Fuzzy Systems: Theory and Applications

Fuzzy Systems: Theory and Applications

Editor: Joshua Hawk

NY Research Press
New York

Published by NY Research Press
118-35 Queens Blvd., Suite 400,
Forest Hills, NY 11375, USA
www.nyresearchpress.com

Fuzzy Systems: Theory and Applications
Edited by Joshua Hawk

International Standard Book Number: 978-1-63238-853-7 (Hardback)

Cataloging-in-Publication Data

Fuzzy systems : theory and applications / edited by Joshua Hawk.
p. cm.
Includes bibliographical references and index.
ISBN 978-1-63238-853-7
1. Fuzzy systems. 2. Fuzzy logic. 3. System analysis. I. Hawk, Joshua.
QA402 .F89 2022
511.313--dc23

Contents

Preface

A control system which operates on fuzzy logic is known as a fuzzy system or a fuzzy control system. Fuzzy logic is a mathematical system which does not operate on absolute binary values of 0 or 1, but instead analyzes analog input values in terms of logical variables that take on continuous values between 0 and 1. Fuzzy systems have found a variety of uses in different fields, from vacuum cleaners to autofocusing cameras and air conditioners. The design of the fuzzy control system is based on empirical methods, which is basically a methodical approach to trial and error. Fuzzy control systems is an upcoming field of science that has undergone rapid development over the past few decades. The extensive content of this book provides the readers with a thorough understanding of the subject.

This book unites the global concepts and researches in an organized manner for a comprehensive understanding of the subject. It is a ripe text for all researchers, students, scientists or anyone else who is interested in acquiring a better knowledge of this dynamic field.

I extend my sincere thanks to the contributors for such eloquent research chapters. Finally, I thank my family for being a source of support and help.

Editor

1

Distribution Network Risk Assessment Using Multicriteria Fuzzy Influence Diagram

Aleksandar Janjic

University of Nis, Faculty of Electronic Engineering, 18 000 Nis, Serbia

Correspondence should be addressed to Aleksandar Janjic; aleksandar.janjic@elfak.ni.ac.rs

Academic Editor: Qi Zeng

Risk assessment of distribution assets is one of the most important factors in the process of network development or maintenance planning decision-making. The process of decision-making is faced with uncertainties, involving technical, financial, safety, environmental, and other operational issues that make standard risk assessment techniques insufficient. Probabilistic uncertainties require appropriate mathematical modeling and quantification when predicting future state of the nature or the value of certain parameters. The paper is proposing a new methodology for the multicriteria risk assessment of the distribution network assets, based on influence diagrams and fuzzy probabilities. Influence diagram has been used to determine all relevant factors concerning risks and their interdependencies are depicted. Fuzzy probabilities are represented by triangular fuzzy numbers with constraints on feasibility of elicited probabilities. This methodology enables the decision process in uncertain environment, with the impact evaluation of each particular distribution asset, or the asset component. The methodology is illustrated on the example of a distribution substation circuit breaker maintenance strategy selection.

1. Introduction

Maintenance planning, development, and reconstruction of distribution networks are playing the crucial role in the asset management of distribution networks [1, 2]. One of the main problems of asset management in distribution companies is to find the best maintenance strategy out of following actions: do nothing and repair only after the breakdown, overhaul, or do the complete replacement of asset. Some activities, like minor or major maintenance, can be performed in a regular time interval, or depending on condition of an asset, but the problem is becoming more complex as alternatives must be evaluated on the basis of several criteria [3]. Some of them are easy to measure (costs and profit), while others can be very difficult to evaluate (public opinion, consequences of outages).

Risk and uncertainties are also present in the process of decision-making, whether it is in presupposed data (consumption increase rate, prices, and preferences) or in decision factors of business environment that affect the process of decision-making. One of the latest approaches is the risk management based maintenance, which evaluates the risk of equipment failure and consequences such failure can produce on the system [4–6]. With the quantification of risk, the most efficient strategy and the optimal risk level for distribution networks assets management can be obtained [6]. In all these approaches, risk is defined as a combination of probability indices and the consequences of failure in the network.

Decision about the optimal level of maintenance depends on several criteria of different nature:

(i) technical criteria

(ii) economic criteria

(iii) health and safety criteria

(iv) environmental impact

(v) public opinion and customer satisfaction

(vi) regulatory requirements

The number and structure of these categories is changing, depending on particular conditions (legislative, regulatory requirements, etc.) but these are basic attributes out of which the others can be derived. Furthermore, the asset management problem is facing the probabilistic uncertainty and imprecision when modeling problem structural parameters, including the required goals, constraints, and external influences.

Various theories of imprecise probability include the Dempster-Shafer evidence theory [7, 8], the coherent lower prevision theory [9], probability bound analysis [10], and the fuzzy probability [11]. Stochastic nature of parameters and subjective probabilities are often described with interval probabilities or fuzzy sets. Interval and fuzzy probabilities are used when it is hard to model uncertainty by point value probabilities: when little or no information to evaluate them is available, or when several information sources (sensors, individual experts in group decision-making) are combined [12]. Fuzzy modeling can be understood as an extension to interval modeling, and fuzzy probabilities can be characterized by a possibility distribution of probability, representing degree of confidence in that probability expressed by an individual [13, 14].

Bayesian networks and Influence diagrams are used as a convenient tool for the large class of engineering problems, while the inherent uncertainty has been modeled by the fuzzification of random variables, and/or prior and conditional probabilities. A comprehensive review of development dealing with imprecise probabilities for the solution of various engineering problems is given in [15]. Fuzzy probabilities are treated as an extension of interval probabilities, emphasizing the correspondence between different α-levels and probability boxes. Various engineering analyses are then enabled using min–max operator and extension principle as the basis for the processing of fuzzy information.

In Bayesian networks, uncertainty embodies both sources: aleatoric (random events or uncontrollable variation) and epistemic (as the absence of complete knowledge). Furthermore, fuzzy probabilities, grouped in several fuzzy sets, can be denoted with linguistic terms: "extremely low", "very low", "medium", etc. [16–19]. These terms represent the information granules that are in great extent influenced by the psychological profile of the decision-maker.

In the deterministic case, alternatives and consequences are directly related in terms of criteria. In the presence of uncertainties, there may exist many possible outcomes that can be described quantitatively or qualitatively (through verbal descriptions).

Approaches like Bayesian networks, fault, and events trees are often used to understand and model random events and outcomes, but issues like interdependencies of different criteria in the decision-making process require further attention. New form of description, the influence diagram, that is both a formal description of the problem that can be treated by computers and a simple, easily understood representation is presented in this paper. The formal theory of Influence diagram is given in [20, 21], with the evaluation, or solving of influence diagram based on Bayesian networks.

This work introduces a new methodology for the risk assessment in distribution network based on the extension of Influence diagrams with the fuzzy probabilities and different consequence evaluation. Risk assessment is performed in two steps. In the first step, influence diagram has been used to determine all relevant factors influencing risks with the depiction of their interdependencies, together with all possible alternative decisions. In the second step, the set of each particular risk values is calculated as the combination of risk factor occurrence and their consequences. Subjective probabilities are represented as information granules described by linguistic terms and modeled as triangular fuzzy numbers.

In the next section of this paper, both steps of a risk assessment methodology using Bayesian networks and Multicriteria Influence diagram are presented. Building of an influence diagram and the way of solving it are presented. Using joint probability rule, the risk of particular event, for different risk categories, is calculated. In Section 3, the notion of fuzzy probability has been explained and in Section 4 the methodology is illustrated on the case study of the choice of circuit breakers maintenance strategy in one transformer substation.

2. Risk Assessment Using Influence Diagrams

2.1. Risk Assessment. Risk assessment, as the first step in the risk management process, attempts to identify possible failure events, evaluate their consequences, determine the probability of their future occurrence, and reduce the detrimental consequences. The usual definition of risk associated with an event E is defined as the product of event probability $p(E)$ and its consequence $Cons(E)$ [22, 23]:

$$Risk(E) = p(E) \cdot Cons(E) \quad (1)$$

More complex relationships between values introducing empirical scaling parameters x, y, and w are presented in the following [24]:

$$Risk(E) = p(E)^{y} \cdot w \cdot Cons(E)^{x} \quad (2)$$

Calculated value of risk became a crucial factor when deciding about the actions to be performed on distribution asset. However, decisions have to be made in a very uncertain environment. In this paper, a new graphical tool based on Bayesian networks—influence diagrams for risk assessment and decision-making under uncertainty—is proposed. The definition of Bayesian networks is given in the sequel, before proceeding to the risk assessment methodology.

2.2. Bayesian Networks. Bayesian network (BN) is a directed acyclic graph represented with pairs $N = \{(V, E), P\}$. Node V represents random variables (events) and links E between nodes represent a causal dependency. A link from variable X to variable Y indicates that X can cause Y, or, in BN terminology, X is a parent of Y, and Y is a child of X. P is a probability distribution over V. Discrete random variables $V = \{X_1, X_2, \ldots, X_n\}$ are assigned to the nodes variables representing a finite set of mutually exclusive states and annotated with a Conditional Probability Table (CPT) that represents the conditional probability of the variable given the values of its parents in the graph.

The simple Bayes net is presented in Figure 1 with two independent variables, X_1 and X_2, and dependent variable Y with appropriate CPT representing probabilities for each

TABLE 1: Consequences grading scale.

	Safety consequences
Grade	Description
1	No harmful consequences
2	Minor: failure results in minor system damage but does not cause injury to personnel or allow any kind of exposure to operational or service personnel
3	Major: failure results in a low level of exposure to Personnel, or activates facility alarm system
4	Critical: failure results in minor injury to personnel
5	Catastrophic: failure results in major injury or death of personnel

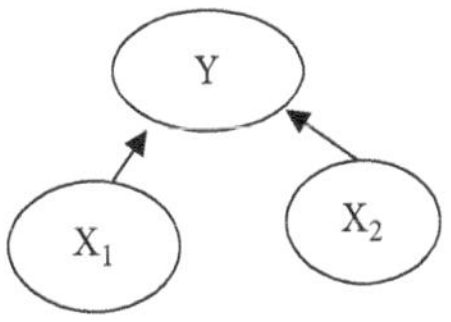

FIGURE 1: Bayes net with two independent variables.

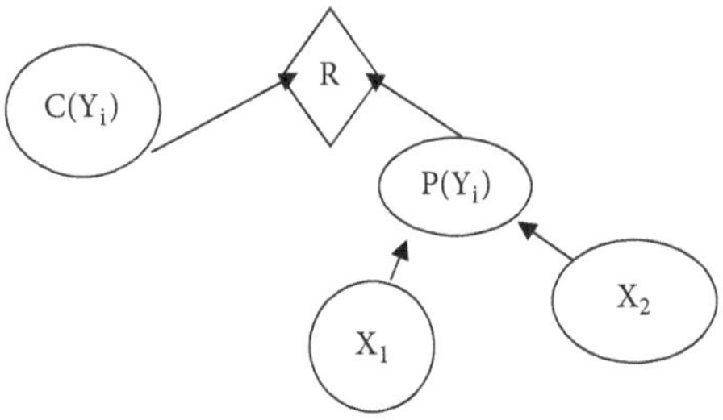

FIGURE 2: Extension of the Bayes net for the risk assessment.

possible state of the nature of variable Y, or event, in the risk assessment terminology.

The solving of BN is based on four rules, including conditional independence, joint probability, marginalization, and Bayesian rule, presented respectively in the following expressions [21, 22].

Conditional Independence

$$P(X_1, X_2, \ldots, X_n) = \prod_{i=1}^{n} P(X_i/Parents(X_i)) \quad (3)$$

Joint Probability

$$P(Y = y_i, X = x_i) = P(X = x_i) \cdot P(Y = y_i/X = x_i) \quad (4)$$

Marginalization Rule

$$P(Y = y_i) = \sum_i P(X = x_i) \cdot P(Y = y_i/X = x_i) \quad (5)$$

Bayesian Rule

$$P(X = x_i/Y = y_j) = \frac{P(X = x_i) \cdot P(Y = y_j/X = x_i)}{P(Y = y_j)} \quad (6)$$

Using expressions (3)–(5), the probability of each possible state j out of n possible states of variable Y can be determined. After the calculation of probabilities, the following step in the methodology is to calculate the risk, using expressions (1) and (2), or more generally:

$$R_i = f(C(Y_i), P(Y_i)) \quad (7)$$

In order to incorporate the risk, BN is extended by two more nodes: consequence node *C(Y)* and risk value node *R* (Figure 2).

For instance, consequences grades for personal safety can be expressed by numerical grades, described in Table 1.

Risk value node is represented by the n-dimensional array, with elements calculated from (7).

2.3. Influence Diagrams. A generalization of a BN is the Influence diagram (ID), proposed by Howard and Matheson [21], as a tool to simplify modeling and analysis of decision trees, allowing not only probabilistic inference but also the graphical representation of decision-making problems. Like in BN, the input and output values of a node in an ID are based on the Bayesian theorem, allowing a user to make inferences with limited available information. Besides the chance nodes of BN, ID also contains decision nodes and utility nodes, depicting available information at the time of making a decision, and the degree of influence of each variable on other variables and decisions. Unlike a decision tree that shows more details of possible paths, ID shows dependencies among variables more clearly. IDs are particularly useful in creating computer-based models that describe a system or as descriptions of decision maker's mental models to assess the impact of their actions.

ID tries to capture system representation in a form that can be communicated to others, through several graphical symbols. A circle depicts an exogenous variable (an external influence) whose values are not affected by previous decisions. A rectangle depicts a decision, while intermediate variables depict an endogenous variable whose values are computed as functions of decision and other variables. Chance node (an ellipse) represents a random variable defined by discrete probability distribution. Arrow shows the influence between variables, and dotted arrow shows information being communicated between elements. Finally, value node (a diamond) is a quantitative criterion representing the subject of optimization. The simple example of an ID is presented in Figure 3.

Methods for evaluating and solving IDs are based on Bayes theorem and can be grouped in several categories. They can be (i) converted to decision trees and solved, (ii) solved by

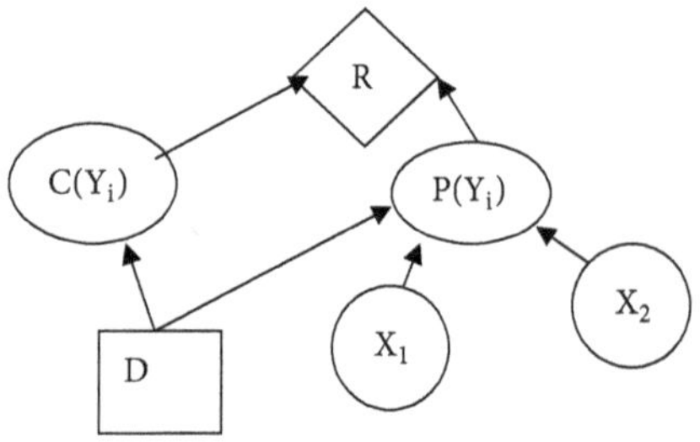

FIGURE 3: Influence diagram for the risk assessment.

the variable elimination algorithms, or (iii) solved by efficient algorithms using graphical structures, like the junction trees [25].

When nodes in the diagram are represented by appropriate fuzzy sets, ID can be solved using fuzzy reasoning [26–28]. A commonly used technique for combining fuzzy sets is fuzzy inference system, like the Mamdani type inference. The final step for building an ID model for the risk assessment is to aggregate different risk factors into one diagram. An illustration for the risk assessment ID with two criteria: safety risk (S) and economic risk (E), is presented in Figure 4.

3. Fuzzy Probability

Uncertain and subjective probabilities can be granulated in different terms like: improbable or doubtful, but we will draw our attention to special form of probability granulation, focusing on point value with inherent uncertainty: around 30%, around 75%, etc. These probabilities are defined starting from previous works on linguistic probability [29–32] defining similar probability measure for fuzzy probabilities.

Klir [31, 32] introduced a notion of a fuzzy interval defined on $[0,1]$ as a probability granule, that is, a normal fuzzy set on $[0,1]$ with α cuts for all as closed subintervals on $[0,1]$. These fuzzy probabilities will be denoted as $P_1, P_2, \ldots, P_n$ and they may be expressed by the canonical form:

$$P_i(x) = \begin{cases} f_i(x), & x \in [0, b) \\ 1, & x \in [b, c] \\ g_i(x), & x \in (c, d] \\ 0, & otherwise \end{cases} \tag{8}$$

where $a, b, c, d \in [0,1]$, $a \le b \le c \le d$, and f_i and g_i are strictly increasing right continuous and strictly decreasing left continuous real valued function, respectively. The α-cuts of P_i are expressed for all $a \in [0,1]$:

$$P_i^\alpha = \begin{cases} \left[f_i^{-1}(\alpha), g_i^{-1}(\alpha)\right], & \alpha \in (0,1) \\ [b, c], & \alpha = 1 \end{cases} \tag{9}$$

In this paper, we will investigate the elicitation of triangular fuzzy set support—left and right bounds of triangular fuzzy numbers. The support of fuzzy set A is the set of all points x in X such that $\mu_{A(x)} > 0$.

Consider a discrete random variable X with values in the set $X = \{x_i, i \in N_n\}$. We will assume that the probabilities of this random variables $P(x_i)$ are assessed approximately, by a triangular fuzzy number:

$$\mu_{P_i}(\mathrm{x}) = \begin{cases} 0, & x < a_i \\ \dfrac{x - a_i}{b_i - c_i}, & a_i < x < b_i \\ \dfrac{c_i - x}{c_i - b_i}, & b_i < x < c_i \\ 0, & x > c_i \end{cases} \tag{10}$$

We can interpret these fuzzy numbers as fuzzy probabilities as follows.

Definition 1. Fuzzy numbers $P_i = [a_i, b_i, c_i]$, $i = 1, \ldots, n$ are called fuzzy probabilities of X if there are $x_1 \in [a_1, c_1], \ldots, x_i \in [a_i, c_i], \ldots, x_n \in [a_n, c_n]$ such that:

$$\sum_{i=1}^{n} x_i = 1, \qquad \sum_{i=1}^{n} b_i = 1 \tag{11}$$

The set of fuzzy numbers P satisfies (11) if and only if the following conditions hold [33]:

$$\begin{aligned} c_i + a_1 + \ldots + a_{i-1} + a_{i+1} + \ldots + a_n \ge 1, \quad \forall i \\ a_i + c_1 + \ldots + c_{i-1} + c_{i+1} + \ldots + c_n \ge 1, \quad \forall i \end{aligned} \tag{12}$$

If there are only two fuzzy probabilities $[a_1, b_1, c_1]$ and $[a_2, b_2, c_2]$, then $a_1 + c_2 = 1$, $a_2 + c_1 = 1$ and $b_1 + b_2 = 1$.Let us consider a set of fuzzy numbers $FP = \{FP_i = [a_i, b_i, c_i], i = 1, \ldots, n\}$. The interval of probability values for every α-cut will be denoted as $[a_{\alpha,i}, c_{\alpha,i}]$. We can interpret these fuzzy numbers as fuzzy probabilities as follows.

Definition 2. Fuzzy numbers $FP_i = [a_i, b_i, c_i]$ are called fuzzy probabilities of X if for $\forall \alpha \in [0,1]$ and $\forall x_i \in [a_{\alpha,i}, c_{\alpha,i}]$ there are $x_1 \in [a_{\alpha,1}, c_{\alpha,1}], \ldots, x_{i-1} \in [a_{\alpha,i-1}, c_{\alpha,i}], x_{i+1} \in [a_{\alpha,i+1}, c_{\alpha,i+1}], \ldots, x_n \in [a_{\alpha,n}, c_{\alpha,n}]$ such that:

$$\sum_{i=1}^{n} x_i = 1 \tag{13}$$

Lemma 3. *The set of fuzzy numbers FP satisfies (13) if and only if the following conditions hold:*

$$\begin{aligned} c_{\alpha,i} + a_{\alpha,1} + \ldots + a_{\alpha,i-1} + a_{\alpha,i+1} + \ldots + a_{\alpha,n} \le 1, \quad \forall \alpha, \forall i \\ a_{\alpha,i} + c_{\alpha,1} + \ldots + c_{\alpha,i-1} + c_{\alpha,i+1} + \ldots + c_{\alpha,n} \ge 1, \quad \forall \alpha, \forall i. \end{aligned} \tag{14}$$

Proof.

Sufficient Conditions. If the first part of Lemma 3 holds, then:

$$\forall \alpha, \forall i$$

$$x_i + a_{\alpha,1} + \ldots + a_{\alpha,i-1} + a_{\alpha,i+1} + \ldots + a_{\alpha,n} \le c_{\alpha,i} + a_{\alpha,1} + \ldots + a_{\alpha,i-1} + a_{\alpha,i+1} + \ldots + a_{\alpha,n} \le 1$$

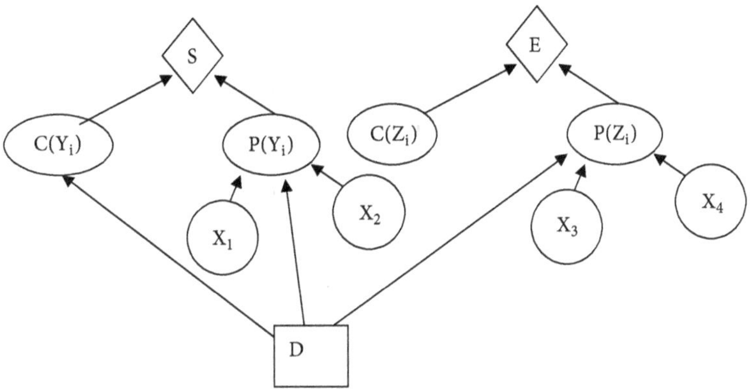

FIGURE 4: Complete influence diagram with different risk factors.

$$\forall\alpha, \forall i$$

$$x_i + c_{\alpha,1} + \ldots + c_{\alpha,i-1} + c_{\alpha,i+1} + \ldots + c_{\alpha,n} \geq a_{\alpha,i} + c_{\alpha,1} + \ldots + c_{\alpha,i-1} + c_{\alpha,i+1} + \ldots + c_{\alpha,n} \geq 1. \tag{15}$$

Then, the following expression holds:

$$x_i + a_{\alpha,1} + \ldots + a_{\alpha,i-1} + a_{\alpha,i+1} + \ldots + a_{\alpha,n} \leq 1 \leq x_i + c_{\alpha,1} + \ldots + c_{\alpha,i-1} + c_{\alpha,i+1} + \ldots + c_{\alpha,n}. \tag{16}$$

The expression shows that there exist $a_{\alpha,j} \leq x_j \leq c_{\alpha,j}, j \in \{1,\ldots n\}, j \neq i$ that satisfies (13).

Necessary Conditions. If the first part of Lemma 3 does not hold, then:

$$\begin{gathered}\forall\alpha, \exists i \\ c_{\alpha,i} + a_{\alpha,1} + \ldots + a_{\alpha,i-1} + a_{\alpha,i+1} + \ldots + a_{\alpha,n} > 1 \\ \forall\alpha, \exists i \\ a_{\alpha,i} + c_{\alpha,1} + \ldots + c_{\alpha,i-1} + c_{\alpha,i+1} + \ldots + c_{\alpha,n} < 1.\end{gathered} \tag{17}$$

Then, taking x_i as $a_{\alpha,i}$ or $c_{\alpha,i}$ (13) cannot hold.

An alternative definition of fuzzy probabilities can be formulated from two extreme cases of $\alpha = 0$ and $\alpha = 1$.

Definition 4. Fuzzy numbers $FP_i = [a_i, b_i, c_i]$ are called fuzzy probabilities of X if for and $\forall x_i \in [a_i, c_i]$ there are $x_1 \in [a_1, c_1], \ldots, x_{i-1} \in [a_{i-1}, c_i], x_{i+1} \in [a_{i+1}, c_{i+1}], \ldots, x_n \in [a_n, c_n]$ such that:

$$\sum_{i=1}^{n} x_i = 1, \qquad \sum_{i=1}^{n} b_i = 1 \tag{18}$$

Bayesian networks with fuzzy numbers replacing point value probabilities are proposed [30] defining "Bayesian fuzzy probability" as convex, normal fuzzy set of $[0,1]$. Complementation law has been relaxed in order to extend a partially defined linguistic probability measure, and this method has been successfully used in forensic statistics [29] and risk analysis [18]. More possible scenarios for fuzzifying the Bayesian approach are presented in [15] using nonfuzzy algorithmically efficient reformulation of the Bayesian formula. Although time-consuming, we will implement the corresponding fuzzy version of Bayesian formulas.

The fuzzy counterparts to the standard arithmetic operators are defined using the extension principle. It is possible to derive these operators by examining the effects of interval based calculations at each α–cut. The extended operators are defined by (19), using a circled arithmetic operator symbol for the extension of a real arithmetic operator.

Definition 5. For all $a, b \in R^F$a, the extended operators are defined by

$$\begin{aligned} \mu_{A\oplus B}(z) &= \sup_{x+y=z} \min\left(\mu_A(x), \mu_B(y)\right) \\ \mu_{A\otimes B}(z) &= \sup_{xy=z} \min\left(\mu_A(x), \mu_B(y)\right) \\ \mu_{A-B}(z) &= \sup_{x-y=z} \min\left(\mu_A(x), \mu_B(y)\right) \\ \mu_{A\oslash B}(z) &= \sup_{x/y=z} \min\left(\mu_A(x), \mu_B(y)\right) \end{aligned} \tag{19}$$

From previous definition, two fuzzy Bayes rules analogue to classical crisp number relations are formulated. Operator "≅" stands for "=" operator.

Fuzzy Joint Probability

$$P\left(Y = y_j, X = x_i\right) \cong P\left(X = x_i\right) \otimes P\left(Y = y_j \setminus X = x_i\right) \tag{20}$$

Table 2: Fuzzy probabilities.

Triangular fuzzy probability number	Description	Notation
[5 10 15]	Extremely low probability	EL
[15 20 25]	Low probability	L
[25 30 35]	Low to medium	LM
[35 40 45]	Medium to low	ML
[45 50 55]	Medium probability	M
[55 60 65]	Medium to high	MH
[65 70 75]	High to medium	HM
[75 80 85]	High probability	H
[85 90 95]	Extremely high probability	EH

Table 3: Prior probability of weather states.

States	Description	Probability
Bad	Severe weather conditions	MM
Medium	No extreme temperatures below – 20 degree	LM
Good	Good weather conditions, no extreme temperatures below -10 degree	L

Fuzzy Bayes Rule

$$P\left(X = x_i \setminus Y = y_j\right) \cong \frac{P\left(X = x_i\right) \otimes P\left(Y = y_j \setminus X = x_i\right)}{P\left(Y = y_j\right)} \quad (21)$$

Based on the law of total probability another rule for the fuzzy marginalization can be added, represented by (22).

Fuzzy Marginalization Rule

$$P\left(Y = y_j\right) \cong \sum_i P\left(X = x_i\right) \otimes P\left(Y = y_j \setminus X = x_i\right) \quad (22)$$

Finally, risk can be calculated from

$$R = \sum_i^n P\left(C_i\right) \otimes C_i \quad (23)$$

It is possible to use any other form of additive, multiplicative, or tabular risk aggregation function. The influence diagram with fuzzy probabilities will be illustrated on a simple case study of maintenance strategy selection.

4. Case Study

Risk assessment methodology is illustrated on substation with low oil circuit breakers. The decision has to be made about three possible alternatives: do nothing, perform minor interventions, or do the overhaul and major repair of circuit breakers. The alternatives will be assessed by the risk assessment of two criteria: safety and environment. Both criteria will be evaluated by their risk and then aggregated in the one influence diagram value node.

Two failure modes and normal operating condition of a circuit breaker are taken into account: breaker is in operating conditions (OK), failure to close (Close), when breaker does not close the circuit to conduct current in one or more poles, and failure to open (FO), when breaker does not open the circuit to interrupt current. In the case of the bad weather conditions in the following year, the network condition will worsen due to the increased number of failure, the network loading will increase, and the breaker will be exposed to more severe operation conditions. Due to the uncertainty about the weather forecast, and consequently network technical condition, network maximal demand power (loading), and possible failure modes, probabilities elicited by experts are also uncertain. According to the definition of the fuzzy probabilities, possible probability grades are represented in Table 2.

Prior fuzzy probabilities of ambient conditions and global weather forecast for the next year are given in Table 3.

Conditional probability tables for network condition, circuit breaker failure modes, loading levels, and consequences are represented in Tables 4, 5, 6, and 7, respectively.

Safety and environment criteria evaluations are expressed in numerical grades (from 1 to 5) and represented in Table 8.

For $\alpha = 1$, the influence diagram becomes the deterministic influence diagram with crisp probability values. The solved diagram with calculated values for three different scenarios and two criteria is represented in Figure 5.

Risk is calculated using the following consequences grades: 1 for no consequences, 2 for minor, and 3 for major consequences, and aggregation of two risks is presented in Table 9.

It is visible that risk calculated values do not show great variance, and that decision about the future maintenance strategy cannot be easily determined. Therefore, the problem is solved again using fuzzy probabilities and expressions (19)–(22). Results are presented on Figures 6, 7, and 8.

Calculated values for risks for both alternatives with crisp values (Figure 5) and fuzzy probability values

Table 4: Conditional probabilities of network conditions.

Weather	States	
	Bad conditions on MV side—no tree trimming, no maintenance, increased number of failures	Good conditions on MV network, no increase in failure rate
B	MH	ML
M	MM	MM
G	ML	MH

Table 5: Conditional probabilities of failure modes.

Decision	NC	Ok	Close	FO
Minor		HM	L	EL
Minor	Good	H	EL	EL
Major	Bad	H	EL	EL
Major	Good	EH	EL	EL
Do nothing	Bad	MH	L	L
Do nothing	Good	MH	L	L

Table 6: Conditional probabilities of network loading levels.

Weather	Low Loading	Medium Loading	High Loading
Bad	EL	LM	MH
Medium	LM	MM	L
Good	MH	LM	EL

Table 7: Conditional probabilities of consequences.

Loading	Failure mode	Safety risk			Environmental risk		
Low loading	OK	EH	EL	Impossible	H	EL	EL
	Failure to close	H	EL	EL	H	EL	EL
	Failure to open	HM	L	EL	HM	L	EL
Medium loading	OK	H	EL	EL	H	EL	EL
	Failure to close	HM	L	EL	MM	LM	L
	Failure to open	MH	LM	EL	MH	LM	EL
High loading	OK	HM	L	EL	HM	L	EL
	Failure to close	MH	L	L	MH	L	L
	Failure to open	MM	LM	L	MM	LM	L

Table 8: Safety and environment criteria grades.

Grade	Safety risk	Environmental risk
1	No harmful consequences	No harmful consequences, Failure does not allow any release of chemicals into the environment
2	Minor: failure results in minor system damage but does not cause injury to personnel	Personnel exposure to harmful chemicals or radiation or fire
3	Critical: failure results in minor injury to personnel	A release of chemical to the environment

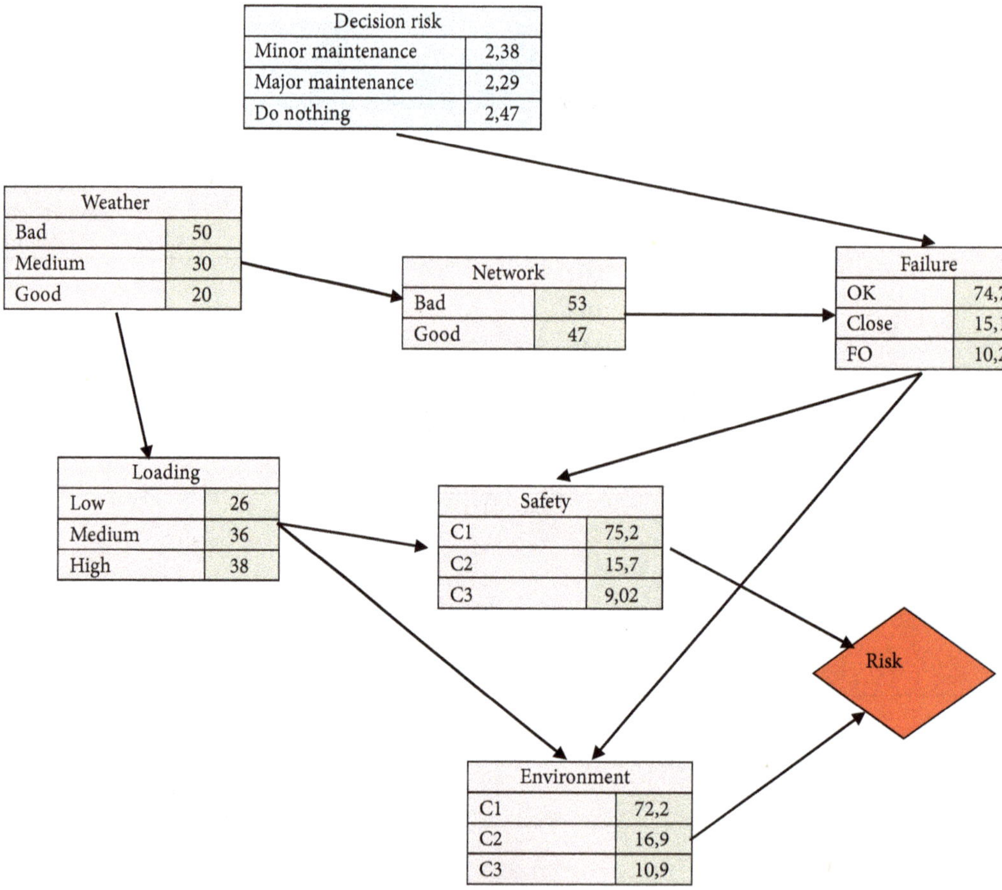

FIGURE 5: Influence diagram with crisp probabilities.

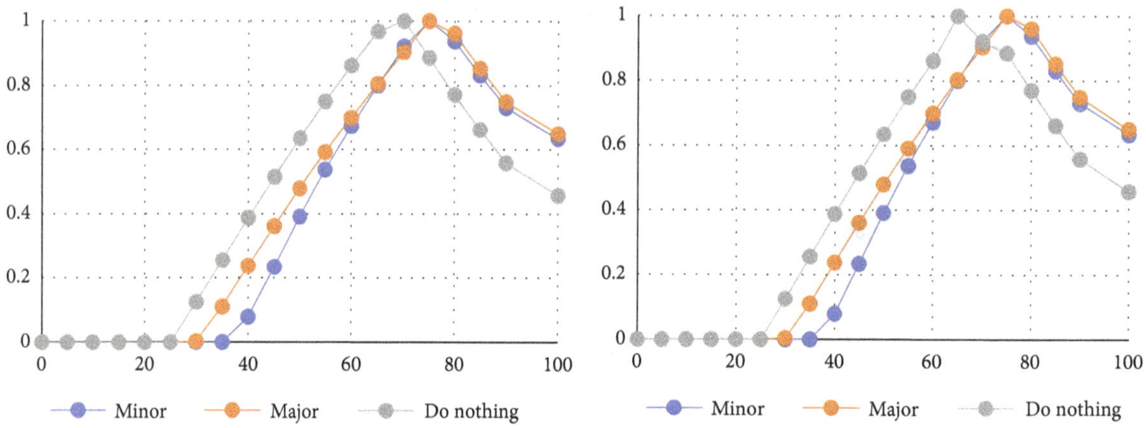

FIGURE 6: Fuzzy probabilities for no consequences state for (a) safety risk and (b) environmental risk.

defuzzified using the centroid method are given in Table 10.

Unlike the crisp calculated values for the risk node, the fuzzy probabilities of possible consequences are highlighting the whole range of future scenarios consequences, facilitating the decision-making process. From Figure 6 it is visible that the "doing nothing" strategy almost doubles the risk for minor and major consequences, while the minor and major maintenance practically do not make any difference.

5. Conclusions

Decision-making in maintenance planning is always confronted with several aspects: technical, financial, safety, environmental, and operational ones. Each of these factors can be modeled by the appropriate risk of possible harmful consequences. This paper introduces a new methodology for the optimization of maintenance activities in distribution network, based on the calculation of the risk of the particular

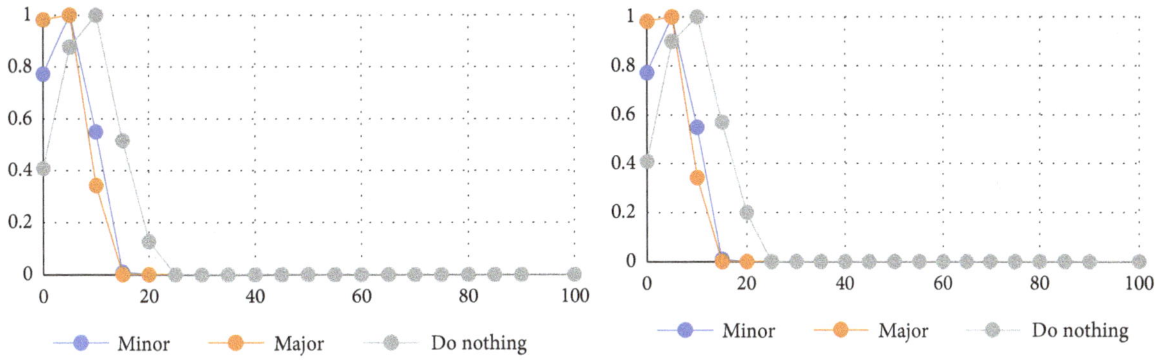

FIGURE 7: Fuzzy probabilities for minor consequences state for (a) safety risk and (b) environmental risk.

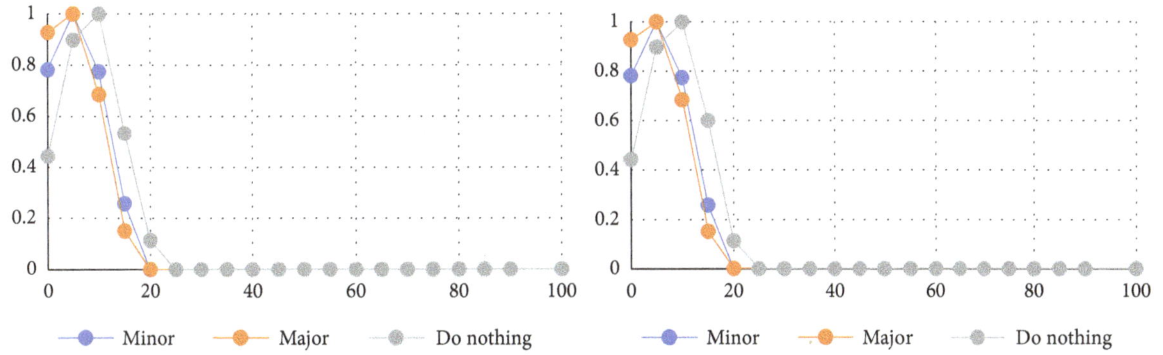

FIGURE 8: Fuzzy probabilities for critical consequences state for (a) safety risk and (b) environmental risk.

TABLE 9: Aggregated risks.

Safety	1	2	3	1	2	3	1	2	3
Environment	1	1	1	2	2	2	3	3	3
Aggregated risk	1	2	3	4	5	8	7	8	10

TABLE 10: Aggregated risk comparison.

	Crisp probability	Fuzzy probability
Minor	2,38	2,40
Major	2,29	2,65
Do nothing	2,48	4,12

component, or the overall risk of the distribution object. For this purpose, the Fuzzy Influence diagram has been used, in order to determine all relevant factors influencing risks, and depicting their interdependencies. Probabilistic uncertainties require appropriate mathematical modeling and quantification when predicting future state of the nature or the value of certain parameters and the proposed model enables evaluation of the impact of each particular component of maintenance decision process. Bayesian networks and Influence diagrams are used as a convenient tool for the large class of engineering problems, while the inherent uncertainty has been modeled by the fuzzification of random variables, and/or prior and conditional probabilities. The methodology is illustrated on the example of the choice of circuit breakers maintenance in one distribution transformer substation. This methodology is especially suited for systems with great number of unobservable components, in the presence of uncertainty and lack of operational data, like power distribution system is. The main challenges for the future research are the introduction of fuzzy multi-criteria decision analysis for the choice of activity minimizing the overall risk and more flexible representation of fuzzy subjective probabilities, as information granules described by linguistic terms and modeled as triangular fuzzy numbers.

Conflicts of Interest

The author declares that he has no conflicts of interest.

References

[1] G. J. Anders, J. Endrenyi, and C. Yung, "Risk-based planner for asset management," *IEEE Computer Applications in Power*, vol. 14, no. 4, pp. 718–723, 2001.

[2] S. Natti and M. Kezunovic, "A risk based decision approach for maintenance scheduling strategies for transmission system equipment," in *Proceedings of the 10th International Conference on Probabilistic Methods Applied to Power Systems, PMAPS '08*, pp. 1–6, 2008.

[3] Y. Jiang, J. D. McCalley, and T. Van Voorhis, "Risk-based resource optimization for transmission system maintenance," *IEEE Transactions on Power Systems*, vol. 21, no. 3, pp. 1191–1200, 2006.

[4] S. R. Yeddanapudi, Y. Li, J. D. McCalley, A. A. Chowdhury, and W. T. Jewell, "Risk-based allocation of distribution system maintenance resources," *IEEE Transactions on Power Systems*, vol. 23, no. 2, pp. 287–295, 2008.

[5] A. D. Janjic and D. S. Popovic, "Selective maintenance schedule of distribution networks based on risk management approach," *IEEE Transactions on Power Systems*, vol. 22, no. 2, pp. 597–604, 2007.

[6] W. Li, *Risk Assessment of Power Systems: Models, Methods, and Applications*, IEEE Press, 2005.

[7] A. P. Dempster, "Upper and lower probabilities induced by a multivalued mapping," *Annals of Mathematical Statistics*, vol. 38, pp. 325–339, 1967.

[8] G. Shafer, *A Mathematical Theory of Evidence*, Princeton University Press, Princeton, NJ, USA, 1976.

[9] P. Walley, *Statistical Reasoning with Imprecise Probabilities*, Chapman and Hall, London, UK, 1991.

[10] S. Dehaene and J. Mehler, "Cross-linguistic regularities in the frequency of number words," *Cognition*, vol. 43, no. 1, pp. 1–29, 1992.

[11] B. Möller and M. Beer, *Fuzzy Randomness – Uncertainty in Civil Engineering and Computational Mechanics*, Springer, Berlin, Germany, 2004.

[12] A. Cano and S. Moral, "Using probability trees to compute marginals with imprecise probabilities," *International Journal of Approximate Reasoning*, vol. 29, no. 1, pp. 1–46, 2002.

[13] J. J. Buckley, *Fuzzy Probabilities. New Approach and Applications*, Springer Science & Business Media, Heidelberg, Germany, 2005.

[14] Y. Pan and G. J. Klir, "Bayesian inference based on interval-valued prior distributions and likelihoods," *Journal of Intelligent & Fuzzy Systems: Applications in Engineering and Technology*, vol. 5, no. 3, pp. 193–203, 1997.

[15] M. Beer, S. Ferson, and V. Kreinovich, "Imprecise probabilities in engineering analyses," *Mechanical Systems and Signal Processing*, vol. 37, no. 1-2, pp. 4–29, 2013.

[16] C. C. Yang, "Fuzzy Bayesian inference," in *Proceedings of the IEEE International Conference on Systems, Man, and Cybernetics, Computational Cybernetics and Simulation*, pp. 2707–2712, Orlando, Fla, USA, 1997.

[17] J. Liu, L. M. López, J. Yang, and J. Wang, "Linguistic Assessment Approach for Hierarchical Safety Analysis and Synthesis," in *Intelligent Decision and Policy Making Support Systems*, pp. 211–230, Springer, 2008.

[18] J. Ren, J. Wang, I. Jenkinson, D. L. Xu, and J. B. Yang, "Bayesian network approach for offshore risk analysis through linguistic variables," *China Ocean Engineering*, vol. 21, no. 3, pp. 371–388, 2007.

[19] J. Talasova and O. Pavlicek, "Fuzzy Probability Spaces and Their Applications in Decision Making," *Austrian Journal of Statistics*, vol. 35, no. 2-3, pp. 347–356, 2006.

[20] F. V. Jensen and T. D. Nielsen, *Bayesian Networks and Decision Graphs*, Springer, 2007.

[21] R. A. Howard and J. E. Matheson, "Influence Diagrams," *Decision Analysis*, vol. 2, no. 3, pp. 127–143, 2005.

[22] A. Janjić, M. Stanković, and L. Velimirović, "Multi-criteria Influence Diagrams – A Tool for the Sequential Group Risk Assessment," in *Granular Computing and Decision-Making*, W. Pedrycz and S. M. Chen, Eds., vol. 10 of *Studies in Big Data*, pp. 165–193, Springer, 2015.

[23] A. Janjic, Z. Stajic, and I. Radovic, "A Practical Inference Engine for Risk Assessment of Power Systems based on Hybrid Fuzzy Influence Diagrams," in *Latest Advances in Information Science, Circuits and Systems*, 2011.

[24] Z. Zhang, Y. Jiang, and J. McCalley, "Condition based failure rate estimation for power transformers," in *Proceedings of the 35th North American Power Symposium*, Rolla, Miss, USA, 2003.

[25] F. Jensen, V. Jensen, and S. Dittmer, "From influence diagrams to junction trees," in *Uncertainty in Artificial Intelligence (UAI)*, pp. 367-363, 1994.

[26] N. H. Mateou, A. P. Hadjiprokopis, and A. S. Andreou, "Fuzzy Influence Diagrams: An Alternative Approach to Decision Making Under Uncertainty," in *Proceedings of the 2005 International Conference on Computational Intelligence for Modelling, Control and Automation, (CIMCA-IAWTIC '05)*, 2005.

[27] N. An, J. Liu, and Y. Bai, "Fuzzy influence diagrams: An approach to customer satisfaction measurement," in *Proceedings of the 4th International Conference on Fuzzy Systems and Knowledge Discovery*, 2007.

[28] L. Hui and X. Y. Ling, "The Traffic Flow Study Based on Fuzzy Influence Diagram Theory," in *Proceedings of the 2009 Second International Conference on Intelligent Computation Technology and Automation*, 2009.

[29] J. Halliwell, J. Keppens, and Q. Shen, "Linguistic Bayesian Networks for reasoning with subjective probabilities in forensic statistics," in *Proceedings of the 5th International Conference of AI and Law*, pp. 42–50, Edinburgh, Scotland, 2003.

[30] J. Halliwell and Q. Shen, "Towards a linguistic probability theory," in *Proceedings of the 11th International Conference on Fuzzy Sets and Systems (FUZZ-IEEE '02)*, pp. 596–601, Honolulu, Hawaii, USA, 2002.

[31] G. J. Klir, "Basic Issues of Computing with Granular Probabilities," in *Data Mining, Rough Sets and Granular Computing*, T. Y. Lin, Y. Y. Yao, and L. A. Zadeh, Eds., Springer Verlag, Germany, 2002.

[32] G. Klir, *Uncertainty and Information: Foundations of Generalized Information Theory*, John Wiley & Sons, 2005.

[33] K. Weichselberger and S. Pohlmann, *A Methodology for Uncertainty in Knowledge-Based Systems*, Springer-Verlag, Germany, 1990.

2

Search of Fuzzy Periods in the Works of Poetry of Different Authors

Artur Nor[1] **and Eugene Korotkov**[1,2]

[1]*National Research Nuclear University "MEPhI", Kashirskoe Highway, 31, 115409, Moscow, Russia*
[2]*Institute of Bioengineering, Research Center of Biotechnology of the Russian Academy of Sciences, Leninsky Ave. 33, bld. 2, 119071, Moscow, Russia*

Correspondence should be addressed to Eugene Korotkov; genekorotkov@gmail.com

Academic Editor: Ferdinando DiMartino

We applied a new method for the identification of fuzzy periods and the insertion and deletion of characters were taken into consideration while studying the works of poetry. The technique employs genetic algorithm, dynamic programming, and the Monte Carlo method. In the present work, the technique was applied to poems written by the famous Russian and foreign classics. A total of 95 poems were studied; and fuzzy periods possessing high statistical significance were identified with more than half of the poems under study. The existence of correlation between the stressed vowel letters in a poem with the position of the fuzzy periods was shown. The present study shows that a work of poetry contains both semantic component and fuzzy periods of letters; hence a poem could have psychological impact on the audience.

1. Introduction

Works of poetry could be considered as a superposition of the semantic content and of the acoustic wave determined by a certain sound alternation periodicity. In relation to this, a poet is capable of combining the semantic content with a certain acoustic wave in a work of poetry. A certain periodicity of sounds alternation in a work of poetry is understood as an acoustic wave. If the meaning of a poetic text is easily understood by each person, the acoustic wave embedded in a work of poetry will be perceived rather intuitively, as some musicality, often fascinating the listeners and exposing them to a certain psychological impact [1]. In order to understand the mechanism of the acoustic wave impact on listeners, it would be very interesting to attempt quantitatively identifying and studying the acoustic wave embedded in a work of poetry, in the form of a certain periodicity of the poetic text [2, 3]. To solve this problem, it seems important to develop and apply new mathematical methods that could quantitatively demonstrate the existence of an acoustic wave in a work of poetry in the form of fuzzy periods and provide the quantitative characteristics of the periodicity found. This task seems to be important, since the quantitative determination of acoustic waves would ensure the classification of existing acoustic waves in the works of poetry. Thus, we could correlate a certain type of acoustic wave and its impact on a listener. After introducing such an important concept as fuzzy periods [4, 5], we could illustrate it with an example. Under the fuzzy periods, we shall obtain the mean of such periods, where the similarity between individual periods is insignificant or is missing at all; and the periodicity becomes statistically significant only on a certain set of periods (more than 2) [6]. Fuzzy periods could be demonstrated with an example. Let us consider a sequence in the following form:

(qzwrt)(qzwrt)(qzwrt)(qzwrt)(qzwrt)(qzwrt)(qzwrt)...

The given sequence is characterized by a perfect periodicity consisting of 5 letters. In this study, each period is highlighted in parentheses, for clarity. There is absolute similarity between the separate periods and it is easily identified using the techniques described previously. Considering a case in the position of each period, a definite and limited set of alphabet letters could be found; for example, such set of letters for each period position is shown as follows: {q,i,u,s,t}; {u,c,i,a,s,r}; {o,p,f,g,l,k,w}; {a,b,n,m,v}; {p,f,g,h,t,j,r}.

Now, let us create a character sequence taking from each set a letter with the use of random technique and corresponding to the period position; then, the sequence can be obtained in the following form:

(iroap)(tufng)(sslmt)(uawaj)(qcgbf)(siknh)(sipvr) . . .

The resulting character sequence lacks absolute periodicity. However, it should be noted that given the sufficient length of this sequence, it could be seen that, in the position of each period, only certain alphabet letters are located. Such a sequence is characterized by fuzzy periods, which could not be identified by pairwise comparison of any two periods but could be detected using a certain set of periods (more than 2).

Nowadays, several mathematical techniques are employed for the detection of fuzzy periods in character and numerical sequences. These include the wavelet transform [7] and the Fourier transform [8]. Previously, the information decomposition (ID) technique was developed [4]. The difference between the ID technique and the Fourier transform lies in the fact that the ID technique could be used for character sequence analysis without recoding it into a numerical series. Such a method of analysis makes it possible to obtain results that are unattainable with the Fourier transform. This allowed the fuzzy periods in DNA sequences [5], amino acid sequences [6], and of several works of poetry to be revealed [4]. However, the ID technique, like other methods previously discussed, does not allow the finding of a statistically significant fuzzy period with insertions and deletions of characters, which in case of literary works could be registered in connection with pronunciation peculiarities. For example, certain sounds may not be pronounced at all or may be pronounced with a certain accent. Consequently, most of the fuzzy periods contained in the sequence could not be determined using the previously developed methods.

As of today, there are mathematical approaches based on dynamic programming that allow the accurate identification of fuzzy periods of time series or character sequence in the presence of characters insertion or deletion [9, 10]. All these techniques are used to construct the multiple alignment of periods; and they are based either on performing the pairwise alignment of periods, followed by the subsequent creation of a guide tree, or on the search for embryos or common words in periods. Thereafter, the initial multiple alignment of periods is provided; and the optimization thereof is carried out in one way or another, including the use of hidden Markov models, iterative procedures, and some other techniques [10–12]. However, all the developed approaches do not ensure construction of the multiple alignment, if the statistically significant pair alignment is missing in the analyzed sequences. It does not allow the creation of a statistically significant guide tree for the progressive alignment; or the sequences are that different that they do not provide searching for the statistically significant embryos or common words. It turns out that nowadays, it is impossible to construct a multiple alignment for significantly different sequences (periods). In this case, it could be argued that all the developed approaches are "blind" and will not identify a statistically significant multiple alignment in the significantly different sequences (periods). Such an alignment could be found, if it would be possible to construct a multiple alignment through the direct application of dynamic programming for all the analyzed sequences. But this is the so-called NP-complete problem [13, 14]; and such an approach requires gigantic computer resources that are not available at present; and it is difficult to think about its creation in the nearest future.

Previously, a new technique was developed for identifying the fuzzy periods in character sequences, which took into consideration the insertions and deletions of characters [15, 16]. This technique is based on the new solution of the NP-complete problem regarding the sequences (periods) multiple alignment. This method employs genetic algorithm, techniques aimed at optimizing weight matrices, dynamic programming, and the Monte Carlo method. It enables identification of the fuzzy periods of a character sequence with insertions and deletions in previously unknown positions. It is important to note that this analysis requires only the symbolic sequence itself (the text of the poetic work) and other information about the poetic work, including the placement of stresses and features of pronunciation, are not required. In the given work, this approach was applied while searching for fuzzy periods in the poems of famous Russian and English-speaking poets. We showed that, in more than half of the works of poetry, it is possible to find fuzzy periods. This study shows that a work of poetry contains both semantic component and fuzzy periods, which could be responsible for the psychological impact of a poem on the audience. Fuzzy periods can be a reflection of the sound "wave" which exists in a poetic work.

2. Fuzzy Periods Search Technique Algorithm Used with Consideration of Characters' Insertions and Deletions

At the beginning of the work, the poetics is transformed in such a way that all the spaces are deleted, uppercase letters are changed to lowercase, and punctuation marks are changed to spaces (Figure 1, Paragraph 1). Thus, the character sequence is created on the basis of the transformed work of poetry for further evaluation. In Figure 1, Paragraph 2, the $Q(n)$ set of random matrices having the $k\times n$ dimension is generated, where n is the period length and k is the size of the original alphabet sequence. In Figure 1, Paragraph 3, modification and optimization of random matrices are performed, which is required for constructing the S sequence alignment.

Then, in Figure 1, Paragraph 4, a search was conducted for a matrix that possesses the greatest value of the similarity function, when the S sequence is aligned. For this purpose, genetic algorithm and dynamic programming are applied. At each phase of the genetic algorithm and for each matrix and the S sequence, we calculate the maximum value of the E_{max} similarity function using dynamic programming. In this case, E_{max} appears to be a fitness function; and each matrix becomes a genotype. Then, to the $Q(n)$ set of matrices we apply the genetic algorithm, which causes the mutation, multiplication, and destruction of matrices. As a result, we

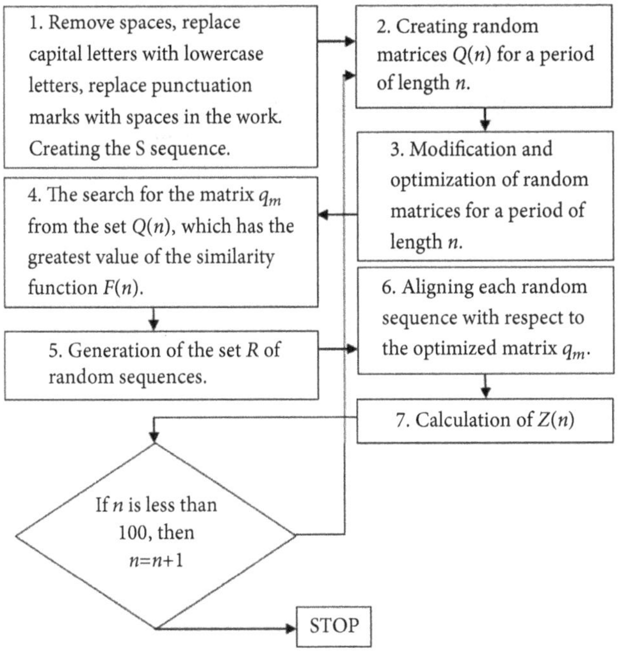
1. Remove spaces, replace capital letters with lowercase letters, replace punctuation marks with spaces in the work. Creating the S sequence.
2. Creating random matrices $Q(n)$ for a period of length n.
3. Modification and optimization of random matrices for a period of length n.
4. The search for the matrix q_m from the set $Q(n)$, which has the greatest value of the similarity function $F(n)$.
5. Generation of the set R of random sequences.
6. Aligning each random sequence with respect to the optimized matrix q_m.
7. Calculation of $Z(n)$
If n is less than 100, then $n=n+1$
STOP

FIGURE 1: The main stages of the algorithm mused for calculation $Z(n)$ of the analyzed sequence S.

find such M_{max} matrix that possesses the greatest E_{max} value; and we denote it as mE_{max}. In order to estimate the statistical significance of mE_{max}, we generate the R random sequences set (Figure 1, Paragraph 5). This set is generated using S sequence random mixing. Then, for each sequence out of the R set, the maximum value of the E_{max} similarity function for the M_{max} matrix is determined (Figure 1, Paragraph 6). This makes it possible to calculate the average value and variance for the E_{max} with the R set and then calculate $Z(n)$ (Figure 1, Paragraph 7). These calculations were performed for n values from 2 to 100. As a result of the algorithm, the dependence of Z on n was obtained and was denoted as $Z(n)$. Let us consider Paragraphs 1-7 in more details.

3. Main Phases of the Technique and the Algorithms Used

3.1. Spaces Removal, Replacing Uppercase Letters with Lowercase Letters and Replacing Punctuation Marks with Spaces. We submitted a poem, which we would like to study for the presence of fuzzy periods, to the program input. The program replaces all uppercase letters with lowercase letters and deletes all spaces (Figure 1, Paragraph 1). Next, the program replaces punctuation marks such as a dot, comma, dash, colon, interrogative and exclamation marks, and also the end of each line to the space character. The space plays the role of a pause [17]. A space is included in the alphabet as an additional character and, thus, an alphabet of the k characters size is used ($k = 34$ for Russian works and $k = 27$ for works in English). At the program output, a transformed work consisting of only the characters of the given alphabet is obtained. This work is regarded as the S sequence with the N length.

The method of preparing the text can introduce certain distortions into the periodicity, which is in the poetic works. However, an incorrect method can only worsen the statistical importance of periodicity, since the shortcomings in the preparation of the text are compensated by the creation of additional insertions or deletions. This means that, for this text preparation, we may find that not all periods are significant. However, those that are discovered exist in the analyzed poems.

3.2. Creating Random Matrices for the n Length Period. The $k \times n$ dimension random matrices, where k is the size of the alphabet and n is the period length, are generated as follows. Each element of $m(i, j), i=1,\ldots,k, j=1,\ldots n$ matrix was randomly filled with equal probability of either 0 or 1. A total of 10^5 of such matrices are created for each n length period, where n varies consecutively in steps of 1 from 2 to 100. Out of the created random matrices and for each n length period, we selected only 10^3 of such matrices, which in the $k \times n$ dimension space were located at a certain value from each other. For this purpose, the matrix in the $k \times n$ space is considered as a point; and we should take only those points that are located at a distance not less than D_0 from each other. The distance between the points is calculated as

$$D = \sqrt{\sum_{i}^{k} \sum_{j}^{n} \left(m_1(i, j) - m_2(i, j)\right)^2} \quad (1)$$

Here $m_1(i, j)$ is the M_1 random matrix element, and $m_2(i, j)$ is the M_2 random matrix element. The matrix (point in the $k \times n$ space) was added to the $Q(n)$ set, if the D distance between it and every already included matrix (point) in this set was greater than the D_0 value. The first generated random matrix was immediately included in the $Q(n)$ set (Figure 1, Paragraph 2).

3.3. Modification and Optimization of Random Matrices for the n Length Period. Then, a modification of the generated random matrices of the $Q(n)$ set (Figure 1, Paragraph 3) was performed. This was done with the goal of ensuring that the mE_{max} distribution functions from different matrices of the $Q(n)$ set and at the R random sequences set were identical. For this purpose, the algorithm described in [15] was used. To do this, each matrix was modified, so that the R^2 and K_d values would be identical for all the matrices.

$$R^2 = \sum_{i=1}^{k} \sum_{j=1}^{n} m(i, j)^2 \quad (2)$$

$$K_d = \sum_{i=1}^{k} \sum_{j=1}^{n} m(i, j)\, p(i, j) \quad (3)$$

where $m(i, j)$ is the matrix element and $p(i, j) = f(i)t(j)$, while $\sum_{i,j} p(i, j) = 1$, $f(i) = b(i)/N$, where $b(i)$ is the number of the i type characters in the S sequence with $\sum_i b(i) = N$ and $t(j) = 1/n$ for any j. Equation (3) is the equation of a sphere in the $k \times n$ space with R radius. Equation (4) is the equation of a plane in the $k \times n$ space. Then, the modified matrix was optimized using genetic algorithm and dynamic programming [15] (Figure 1, Paragraph 3). The genetic algorithm was applied immediately to the entire $Q(n)$ set of matrices, in order to create such an M_{max} matrix and such a S subsequence that would have the greatest E_{max}. E_{max} is the maximum similarity function when searching for local alignment [18] of the S sequence, in respect to a certain matrix of the $Q(n)$ set [19]. Each matrix in the genetic algorithm appears to be the genotype, and E_{max} here acts as the fitness function. This procedure has already been described in [15]. As a result of this algorithm operation, we obtained the M_{max} matrix (let us call it the mM_{max}), as well as a fragment of the S sequence (let us call it S'), which possessed the maximum value of the similarity function, when it was aligned with the mM_{max} matrix.

3.4. Generation of the R Random Sequences Set. Random sequences were generated using the S' subsequence (Figure 1, Paragraph 5). The R random sequences set was created using the random mixing of characters in the original S' subsequence. The size of this set contains 200 sequences. The random sequence was created on the basis of the original S' subsequence, by randomly mixing the sequences. For this purpose and using the random numbers sensor, the r sequence was generated with a length of N', where N' is the length of the initial S' subsequence. Then, the r sequence was regularized in ascending order and the permutations made were memorized. Thereafter, the permutations made in the r sequence were applied to the S' subsequence. In total, 200 random sequences were created and were included in the R set.

3.5. Random Sequences Alignment in respect to the Optimized Matrix. For the obtained optimized M_{max} matrix (Section 3.3), the E_{max} average value and value variance were calculated. To do this, we constructed the local alignment of each random sequence out of the R set in respect to the optimized M_{max} matrix (Figure 1, Paragraph 6) [20]. Using this algorithm, we searched for the best local alignment between each sequence out of the R set and the sequence of column numbers of the optimized M_{max} matrix. For this purpose, the matrix for the E similarity function was filled using the optimized $m_{max}(i, j)$ matrix:

$$E(i,j) = \max \begin{cases} 0 \\ E(i-1, j-1) + m_{max}(s(i), l) \\ E(i, j-1) - d \\ E(i-1, j) - d \end{cases} \quad (4)$$

where $s(i)$ is the character sequence element and d is the price for the character insertion or deletion from the alphabet in the S character sequence. Here, i and j vary from 1 to N, $l = j - n \cdot \text{int}((j-1)/n)$. This means that the matrix column with the l number always corresponds to the j index. The E matrix has the $N \times N$ dimension, where N is the length of the character sequence. After filling the F matrix, the following value was used: $E_{max} = E(N, N)$.

3.6. $Z(n)$ Calculation. As a result of calculations using formula (5), the value for each random sequence out of the R set was found. Then, the $\overline{E_{max}}$ average value and the $D(E_{max})$ variance were calculated using the E_{max} set obtained for the $Z(n)$ random sequences as follows:

$$Z(n) = \frac{E_{max} - \overline{E_{max}}}{\sqrt{D(E_{max})}} \quad (5)$$

All calculations were performed for the n period length from 2 to 100.

4. Constructing Multiple Alignment

After completing the algorithm operation, a specific n period length was selected that possessed the greatest Z value. For this period length, a local alignment was constructed, which consisted of two sequences located one below the other. The first sequence was a sequence of indices that periodically varied from 1 to n (denoted as *index S*). The second sequence is the character sequence, and, namely, the transformed poem (denoted as *symbol S*). Local alignment was employed to construct multiple alignment in the following way. The local alignment was divided into short fragments as follows: if the index in the *index S* sequence reached the n value, then the alignment fragment was cut from the entire local alignment and so on until the very end of the local alignment. Thus, a set of the alignment fragments was obtained. It should be noted that both in the *index S* and in the *symbol S* sequences the insertion character or the "*" deletion character could be contained. Then, the multiple alignment was constructed using the obtained alignment fragments. In details, the process of constructing multiple alignment is described in [21].

5. Calculating the Chi-Square Distribution Using Multiple Alignment and Its Transfer into Normal Arguments

For the multiple alignment columns, the chi-square distribution was calculated. The column numbers were the period positions and the "*" character was not involved in the calculation. The number of letters in the entire local alignment was counted and denoted by L. The number of I type letters was also counted in the entire local alignment and was denoted as $u(i)$, where I is the letter from the alphabet. Then the i type letter probability was calculated within the entire local alignment, as $p(i) = u(i)/L$. The number of letters was counted without taking in to consideration the "*" in the j column, and it was denoted as $V(j)$, where j varied from 1 to n. The $f(i, j)$ value denoted the number of i letters in the j column. As a result, the chi-square value was calculated for each column with the (n-1) degree of freedom:

$$x^2(j) = \sum_i \frac{(f(i,j) - p(i) \cdot V(j))^2}{p(i) \cdot V(j)} \quad (6)$$

Thereafter, the obtained chi-square distribution was transformed into the arguments of normal distribution. For this purpose, the Wilson-Hilferty approximation was used, based on the fact that the $(x_v^2/v)^{1/3}$ distribution and the increasing v were approaching normal distribution with the $\mu = 1-2/(9\cdot v)$ mathematical expectation and the $\sigma = \sqrt{2/(9\cdot v)}$ variance [22]. As a result, we obtained a formula for converting to a normal distribution with the number of degrees of freedom equal to $(n\text{-}1)$:

$$w\left(x^2(j)\right) = \frac{\left(x^2(j)/(n-1)\right)^{1/3} - (1-2/(9\cdot(n-1)))}{(2/(9\cdot(n-1)))^{1/2}} \quad (7)$$

where n is the period length.

6. Calculating Mutual Information

In order to check the relationship between stresses in a poem and period positions, we calculated the mutual information between the stressed vowels and the period positions. In the multiple alignment, the stressed vowel letters were replaced by 1 and other letters were replaced by 0. Then, the number of zeros (0s) and ones (1s) in each multiple alignment column was counted. As a result, the $2\times n$ dimension table was constructed; and the first line indicated the number of zeros (0s) in each column, whereas the second line indicated the number of ones (1s) in each column. In the end, mutual information was calculated according to the following formula [23]:

$$I = \sum_{i=1}^{r}\sum_{j=1}^{n} a_{ij}\ln\left(a_{ij}\right) - \sum_{i=1}^{r} a_{i\cdot}\ln\left(a_{i\cdot}\right) - \sum_{j=1}^{n} a_{\cdot j}\ln\left(a_{\cdot j}\right) + N\ln(N) \quad (8)$$

where $a_{i\cdot} = \sum_j^n a_{ij}$, $a_{\cdot j} = \sum_i^r a_{ij}$, $N = \sum_i^r \sum_j^n a_{ij}$, and a_{ij} are the table elements. The $2I$ value was distributed as the chi-square with $(r\text{-}1)(n\text{-}1)$ degree of freedom. In order to convert the normal distribution, formula (7) was used. Thus, the w value reflects the correlation of the stressed vowel letters and all the other letters with the period positions. If a correlation is present, the $w(2I)$ values should be greater than 4.0.

7. Study of Artificial Sequences

The developed algorithm was first applied to the study of artificial sequences, one of which was a sequence in the following form: $[abcdefg]_{45}$ (the set of *abcdefg* letters was repeated 45 times); the sequence had an alphabet of 7 letters. Then random substitutions were introduced to this sequence (the number of random substitutions was indicated in % of the initial sequence length). Figure 2 shows that the application of the developed algorithm to an artificial sequence randomly changed by 50%. It could be seen that the Z value takes the maximum value at the 7 letters period length, while the maxima are significant for length periods that are multiples of 7, but these maxima gradually decrease.

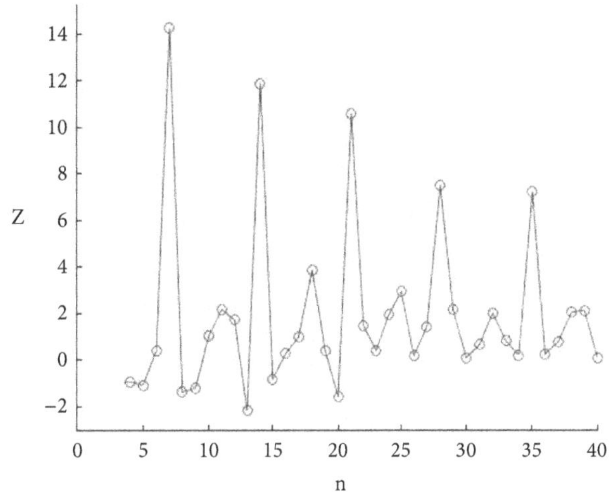

FIGURE 2: Graph of $Z(n)$ for an artificial sequence randomly changed to 50%.

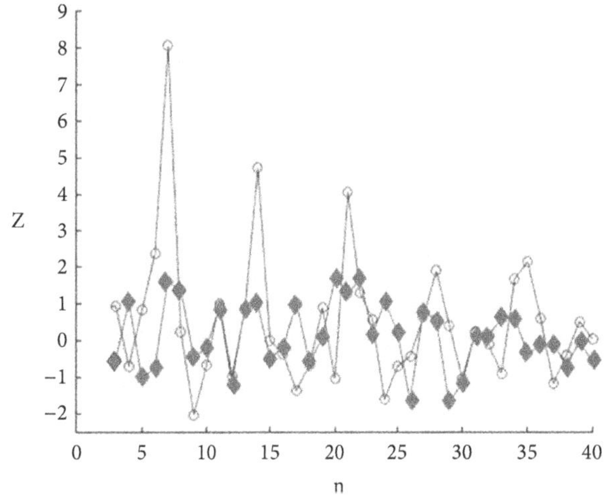

FIGURE 3: Graph $Z(n)$ for an artificial sequence containing 60% random base substitutions with the addition of 8 inserts and 12 deletions (circles) and $Z(n)$ for a random sequence (rhombus).

Figure 3 shows that the result for the artificial sequence is randomly changed by 60% and involves the addition of 8 inserts and 12 deletions of letters, as well as for the random sequence (obtained by random mixing of the initial sequence). It could be seen that there are no fuzzy periods in the random sequence. The test results show that the technique confidently identifies fuzzy periods in the presence of insertions and deletions of characters, as well as of the substitution of random characters.

8. Searching for Periodicity in Works of Poetry

Afterwards, the developed algorithm was applied to identify fuzzy periods in the works of poetry written in Russian and English languages. Certain results were presented which contained the discovered fuzzy periods in the poems of famous classics. The poem by A. Pushkin, the Russian classic,

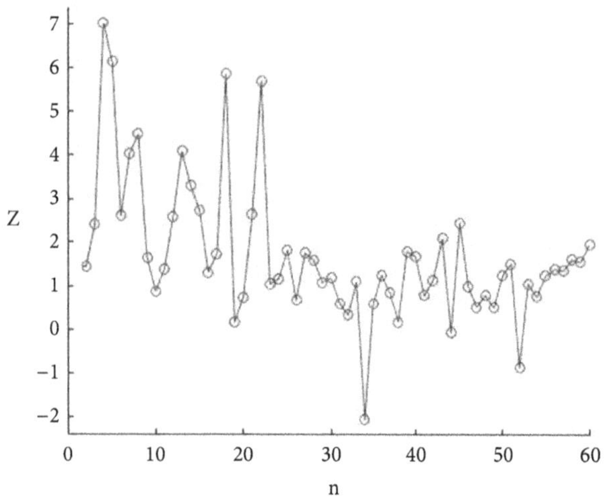

FIGURE 4: Graph Z (n) for the poem A.S. Pushkin "I remember a wonderful moment..."

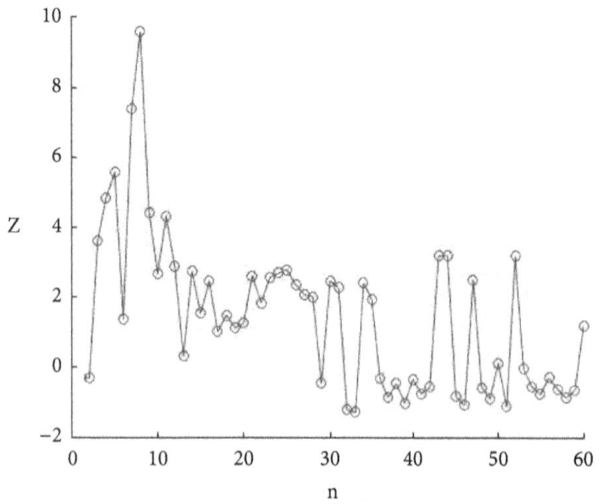

FIGURE 5: Graph $Z(n)$ for William Blake's poem "Spring."

entitled "I remember a wonderful moment. . ." was written with the iambic tetrameter (iamb is a two-syllable verse meter with stress on the second syllable in the foot; the foot in this case consists of two syllables). After applying a set of programs to the poem, a graph of the Z dependence upon the n period length was obtained (Figure 4). Z=7.04 and the maximum value exceeds the Z_0 threshold value and is reached at n=4, while the fuzzy period length is equal to 4. The Z_0 threshold value was determined experimentally by calculation based upon the random sequences obtained from the converted poem (initial sequence) by adding a large number of random substitutions to it. It was calculated that after taking into consideration the probability of the Z>6.0, accidental occurrence was less than 5% for all the analyzed poems.

William Blake's poem entitled "Spring" was written in the two-legged trochee (trochee is the two-syllable verse meter with stress on the first syllable in the foot). The Z=9.58 maximum value (Figure 5) is reached at n=8, which means

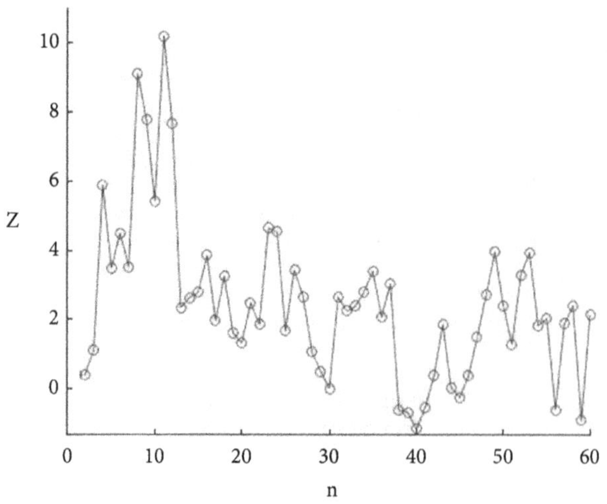

FIGURE 6: Graph $Z(n)$ for William Blake's poem "The Lamb."

that the length of the fuzzy period is equal to 8. It should be noted that, in some cases and for lengths close to a period with the maximum Z, for example, at n=7, a sufficiently great value of Z is registered. This is due to the addition of superfluous inserts and deletions and, thus, periodicity close to that found is being simulated. The $w(2I)$=4.68 mutual information value was calculated, which indicated the presence of a correlation between the stressed vowels and the period positions. An evaluation of the multiple alignment positions (Section 5) shows that the 6th period position is the most significant.

The poetic meter of the famous poem by William Blake called "The Lamb" is the base trochee. The Z=10.17 maximum value (Figure 6) is reached at n=11; i.e., the length of the fuzzy period is equal to 11. It is interesting to note that the Z values of the length period are greater and are close to n=11. However, this happens with the additional inserts and deletions, thus simulating the main period. The mutual information value for this poem is equal to $w(2I)$ = 7.96, which indicates a strong relationship between stresses in the poem and the period positions. A study of the multiple alignment positions (Section 5) shows that the 10th period position is the most significant.

The poem "Fire and Ice" by Robert Frost was studied; the poem was written with the iambic tetrameter with a variable number of feet stops in the lines, either 4 or 8. The Z=7.55 maximum value (Figure 7) was reached at n=13; i.e., the length of the fuzzy period is equal to 13 letters.

The mutual information value is $w(2I)$=6.00. After calculating the chi-square distribution, it turned out that the most significant is the 4th period position in the multiple alignment. Table 2 presents the multiple alignment. By selecting the most common letter in each period position in the multiple alignment, the following set of letters will be received: *iresoleshith*. After substituting all the stressed letters in the poem with 1s and all the remaining letters with 0s, it becomes absolutely evident that there is a relation between the stresses in the poem and the period positions in the multiple alignment, because the first period position practically consists of 1s (Table 3).

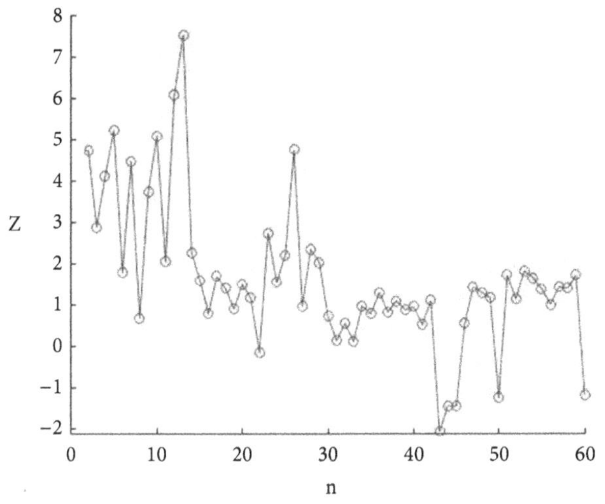

FIGURE 7: Graph Z (n) for the poem by Robert Frost "Fire and ice."

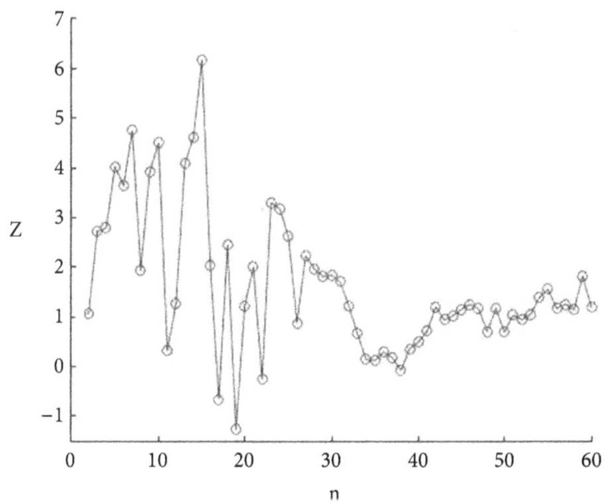

FIGURE 8: Graph $Z(n)$ for the poem by George Gordon Byron "Remember thee."

In an additional study, we considered the possibility of the existence of an interrelation between the positions of the fuzzy period and the stressed letters in the poem. For example, the result is given for the poem "Fire and Ice" by Robert Frost. In this example, the placement of stresses was done manually. To do this, all the percussive letters in the poem were replaced by 1 and all the other letters by 0. After this, it became evident that there is a relationship between the stresses in the poem and the positions of the period in the multiple alignment, so the first position of the period consists of almost only 1 (Table 3).

The poem by George Gordon Byron "Remember Thee" was written with the iambic tetrameter. The Z=6.16 maximum value (Figure 8) was obtained at n=15; i.e., the length of the fuzzy period is equal to 15. The mutual information value is $w(2I)$=4.90; and the most significant is the 11th position of the period in the multiple alignment, which practically consists of the letter "h." If the most popular letter is selected in the position of each period of the multiple alignment and in case

TABLE 1: Lengths of fuzzy periods found in 95 works of poetry of different authors.

Author	Number of analyzed works of poetry	Lengths of fuzzyperiods found in the works of poetry
Pushkin A.	15	2,4,7,10
YeseninS.	10	2,6,8,11
Blok A.	5	3,5,10,11
Tutchev F.	5	2,7,8,15
Fet A.	5	2,4,6,9
Mayakovsky V.	7	2,3,5,8
Shakespeare W.	5	10,15,18,37
Byron D.	5	8,16,28
Frost P.	20	5,8,13,15,20
Blake W.	18	4,6,8,10,11,23

of the same number of certain letters in the column, the one used most rarely in this poem is selected, then the following set will be obtained: *lremombertheea*.

It should be noted that, in English, it is not the letter that strikes but the sound. Since the sound can consist of several vowel letters, then for the sake of certainty, the first letter in the sound was considered (marked) by the stressed letter. Concerning the very arrangement of accents in this poem, we arranged them according to the poetic size (iambic, trochee, etc.). Therefore, in the case of chorea, which is characterized by an alternating sequence, a shock and then an unstressed sound, the poem was placed stress. However, to search for the periods themselves, as earlier noted, only the text itself is used and no other information is required.

In total, 95 poems by Russian and foreign poets were studied. In more than half of the poems studied, fuzzy periods with the Z>6.0 value were found. The other half also had fuzzy periods, but the Z level is lower than 6.0. These results could be explained by the fact that not all poems have a "clear structure" and rhyme. They also combine poetic dimensions that make it difficult to detect the periodicity that is often used. There is another explanation, which is connected with the fact that in many cases a large number of insertions or deletions of symbols in the text are required to notice fuzzy periods. Such causes can lead to a relatively low level of statistical significance of fuzzy periods.

Table 1 shows that the short lengths of the fuzzy period are mostly often encountered in Russian poems, whereas in English poems the lengths of the fuzzy period are longer. This can be explained by the structure of the language. For example, in the Russian language there is a frequent alternation of vowel and consonant letters, and in the English language a case is more widely spread, where several consonant or vowel letters are consecutive, which in turn prolongs the period.

9. Conclusions

The present study aimed at evaluating the efficiency of a new technique in searching for fuzzy periods which are accompanied by insertions and deletions [15] in the texts of

Table 2: Multiple alignment of Robert Frost's poem "Fire and Ice." The zero line shows the positions of the period.

0)	*	*	1	2	*	3	*	4	*	5	6	7	8	9	*	10	*	*	11	12	13
1)	*	*	*	*	*	*	*	*	*	s	o	m	e	s	*	a	*	*	y	t	h
2)	e	w	o	r	l	d	*	w	*	i	l	l	e	n	*	d	*	*	i	n	f
3)	*	*	i	r	*	e	*		*	s	o	m	e	s	*	a	*	*	y	i	n
4)	*	*	i	c	*	e	*		*	f	r	o	m	w	*	h	*	*	a	t	*
5)	*	*	i	v	*	e	t	a	*	s	t	e	d	o	*	f	*	*	d	e	s
6)	*	*	i	r	*	e	*		i	h	o	l	d	w	i	t	*	*	h	t	h
7)	*	*	o	s	*	e	*	w	*	h	o	f	a	*	*	v	*	*	o	r	f
8)	*	*	i	r	*	e	*		*	*	b	u	t	i	*	f	*	*	i	t	h
9)	*	*	a	*	*	d	*	*	*	t	o	p	e	r	*	i	s	*	h	t	w
10)	*	*	i	c	*	e	*		*	i	t	h	i	n	*	k	*	*	i	k	n
11)	*	*	o	w	*	e	*	*	*	n	o	*	u	g	*	h	*	*	o	f	h
12)	*	*	a	t	*	e	*		*	t	o	s	a	y	t	h	*	*	a	t	f
13)	*	*	o	r	*	d	*	e	*	s	t	r	u	c	*	t	*	*	i	o	n
14)	*	*	i	c	*	e	*		i	s	a	l	s	o	*	g	r	e	a	t	
15)	*	*	a	n	*	d	*	w	*	o	u	l	d	s	*	*	*	*	u	f	f
16)	*	*	i	c	*	e	*		*	*	*	*	*	*	*	*	*	*	*	*	*

Table 3: Multiple alignment with the replacement of letters by 0 and 1 for the work of Robert Frost "Fire and Ice." The null line shows the positions of the period.

0)	*	*	1	2	*	3	*	4	*	5	6	7	8	9	*	10	*	*	11	12	13
1)	*	*	*	*	*	*	*	*	*	0	0	0	0	0	*	1	*	*	0	0	0
2)	0	0	1	0	0	0	*	0	*	0	0	0	1	0	*	0	*	*	0	0	0
3)	*	*	1	0	*	0	*	0	*	0	0	0	0	0	*	1	*	*	0	0	0
4)	*	*	1	0	*	0	*	0	*	0	0	0	0	0	*	0	*	*	1	0	*
5)	*	*	0	0	*	0	0	1	*	0	0	0	0	1	*	0	*	*	0	0	0
6)	*	*	1	0	*	0	*	0	0	0	1	0	0	0	0	0	*	*	0	0	0
7)	*	*	1	0	*	0	*	0	*	0	0	0	1	*	*	0	*	*	0	0	0
8)	*	*	1	0	*	0	*	0	*	*	0	0	0	1	*	0	*	*	0	0	0
9)	*	*	1	*	*	0	*	*	*	0	0	0	1	0	*	0	0	*	0	0	0
10)	*	*	1	0	*	0	*	0	*	0	0	0	1	0	*	0	*	*	0	0	0
11)	*	*	1	0	*	0	*	*	*	0	1	*	0	0	*	0	*	*	0	0	0
12)	*	*	1	0	*	0	*	0	*	0	0	0	1	0	0	0	*	*	0	0	0
13)	*	*	1	0	*	0	*	0	*	0	0	0	1	0	*	0	*	*	0	0	0
14)	*	*	1	0	*	0	*	0	0	0	1	0	0	0	*	0	0	1	0	0	0
15)	*	*	0	0	*	0	*	0	*	1	0	0	0	0	*	*	*	*	0	0	0
16)	*	*	1	0	*	0	*	0	*	*	*	*	*	*	*	*	*	*	*	*	*

poems. The applied mathematical method analyzes the text of a poetic work and does not require any other information. It should also be noted that we were unable to find a statistically significant periodicity in ordinary novels both in the Russian and in the English languages. This shows that the fuzzy periods within the linguistic texts are observed exclusively in works of poetry. In general, the results of the present study suggest that a poet uses certain acoustic waves, when writing a poem. It could be noted that poets use a fairly diverse set of acoustic wave lengths, when creating a poem. Probably the fuzzy periodicity of the text is reflection of such acoustic waves. It could be assumed that the acoustic wave is rather important to ensure the psychological impact on the audience, when reciting a poem.

Conflicts of Interest

The authors declare that they have no conflicts of interest.

Acknowledgments

In this work, we used the resources of the high-performance computing center of the National Nuclear Research University "MEPhI."

Supplementary Materials

Guide for launching programs: (1) Generation and selection of random matrices for each length of the period n (n = 2..100) by the program ENG_ComparisonMatrix. The program code in the file "ENG_ComparisonMatrix.c". This program must be run alternately for each length of the period n, which changes in the program code in the variable "n." For example, in order to generate a plurality of matrices for a period length of 2, it is necessary to (a) open the file "ENG_ComparisonMatrix.c" and (b) assign a value of 2 to the variable "n"; it should be "n = 2;". (c) In the program code at the end of the filename to store the matrices to make 2, the following should turn out: FILE ∗ fp1 = fopen ("matrix_n_2.txt", "w"); (d) save changes to the program file and compile. As a result of the program, we get the file "matrix_n_2.txt", which will contain the matrix. And the file "report_number_matrix.txt", in which the number of generated matrices will be stored (this amount is useful in the program ENG_POEM). After starting the program for each length of the period, we get a set of files of the form: "matrix_n_2.txt", "matrix_n_3.txt", "matrix_n_4.txt", and so on to "matrix_n_100.txt". In the program file "ENG_POEM.c" in the array int Number [101] add the value of the number of matrices for each length of the period. For example, for the length of the period n = 2, 1062 matrices were obtained, and then in the array it is necessary to fill the third position and then get: int Number [101] = {0, 0, 1062, ...}. (2) Choosing a poem for research: we go in the folder with poems "95 poem" and open the folder with the desired author and then copy the text from the file with the poem to the file "poem.txt" in the folder with the program Changetext. (3) Changetext program converts the original poem into a sequence for research (removes spaces, punctuation marks replaces with spaces). The program code in the file "Changetext.c". The input file of the program is a file with a poem "poem.txt". The output file of the program is the file "poemSequence.txt". (4) The mixingSeq program generates 200 random sequences from the original sequence in the file "poemSequence.txt". The program code in the file "mixingSeq.c". The input file of the program is the file "poemSequence.txt". The output file of the program is the file "Sequences.txt". (5) The ENG_POEM program checks poems in English for periodicity. The program code in the file "ENG_POEM.c". This program must be run alternately for each length of the period n, which changes in the program code in the variable "n" with the parameters d and Kd given for each poem. Parameters d and Kd are indicated for each poem in the file "d_Kd_poem.txt". Taking the values of these parameters, it is necessary to fill them in the code of the program "ENG_POEM.c". For example, in order to run the program for the length of the period n = 2, with the parameters d = 5, Kd = -0.2, it is necessary to (a) open the file "ENG_POEM.c" and (b) assign a value of 2 to the variable "n", and it should be "n = 2;" and (c) in the code of the program at the end of the file name to store the matrices to make 2, the following should turn out: ∗ Matrix = fopen ("matrix_n_2.txt", "r"); (d) in the program code at the end of the file name for the report to make 2, the following should turn out: ∗ Report = fopen ("report_n_2.txt", "a"); (i) assign a value of 5 to the program variable "d" and it should be "d = 5;". (f) Assign the value -0.2 to the variable "Kd" and it should be "Kd = -0.2;" and (g) save changes to the program file and compile. Input files of the program are the following files: "Sequences.txt" and "matrix_n_2.txt" ("matrix_n_2.txt", in this case n = 2, and if another length of periodicity is investigated, then it is necessary to replace 2 with another value). The output file of the program is the file "report_n_2.txt", and this file stores the local and multiple alignment of the investigated sequence, as well as local alignments for random sequences. At the end of this file, a value of statistical significance for the period n = 2 is indicated in a line of type Z (2) = 4.800792. Carrying out the calculations for each length of the period n we get a set of files of the form: "report_n_2.txt", "report_n_3.txt", "report_n_4.txt",..... "report_n_100.txt". Then a graph is constructed from the values of Z (2), Z (3), Z (4), ... Z (100).

References

[1] H. Wang, P. Mok, and H. Meng, "Capitalizing on musical rhythm for prosodic training in computer-aided language learning," *Computer Speech and Language*, vol. 37, pp. 67–81, 2016.

[2] F. Orsucci, A. Giuliani, C. Webber Jr., J. Zbilut, P. Fonagy, and M. Mazza, "Combinatorics and synchronization in natural semiotics," *Physica A: Statistical Mechanics and its Applications*, vol. 361, no. 2, pp. 665–676, 2006.

[3] O. A. Rosso, H. Craig, and P. Moscato, "Shakespeare and other English Renaissance authors as characterized by Information Theory complexity quantifiers," *Physica A: Statistical Mechanics and its Applications*, vol. 388, no. 6, pp. 916–926, 2009.

[4] E. V. Korotkova, M. A. Korotkovaa, and N. A. Kudryashova, "Information decomposition method to analyze symbolical sequences," *Physics Letters Section A: General, Atomic and Solid State Physics*, vol. 312, no. 3-4, pp. 198–210, 2003.

[5] E. V. Korotkov, M. A. Korotkova, and J. S. Tulko, "Latent sequence periodicity of some oncogenes and DNA-binding protein genes," *Computer Applications in the Biosciences*, vol. 13, no. 1, pp. 37–44, 1997.

[6] V. P. Turutina, A. A. Laskin, N. A. Kudryashov, K. G. Skryabin, and E. V. Korotkov, "Identification of amino acid latent periodicity within 94 protein families," *Journal of Computational Biology*, vol. 13, no. 4, pp. 946–964, 2006.

[7] Z. R. Struzik, "Wavelet methods in (financial) time-series processing," *Physica A: Statistical Mechanics and its Applications*, vol. 296, no. 1-2, pp. 307–319, 2001.

[8] V. Afreixo, P. J. S. G. Ferreira, and D. Santos, "Fourier analysis of symbolic data: a brief review," *Digital Signal Processing*, vol. 14, no. 6, pp. 523–530, 2004.

[9] G. Benson, "Tandem repeats finder: a program to analyze DNA sequences," *Nucleic Acids Research*, vol. 27, no. 2, pp. 573–580, 1999.

[10] M. Pellegrini, "Tandem repeats in proteins: Prediction algorithms and biological role," *Frontiers in Bioengineering and Biotechnology*, 2015.

[11] J. Jorda and A. V. Kajava, "T-REKS: identification of Tandem REpeats in sequences with a K-meanS based algorithm," *Bioinformatics*, vol. 25, no. 20, pp. 2632–2638, 2009.

[12] S. Kurtz, J. V. Choudhuri, E. Ohlebusch, C. Schleiermacher, J. Stoye, and R. Giegerich, "REPuter: The manifold applications of repeat analysis on a genomic scale," *Nucleic Acids Research*, vol. 29, no. 22, pp. 4633–4642, 2001.

[13] I. Elias, "Settling the intractability of multiple alignment," *Journal of Computational Biology: A Journal of Computational Molecular Cell Biology*, vol. 13, no. 7, pp. 1323–1339, 2006.

[14] H. T. Wareham, "A Simplified Proof of the NP- and MAX SNP-Hardness of Multiple Sequence Tree Alignment," *Journal of Computational Biology*, vol. 2, no. 4, pp. 509–514, 1995.

[15] V. M. Pugacheva, A. E. Korotkov, and E. V. Korotkov, "Search of latent periodicity in amino acid sequences by means of genetic algorithm and dynamic programming," *Statistical Applications in Genetics and Molecular Biology*, vol. 15, no. 5, pp. 381–400, 2016.

[16] F. E. Frenkel, M. A. Korotkova, and E. V. Korotkov, "Database of Periodic DNA Regions in Major Genomes," *BioMed Research International*, vol. 2017, Article ID 7949287, 9 pages, 2017.

[17] A. Oras, *Pause Patterns in Elizabethan and Jacobean Drama: An Experiment in Prosody*, 1960.

[18] T. F. Smith and M. S. Waterman, "Identification of common molecular subsequences," *Journal of Molecular Biology*, vol. 147, no. 1, pp. 195–197, 1981.

[19] A. A. Laskin, E. V. Korotkov, M. B. Chaley, and N. A. Kudryashov, "The locally optimal method of cyclic alignment to reveal latent periodicities in genetic texts. The NAD-binding protein sites," *Molekularna Biologija*, vol. 37, no. 4, pp. 663–673, 2003.

[20] S. B. Needleman and C. D. Wunsch, "A general method applicable to the search for similarities in the amino acid sequence of two proteins," *Journal of Molecular Biology*, vol. 48, no. 3, pp. 443–453, 1970.

[21] F. Sievers and D. G. Higgins, *Multiple Sequence Alignment Methods*, 2014.

[22] E. B. Wilson and M. M. Hilferty, "The Distribution of Chi-Square," *Proceedings of the National Acadamy of Sciences of the United States of America*, vol. 17, no. 12, pp. 684–688, 1931.

[23] S. Kullback, *Information Theory and Statistics*, Dover, New York, NY, USA, 1997.

3

Implementing Fuzzy TOPSIS in Cloud Type and Service Provider Selection

Aveek Basu[1] **and Sanchita Ghosh**[2]

[1]*Department of Management, BIT Mesra Kolkata Campus 700107, India*
[2]*Department of Computer Science, BIT Mesra Kolkata Campus 700107, India*

Correspondence should be addressed to Aveek Basu; aveekbasu@gmail.com

Guest Editor: Yiyi Zhang

Cloud computing can be considered as one of the leading-edge technological advances in the current IT industry. Cloud computing or simply cloud is attributed to the Service Oriented Architecture. Every organization is trying to utilize the benefit of cloud not only to reduce the cost overhead in infrastructure, network, hardware, software, etc., but also to provide seamless service to end users with the benefit of scalability. The concept of multitenancy assists cloud service providers to leverage the costs by providing services to multiple users/companies at the same time via shared resource. There are several cloud service providers currently in the market and they are rapidly changing and reorienting themselves as per market demand. In order to gain market share, the cloud service providers are trying to provide the latest technology to end users/customers with the reduction of costs. In such scenario, it becomes extremely difficult for cloud customers to select the best service provider as per their requirement. It is also becoming difficult to decide upon the deployment model to choose among the existing ones. The deployment models are suitable for different companies. There exist divergent criteria for different deployment models which are not tailor made for an organization. As a cloud customer, it is difficult to decide on the model and determine the appropriate service provider. The multicriteria decision making method is applied to find out the best suitable service provider among the top existing four companies and choose the deployment model as per requirement.

1. Introduction

Cloud computing (CC) provides service to users adopting the distributed computing model. It provides computing resources and service to the users as per demand. Cloud computing enhances user's opportunity who can access infrastructure and software applications in a ubiquitous manner [1]. Hardware and licensing costs can be leveraged by utilizing cloud computing and customers can be served in an efficient manner with the aid of scalability attribute. Service offerings in cloud are complex and are constantly evolving. On-demand resource provisioning, broad network access, resource pooling, rapid elasticity, and measured services are some of the key characteristics in cloud computing. Various organizations are trying to adopt cloud from their existing IT infrastructure. The scalability and potential cost effectiveness are attracting various organizations to shift to cloud environment. Recent surveys have revealed that various organizations are willing to transfer their applications to cloud to avail the diverse advantages it offers. The cloud computing market has been growing over the years and the service providers are trying to gain foot hold in the market with various offers in terms of services [2]. There are several cloud service providers in current scenario who are providing services almost identical in nature but with variation in characteristics and offerings. The consumers often face difficulty in selecting the best cloud provider as per their requirement. Cloud providers including Amazon Web Services (AWS) and Microsoft give customers the choice to deploy their applications over a pool of virtual services with practically no upfront investment and with an operating cost proportional to their actual usage [3]. The cloud service providers help the companies to concentrate on their core business areas, but there are certain factors and parameters which customers need to consider during choice of service [4]. Cloud has different deployment models (Public, Private, and Hybrid) and different service models like SaaS, PaaS, and IaaS. Big IT organizations like Google, IBM, Microsoft, Amazon, etc., are offering various cloud services to users. It becomes an uphill task for a cloud customer or user to determine which company to choose [5, 6]. Also it becomes complex to decide on the deployment model. Customers are lacking relevant experience and information to assess the service providers capability in various occasions.

This paper analyzes the different criteria for choosing the suitable service provider along with the deployment model using the Multi Criteria Decision Making (MCDM concept). The evaluation will be done using the Technique for Order Preference by Similarity to an Ideal Solution (TOPSIS) method [7]. MCDM method helps decision makers (DMs) in integrating objective measurements with value judgments that are based on collective group ideas instead of individual opinions.

The best alternative is deduced based on the shortest distance from the fuzzy positive ideal solution (FPIS) and farthest distance from the fuzzy negative ideal solution (FNIS). FPIS refers to maximization of benefit criteria while minimizing cost criteria whereas FNIS will maximize cost criteria and minimize benefit criteria. Utilizing the concept of Fuzzy TOPSIS, FPIS, and FNIS was defined and distance from each alternative from FPIS and FNIS was calculated. In final stage the closeness coefficient will help in determining the ranking order of the alternatives [6].

The current research work deals with the application of TOPSIS in the two most critical areas of concern, viz., selection of the suitable cloud service provider from the top 3 in current fiercely competitive cloud industry and most suitable cloud based on its type. Section 2 deals with related works. Section 3 describes the different cloud service providers and cloud types. Section 4 describes the MCDA techniques. Section 5 deals with fuzzy TOPSIS. Section 6 has two parts dealing with cloud service provider selection and cloud type selection using TOPSIS. Section 7 concludes the paper.

MCDA technique has found its application in several research areas to determine the best alternative among numerous alternatives with different set of criteria. In the current scenario there are multiple cloud service providers offering numerous attractive benefits to customers. Similarly, it is very difficult to determine the suitable cloud type for an organization. Fuzzy TOPSIS has been applied in this paper to determine the most suitable service provider and also the cloud type for an organization.

2. Related Work

In recent years there had been numerous studies on cloud service provider selection and cloud type selection. There are top cloud service providers offering plethora of services at different rate and multiple features. It becomes extremely difficult for a company to decide the best service provider and also the type of cloud to choose [8]. Kumar and Rai (2016) have studied IaaS with 3 different sets of criteria and provided a framework on cloud simulation. Costa (2013) has worked on selection of cloud service providers using MACBETH

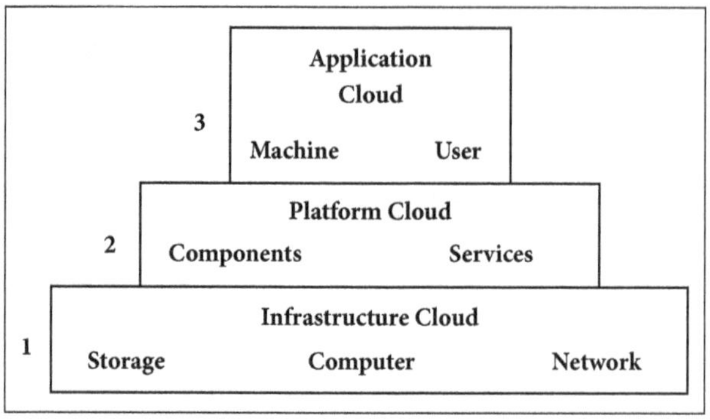

FIGURE 1: Cloud computing models.

MCDA technique. Park and Jeong (2013) proposed a new MCDM approach and applied the same on SaaS based ERP. Rad et al. have studied cloud service platforms and its salient features. Li et al. have worked on the issues related to cloud application performances. Peng et al. have done survey on cloud middleware.

Chen et al. applied constraint programming in cloud provider selection and provided inputs on enterprise policies and its conflicts with users expectations. Chung and Seo (2015) applied ANP technique while working on evaluation on cloud services. Lee and Seo (2013) applied AHP in their research on cloud IaaS.

Godse and Mulik (2009) applied MCDA technique on 3 companies for comparison.

3. Cloud Computing and Cloud Service Providers

3.1. Cloud Computing Overview. Cloud computing refers to storing of data in a remote place and accessing it via Internet instead of doing it in the local machine. So, the greatest advantage is that we need not require a hard drive or dedicated network for data storage and access. One well-known application is Office 365 by which user can store, access, and edit their MS Office documents online without the installation of software in their local machine. The architecture of cloud computing mainly comprises front-end device, back-end platform, cloud-based delivery, and network. The storage in cloud includes three options like public, private, and hybrid. In case of public cloud, it is available to the general public whereas infrastructure is owned and operated by service providers like Google and Microsoft. For private cloud, it is dedicated to a specific organization which can use it for storing organization's data, hosting business application, etc. Other organizations are not able to access the same. Advantages of both public and private cloud are present in hybrid cloud. Organizations can utilize private clouds for sensitive application, while public clouds are meant for nonsensitive applications.

3.2. Cloud Computing Models. Cloud computing models can be mapped against the layers of business value pyramid. Figure 1 depicts the same.

(i) SaaS. The top most layer of the above pyramid is SaaS or functional layer. This specific cloud type is responsible for delivering a single application with the help of a browser to various users through multitenant architecture. It is basically a "pay-as-you-go" model where provider sells an application based on license. The users need not have to take the hazards of maintaining servers or any software which basically reduces the cost. Service providers can also handle it easily as one application needs to be maintained here. Thus, it is cost effective for both sides, users and providers. Few well-known applications are Salesforce.com, SRM, ERP, etc. Few major characteristics of SaaS are listed in the following:

(I) Centralized web-based access to company and commercial software

(II) Providing superior services to client

(III) No software maintenance required from user's perspective

(IV) Integration with different applications possible through Application Programming

(V) Interfaces (APIs)

(ii) PaaS. PaaS or Platform as a service delivers development or operating environments as a service. It is a combination of tools and services designed for coding and deploying the applications in an effective and efficient manner. The major difference with SaaS model is that PaaS is a platform for development/deployment of the software instead of readymade software delivered over the Internet. Few major examples include Salesforce.com's Force.com, Azure from Microsoft, and Google App Engine. The major characteristics are the following:

(a) A one stop solution for developing, testing, deploying, hosting, and maintaining applications

(b) Web-based UI designing tools to create, modify, test, and deploy different UI scenarios

(c) Multitenant architecture facilitating concurrent users

(d) Load balancing, security, and failover capabilities for application to be deployed

(e) OS and cloud programming APIs to create new apps for cloud or to cloudify the current apps

(f) Tools to handle billing and subscription

(iii) IaaS. The infrastructure cloud is responsible for storage and compute resources as a service which is basically used by various IT organizations for providing business solutions. Complete flexibility is provided in this approach to the user; users can choose among desktops, servers, and network resources. The entire infrastructure package can be customized by choosing anything from the list of CPU hours, storage space, bandwidth, etc. This cloud type has different categories like private, public, and hybrid. Public cloud consists of shared resources whereas private cloud is responsible for providing secure access to the resources and is managed by the organization it serves [9]. This type of cloud is maintained by both internal and external providers. Some notable characteristics are the following:

(a) Resources distributed as a service

(b) Dynamic, on-demand scaling of resources

(c) Utility based pricing model

(d) Concurrent users on a single piece of hardware

3.3. Cloud Computing Benefits. Cloud computing provides different benefits. Cloud services offer scalability. Dynamic allocation and deallocation of resources happen based on demand. Cost savings are another major advantage which happens due to cost reduction in capital infrastructure. Applications can be accessed across the globe and without the hardware configuration in the local machine also. Network is simplified, and client can access the application without buying license for individual machine. Storing data on cloud is more reliable as it is not lost easily.

3.4. Challenges behind Cloud Services. Cloud services cover various issues along with its advantages. Few such concerns are listed in the following:

(a) Security and Privacy

(b) Interoperability and Portability

(c) Reliability and Availability

(d) Performance and Bandwidth Cost

3.5. Cloud Service Providers. Cloud service providers refers to different organizations that offer infrastructure, network services, software, hardware components, etc. to different customers and business entities. Cisco, Citrix, IBM, Google, Microsoft, Rackspace, etc. are examples of cloud service providers. In the paper we have considered currently, the top cloud service providers in market are like Amazon Web Services, IBM Bluemix, and Google Cloud Compute. Evaluating the cloud service provider is not an easy activity, but it requires thorough analysis. This has been dealt with in this research article in detail. Cost cannot be the single criteria for selecting a service provider, but different offerings should also be considered in detail. The different fine prints in the agreement need to be analyzed by customers before selecting the provider.

3.6. Public Cloud. In a public cloud a service provider manages resources such as infrastructure, application, and storage and makes it available to cloud consumers via Internet. The service providers like Microsoft, Amazon, Google, etc. own and operate their infrastructure from their own data centers [10]. With the increase in demand of service, users do not need to purchase hard ware as public cloud providers manage the infrastructure. Public clouds are owned by third party organizations and are made available to organizations. Google, Amazon, and Microsoft are notable examples of public cloud vendors.

Some advantages of public cloud are

(i) seamless data availability,

(ii) all round technical support,

(iii) scalability on demand,

(iv) limited investment,

(v) proper resource utilization.

Limitations of public cloud are

(i) data security and privacy.

3.7. Private Cloud. Private cloud as the name suggests refers to infrastructure which is linked to a concern either managed by an organization or third party. It may be present on premise or off site. In private cloud the service is offered to a specific organization and is not meant for public use. In terms of security private clouds are providing highest amount of security service. Private clouds can be built and managed by companies own infrastructure or by cloud service provider.

Some advantages of public cloud are

(i) control over data and information assets,

(ii) high level security,

(iii) superior performance due to intranet and network performance,

(iv) easier to achieve compliance.

Limitations of private cloud are

(i) underutilization of resources

(ii) costliness

3.8. Hybrid Cloud. Hybrid cloud deployment model involves composition of two or more clouds like private, public, etc. The combination of public cloud provider and private cloud platform can also be referred to as a hybrid cloud where they operate independently. Organizations can store sensitive data on private cloud environment and leverage the computational services from public cloud. The hybrid environment ensures minimum data exposure while taking advantage of public cloud platform. Some advantages of public cloud are

(i) private infrastructure to ensure easy accessibility,

(ii) reduction of access time and efficient resource utilization,

(iii) advantage of using computational infrastructure.

Limitations of hybrid cloud are

(i) higher cost,

(ii) security aspects,

(iii) compatibility issues.

4. Multicriteria Decision Analysis (MCDA)

4.1. Background of MCDA. Multicriteria Decision Analysis (MCDA) or Multi Criteria Decision Making is a subbranch of operational research which helps in decision making where several decision making criteria exist. Finding out the best option from the available alternatives is known as decision making. In real world scenario decision making is difficult where there are conflicting goals, different constraints, and unpredictable end results [11]. Here the fuzzy set theory can be used where we are unable to conclude precisely. In 1951 the vector maximum problem was first introduced by Harold William Kuhn and Albert William Tucker. This can be considered as the basics of MCDA. Later in 1972 "Multiple Criteria Decision Making" conference was held in Columbia University. MCDA has been growing in rapid space in the following decades since then.

The MCDA uses the mathematical and computational tools in selection of the best alternative among different choices which may have conflicting criteria. MCDA helps in finding the best alternative among different available choices with respect to specific criteria by decision maker.

We human beings face difficulty in finding the best alternative if there exists multiple criteria and in such situation MCDA can guide in proper decision making. As an example we may consider our current scenario where we have different cloud providers. All the cloud providers are competing against each other to gain the top position and have been trying to draw customers by providing different attractive and cost competitive features. There are distinctive features like control interface features, support services availability, and server OS types which are being offered by the cloud service providers. A customer needs to take decision on the distinctive features being offered by the cloud providers and select the one which is the best alternative among them. MCDA is developed based on the human thinking and their approach in decision making. There are several MCDA methods and techniques available, but the basic methodology is similar based on existing diverse set of criteria and decision making. MCDA consists of methodologies, application of theories, and techniques aiding and dealing with decision making problems. Decision making theory has been applied to solve various real-life problems where multiple conflicting criteria can exist.

4.2. MCDA Methods. MCDA is part of operational research which aims to select the suitable or best alternative among several options with the aid of mathematical and computational tools. It consists of two main categories: Multiattribute Decision Making (MADM) and Multiobjective Decision Making (MODM). MCDA can also be categorized into 2 types, viz., (a) Multiattribute Utility Theory (MAUT) and (b) outranking methods. Using MAUT we try to find a function which determines the utility or usefulness of an alternative. Every action is linked with a marginal utility and a real number will represent the preference in the considered action. The resultant utility represents the addition of the marginal utilities. Outranking method helps

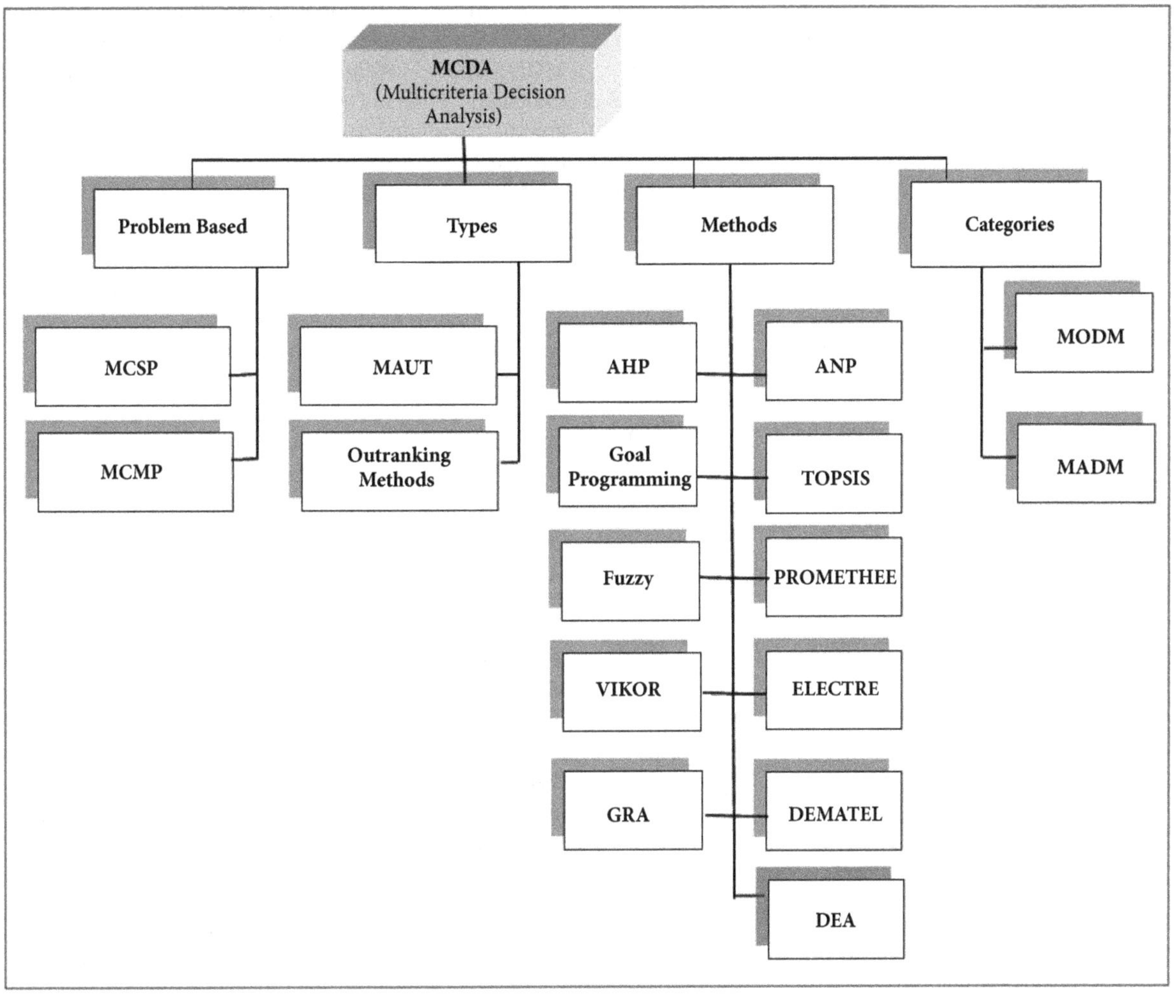

FIGURE 2: Different branches of MCDA.

in finding the alternative which is ranked higher when compared pairwise. Figure 2 shows the different branches of MCDA.

4.2.1. Analytic Hierarchical Process (AHP). Analytic Hierarchical Process (AHP) was introduced by Thomas L Satty in 1980. This is a popular and widely used method for MCDA. Complex MCDM problems are divided into system of hierarchies. In final stage AHP deals with an M X N matrix where M refers to number of alternatives and N represents number of criteria. The matrix is formed considering the relative importance of alternatives against each criterion. Both qualitative and quantitative criteria are used in AHP to find the alternatives and attributes are not entirely independent of each other [12]. Pair wise comparison is used in AHP and the attributes are structured into a hierarchical relationship. Hierarchy starts from top level and then proceeds towards the goal. Criteria, subcriteria, etc., represent the lower levels. The process execution in hierarchy tree initiates from the leaf nodes and it proceeds to the top level. Output level represents hierarchy related to the weight or the influence of different branches which originated at that level. In final stage the comparison is done and best alternative against each attribute is selected.

4.2.2. Analytic Network Process (ANP). Analytic Network Process (ANP) can be referred to as an extension or generalization of Analytic Hierarchy Process (AHP). ANP decision making technique is designed using unidirectional hierarchical relationships between different levels and taking upon the problem of dependence and feedback on different criteria. ANP considers interrelationships within decision levels and attributes using unidirectional hierarchical relationships. It models the decision problem by implementing ratio scale measurements based upon pair wise compare. The interdependence between elements is effectively handled by ANP using composite weights and "super matrix". In many real world scenarios of decision making, ANP has been successfully applied. It has been observed that many decision making problems cannot be hierarchically structured as there is involvement of interaction and dependence between higher and lower level elements [13]. Thus ANP is represented as a network instead of hierarchy. The feedback structure is devoid of the top-to-bottom form in hierarchy. It rather looks like a network with cycles connecting its component of elements which cannot be referred as levels and it loops to connect a component to itself. ANP has sources and sinks. Source node is the origin of paths of influence and is not the destination of paths. Sink node is a destination of paths of influence and is not an origin of paths. A full network may consist of source nodes, intermediate nodes which appear on the paths from source nodes and lie or fall on path to sink nodes and finally sink nodes.

4.2.3. Technique for Order of Preferences by Similarity to Ideal Solutions (TOPSIS). In multicriteria decision making (MCDM) methods we know the ratings and weights of the criteria. TOPSIS was first developed by Hwang and Yoon for solving issues where multicriteria exist and decision making becomes a complex affair. In TOPSIS the performance ratings and weights of the criteria are provided with crisp values. C.T. Chen developed TOPSIS methodology further in solving multiperson and multicriteria decision issues in real world environment where fuzzy exists. Linguistic variables are used to determine weights of all existing criteria and ratings given on each alternative linked to each criterion as there exists fuzziness in decision data and group decision.

In Fuzzy TOPSIS we define the Fuzzy Positive Ideal Solution (FPIS) and Fuzzy Negative Ideal Solution (FNIS). Then calculation is done on distance of each alternative from FPIS and FNIS. Finally ranking order of alternatives is determined using closeness coefficient.

4.2.4. Elimination and Choice Expressing Reality (ELECTRE). Elimination and Choice Expressing Reality (ELECTRE) was introduced initially in 1966. This deals with "outranking relations" by performing pairwise comparison among alternatives under each criterion separately. Later several versions were developed like ELECTRE I, ELECTRE II, ELECTRE III, ELECTRE IV, and so on. ELECTRE belongs to the class of outranking methods and it involves up to 10 steps. Pairwise comparison is done between alternatives to find out the outranking relationships. The relationships in turn help in identifying and removing the alternatives which are dominated by others, resulting in a smaller set of alternatives.

ELECTRE method handles discrete criteria that are both qualitative and quantitative and provides ordering of alternatives. Ranking of alternatives is obtained by using graphs in an iterative procedure. This method starts comparing pair wise of alternatives under each criterion. The ELECTRE method finds a whole system of binary outranking relations among the alternatives. ELECTRE method at times is unable to identify the preferred alternative since the systems are not necessarily complete ones. It yields the core of leading alternatives. This method eliminates the less favorable ones thus giving a clear understanding of the alternatives. In cases where we need to deal with few criteria and large alternatives, this ELECTRE method will be useful.

4.2.5. Fuzzy. Fuzzy set theory has been initially proposed by Zadeh in 1965 and is applied in areas of uncertain data or there is lack of precise information. Fuzzy can help in multicriteria decision making where there exist several uncertainties in available information. The decision pools help in finding selected alternative criteria using the fuzzy MCDA model. Weights are assigned to criteria which are evaluated in terms of linguistic values. Linguistic values are then assigned fuzzy numbers. Inside fuzzy set, fuzzy terms are described by linguistic variables which in turn are used to map the linguistic variables to numeric variables [14].

4.2.6. Goal Programming. Goal Programming is a MODM tool proposed by Charnes in 1955. In areas of multiple conflicting objects the Goal Programming is applied. This is an extension of Linear Programming. Multiple conflicting objective measures can be handled by the Goal Programming optimization procedure. Mathematical programming is combined with the logic of optimization in order to take decisions involving several objectives in different multicriteria decision making problems.

4.3. Motivations in Selecting TOPSIS Method. TOPSIS is one of the most popular multicriteria decision making (MCDM) methods. It deals with the shortest distance from the positive ideal solution and the farthest distance from the negative ideal solution while determining the best alternative. TOPSIS is a well-known method due to the following reasons: (a) theoretical stringency, (b) effective usage of human thinking in selection process, (c) guides in decision making using rank alternatives in fuzzy environment, (d) proper implementation of subjective and objective criteria, (e) crisp values assigned to performance ratings and also to the weights of the criteria which helps in dealing with MCDM problems.

5. Brief Overview of TOPSIS Method

TOPSIS stands for Technique for Order Preference by Similarity to Ideal Solution. Here two artificial alternatives are hypothesized which are Ideal Alternative and Negative Ideal Alternative. Ideal Alternative is the one which has the best attribute values like maximum benefit attributes and minimum cost attributes. Similarly Negative Ideal Alternative includes the worst attribute values like minimum benefit attributes and maximum cost attributes. The TOPSIS method chooses the alternative which is nearest to the ideal solution and farthest from the negative ideal solution [15, 16]. The outline of the TOPSIS method is presented in the following.

Step 1. Evolution matrix is formed of m alternatives and n criteria, using the intersection of each alternative and criteria given as x_{ij}, and then we have a matrix $(x_{ij})_{\mathrm{m\ x\ n}}$

Step 2. The matrix $(x_{ij})_{mxn}$ is then normalized to form the matrix.

R $= (r_{ij})_{m\ x\ n}$ using the normalization method $r_{ij} = X_{ij}/\sqrt{\sum_{i=1}^{m} x_{ij}^2}$, $\mathrm{i} = 1, 2, \ldots\ldots \mathrm{m}, \mathrm{j} = 1, 2, \ldots., \mathrm{n}$

Step 3. Calculate the weighted normalized decision matrix

$t_{ij} = r_{ij} \cdot w_j, \mathrm{i} = 1, 2, \ldots., \mathrm{m}, \mathrm{j} = 1, 2 \ldots\ldots, \mathrm{n}$

where $w_j = W_j / \sum_{j=1}^{n} W_j$, j = 1, 2......n so that $\sum_{j=1}^{n} w_j =$ 1 and W_j is the original weight given to the indicator v_j, j = 1, 2.....n

Step 4. Determine the worst alternative (A_w) and the best alternative (A_b)

$A_w = \{(\max(t_{ij} \mid i = 1, 2\ldots, m) \mid j \in J_-), (\min(t_{ij} \mid i = 1, 2\ldots m) \; j \in J_+)\} \equiv \{t_{wj} \mid j = 1, 2\ldots n\}$,

$A_b = \{(\min(t_{ij} \mid i = 1, 2\ldots, m) \mid j \in J_-), (\max(t_{ij} \mid i = 1, 2\ldots m) \; j \in J_+)\} \equiv \{t_{bj} \mid j = 1, 2\ldots n\}$,

where

$J_+ = \{j = 1, 2, \ldots, n \mid$ j associated with the criteria having a positive impact and

$J_- = \{j = 1, 2, \ldots, n \mid$ j associated with the criteria having a negative impact

Step 5. Calculate the L2 - distance between the target alternative i and the worst condition A_w

$$d_{iw} = \sqrt{\sum_{j=1}^{n} \left(t_{ij} - t_{wj}\right)^2}, \quad i = 1, 2 \ldots\ldots, m \qquad (1)$$

and the distance between the alternative i and the best condition A_b

$$d_{ib} = \sqrt{\sum_{j=1}^{n} \left(t_{ij} - t_{bj}\right)^2}, \quad i = 1, 2 \ldots\ldots, m \qquad (2)$$

where d_{iw} and d_{ib} are L2 - norm distances from the target alternative i to the worst and the best conditions, respectively.

Step 6. Calculate the similarity to the worst condition:

$s_{iw} = d_{iw} / (d_{iw} + d_{ib}), 0 \le s_{iw} \le 1, i = 1, 2 \ldots\ldots m.$

$s_{iw} = 1$ if and only if the alternative solution has the best condition.

$s_{iw} = 0$ if and only if the alternative solution has the worst condition.

Step 7. Rank the alternative according to s_{iw} (i = 1, 2.....m).

6. Applying MCDM Topsis in Cloud

6.1. Evaluation of Cloud Service Provider Using TOPSIS. Three experts evaluate three types of cloud service providers A, I, G and find their evaluations in linguistic variables with respect to objectives, i.e., criteria $C_1 \ldots\ldots C_9$.

The decision makers use seven point scale linguistic variables which are represented by triangular fuzzy numbers to express importance of weight/priority to *Nine* criteria given by Box 1

The criteria are assessed by decision makers which are represented in Table 1.

The three different decision makers are represented in Table 1 by D1, D2, and D3.

Very Low (VL)	(0,0,0.1)
Low (P)	(0,0.1,0.3)
Medium Low (ML)	(0.1,0.3,0.05)
Medium (M)	(0.3,0.5,0.7)
Medium High (MH)	(0.5,0.7,0.9)
High (H) (0.7,0.9,1.0)	
Very High (VH)	(0.9,1.0,1.0)

Box 1

TABLE 1: Criteria assessed by decision makers.

Feature Name	D1	D2	D3
Business Size Support	H	VH	VH
Support for Versatile Industries	VH	H	H
Control Interface Features	H	H	H
Availability of Support Services	VH	VH	VH
Server OS Types	H	H	VH
Preconfigured Operating Systems	MH	MH	MH
Available Runtimes	MH	H	MH
Middleware	H	MH	MH
Native Databases	VH	VH	H

As per above assessment and based on the values of linguistic variables, the fuzzy weight of each criteria j is found as

$$\widetilde{w_j} = \frac{1}{3}\left[w_j^{(1)} + w_j^{(2)} + w_j^{(3)}\right] \qquad (3)$$

Thus

$$\begin{aligned} \widetilde{w_1} &= \frac{1}{3}[G + VG + VG] \\ &= \frac{1}{3}[G + VG + VG] \\ &= \frac{1}{3}[(0.7, 0.9, 1.0) + (0.9, 1.0, 1.0) + (0.9, 1.0, 1.0)] \\ &= \frac{1}{3}[2.5, 2.9, 3] \\ &= (0.83, 0.97, 1) \end{aligned} \qquad (4)$$

Similarly we can obtain the values of $\widetilde{w_2}, \widetilde{w_3} \ldots\ldots \widetilde{w_9}$

In Table 2 features of different cloud service providers are given along with the reason for the different weightage and motivation behind the weightage.

The three cloud companies are evaluated by three decision makers on a seven point linguistic scale comprising the values in Box 2.

The decision makers' opinion is considered for each criterion in Table 3. The fuzzy decision matrix of 3 cloud service providers is given by the following.

For cloud provider AWS, under the feature F_1, the evaluation is

$$\widetilde{x}_{11} = \frac{1}{3}[G + VG + VG]$$

TABLE 2: Cloud service providers and feature compare.

Feature Name	Cloud Service Providers					
	Amazon Web Services(AWS)	**Major Motivators for Weight Assignment**	**IBM Bluemix (IB)**	**Major Motivators for Weight Assignment**	**Google Compute Engine (GCE)**	**Major Motivators for Weight Assignment**
Business Size Support	Good	Supporting Small-Medium Business	Very Good	Supporting Large - Small-Medium Business	Very Good	Supporting Large - Small-Medium Business
Support for Versatile Industries	Good	Supporting medium range of industries	Very Good	Supporting large set of industries	Poor	Supporting very few industries
Control Interface Features	Very Good	Supporting API, GUI, Web Based Application/Control Panel and Command Line	Poor	Supporting Web Based Application/Control Panel and Command Line	Good	Supporting API, Web Based Application/Control Panel and Command Line
Availability of Support Services	Very Good	Supporting Live Chat, Phone, 24/7, Forums, Online/Self-Serve Resources	Good	Supporting 24/7, Forums, Online/Self-Serve Resources	Good	Supporting 24/7, Forums, Online/Self-Serve Resources
Server OS Types	Very Good	Support Linux and Windows	Good	Supporting Windows	Very Good	Supporting Linux and Windows
Preconfigured Operating Systems	Very Good	Supporting Amazon Linux, Cent OS, Debian, Oracle Enterprise Linux, Red Hat Enterprise Linux, SUSE Enterprise Linux, Ubuntu, Windows Server	Poor	Supporting None	Good	Supporting Cent OS, Debian, Red Hat Enterprise Linux, Ubuntu, FreeBSD, openSUSE Linux
Available Runtimes	Good	Supporting NET, Java, PHP, Python and Ruby	Very Good	Supporting Go, Node, Java, PHP, Python and Ruby	Poor	Supporting None
Middleware	Good	Supports Tomcat	Very Good	Supports Jboss, Tomee	Poor	Supports None
Native Databases	Very Good	Supports CouchDB, Microsoft SQL, MongoDB, MySQL	Good	Supports MySQL and PostGreSQL	Poor	Supports None

Very Poor (VP)	(0,0,1)
Poor (P)	(0,1,3)
Medium Poor (MP)	(1,3,5)
Fair (F)	(3,5,7)
Medium Good (MG)	(5,7,9)
Good (G)	(7,9,10)
Very Good (VG)	(9,10,10)

Box 2

$$= \frac{1}{3}\left[(7,9,10)+(9,10,10)+(9,10,10)\right]$$
$$= \frac{1}{3}(25,29,30) = (8.3,9.6,10) \tag{5}$$

Under Feature F_2,

$$\begin{aligned}\widetilde{x}_{12} &= \frac{1}{3}\left[G+MG+MG\right]\\ &= \frac{1}{3}\left[(7,9,10)+(5,7,9)+(5,7,9)\right]\\ &= \frac{1}{3}(17,23,28) = (5.6,7.6,9.3)\end{aligned} \tag{6}$$

Likewise, evaluation is done for AWS for remaining features.

Similarly for other 2 cloud service providers, viz., IB & GCE under 9 Features ($F_1, F_2 \ldots F_9$) the evaluations are done.

Normalized decision matrix for each 9 features is determined against the 3 cloud service providers. Normalized fuzzy decision matrix $\widetilde{v} = (\widetilde{v}_{ij})$ where $\widetilde{v}_{ij} = (\widetilde{r}_{ij})(.)(\widetilde{w}_j)$.

TABLE 3: Cloud service provider features and decision makers analysis.

Feature Name	Cloud Providers	Decision Makers D_1	D_2	D_3
Business Size Support (F_1)	AWS	G	VG	VG
	IB	VG	G	G
	GCE	VG	VG	BG
Support for Versatile Industries (F_2)	AWS	G	MG	MG
	IB	VG	G	VG
	GCE	P	F	MP
Control Interface Features (F_3)	AWS	VG	VG	G
	IB	P	F	MP
	GCE	G	G	MG
Availability of Support Services (F_4)	AWS	VG	G	VG
	IB	G	G	MG
	GCE	G	G	G
Server OS Types (F_5)	AWS	VG	VG	VG
	IB	G	MG	G
	GCE	VG	VG	VG
Preconfigured Operating Systems (F_6)	AWS	VG	G	G
	IB	P	MG	MP
	GCE	G	G	G
Available Run Times (F_7)	AWS	G	G	VG
	IB	VG	G	G
	GCE	P	F	P
Middleware (F_8)	AWS	G	MG	MG
	IB	VG	G	VG
	GCE	P	MP	F
Native Databases (F_9)	AWS	VG	VG	VG
	IB	G	G	G
	GCE	P	F	F

Weighted normalized fuzzy decision matrix is determined next.

The fuzzy positive and fuzzy negative ideal solutions are

$$P^* = (\widetilde{V}_1^{\ *}, \widetilde{V}_2^{\ *} \ldots . \widetilde{V}_9^{\ *})$$

$\overline{N} = (\overline{\overline{V}}_1, \overline{\overline{V}}_2 \ldots\ldots \overline{\overline{V}}_9)$ respectively such that

$$\widetilde{V}_j^{\ *} = (1, 1, 1) \text{ and } \overline{\overline{V}}_j = (0, 0, 0)$$

The distance of the alternatives from B_i from positive solution is calculated by

$$d_i^+ = \sum_{j=1}^{n} d\left(V_{ij}, V_j^*\right) \tag{7}$$

This is done for all the 3 cloud service providers.

Similarly, the distance from the alternatives from (0,0,0) is calculated.

The separation measures from positive ideal solution and negative ideal solution are calculated [17]. Table 4 depicts the same.

TABLE 4: Separation measures.

Cloud Providers	d_1^+	d_1^-
AWS	3.6759	6.0917
IB	4.285	5.56645
GCE	3.78625	6.0728

In Table 4 the separation measures are provided. The closeness coefficient will be calculated based on the separation measures obtained in Table 4.

The closeness coefficient CC_i is given by $d_i^-/(d_i^+ + d_i^-)$

$$CC_1 = \frac{6.0917}{(3.6759 + 6.0917)} = 0.6237$$
$$CC_2 = \frac{5.56645}{(4.285 + 5.56645)} = 0.5650 \tag{8}$$
$$CC_3 = \frac{6.0728}{(3.78625 + 6.0728)} = 0.6159$$

Very Low (VL)	(0,0,0.1)
Low (P)	(0,0.1,0.3)
Medium Low (ML)	(0.1,0.3,0.05)
Medium (M) (0.3,0.5,0.7)	
Medium High (MH)	(0.5,0.7,0.9)
High (H)	(0.7,0.9,1.0)
Very High (VH)	(0.9,1.0,1.0)

Box 3

TABLE 5: Assessment criteria by decision makers.

Feature Name	D1	D2	D3
Cloud environment	H	VH	H
Data center location	VH	H	H
Resource sharing	H	H	H
Cloud storage	VH	VH	VH
Scalability	H	H	VH
Pricing structure	MH	MH	MH
Cloud security	MH	H	MH
Performance	H	MH	MH

The ranking order is now determined based on the closeness coefficient and its found AWS>GCE>IB. Hence the best alternative cloud service provider is AWS, i.e., Amazon Web Services.

6.2. Evaluation of Suitable Cloud Types Based on Notable Features. Evaluations are done in linguistic variables by cloud experts to evaluate suitable cloud platforms with respect to the different features like cloud environment, data center location, resource sharing, cloud storage, scalability, pricing structure, cloud security, and performance [18, 19].

Cloud experts use seven points linguistic variable scale based on the triangular fuzzy numbers and express the weightage/priority to 8 unique features (Box 3).

A committee is formed with decision makers to identify the evaluation criteria, which is shown in following Table 5. The committee of decision makers is represented by D1, D2, and D3 and assessment of criteria importance is shown in Table 5.

The fuzzy weight of each criterion j is determined with the help of given values of linguistic variables. These are provided below.

$$\widetilde{w_j} = \frac{1}{3}\left[w_j^{(1)} + w_j^{(2)} + w_j^{(3)}\right] \tag{9}$$

Thus

$$\begin{aligned}\widetilde{w_1} &= \frac{1}{3}[\mathrm{H} + \mathrm{VH} + \mathrm{H}] \\ &= \frac{1}{3}[\mathrm{H} + \mathrm{VH} + \mathrm{H}] \\ &= \frac{1}{3}[(0.7, 0.9, 1.0) + (0.9, 1.0, 1.0) + (0.7, 0.9, 1.0)] \\ &= \frac{1}{3}[2.3, 2.8, 3] \\ &= (0.77, 0.93, 1)\end{aligned} \tag{10}$$

Very Poor (VP)	(0,0,1)
Poor (P)	(0,1,3)
Medium Poor (MP)	(1,3,5)
Fair (F)	(3,5,7)
Medium Good (MG)	(5,7,9)
Good (G)	(7,9,10)
Very Good (VG)	(9,10,10)

Box 4

Similarly, we can obtain the values of $\widetilde{w_2}, \widetilde{w_3} \ldots\ldots \widetilde{w_9}$

The three cloud platforms are evaluated by three decision makers on a seven point linguistic scale comprising the values in Box 4

The decision makers' opinion is combined for each criterion in Table 6. The fuzzy decision matrix of 3 cloud platforms is given by

For Cloud Platform Public, under the feature CE, the evaluation is

$$\begin{aligned}\tilde{x}_{11} &= \frac{1}{3}[\mathrm{G} + \mathrm{VG} + \mathrm{G}] \\ &= \frac{1}{3}[(7, 9, 10) + (9, 10, 10) + (7, 9, 10)] \\ &= \frac{1}{3}(23, 28, 30) = (7.6, 9.6, 10)\end{aligned} \tag{11}$$

Under feature DC,

$$\begin{aligned}\tilde{x}_{12} &= \frac{1}{3}[\mathrm{G} + \mathrm{G} + \mathrm{MG}] \\ &= \frac{1}{3}[(7, 9, 10) + (7, 9, 10) + (5, 7, 9)] \\ &= \frac{1}{3}(19, 25, 29) = (6.3, 8.3, 9.6)\end{aligned} \tag{12}$$

Likewise, evaluation is done for public cloud for remaining features.

Similarly for the other 2 cloud platforms, viz., Private and Hybrid under 8 features (CE, DC...PR) the evaluations are done.

Normalized decision matrix for each 8 features is determined against the 3 cloud platforms.

Normalized fuzzy decision matrix $\tilde{v} = (\tilde{v}_{ij})$

where $\tilde{v}_{ij} = (\tilde{r}_{ij})(.)(\widetilde{w}_j)$.

Weighted normalized fuzzy decision matrix is determined next.

The fuzzy positive and fuzzy negative ideal solutions are $\mathrm{P}^* = (\widetilde{V}_1{}^*, \widetilde{V}_2{}^* \ldots\ldots \widetilde{V}_9{}^*)$

TABLE 6: Assessment on different platforms by decision makers.

Feature Name	Cloud Platforms	Decision Makers		
		D_1	D_2	D_3
Cloud Environment CE	Public	G	VG	G
	Private	MG	F	MG
	Hybrid	VG	VG	VG
Data Center Location DC	Public	G	G	MG
	Private	MG	MG	F
	Hybrid	G	VG	G
Resource Sharing RS	Public	VG	G	VG
	Private	MG	MG	F
	Hybrid	G	G	G
Cloud Storage CS	Public	G	VG	VG
	Private	MG	G	G
	Hybrid	MG	G	G
Scalability SC	Public	VG	VG	VG
	Private	F	G	G
	Hybrid	G	VG	VG
Pricing Structure PS	Public	VG	G	VG
	Private	F	MG	F
	Hybrid	G	MG	G
Cloud Security SE	Public	MG	F	F
	Private	VG	VG	VG
	Hybrid	G	G	G
Performance PR	Public	F	F	MG
	Private	VG	G	VG
	Hybrid	G	VG	G

TABLE 7: Separation measures.

Cloud Types	d_1^+	d_1^-
Public	1.413	3.378
Private	1.645	2.914
Hybrid	2.78625	4.56

$\overline{N} = (\overline{\overline{V}}_1, \overline{\overline{V}}_2 \ldots\ldots \overline{\overline{V}}_9)$ respectively such that
$\widetilde{V}_j^{\,*} = (1,1,1)$ and $\overline{\overline{V}}_j = (0,0,0)$

The distance of the alternatives from B_i from positive solution is calculated by

$$d_i^+ = \sum_{j=1}^{n} d\left(V_{ij}, V_j^*\right) \tag{13}$$

This is done for all the 3 cloud platforms.

Similarly, the distance from the alternatives from (0,0,0) is calculated.

The separation measures from positive ideal solution and negative ideal solution are calculated [20]. This is given in Table 7.

The closeness coefficient CC_i is given by $d_i^- / (d_i^+ + d_i^-)$ based on the separation measures obtained in Table 7. The separation measure in Table 7 is determined based upon the FPIS and FNIS.

The ranking order is determined from the closeness coefficient matrix and it was found Hybrid>Public>Private. The best alternative cloud type is Hybrid.

7. Conclusion

In today's smart era, competition is gradually increasing among the Cloud service providers in the market. It is getting steeper day by day as new entrants are joining in the service provider pool. Top cloud service providers are changing their strategies to retain their position in this volatile market. Hence they are very keen on selection of features which they are providing to the customers. So every provider offers a set of specific features which differ from those of the others. Now it is the client's responsibility to choose the appropriate vendor from the available ones based on their need. This vendor selection requires understanding and analyzing the features in deep, which is quite tedious if done manually. So there is a crying need of some technique which can perform this analysis automatically. This paper deals with TOPSIS methodology which helps us to select the most suitable service provider by analyzing its available offerings and features. It also studied in detail the different MCDA methods available along with the TOPSIS methodology. The TOPSIS technique is applied in selecting the suitable cloud for an organization which is embracing cloud from on-premise architecture. However, the detailed study will help cloud consumers in selecting the best service provider and cloud service from a set of different offerings and cloud features.

Conflicts of Interest

The authors declare that they have no conflicts of interest.

Authors' Contributions

Aveek Basu carried out the research work. Sanchita Ghosh participated as the reviewer and research guide. All authors read and approved the final manuscript.

References

[1] R. Buyya, C. S. Yeo, S. Venugopal, J. Broberg, and I. Brandic, "Cloud computing and emerging IT platforms: vision, hype, and reality for delivering computing as the 5th utility," *Future Generation Computer Systems*, vol. 25, no. 6, pp. 599–616, 2009.

[2] A. Alkhalil, R. Sahandi, and D. John, "Migration to Cloud Computing: A Decision Process Model," in *Proceedings of the Central European Conference on Information and Intelligent Systems - CECIIS-, 2014, Varadin*, pp. 154–163, 2014.

[3] M. Zhang, R. Ranjan, A. Haller, D. Georgakopoulos, and P. Strazdins, "Investigating decision support techniques for automating Cloud service selection," in *Proceedings of the 2012 4th IEEE International Conference on Cloud Computing Technology and Science, CloudCom 2012*, pp. 759–764, Taiwan, December 2012.

[4] M. Firdhous, S. Hassan, and O. Ghazali, "A comprehensive survey on quality of service implementations in cloud computing," *International Journal of Scientific & Engineering Research*, 2013.

[5] M. Whaiduzzaman, A. Gani, N. B. Anuar, M. Shiraz, M. N. Haque, and I. T. Haque, "Cloud service selection using multicriteria decision analysis," *The Scientific World Journal*, vol. 2014, Article ID 459375, 10 pages, 2014.

[6] C.-C. Lo, D.-Y. Chen, C.-F. Tsai, and K.-M. Chao, "Service selection based on fuzzy TOPSIS method," in *Proceedings of the 24th IEEE International Conference on Advanced Information Networking and Applications Workshops, WAINA 2010*, pp. 367–372, Australia, April 2010.

[7] B. Ashtiani, F. Haghighirad, A. Makui, and G. A. Montazer, "Extension of fuzzy TOPSIS method based on interval-valued fuzzy sets," *Applied Soft Computing*, vol. 9, no. 2, pp. 457–461, 2009.

[8] L. Sun, H. Dong, F. K. Hussain, O. K. Hussain, and E. Chang, "Cloud service selection: state-of-the-art and future research directions," *Journal of Network and Computer Applications*, vol. 45, pp. 134–150, 2014.

[9] S. Soltani, *IaaS Cloud Service Selection using Case-Based Reasoning [Doctoral dissertation]*, 2016.

[10] A. Li, X. Yang, S. Kandula, and M. Zhang, "Comparing public-cloud providers," *IEEE Internet Computing*, vol. 15, no. 2, pp. 50–53, 2011.

[11] R. K. Gavade, "Multi-Criteria Decision Making: An overview of different selection problems and methods," *International Journal of Computer Science and Information Technologies*, vol. 5, no. 4, pp. 5643–5646, 2014.

[12] Y. Lee, "A Decision Framework for Cloud Service Selection for SMEs: AHP Analysis," *SOP Transactions on Marketing Research*, vol. 1, no. 1, pp. 51–61, 2014.

[13] B. Do Chung and S. Kwang-Kyu, "A Cloud Service Selection Model based on Analytic Network Process," *Indian Journal of Science and Technology*, vol. 8, no. 18, 2015.

[14] H. Zimmermann, *Fuzzy Set Theory—and Its Applications*, Springer, Dordrecht, Netherlands, 1996.

[15] M. Behzadian, S. K. Otaghsara, M. Yazdani, and J. Ignatius, "A state-of the-art survey of TOPSIS applications," *Expert Systems with Applications*, vol. 39, no. 17, pp. 13051–13069, 2012.

[16] J. Papathanasiou, V. Kostoglou, and D. Petkos, "A comparative analysis of cloud computing services using multicriteria decision analysis methodologies," *International Journal of Information and Decision Sciences*, vol. 7, no. 1, pp. 51–70, 2015.

[17] S. K. Garg, S. Versteeg, and R. Buyya, "A framework for ranking of cloud computing services," *Future Generation Computer Systems*, vol. 29, no. 4, pp. 1012–1023, 2013.

[18] G. Shivi and T. Narayanan, "A review on matching public, private, and hybrid cloud computing options," *International Journal of Computer Science and Information Technology Research*, vol. 2, no. 2, pp. 213–216, 2014.

[19] S. Singh and T. Jangwal, "Cost breakdown of public cloud computing and private cloud computing and security issues," *International Journal of Computer Science & Information Technology*, vol. 4, no. 2, p. 17, 2012.

[20] M. Pastaki Rad, A. Sajedi Badashian, G. Meydanipour, M. Ashurzad Delcheh, M. Alipour, and H. Afzali, "A Survey of Cloud Platforms and Their Future," in *Computational Science and Its Applications - ICCSA 2009*, vol. 5592 of *Lecture Notes in Computer Science*, pp. 788–796, Springer Berlin Heidelberg, Berlin, Heidelberg, 2009.

4

Mobility Load Balancing in Cellular System with Multicriteria Handoff Algorithm

Solomon T. Girma[1] **and Abinet G. Abebe**[2]

[1] *Department of Electrical Engineering, Pan African University Institute of Sciences, Technology and Innovations, Addis Ababa, Ethiopia*
[2] *Ethio Telecom, Radio Access Network Rollout Department, Addis Ababa, Ethiopia*

Correspondence should be addressed to Solomon T. Girma; solomon.tshm@gmail.com

Academic Editor: Antonella Petrillo

Efficient traffic load balancing algorithm is very important to serve more mobile users in the cellular networks. This paper is based on mobility load balancing handoff algorithm using fuzzy logic. The rank of the serving and the neighboring Base Transceiver Stations (BTSs) are calculated every half second with the help of measurement report from the two-ray propagation model. This algorithm is able to balance load of the BTS by handing off some ongoing calls on BTS's edge of highly loaded BTS to move to overlapping underloaded BTS, such that the coverage area of loaded BTS virtually shrunk towards BTS center of a loaded sector. In case of low load scenarios, the coverage area of a BTS is presumed to be virtually widened to cover up to the partial serving area of neighboring BTS. This helps a highly loaded neighboring BTS or failed BTS due to power or transmission. Simulation shows that new call blocking and handoff blocking using the proposed algorithm are enhanced notably.

1. Introduction

Load balancing is a mechanism whereby overloaded BTSs distribute some of their traffic to less loaded neighbors in order to make the radio resource more efficient [1].

Telecommunication infrastructures originally designed to carry a defined amount of traffic are often congested by an overwhelming request of resources. A naive solution is to expand the infrastructure to match the increasing demand. Due to limitations in available space and shortness of resources, this solution is not possible. The best solution is to carefully tune the parameters of the existing system to accommodate the new traffic demands [2].

There are a number of methods to balance traffic load among BTS [3]. One of the most important methods is based on antenna down tilting [4]. With the down tilt, one directs the antenna radiation further down to the ground. The down tilt is advisable when one wishes to decrease interference and coverage in some specific areas, each BTS to meet only its designed area. When selecting the optimum tilt angle, the goal is to have as high signal strength as possible in the area where the BTS should be serving traffic [4]. Beyond the serving area of the BTS, the signal strength should be as low as possible. A too aggressive down tilting strategy however leads to an overall loss of coverage and creates coverage holes which eventually lead to call drop.

The second method is to adjust the BTS coverage by varying transmitter power. A minimal transmitter power effectively disallows more distant mobiles to access the BTS, thereby decreasing the coverage area and prohibiting distant mobile users to access the BTS [5]. Hence, reducing BTS coverage area runs at the danger of creating coverage holes. Having the coverage holes on the cellular system adversely affect the performance of cellular system leading to call drop. Conversely maximum transmitter power effectively allows mobile at distance to access the BTS, at the cost of producing interference on those BTSs using the same frequency [5].

The antenna height is fundamental to BTS coverage area. If the antenna height is increased, path-loss will reduce. The relation between antenna height and coverage area is stated on two-ray model [6]. If antenna height is doubled, then coverage will be increased by 6 db [6]. To prohibit the distant mobile user from accessing BTS, reducing the height of antenna is good alternative.

Mobile station (MS), in the presence of random networks with overlapping cellular coverage, can connect to any of BTSs. MS in a wireless network switches its current Point of Attachment (PoA) to a new wireless network using a process called handoff [7]. To have global connectivity, handoffs are extremely important cellular communication because of the cellular architecture employed to maximize spectrum utilization. When a mobile terminal moves away from a base station, the signal level drops and there is a need to switch the communications channel to another base station. That time there is a need for a handoff to be executed. Handoff is the process of changing the communications channel associated with the current ongoing connection while a call is in progress [8].

Therefore, by adjusting the handoff regions between neighboring BTSs, it is possible to cause cell edge users in overloaded BTS to migrate to less loaded neighboring BTS. Such an approach is referred to as mobility load balancing, thereby increasing the efficiency of resource utilization [9].

Many metrics have been used to support handoff decisions, including received signal strength (RSS), signal to noise ratio (SIR), power budget, and distance between the mobile station and BTS, traffic load, and mobile velocity, among others. The single criteria handoff decision compares one of the metrics from the serving BTS with that from one of the neighboring BTSs, using a constant handoff threshold value. The selection of the threshold is important to handoff performance. If the threshold is too small, a lot of unnecessary handoffs may take place. On the contrary, the quality of service (QoS) could be low and calls could be dropped if the threshold is too large.

However, all the above techniques of load balancing are not self-adaptive. The problem of network congestion control remains a critical issue and a high priority, mainly given the growing size, demand, and speed of the networks. Therefore, network congestion is becoming a real threat to the growth of existing real time networks (circuit switching). It is a problem that cannot be ignored.

A multicriteria handoff algorithm can provide better performance than a single criterion handoff algorithm due to the extra number of evaluation parameters and the greater potential for achieving the desired balance among different system characteristics. This multicriteria nature of the algorithm allows simultaneous consideration of several significant aspects of the handoff procedure in order to enhance the system performance [8].

Justification of Using Fuzzy Logic in Handoff Process. Network parameters like RSS, power budget, SNR, and traffic load are intrinsically imprecise, vague, and uncertain [10]. Due to this nature of fuzziness, the accurate measurement of these network parameters in a wireless environment is a difficult task. A fuzzy logic method seems to produce better results when used for system design in such condition [10].

The fact that fuzzy logic can mimic human expert reasoning and that many of the terms used to describe a radio signal (weak, far, strong, and congested) are fuzzy in nature makes fuzzy logic a strong candidate for performing handoff decisions. Fuzzy logic can adapt easily to these decisions as it can overcome radio environment uncertainty and fluctuations and can deal with heterogeneous intersystem parameters (shadowing effect, traffic variations, etc.) [11].

2. System Model

2.1. Network Model. We consider a GSM cellular network in which cells are omnidirectional that each cell is served by central BTS and has 100 traffic channels. Mobile users inside the coverage area of a BTS are assumed to be log-normal distribution. BTSs are connected to a single BTS Controller (BSC) where the proposed algorithm is to be located. A mobile station measures the received signal strength, the path-loss, and the received signal to interference ratio from the current serving BTS and all the neighboring BTS. This measurement report is done by mobile station every half second and report to BTS through common control channel. The proposed algorithm ranks the index of all the BTSs in the whole cluster and determines if handoff has to take place.

In this paper, we are assuming a suburban area environment. In instances of low load scenarios, the coverage area of a cell sector is assumed to be virtually extendable to cover up to the partial serving area of a nearby BTS when the BTS has low or medium load. This helps a nearby highly loaded or failed BTS due to power or transmission. Figure 1 shows a cluster of seven cells, whereby every BTS has gotten six adjacent cells. In case of congestion, the adjacent cell will cover its immediate cell.

3. Propagation Model

The two-ray model is used when a single ground reflection dominates the multipath effect. The received signal consists of two components: the line of sight component which is just the transmitted signal propagating through free space and a reflected component which is the transmitted signal reflected off the ground [6]. Equation (1) indicates the received power falls off inversely the fourth power of d and independent of wavelength.

$$\begin{aligned}\text{Pr dbm} &= \text{pt dbm} + 20\log_{10}(\text{ht hr}) - 40\log_{10} d + \text{Gr} \\ &\quad + \text{Gt},\end{aligned} \tag{1}$$

where Pr is the received power in dbm, pt is transmitted power in dbm, Gr is receiver antenna gain in db, Gt is transmitter antenna gain in db, ht is transmitter height in meters, hr is received antenna height in meters, and d is the distance between transmitter and receiver in meters.

Figure 2 shows the received signal strength (RSS) from two neighboring BTSs. It is assumed that the RSS averaged over time, so the fluctuations due to multipath nature of the radio environment can be eliminated [13]. Figure 2 shows a MS moving from BTS I to BTS J. The RSS of BTS I decrease as MS moves away from the BTS and from BTS J increases as it approaches. With conventional algorithm, looking at the variation of RSS from either BTS; it is possible to tell

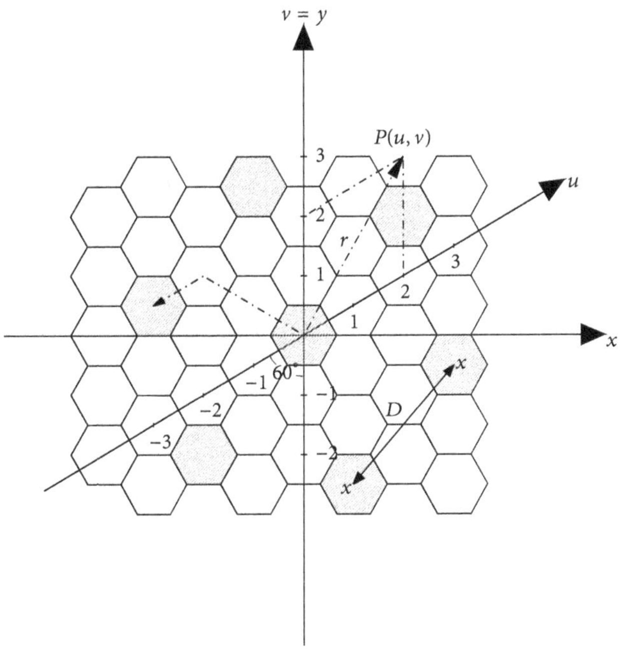

FIGURE 1: GSM system.

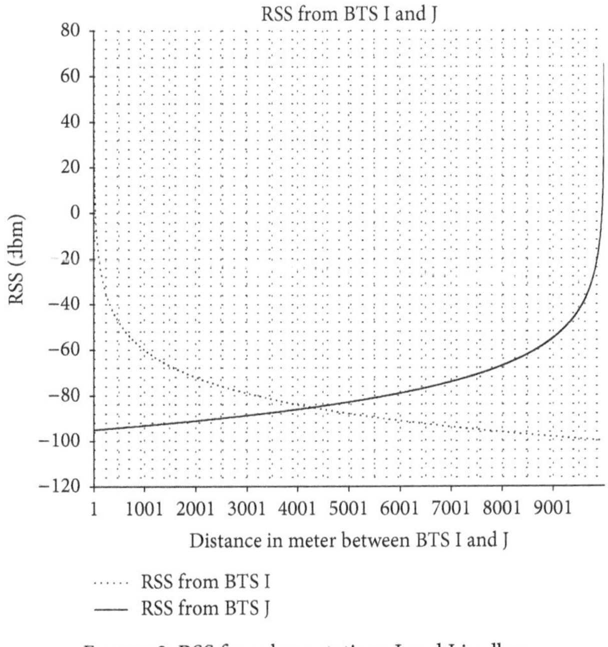

FIGURE 2: RSS from base stations I and J in dbm.

that 4500 m is optimum area where handoff can take place. However, the conventional algorithm with hysteresis allows a mobile station (MS) to make handoff decision only if the RSS received from the neighboring BTS is sufficiently stronger than the current one by the specified hysteresis margin, provided a certain minimum signal level is assured.

Load balancing can be achieved with this method by handing off cell edge users in overloaded cells to migrate to underloaded neighboring BTS. But this type of load balancing is manual. Optimization engineers have to tune the hysteresis.

3.1. Path-Loss. Link budget is calculation of all the gains and losses in a transmission system. It looks at the elements that will decide the signal strength incoming at the receiver. It is necessary to determine link budget in the whole design of radio communication system. Link budget calculations are used for calculating the power levels wanted for cellular communications systems and for obtaining the BTS coverage [14].

To determine a link budget equation, it is necessary to look into all the areas where gains and losses may take place between the transmitter and the receiver. The calculation of the basic link budget is as follows.

$$\text{Received power} = \text{Transmitted power} + \text{gains} - \text{losses}. \quad (2)$$

In the basic calculation of link budget equation it is assumed that the power spreads out equally in all directions from the transmitter antenna source. This is good for theoretical calculations, but not for practical calculations [14].

$$\text{Pr} = \text{Pt} + \text{Gt} + \text{Gr} - \text{LT} - \text{LFS} - \text{LFM} - \text{LR}, \quad (3)$$

where

Pr is received power in dbm,

Pt is transmitter output power in dbm,

Gt is transmitter antenna gain in db,

Gr is receiver antenna gain in db,

LT is transmit feeder and associated losses (feeder, connectors, etc.) in db,

LFS is free space loss or path-loss,

LFM is many-sided signal propagation losses (these include fading margin, polarization mismatch, and losses associated with medium through which signal is travelling),

LR is receiver feeder losses (feeder, connectors, etc.) in db.

Figure 3 shows path-loss from two neighboring BTSs. In cellular networks, the handoff process takes on the responsibility of ensuring that any MS is always connected to the most suitable BTS.

Power budget (PBGT) handover assures that, under normal conditions, any MS is served by BTS that provides minimum path-loss and hence handoff due to path-loss takes place at about 4700 m. However, the conventional algorithm with hysteresis allows a mobile station (MS) to make handoff decision only if the path-loss received from the neighboring BTS is better than the current one by the specified hysteresis margin, provided a certain threshold path-loss is assured. Same as in RSS based handoff algorithm, load balancing can be achieved manually by handing off cell edge users in overloaded cells to migrate to underloaded neighboring BTS.

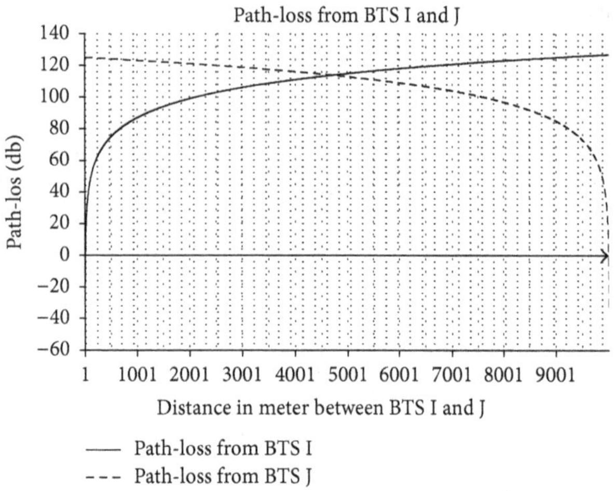

FIGURE 3: Path-loss from base stations I and J in db.

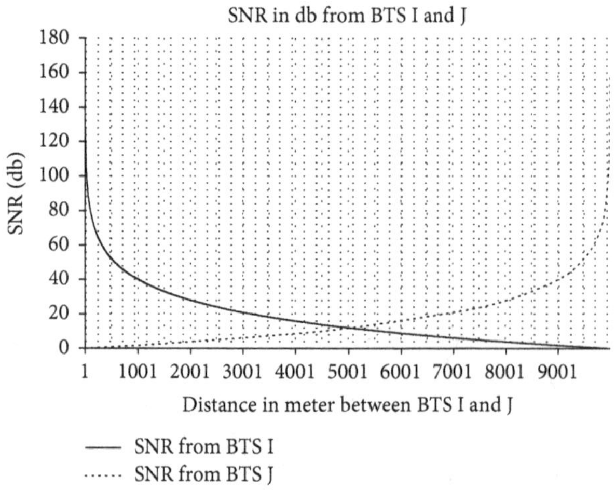

FIGURE 4: SNR from base stations I and J db.

4. Signal to Noise Ratio

It is worth mentioning that in cellular system the cochannel interference is actually the limiting factor in their efficiency and performance and not the total in-band noise in the system. This is because the unwanted signal power is very much higher than the total in-band noise (thermal, man-made) power in the system; hence the noise can be ignored [12].

Mathematically,

$$\text{Signal to noise ratio, SNR} = \frac{\text{RSS}}{I + \text{Ns}}$$
$$\text{Signal to interference ratio, } \frac{C}{I} = \frac{\text{RSS}}{I} \quad \text{for } I \gg \text{Ns}, \tag{4}$$

where RSS is the wanted signal power, I is the unwanted cochannel interfering signal power, and Ns is the total in-band noise power in the system. The signal to interference ratio is a valuable measure of the performance of the modulation technique in the cellular system and it can indeed influence its spectral efficiency.

Figure 4 shows the C/I from two neighboring BTSs. In cellular networks, the handoff process takes on the responsibility of ensuring that any MS is always connected to BTS having a better C/I. Therefore, handoff due to interference takes place at about 5000 m to ensure better C/I.

Constant hysteresis can also be introduced in interference based handoff algorithm, to fasten handoff process when BTS I is congested. This time MS on the edge of BTS I can migrate to BTS J and the load on BTS I can be decongested. If BTS J is congested next time, the optimization engineers have to tune hysteresis to decongest BTS J. This task shows how load is manually shared between neighboring BTSs and it cannot be effective.

4.1. Geographical Model with Second Tier. This model, as indicted on Figure 5, is built depending on the relative geographical location of the serving and interfering BTS with respect to mobile station [12]. The model accounts for the signal path-loss due to free space and propagation loss over the flat earth.

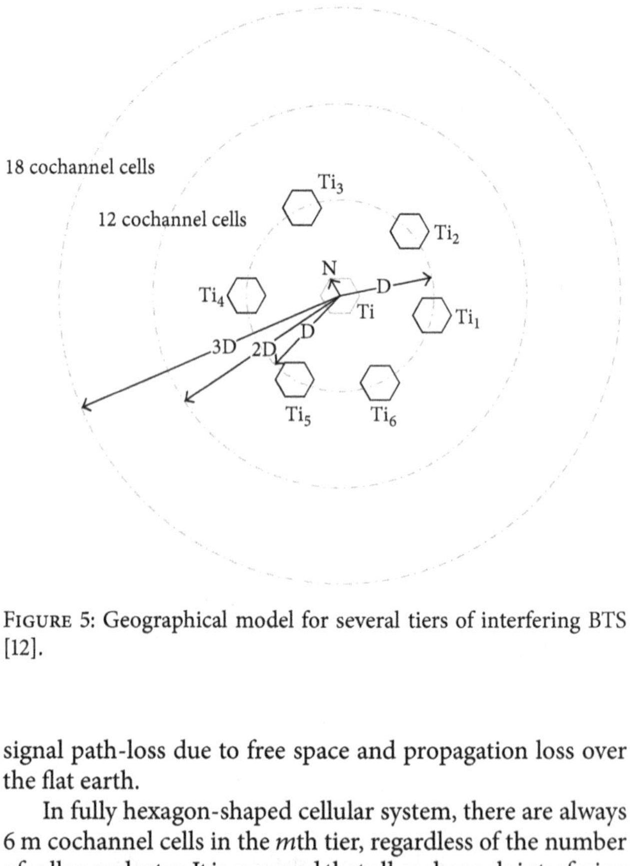

FIGURE 5: Geographical model for several tiers of interfering BTS [12].

In fully hexagon-shaped cellular system, there are always 6 m cochannel cells in the mth tier, regardless of the number of cells per cluster. It is assumed that all cochannels interfering BTS, up to the mth tier considered, are active as in a busy hour situation [12]. It is also assumed that the interference from the BTS higher order tiers (i.e., the $(m+1)$th tier onwards) is negligible. Mobile user in the center BTS can be interfered by six BTSs on the first tier, 12 BTSs on the second tier, 18 BTSs on the third tier, and so on.

At the MS user, being served by BTS on the center of the cluster, the RSS is given by [12]

$$\text{RSS} \propto \frac{1}{R^{\alpha}}, \tag{5}$$

where R is distance between a MS and BTS in meters and σ path-loss constant.

The total interference from the all cochannel cells is given by [12]

$$I \propto \left[\frac{6}{D^{\alpha}} + \frac{12}{(2D)^{\alpha}} + \frac{18}{(3D)^{\alpha}} + \frac{24}{(4D)^{\alpha}} + \cdots\right] \quad (6)$$

$$I \propto \left[\frac{6\times 1}{D^{\alpha}} + \frac{6\times 2}{(2D)^{\alpha}} + \frac{6\times 3}{(3D)^{\alpha}} + \frac{6\times 4}{(4D)^{\alpha}} + \cdots\right], \quad (7)$$

where D is the frequency reuse distance in meters.

Hence,

$$I \propto \sum_{M=1}^{T} \frac{6m}{(mD)^{\alpha}}, \quad (8)$$

where T is the number of tiers and σ path-loss constant

$$\frac{\text{RSS}}{I} = \left(\frac{D}{R}\right)^{\alpha} \frac{1}{6\sum_{m=1}^{T}\left(1/m^{\alpha-1}\right)}. \quad (9)$$

For second tier cochannel cell the above equation will reduce to the following and the third and higher tier can be ignored because (9) is Taylor series.

$$\text{SNR} = \frac{\text{RSS}}{I} = \frac{(D/R)^4}{6.75}. \quad (10)$$

In db the same equation will be reduced to

$$\text{SNR} = \frac{\text{RSS}}{I} = 40\log_{10}(D) - 40\log_{10}R - 10\log_{10}6.75. \quad (11)$$

4.2. Mathematical Justification of Geographical Model. The analytical results for propagation over a plane earth or two-ray models have been derived [6]. For the BTS and mobile station elevated heights ht and hr, respectively, above the ground level and separated a distance d apart, the received power Pr is given in terms of the transmitted power pt as

$$\text{RSS} = \text{Pr} = \text{Gt}_s\text{Gr}\left[\frac{\text{ht}(s)\,\text{hr}}{R^2}\right]^2 \text{p}_{t(s)}. \quad (12)$$

Pr is the desired signal power received at the mobile station from the serving BTS, $p_{t(s)}$ is the transmitted power from the serving BTS, ht(s) is height of the serving BTS antenna in meters, hr is height of the mobile station antenna in meters, Gt_s is the serving BTS antenna gain, and Gr is the mobile station antenna gain. R is the distance between the MS and BTS in meters.

Similarly for the interfering signal power from the interfering BTS,

$$I \approx 6.75\text{Gt}_i\text{Gr}\left[\frac{\text{ht}(i)\,\text{hr}}{D^2}\right]^2 \text{p}_{t(i)}. \quad (13)$$

I is the unwanted signal power received at the mobile station from the interfering BTS, $p_{t(i)}$ is the transmitted power from the interfering BTS, ht(i) is height of the interfering BTS antenna in meters, hr is height of the mobile station antenna in meters, Gt_i is the interfering BTS antenna gain, Gr is the mobile station antenna gain, and D is the frequency reuse distance in meters.

Hence combining equations

$$\frac{\text{RSS}}{I} = \text{SNR} = \frac{\text{Gt}_s\text{Gr}\left[\text{ht}(s)\,\text{hr}/d^2\right]^2 \text{p}_{t(s)}}{6.75\text{Gt}_i\text{Gr}\left[\text{ht}(i)\,\text{hr}/d^2\right]^2 \text{p}_{t(i)}}. \quad (14)$$

From (14), it can be shown that SNR can be maximized by maximizing Gt_s, h_{ts}, and $\text{p}_{t(s)}$.

$$\text{SNR} = \frac{\text{RSS}}{I} = \frac{(D/R)^4}{6.75} \text{ same as } (10). \quad (15)$$

Equation (15) agrees with (10) based on the assumption of equal radiated power in the entire clusters; nevertheless (10) remains valid for mixed cell size when $\text{Gt}_s \neq \text{Gt}_i$, $\text{p}_{t(s)} \neq \text{p}_{t(i)}$, $\text{Gt}_s \neq \text{Gt}_i$, and $\text{ht}(s) \neq \text{ht}(i)$.

5. Simulation Using Matlab

5.1. Fuzzification. In the first step of the handoff process, the model would collect the parameters like RSS, path-loss, and SNR and traffic load of the base station and fed into a fuzzifier. The fuzzifier transforms real time measurements into fuzzy sets. In order to improve the reliability and robustness of the system, Gaussian membership functions (MFs) are used as an alternative to the traditional triangular MFS [15]. For instance, if RSS is considered in crisp set, it can only be weak or strong. RSS cannot be both at a time. However, in a fuzzy set the signal can be considered as weak signal and medium at the same time with graded membership. The membership values are obtained by mapping the values obtained for particular parameter into a membership function.

5.2. Fuzzy Inference. The second step of handoff process involves feeding the fuzzy sets into an inference engine, where a set of fuzzy IF-THEN rules are applied to obtain fuzzy decision sets [15]. These sets are mapped to the corresponding Gaussian membership functions. Since there are four fuzzy inputs and each of them has three subsets, there are $3^4 = 81$ rules (e.g., if RSS is strong and SNR is high and path-loss is small and load is low, then there is no handoff). Fuzzy rules can be defined as a set of possible scenarios. For simple understanding, the set (no handoff, wait, be careful, and handoff) is used to represent the fuzzy set of output handoff decision; the range of the decision matrix is from 0 to 1, where 0 is no handoff and 1 is exactly handoff.

5.3. Defuzzification. Finally, the output fuzzy decision sets are aggregated into a single fuzzy set and passed to the defuzzifier to be converted into a precise quantity during the last stage of the handoff decision. The centroid of area method is elected to defuzzify for changing the fuzzy value into the crisp set [11]. Figure 6 shows the structure of the proposed algorithm, with four inputs and one output, fuzzifier and defuzzifier.

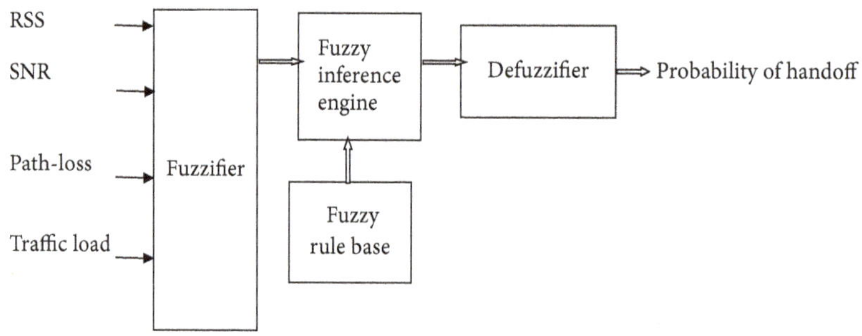

Figure 6: Structure of the proposed algorithm.

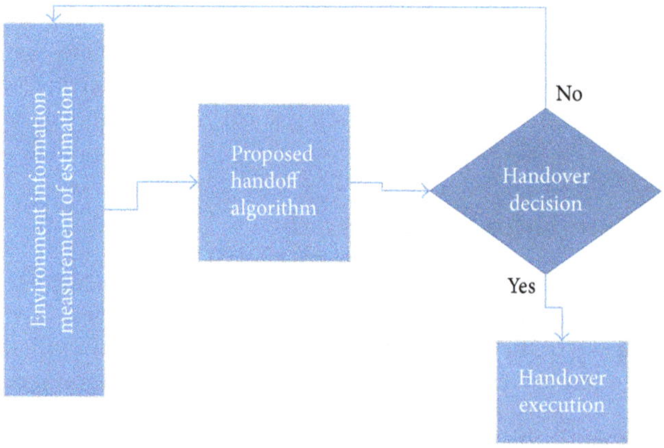

Figure 7: General structure of the proposed multicriteria handoff algorithm.

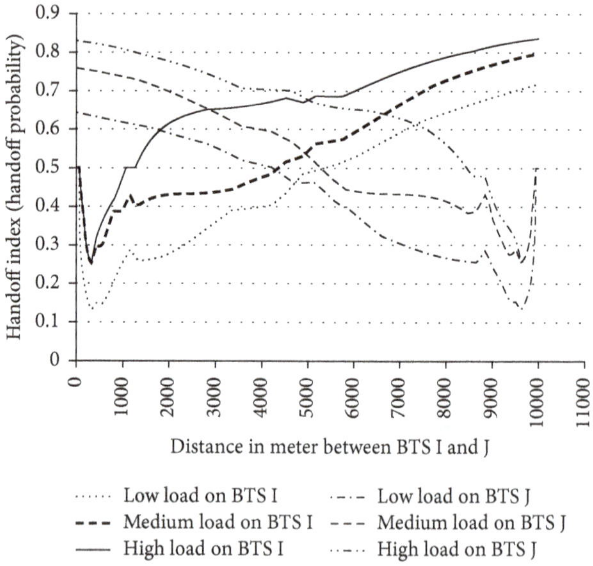

Figure 8: Movement of the mobile station between two adjacent BTSs I and J.

For system simulation Mamdani Fuzzy Inference system is proposed due to fact that Mamdani method is well suited to human input and the nature of wireless is nonlinear [15]. Fuzzy inferences gather the input values of RSS and path-loss and SNR are collected from the propagation model and the load of the BTS from BTS Controller (BSC) and then evaluate them according to the fuzzy interference rules base. The composed and aggregated output of rules evaluation is defuzzified using the centroid of area method and crisp output is obtained [11].

6. Multicriteria Handoff Algorithm

As indicated in Figure 7, fuzzy logic reasoning scheme is applied for mapping of nonlinear data set to scalar output figure. When the problems are in doubt and ambiguous, then to anticipate the correct value among all uncertainties we select fuzzy logic to develop computational techniques that can carry out reasoning and problem solving tasks that require human intelligence [16]. Thus, fuzzy logic is used to choose the optimal serving BTS among given neighboring BTSs for handoff decision founded on the multiple parameters (RSS, path-loss, C/I, and the load of the BTS) as crisp inputs and give the best possible answer to choose the best BTS.

Using single criteria may result in inefficient handover and uneven traffic load distribution. To decide handoff more accurately, more parameters are needed [15]. In our study, we propose four criteria to decide handoff. These input parameters are RSS, path-loss, C/I, and traffic load of the BTS. The first three criteria are user mobility related parameters; meanwhile traffic load on the BTS is network-related parameter [15].

Mobility related parameters are gathered from the wireless environments using propagation models and fed into the fuzzy system along with the traffic load of the BTS [15].

The proposed handoff algorithm gives the handoff index of the serving BTS and all the neighboring BTS and the handoff process are executed once the neighboring BTS is identified better. The output of the fuzzy system is between 0 and 1, where 0 is no handoff and 1 is exactly handoff; the lower the better for MS to be served [15].

7. Result and Discussion

Figure 8 shows a mobile station is moving from one BTS (named BTS I) to another BTS (named BTS J). The handoff index of BTS I increases as a mobile station moves away and decreases as mobile station gets close to the BTS J. Under normal condition the BTS I serves the first 5000 m and BTS J serves the second 5000 m. The handoff index varies depending on the load BTS. The higher the load of BTS the higher the handoff index of the BTS so that the handoff process takes place very fast. For low and medium load on the BTS, the handoff index will be smaller so that the handoff process will be delayed.

The fastest handoff takes place at about 1700 m when BTS I has high load and BTS J has low load, so that the coverage area of a BTS I can virtually be shrunk inwards and mobile on that region is served by the neighboring BTS J. This technique is called dynamic mobility load balancing. The logic behind this method is to adjust the handoff regions by biasing the handoff region depending on the load of the BTSs, causing cell edge users in loaded cells to migrate to less loaded overlapping cells, thereby releasing some traffic channels being occupied by edge user so that the call can proceed. With this process new call blocking can be solved in case the neighboring BTS has low load, satisfying the minimum RSS required. If the neighboring BTS has high load then the handoff takes place on the ideal boundary of the BTS as the conventional handoff as shown in Figure 8.

The most delayed handoff takes place at about 7400 m when the BTS I has low load and BTS J has high load. In this case BTS I coverage area can virtually be expanded towards loaded BTS J. This is due to fact that if no channel is available in the BTS J, the handoff call is blocked which leads a call to terminate prematurely due to dropout. The MS keeps on BTS I as long as RSS, path-loss, and SNR of BTS I are on the recommended threshold. The is a method by which handoff blocking is reduced and avoids call drop due to handover failure as there is no channel available in BTS J [17]. If the neighboring BTS has also low load, then there is no need to delay handoff process. Handoff will take place on the ideal boundary between the base stations.

Figure 3 shows how the mobile in the region between 1700 m and 7400 m swings between BTSs I and J dynamically depending on the load on the BTS, such that the coverage area of a BTS can dynamically be expanded towards a nearby loaded cell or shrunk towards cell center for a loaded sector. Therefore, this mechanism activates a handoff procedure to shift some traffic of a loaded cell towards a lightly loaded cell.

For the same load on either BTS the multicriteria have the same performance as single criteria (RSS based, path-loss based, and SNR based) handoff algorithm; the handoff process takes place at an ideal boundary between the BTSs which is 5000 m.

8. Conclusion

In this paper, we show that it is possible to balance traffic load of cellular network by handing off some ongoing calls on cell edge in highly loaded cells to migrate to overlapping underloaded cells, such that the coverage area of loaded BTS virtually shrunk towards cell center of a loaded sector. In case of low load scenarios, the coverage area of a BTS is presumed to be virtually widened to cover up to the partial serving area of neighboring BTS.

Conflicts of Interest

The authors declare that they have no conflicts of interest.

References

[1] K. Raymond, R. Arnott, R. R. Trivisonno, and M. Kubota, "On mobility load balancing for LTE systems," in *Proceedings of the IEEE 72nd Vehicular Technology Conference Fall*, IEEE, September 2010.

[2] S. Scellato, L. Fortuna, M. Frasca, J. Gómez-Gardeñes, and V. Latora, "Traffic optimization in transport networks based on local routing," *The European Physical Journal B*, vol. 73, no. 2, pp. 303–308, 2010.

[3] P. Muñoz, R. Barco, I. De La Bandera, M. Toril, and S. Luna-Ramírez, "Optimization of a fuzzy logic controller for handover-based load balancing," in *Proceedings of the IEEE 73rd Vehicular Technology Conference*, IEEE, May 2011.

[4] K. A. Akpado, O. S. Oguejiofor, C. O. Ezeagwu, and A. U. Okolibe, "Investigating the impacts of BTS antenna height, tilt and transmitter power on network coverage," *International Journal of Engineering Science Invention*, vol. 2, no. 7, pp. 32–38, 2013.

[5] K. A. Ali, H. S. Hassanein, and H. T. Mouftah, "Directional cell breathing based reactive congestion control in WCDMA cellular networks," in *Proceedings of the 12th IEEE International Symposium on Computers and Communications*, pp. 685–690, July 2007.

[6] A. Goldsmith, *Wireless Communications*, Cambridge University Press, New York, NY, USA, 1st edition, 2005.

[7] S. Srinivas, A. Sahu, and S. K. Jena, "Efficient load balancing in cloud computing using fuzzy logic," *IOSR Journal of Engineering (IOSRJEN)*, vol. 2, pp. 65–71, 2012.

[8] Chandrasekhar and P. K. Behera, "Use of adaptive resonance theory for vertical handoff decision in heterogeneous wireless environment," *International Journal of Recent Trends in Engineering*, vol. 2, no. 3, 2009.

[9] R. Nasri and Z. Altman, "Handoff adaptation for dynamic load balancing in 3GPP long term evolution systems," in *Proceedings of the International Conference on Advancnes in Mobile Computing Multimedia (MoMM '07)*, 2007.

[10] F. Kaleem, A. Mehbodniya, K. K. Yen, and F. Adachi, "A fuzzy preprocessing module for optimizing the access network selection in wireless networks," *Advances in Fuzzy Systems*, vol. 2013, Article ID 232198, 9 pages, 2013.

[11] F. Kaleem, *VHITS: Vertical Handoff Initiation and Target Selection in a Heterogeneous Wireless Network*, Florida International University, 2012.

[12] H. Hammunda, *Cellular Mobile Radio Systems: Designing Systems for Capacity Optimization*, John Wiley and Sons' Ltd, 1997.

[13] L. Chandra Paul, "Handoff/handover mechanism for mobility improvement in the wireless," *The Global Journal of Researches in Engineering*, vol. 13, 2013.

[14] P. K. Sharma and R. K. Singh, "Cell coverage area and link budget calculations in GSM system," *International Journal of Modern Engineering Research (IJMER)*, vol. 2, no. 2, pp. 170–176, 2012.

[15] T. G. Solomon, D. B. O. Konditi, and E. N. Ndungu, "Real time traffic balancing in cellular network by multi-criteria handoff algorithm using fuzzy logic," *International Journal of Emerging Technology and Advanced Engineering*, vol. 4, no. 4, 2014.

[16] G. P. Pollini, "Trends in Handover design," *IEEE Communications Magazine*, vol. 34, no. 3, pp. 82–90, 1996.

[17] Y. Iraqi and R. Boutaba, "handoff and call dropping probabilities in wireless cellular networks," in *Proceedings of the International Conference on Wireless Networks, Communications and Mobile Computing*, June 2005.

5

On (L, M)-Double Fuzzy Filter Spaces

A. A. Abd El-Latif,[1,2] H. Aygün,[3] and V. Çetkin[3]

[1]*Department of Mathematics, Faculty of Science and Arts at Balgarn, P.O. Box 60, University of Bisha, Sabt Al-Alaya 61985, Saudi Arabia*
[2]*High Institute of Computer King Marriott, P.O. Box 3135, Alexandria, Egypt*
[3]*Department of Mathematics, Kocaeli University, 41380 Kocaeli, Turkey*

Correspondence should be addressed to V. Çetkin; vcetkin@gmail.com

Academic Editor: Kemal Kilic

We give in this paper the definitions of (L, M)-double fuzzy filter base and (L, M)-double fuzzy filter structures where L and M are strictly two-sided commutative quantales, and we also investigate the relations between them. Moreover, we propose second-order image and preimage operators of (L, M)-double fuzzy filter base and study some of its fundamental properties. Finally, we handle the given structures in the categorical aspect. For instance, we show that the category (L, M)-**DFIL** of (L, M)-double fuzzy filter spaces and filter maps between these spaces is a topological category over the category **SET**.

1. Introduction

Kubiak [1] and Šostak [2] introduced the notion of L-fuzzy topological space as a generalization of L-topological spaces introduced by Chang [3]. At the bottom of it lies the degree of openness of an L-fuzzy set. A general approach to the study of topological-type structures on fuzzy powersets was developed in [4–14].

On the other hand, Atanassov [15] introduced the idea of intuitionistic (double graded) fuzzy set. Çoker and his coworker(s) [16, 17] introduced the idea of topology of intuitionistic fuzzy sets. Recently, Mondal and Samanta [18] introduced the notion of intuitionistic gradation of openness which is a generalization of both fuzzy topological spaces [2] and the topology of intuitionistic fuzzy sets [16].

Working under the name "intuitionistic" did not continue because doubts were thrown about the suitability of this term, especially when working in the case of complete lattice L. These doubts were quickly ended in 2005 by Gutiérrez García and Rodabaugh [19]. They argued that this term is unsuitable in mathematics and applications. They concluded that they work under the name "double."

The notion of L-filter was introduced by Höhle and Šostak [7] as an expansion of fuzzy filter [20–25]. In recent years, L-filters were used to introduce many kinds of lattice-valued convergence spaces [26–28]. L-filter is an important tool to study L-fuzzy topology [29, 30] and L-fuzzy uniform space [26]. The structure of this paper is as follows. In Section 2, we recall some fundamental definitions related to quantale lattice by giving illustrative examples and also recall some definitions necessary for the main sections. In Section 3, we define (L, M)-double fuzzy filter and (L, M)-double fuzzy filter base and then study relations between them. In the next two sections, we consider two types of second-order Zadeh image and preimage operators of (L, M)-double fuzzy filter base and examine their characteristics by giving examples.

2. Preliminaries

Throughout this paper, let X be a nonempty set. Let $L = (L, \leq, \vee, \wedge)$ be a complete lattice with the least element 0_L and the greatest element 1_L. For $\alpha \in L$, $\underline{\alpha}(x) = \alpha$ for all $x \in X$. The second lattice belonging to the context of our work is denoted by M and $M_0 = M - \{0_M\}$ and $M_1 = M - \{1_M\}$.

A complete lattice $L = (L, \leq, \wedge, \vee)$ is called completely distributive, if for any family $\{\{a_{i,j} : j \in J_i\} : i \in I\}$ in L the following identity holds:

$$\text{(CD)} \quad \bigwedge_{i\in I}\left(\bigvee_{j\in J_i} a_{i,j}\right) = \bigvee_{\varphi\in\prod_{i\in I}J_i}\left(\bigwedge_{i\in I} a_{i,\varphi(i)}\right). \tag{1}$$

Definition 1 (see [24, 31–33]). A triple $L = (L, \leq, \odot)$ is called a strictly two-sided commutative quantale (stsc-quantale, for short) iff it satisfies the following properties:

(L1) $(L, \odot)$ is a commutative semigroup.

(L2) $x \odot 1_L = x$, for all $x \in L$.

(L3) $\odot$ is distributive over arbitrary joins:

$$x \odot \left(\bigvee_{i\in I} y_i\right) = \bigvee_{i\in I}(x \odot y_i), \quad \forall x \in L,\ \forall \{y_i\}_{i\in I} \subseteq L. \tag{2}$$

An stsc-quantale $L = (L, \leq, \odot)$ is an $\bigwedge$-distributive quantale (or stsc-biquantale [34]) if $\odot$ is distributive over nonempty meets:

$$x \odot \left(\bigwedge_{i\in I} y_i\right) = \bigwedge_{i\in I}(x \odot y_i), \quad \forall x \in L,\ \forall \{y_i\}_{i\in I} \subseteq L. \tag{3}$$

Remark 2 (see [24, 25, 31–33, 35]). (1) A complete lattice satisfying the infinite distributive law is an stsc-quantale. In particular, the unit interval $([0,1], \leq, \wedge, 0, 1)$ is an $\bigwedge$-distributive quantale.

(2) Every left-continuous t-norm T on $[0,1]$, $([0,1], \leq, T)$ is an stsc-quantale.

(3) Every continuous t-norm T on $[0,1]$, $([0,1], \leq, T)$ is an $\bigwedge$-distributive quantale.

(4) Every GL-monoid is an stsc-quantale.

(5) Let $(L, \leq, \odot)$ be an stsc-quantale. For each $x, y \in L$, we define

$$x \longmapsto y = \bigvee \{z \in L \mid x \odot z \leq y\}. \tag{4}$$

Then, it satisfies Galois correspondence; that is,

$$x \odot z \leq y \Longleftrightarrow z \leq x \longmapsto y, \quad \forall x, y, z \in L. \tag{5}$$

Definition 3 (see [1, 7, 24, 29, 31, 33, 35–38]). Let $(L, \leq, \odot)$ be an stsc-quantale. A mapping $\star : L \to L$ is called an order-reversing involution if it satisfies the following conditions:

(1) $x^{\star\star} = x$, for each $x \in L$.

(2) If $x \leq y$, then $y^\star \leq x^\star$, for each $x, y \in L$.

An stsc-quantale is called a Girard monoid [37] if $(x \mapsto 0_L) \mapsto 0_L = x$, $\forall x \in L$.

Hence, in case L is a Girard monoid, residuation $\mapsto$ induces an order-reversing involution $\star : L \to L$. In this paper, we always assume that $(L, \leq, \odot, \oplus, \star)$ (resp., $(M, \leq, \tilde{\odot}, \tilde{\oplus}, \star)$) is a Girard monoid with an order-reversing involution $\star$, and the operation $\oplus$ is defined by

$$x \oplus y = (x^\star \odot y^\star)^\star, \quad x^\star = x \longmapsto 0_L \tag{6}$$

unless otherwise specified, where $\tilde{\odot}, \tilde{\oplus}$ denote the quantale operations on M.

Remark 4 (see [39]). When the underlying lattice L is the unit interval $[0,1]$ of the real numbers, the notion of a Girard monoid coincides with the notion of a left-continuous t-norm with strong induced negation $\star$, $(x^\star = x \mapsto 0)$.

Lemma 5 (see [34]). *Let L be a Girard monoid. For each $x, y, z, x_i, y_i \in L$, one has the following properties:*

(1) If $y \leq z$, then $x \odot y \leq x \odot z$, $x \oplus y \leq x \oplus z$, $x \mapsto y \leq x \mapsto z$, and $y \mapsto x \geq z \mapsto x$.

(2) $x \odot y \leq x \wedge y \leq x \vee y \leq x \oplus y$.

Let L be a complete lattice and $\varphi : X \to Y$ be a function. The Zadeh image and preimage operators $\varphi^{\to} : L^X \to L^Y$ and $\varphi^{\leftarrow} : L^Y \to L^X$ are defined by

$$\varphi^{\to}(\lambda)(y) = \bigvee \{\lambda(x) \mid y = \varphi(x)\}, \quad \varphi^{\leftarrow}(\mu)(x) = \mu(\varphi(x)), \quad \forall x \in X,\ y \in Y. \tag{7}$$

Lemma 6 (see [40]). *Let $(L, \leq, \odot)$ be an stsc-quantale and $\varphi : X \to Y$ be a function. For each $\lambda, \mu \in L^X$ and $\lambda_i \in L^Y$, one has the following properties:*

(1) *$\varphi^{\to}(\lambda \odot \mu) \leq \varphi^{\to}(\lambda) \odot \varphi^{\to}(\mu)$ with equality if φ is injective.*

(2) *$\varphi^{\leftarrow}(\odot_{i=1}^{n}\lambda_i) = \odot_{i=1}^{n}\varphi^{\leftarrow}(\lambda_i)$.*

Definition 7 (see [40]). Basic scheme for second-order image operators: let $\varphi : X \to Y$ be a function.

Case 1. Consider

$$[\varphi_L^{\to}]_L^{\to} : L^{L^X} \longrightarrow L^{L^Y}. \tag{8}$$

This is the Zadeh image operator of the Zadeh image operator. We denote it by $\varphi_1^{\rightrightarrows}$; that is, for all $\mathcal{U} \in L^{L^X}$ and $\mu \in L^Y$,

$$\varphi_1^{\rightrightarrows}(\mathcal{U})(\mu) = [\varphi_L^{\to}]_L^{\to}(\mathcal{U})(\mu) = \bigvee \{\mathcal{U}(\lambda) : \mu = \varphi_L^{\to}(\lambda)\}. \tag{9}$$

Case 2. Consider

$$[\varphi_L^{\leftarrow}]_L^{\leftarrow} : L^{L^X} \longrightarrow L^{L^Y}. \tag{10}$$

This is the Zadeh preimage operator of the Zadeh preimage operator. We denote it by $\varphi_2^{\rightrightarrows}$; that is, for all $\mathcal{U} \in L^{L^X}$ and $\mu \in L^Y$,

$$\varphi_2^{\rightrightarrows}(\mathcal{U})(\mu) = [\varphi_L^{\leftarrow}]_L^{\leftarrow}(\mathcal{U})(\mu) = \mathcal{U} \circ \varphi_L^{\leftarrow}(\mu). \tag{11}$$

Basic scheme for second-order preimage operators: let $\varphi : X \to Y$ be a function.

Case 1. Consider

$$[\varphi_L^{\leftarrow}]_L^{\to} : L^{L^X} \longleftarrow L^{L^Y}. \quad (12)$$

This is the Zadeh image operator of the Zadeh preimage operator. We denote it by $\varphi_1^{\Leftarrow}$; that is, for all $\mathcal{V} \in L^{L^Y}$ and $\lambda \in L^X$,

$$\varphi_1^{\Leftarrow}(\mathcal{V})(\lambda) = [\varphi_L^{\leftarrow}]_L^{\to}(\mathcal{V})(\lambda) = \bigvee \{\mathcal{V}(\mu) : \lambda = \varphi_L^{\leftarrow}(\mu)\}. \quad (13)$$

Case 2. Consider

$$[\varphi_L^{\to}]_L^{\leftarrow} : L^{L^X} \longleftarrow L^{L^Y}. \quad (14)$$

This is the Zadeh preimage operator of the Zadeh image operator. We denote it by $\varphi_2^{\Leftarrow}$; that is, for all $\mathcal{V} \in L^{L^Y}$ and $\lambda \in L^X$,

$$\varphi_2^{\Leftarrow}(\mathcal{V})(\lambda) = [\varphi_L^{\to}]_L^{\leftarrow}(\mathcal{V})(\lambda) = \mathcal{V} \circ \varphi_L^{\to}(\lambda). \quad (15)$$

In this paper, we consider additional operators as follows.

Define the operator $\varphi_1^{*\Leftarrow} : M^{L^X} \leftarrow M^{L^Y}$ as $\varphi_1^{*\Leftarrow}(\mathcal{V})(\lambda) = \bigwedge\{\mathcal{V}(\mu) : \lambda = \varphi_L^{\leftarrow}(\mu)\}$, for all $\mathcal{V} \in M^{L^Y}$ and $\lambda \in L^X$.

Define the operator $\varphi_1^{*\Rightarrow} : M^{L^X} \to M^{L^Y}$ as $\varphi_1^{*\Rightarrow}(\mathcal{V})(\mu) = \bigwedge\{\mathcal{V}(\lambda) : \mu = \varphi_L^{\to}(\lambda)\}$, for all $\mathcal{V} \in M^{L^X}$ and $\mu \in L^Y$.

All algebraic operations on L can be extended pointwise to the sets L^X and M^{L^X} as follows: for all $x \in X, \lambda, \mu \in L^X$, and $\mathcal{U}, \mathcal{V} \in M^{L^X}$;

(1) $\lambda \le \mu$ iff $\lambda(x) \le \mu(x)$.

(2) $(\lambda \odot \mu)(x) = \lambda(x) \odot \mu(x)$.

(3) $\mathcal{U} \le \mathcal{V}$ iff $\mathcal{U}(\lambda) \le \mathcal{V}(\lambda)$.

Definition 8 (see [41]). The pair $(\mathcal{T}, \mathcal{T}^*)$ of maps $\mathcal{T}, \mathcal{T}^* : L^X \to M$ is called an (L, M)-double fuzzy topology on X if it satisfies the following conditions:

(LO1) $\mathcal{T}(\lambda) \le (\mathcal{T}^*(\lambda))^*$, for each $\lambda \in L^X$,

(LO2) $\mathcal{T}(\underline{0}) = \mathcal{T}(\underline{1}) = 1_M$, $\mathcal{T}^*(\underline{0}) = \mathcal{T}^*(\underline{1}) = 0_M$,

(LO3) $\mathcal{T}(\lambda_1 \odot \lambda_2) \ge \mathcal{T}(\lambda_1) \tilde{\odot} \mathcal{T}(\lambda_2)$ and $\mathcal{T}^*(\lambda_1 \odot \lambda_2) \le \mathcal{T}^*(\lambda_1) \tilde{\oplus} \mathcal{T}^*(\lambda_2)$, for each $\lambda_1, \lambda_2 \in L^X$,

(LO4) $\mathcal{T}(\bigvee_{i \in \Delta} \lambda_i) \ge \bigwedge_{i \in \Delta} \mathcal{T}(\lambda_i)$ and $\mathcal{T}^*(\bigvee_{i \in \Delta} \lambda_i) \le \bigvee_{i \in \Delta} \mathcal{T}^*(\lambda_i)$, for each $\lambda_i \in L^X$, $i \in \Delta$.

The triplet $(X, \mathcal{T}, \mathcal{T}^*)$ is called an (L, M)-double fuzzy topological space ((L, M)-dfts, for short). $\mathcal{T}$ and $\mathcal{T}^*$ may be interpreted as gradation of openness and gradation of nonopenness, respectively.

Let $(\mathcal{U}, \mathcal{U}^*)$ and $(\mathcal{T}, \mathcal{T}^*)$ be (L, M)-double fuzzy topologies on X. We say that $(\mathcal{U}, \mathcal{U}^*)$ is finer than $(\mathcal{T}, \mathcal{T}^*)$ ($(\mathcal{T}, \mathcal{T}^*)$ is coarser than $(\mathcal{U}, \mathcal{U}^*)$) if $\mathcal{T}(\lambda) \le \mathcal{U}(\lambda)$ and $\mathcal{T}^*(\lambda) \ge \mathcal{U}^*(\lambda)$ for all $\lambda \in L^X$.

Let $(X, \mathcal{T}, \mathcal{T}^*)$ and $(Y, \mathcal{U}, \mathcal{U}^*)$ be (L, M)-dfts's. A function $\varphi : X \to Y$ is called LF-continuous iff $\mathcal{U}(\lambda) \le \mathcal{T}(\varphi^{\leftarrow}(\lambda))$ and $\mathcal{U}^*(\lambda) \ge \mathcal{T}^*(\varphi^{\leftarrow}(\lambda))$, for all $\lambda \in L^Y$.

Thus, we have the category **(L, M)-DFTOP** where the objects are (L, M)-dfts's and the morphisms are LF-continuous maps between these spaces.

Example 9. Let $X = \{x, y\}$ be a set, $L = M = [0, 1]$ and $x \odot y = \max\{x + y - 1, 0\}$, $x \oplus y = \min\{x + y, 1\}$. Then, $([0, 1], \le, \odot)$ is a left-continuous t-norm (Lukasiewicz t-norm) with strong induced negation $x \mapsto 0 = \min\{1 - x, 1\}$. Let $\mu, \rho \in [0, 1]^X$ be defined as follows: $\mu(x) = 0.6$, $\mu(y) = 0.3$, $\rho(x) = 0.5$, $\rho(y) = 0.7$. Define $\mathcal{T}, \mathcal{T}^* : [0, 1]^X \to [0, 1]$ as follows:

$$\mathcal{T}(\lambda) = \begin{cases} 1, & \text{if } \lambda = \underline{0}, \underline{1}; \\ 0.8, & \text{if } \lambda = \mu; \\ 0.3, & \text{if } \lambda = \rho; \\ 0.7, & \text{if } \lambda = \mu \vee \rho; \\ 0.2, & \text{if } \lambda = \mu \wedge \rho; \\ 0, & \text{otherwise,} \end{cases}$$
$$\mathcal{T}^*(\lambda) = \begin{cases} 0, & \text{if } \lambda = \underline{0}, \underline{1}; \\ 0.2, & \text{if } \lambda = \mu; \\ 0.7, & \text{if } \lambda = \rho; \\ 0.3, & \text{if } \lambda = \mu \vee \rho; \\ 0.8, & \text{if } \lambda = \mu \wedge \rho; \\ 1, & \text{otherwise.} \end{cases} \quad (16)$$

Then, the pair $(\mathcal{T}, \mathcal{T}^*)$ is a $([0, 1], [0, 1])$-dft on X.

Remark 10. (1) If $(L = M = [0, 1], \odot = \wedge, \oplus = \vee)$ with an order-reversing involution $\star$, $(a^\star = 1 - a)$ (L, M)-dfts is the concept of Mondal and Samanta [18].

(2) If L and M are frames with 0 and 1, (L, M)-dfts is the concept of Gutiérrez García and Rodabaugh [19].

(3) If $\odot = \wedge$, (L, M)-dfts is the concept of Abd El-latif [42].

Definition 11 (see [29, 30]). A map $\mathcal{F} : L^X \to L$ is called an L-filter if it fulfills the following conditions:

(LF1) $\mathcal{F}(\underline{0}) = 0_M$ and $\mathcal{F}(\underline{1}) = 1_M$.

(LF2) $\mathcal{F}(\lambda \odot \mu) \ge \mathcal{F}(\lambda) \odot \mathcal{F}(\mu)$, for each $\lambda, \mu \in L^X$.

(LF3) If $\lambda \le \mu$, then $\mathcal{F}(\lambda) \le \mathcal{F}(\mu)$.

The pair, $(X\mathcal{F})$, is called an L-filter space.

3. (L,M)-Double Fuzzy Filters and (L,M)-Double Fuzzy Filter Bases

Definition 12. The pair $(\mathcal{F}, \mathcal{F}^*)$ of maps $\mathcal{F}, \mathcal{F}^* : L^X \to M$ is called an (L, M)-double fuzzy filter (briefly, (L, M)-dff) on X if it fulfills the following axioms:

(DFF1) $\mathcal{F}(\lambda) \le (\mathcal{F}^*(\lambda))^*$, for each $\lambda \in L^X$.

(DFF2) $\mathcal{F}(\underline{0}) = 0_M$, $\mathcal{F}(\underline{1}) = 1_M$ and $\mathcal{F}^*(\underline{0}) = 1_M$, $\mathcal{F}^*(\underline{1}) = 0_M$.

(DFF3) $\mathcal{F}(\lambda \odot \mu) \ge \mathcal{F}(\lambda) \tilde{\odot} \mathcal{F}(\mu)$ and $\mathcal{F}^*(\lambda \odot \mu) \le \mathcal{F}^*(\lambda) \tilde{\oplus} \mathcal{F}^*(\mu)$, for each $\lambda, \mu \in L^X$.

(DFF4) If $\lambda \leq \mu$, then $\mathscr{F}(\lambda) \leq \mathscr{F}(\mu)$ and $\mathscr{F}^*(\lambda) \geq \mathscr{F}^*(\mu)$.

The triplet $(X, \mathscr{F}, \mathscr{F}^*)$ is called an (L, M)-double fuzzy filter space (briefly, (L, M)-dffs).

If $(\mathscr{F}_1, \mathscr{F}_1^*)$ and $(\mathscr{F}_2, \mathscr{F}_2^*)$ are two (L, M)-dffs on X, we say that $(\mathscr{F}_1, \mathscr{F}_1^*)$ is finer than $(\mathscr{F}_2, \mathscr{F}_2^*)$ (or $(\mathscr{F}_2, \mathscr{F}_2^*)$ is coarser than $(\mathscr{F}_1, \mathscr{F}_1^*)$), denoted by $(\mathscr{F}_2, \mathscr{F}_2^*) \leq (\mathscr{F}_1, \mathscr{F}_1^*)$ if and only if $\mathscr{F}_2(\lambda) \leq \mathscr{F}_1(\lambda)$ and $\mathscr{F}_2^*(\lambda) \geq \mathscr{F}_1^*(\lambda)$, for each $\lambda \in L^X$.

Definition 13. Let $(X, \mathscr{F}_1, \mathscr{F}_1^*)$ and $(Y, \mathscr{F}_2, \mathscr{F}_2^*)$ be two (L, M)-dffs's. Then, a map $\varphi : X \to Y$ is said to be

(i) a filter map if and only if $\mathscr{F}_2 \leq \mathscr{F}_1 \circ \varphi_L^{\leftarrow}$ and $\mathscr{F}_2^* \geq \mathscr{F}_1^* \circ \varphi_L^{\leftarrow}$;

(ii) a filter preserving map if and only if $\mathscr{F}_1 \leq \mathscr{F}_2 \circ \varphi_L^{\rightarrow}$ and $\mathscr{F}_1^* \geq \mathscr{F}_2^* \circ \varphi_L^{\rightarrow}$.

Normally, the composition of filter maps (resp., filter preserving maps) is a filter map (resp., filter preserving map).

Hence, we get to the category **(L, M)-DFIL** with objects of all the (L, M)-dffs's and the morphisms are filter maps between these spaces.

Remark 14. (i) Let $\mathscr{F}$ be an L-filter on X and $\mathscr{F}^* : L^X \to L$ defined by $\mathscr{F}^*(\lambda) = (\mathscr{F}(\lambda))^*$. Then, the pair $(\mathscr{F}, \mathscr{F}^*)$ is an (L, L)-dff on X. Therefore, (L, M)-dff is a generalization of L-filter due to Höhle and Šostak [7, 29].

(ii) If $\odot = \wedge$, the definition of (L, M)-dff coincides with the definition of a proper (L, M)-intuitionistic fuzzy filter due to Abd El-latif [42].

Theorem 15. *Each (L, M)-dff $(X, \mathscr{F}, \mathscr{F}^*)$ produces an (L, M)-dfts $(X, \mathscr{T}_{\mathscr{F}}, \mathscr{T}^*_{\mathscr{F}^*})$.*

Proof. When contrasting the axioms of (L, M)-dff and (L, M)-dft, we find (DFF4) implying (DFT4).

Let $\{\lambda_i : i \in \Gamma\} \subseteq L^X$. Then, $\lambda_i \leq \bigvee_{i\in\Gamma}\lambda_i$ for all $i \in \Gamma$; due to (DFF4), we have that $\mathscr{F}(\lambda_i) \leq \mathscr{F}(\bigvee_{i\in\Gamma}\lambda_i)$ and $\mathscr{F}^*(\lambda_i) \geq \mathscr{F}^*(\bigvee_{i\in\Gamma}\lambda_i)$ for all $i \in \Gamma$. So,

$$\mathscr{F}\left(\bigvee_{i\in\Gamma}\lambda_i\right) \geq \mathscr{F}(\lambda_i) \geq \bigwedge_{i\in\Gamma}\mathscr{F}(\lambda_i),$$
$$\mathscr{F}^*\left(\bigvee_{i\in\Gamma}\lambda_i\right) \leq \mathscr{F}^*(\lambda_i) \leq \bigvee_{i\in\Gamma}\mathscr{F}^*(\lambda_i). \tag{17}$$

Then, we can get an (L, M)-dft $(\mathscr{T}_{\mathscr{F}}, \mathscr{T}^*_{\mathscr{F}^*})$ defined by

$$\mathscr{T}_{\mathscr{F}}(\lambda) = \begin{cases} \mathscr{F}(\lambda), & \text{if } \lambda \neq \underline{0} \\ 1_M, & \text{if } \lambda = \underline{0}, \end{cases}$$
$$\mathscr{T}^*_{\mathscr{F}^*}(\lambda) = \begin{cases} \mathscr{F}^*(\lambda), & \text{if } \lambda \neq \underline{0} \\ 0_M, & \text{if } \lambda = \underline{0}. \end{cases} \tag{18}$$

Theorem 16. *Let $(X, \mathscr{F}_1, \mathscr{F}_1^*)$ and $(Y, \mathscr{F}_2, \mathscr{F}_2^*)$ be (L, M)-dffs's. If $\varphi : (X, \mathscr{F}_1, \mathscr{F}_1^*) \to (Y, \mathscr{F}_2, \mathscr{F}_2^*)$ is a filter map, then $\varphi : (X, \mathscr{T}_{\mathscr{F}_1}, \mathscr{T}^*_{\mathscr{F}_1^*}) \to (Y, \mathscr{T}_{\mathscr{F}_2}, \mathscr{T}^*_{\mathscr{F}_2^*})$ is an LF-continuous map.*

Proof. Let $\mu \in L^Y$. If $\mu = 0_Y$ or $\varphi_L^{\leftarrow}(\mu) = 0_X$, then the proof is easy. Let $\mu \neq 0_Y$ and $\varphi_L^{\leftarrow}(\mu) \neq 0_X$. Then, from the definition of double filter map and Theorem 15, we have

$$\mathscr{T}_{\mathscr{F}_1}(\varphi_L^{\leftarrow}(\mu)) = \mathscr{F}_1(\varphi_L^{\leftarrow}(\mu)) \geq \mathscr{F}_2(\mu) = \mathscr{T}_{\mathscr{F}_2}(\mu),$$
$$\mathscr{T}^*_{\mathscr{F}_1^*}(\varphi_L^{\leftarrow}(\mu)) = \mathscr{F}_1^*(\varphi_L^{\leftarrow}(\mu)) \leq \mathscr{F}_2^*(\mu) = \mathscr{T}^*_{\mathscr{F}_2^*}(\mu). \tag{19}$$

Corollary 17. *The function F : **(L, M)-DFIL** $\to$ **(L, M)-DFTOP** defined by $F(X, \mathscr{F}, \mathscr{F}^*) = (X, \mathscr{T}_{\mathscr{F}}, \mathscr{T}^*_{\mathscr{F}^*})$ and $F(\varphi) = \varphi$ is a functor.*

Notation 18. Let $\mathscr{B}, \mathscr{B}^* : L^X \to M$ be two maps and $\lambda \in L^X$. Then, $\langle\mathscr{B}\rangle$ and $\langle\mathscr{B}^*\rangle$ are defined as follows:

$$\langle\mathscr{B}\rangle(\lambda) = \bigvee_{\mu\leq\lambda}\mathscr{B}(\mu),$$
$$\langle\mathscr{B}^*\rangle(\lambda) = \bigwedge_{\mu\leq\lambda}\mathscr{B}^*(\mu). \tag{20}$$

Definition 19. The pair $(\mathscr{B}, \mathscr{B}^*)$ of maps $\mathscr{B}, \mathscr{B}^* : L^X \to M$ is called an (L, M)-double fuzzy filter base (briefly, (L, M)-dffb) on X if it fulfills the following axioms:

(DFFB1) $\mathscr{B}(\lambda) \leq (\mathscr{B}^*(\lambda))^*$, for each $\lambda \in L^X$.

(DFFB2) $\mathscr{B}(\underline{0}) = 0_M$, $\mathscr{B}(\underline{1}) = 1_M$ and $\mathscr{B}^*(\underline{0}) = 1_M$, $\mathscr{B}^*(\underline{1}) = 0_M$.

(DFFB3) $\langle\mathscr{B}\rangle(\lambda \odot \mu) \geq \mathscr{B}(\lambda) \,\tilde{\odot}\, \mathscr{B}(\mu)$ and $\langle\mathscr{B}^*\rangle(\lambda \odot \mu) \leq \mathscr{B}^*(\lambda) \,\tilde{\oplus}\, \mathscr{B}^*(\mu)$, for each $\lambda, \mu \in L^X$.

If $(\mathscr{B}_1, \mathscr{B}_1^*)$ and $(\mathscr{B}_2, \mathscr{B}_2^*)$ are two (L, M)-dffb's on X, we say $(\mathscr{B}_1, \mathscr{B}_1^*)$ is finer than $(\mathscr{B}_2, \mathscr{B}_2^*)$ (or $(\mathscr{B}_2, \mathscr{B}_2^*)$ is coarser than $(\mathscr{B}_1, \mathscr{B}_1^*)$) denoted by $(\mathscr{B}_2, \mathscr{B}_2^*) \leq (\mathscr{B}_1, \mathscr{B}_1^*)$ if and only if $\langle\mathscr{B}_2\rangle(\lambda) \leq \langle\mathscr{B}_1\rangle(\lambda)$ and $\langle\mathscr{B}_2^*\rangle(\lambda) \geq \langle\mathscr{B}_1^*\rangle(\lambda)$, for each $\lambda \in L^X$.

Remark 20. (i) An (L, M)-dffb is a generalization of L-filter base due to Kim and Ko [40].

(ii) If $(\mathscr{F}, \mathscr{F}^*)$ is an (L, M)-dff, then $(\mathscr{F}, \mathscr{F}^*)$ is an (L, M)-dffb with $\langle\mathscr{F}\rangle = \mathscr{F}$ and $\langle\mathscr{F}^*\rangle = \mathscr{F}^*$.

(iii) If $(\mathscr{B}, \mathscr{B}^*)$ is an (L, M)-dffb, then, by (DFFB3), $\lambda \odot \mu = \underline{0}$ implies $\mathscr{B}(\lambda) \,\tilde{\odot}\, \mathscr{B}(\mu) = 0_M$ and $\mathscr{B}^*(\lambda) \,\tilde{\oplus}\, \mathscr{B}^*(\mu) = 1_M$.

Theorem 21. *If $(\mathscr{B}, \mathscr{B}^*)$ is an (L, M)-dffb, then $(\langle\mathscr{B}\rangle, \langle\mathscr{B}^*\rangle)$ is the coarsest (L, M)-dff which satisfies $\mathscr{B} \leq \langle\mathscr{B}\rangle$ and $\mathscr{B}^* \geq \langle\mathscr{B}^*\rangle$.*

Proof. (DFF1) For each $\lambda \in L^X$,

$$\langle\mathscr{B}\rangle(\lambda) = \bigvee_{\mu\leq\lambda}\mathscr{B}(\mu) \leq \bigvee_{\mu\leq\lambda}\mathscr{B}^*(\mu)^*$$
$$= \left(\bigwedge_{\mu\leq\lambda}\mathscr{B}^*(\mu)\right)^* = (\langle\mathscr{B}^*\rangle(\lambda))^*. \tag{21}$$

(DFF2) and (DFF4) are easily checked.

(DFF3) Suppose that there exist $\lambda, \mu \in L^X$ such that

$$\langle\mathscr{B}^*\rangle(\lambda \odot \mu) \nleq \langle\mathscr{B}^*\rangle(\lambda) \,\tilde{\oplus}\, \langle\mathscr{B}^*\rangle(\mu). \tag{22}$$

By the definition of $\langle\mathscr{B}^*\rangle$ and (L4′), there exist $\lambda_1, \mu_1 \in L^X$ with $\lambda_1 \le \lambda$ and $\mu_1 \le \mu$ such that

$$\langle\mathscr{B}^*\rangle(\lambda \odot \mu) \nleq \mathscr{B}^*(\lambda_1) \,\tilde{\oplus}\, \mathscr{B}^*(\mu_1). \tag{23}$$

Since $(\mathscr{B}, \mathscr{B}^*)$ is an (L, M)-dffb,

$$\langle\mathscr{B}^*\rangle(\lambda_1 \odot \mu_1) \le \mathscr{B}^*(\lambda_1) \,\tilde{\oplus}\, \mathscr{B}^*(\mu_1). \tag{24}$$

Since $\lambda_1 \odot \mu_1 \le \lambda \odot \mu$, we have

$$\begin{aligned}\langle\mathscr{B}^*\rangle(\lambda \odot \mu) &\le \langle\mathscr{B}^*\rangle(\lambda_1 \odot \mu_1)\\ &\le \mathscr{B}^*(\lambda_1) \,\tilde{\oplus}\, \mathscr{B}^*(\mu_1).\end{aligned} \tag{25}$$

It is a contradiction. Thus, $\langle\mathscr{B}^*\rangle(\lambda \odot \mu) \le \langle\mathscr{B}^*\rangle(\lambda) \,\tilde{\oplus}\, \langle\mathscr{B}^*\rangle(\mu)$, for each $\lambda, \mu \in L^X$. Similarly, $\langle\mathscr{B}\rangle(\lambda \odot \mu) \ge \langle\mathscr{B}\rangle(\lambda) \,\tilde{\odot}\, \langle\mathscr{B}\rangle(\mu)$, for each $\lambda, \mu \in L^X$.

Let $(\mathscr{F}, \mathscr{F}^*)$ be another (L, M)-dff which is finer than $(\mathscr{B}, \mathscr{B}^*)$, that is, $\mathscr{B} \le \mathscr{F}$ and $\mathscr{B}^* \ge \mathscr{F}^*$. Then, we have

$$\begin{aligned}\langle\mathscr{B}\rangle(\lambda) &= \bigvee_{\mu\le\lambda} \mathscr{B}(\mu) \le \bigvee_{\mu\le\lambda} \mathscr{F}(\mu) = \mathscr{F}(\lambda),\\ \langle\mathscr{B}^*\rangle(\lambda) &= \bigwedge_{\mu\le\lambda} \mathscr{B}^*(\mu) \ge \bigwedge_{\mu\le\lambda} \mathscr{F}^*(\mu) = \mathscr{F}^*(\lambda).\end{aligned} \tag{26}$$

Theorem 22. *$\mathscr{H}, \mathscr{H}^* : L^X \to M$ are maps fulfilling the following conditions:*

(C1) $\mathscr{H}(\lambda) \le (\mathscr{H}^*(\lambda))^*$, for each $\lambda \in L^X$,

(C2) $\mathscr{H}(\underline{1}) = 1_M$ and $\mathscr{H}^*(\underline{1}) = 0_M$ and for each finite index set K, if $\bigodot_{i\in K}\lambda_i = \underline{0}$, then $\widetilde{\bigodot}_{i\in K}\mathscr{H}(\lambda_i) = 0_M$ and $\widetilde{\bigoplus}_{i\in K}\mathscr{H}^*(\lambda_i) = 1_M$.

We define the maps $\mathscr{B}_{\mathscr{H}}, \mathscr{B}^*_{\mathscr{H}^*} : L^X \to M$ as

$$\begin{aligned}\mathscr{B}_{\mathscr{H}}(\lambda) &= \bigvee\left\{\widetilde{\bigodot_{i\in K}} \mathscr{H}(\lambda_i) : \lambda = \bigodot_{i\in K}\lambda_i\right\},\\ \mathscr{B}^*_{\mathscr{H}^*}(\lambda) &= \bigwedge\left\{\widetilde{\bigoplus_{i\in K}} \mathscr{H}^*(\lambda_i) : \lambda = \bigodot_{i\in K}\lambda_i\right\},\end{aligned} \tag{27}$$

where $\bigvee$ and $\bigwedge$ are taken for every finite index set K such that $\lambda = \bigodot_{i\in K}\lambda_i$, respectively. Then, the following properties are satisfied:

(i) $(\mathscr{B}_{\mathscr{H}}, \mathscr{B}^*_{\mathscr{H}^*})$ is an (L, M)-dffb on X.

(ii) If $\mathscr{H} \le \mathscr{B}$, $\mathscr{H}^* \ge \mathscr{B}^*$ and $(\mathscr{B}, \mathscr{B}^*)$ is an (L, M)-dffb on X, then $\langle\mathscr{B}_{\mathscr{H}}\rangle \le \langle\mathscr{B}\rangle$ and $\langle\mathscr{B}^*_{\mathscr{H}^*}\rangle \ge \langle\mathscr{B}^*\rangle$.

Proof. (i) (DFFB1) For each $\lambda \in L^X$, the following is valid:

$$\begin{aligned}\mathscr{B}_{\mathscr{H}}(\lambda) &= \bigvee\left\{\widetilde{\bigodot_{i\in K}} \mathscr{H}(\lambda_i) : \lambda = \bigodot_{i\in K}\lambda_i\right\}\\ &\le \bigvee\left\{\widetilde{\bigodot_{i\in K}} (\mathscr{H}^*(\lambda_i))^* : \lambda = \bigodot_{i\in K}\lambda_i\right\}\\ &= \left(\bigwedge\left\{\widetilde{\bigoplus_{i\in K}} \mathscr{H}^*(\lambda_i) : \lambda = \bigodot_{i\in K}\lambda_i\right\}\right)^*\\ &= (\mathscr{B}^*_{\mathscr{H}^*}(\lambda))^*.\end{aligned} \tag{28}$$

(DFFB2) It is clear by condition (C2).

(DFFB3) For each $\lambda, \mu \in L^X$ and for any two finite index sets K, J with $\lambda = \bigodot_{k\in K}\lambda_k$ and $\mu = \bigodot_{j\in J}\mu_i$, since $\lambda \odot \mu = (\bigodot_{k\in K}\lambda_k) \odot (\bigodot_{j\in J}\mu_i)$, by the definition of $\mathscr{B}_{\mathscr{H}}$ and $\mathscr{B}^*_{\mathscr{H}^*}$, we get

$$\begin{aligned}\langle\mathscr{B}_{\mathscr{H}}\rangle(\lambda \odot \mu) &\ge \mathscr{B}_{\mathscr{H}}(\lambda \odot \mu)\\ &\ge \left(\widetilde{\bigodot_{k\in K}} \mathscr{H}(\lambda_k)\right)\\ &\quad \tilde{\odot}\left(\widetilde{\bigodot_{j\in J}} \mathscr{H}(\mu_j)\right),\\ \langle\mathscr{B}^*_{\mathscr{H}^*}\rangle(\lambda \odot \mu) &\le \mathscr{B}^*_{\mathscr{H}^*}(\lambda \odot \mu)\\ &\le \left(\widetilde{\bigoplus_{k\in K}} \mathscr{H}^*(\lambda_k)\right)\\ &\quad \tilde{\oplus}\left(\widetilde{\bigoplus_{j\in J}} \mathscr{H}^*(\mu_j)\right).\end{aligned} \tag{29}$$

If supremum and infimum are taken over finite index set K, respectively, then by (2) and (3),

$$\begin{aligned}\langle\mathscr{B}_{\mathscr{H}}\rangle(\lambda \odot \mu) &\ge \mathscr{B}_{\mathscr{H}}(\lambda) \,\tilde{\odot}\, \mathscr{B}_{\mathscr{H}}(\mu),\\ \langle\mathscr{B}^*_{\mathscr{H}^*}\rangle(\lambda \odot \mu) &\le \mathscr{B}^*_{\mathscr{H}^*}(\lambda) \,\tilde{\oplus}\, \mathscr{B}^*_{\mathscr{H}^*}(\mu).\end{aligned} \tag{30}$$

Thus, $(\mathscr{B}_{\mathscr{H}}, \mathscr{B}^*_{\mathscr{H}^*})$ is an (L, M)-dffb on X.

(ii) For any finite family $\{\lambda_i : \bigodot_{i\in K}\lambda_i \le \lambda\}$, the following are true:

$$\begin{aligned}\langle\mathscr{B}\rangle(\lambda) &\ge \langle\mathscr{B}\rangle\left(\bigodot_{i\in K}\lambda_i\right) \ge \widetilde{\bigodot_{i\in K}} \langle\mathscr{B}\rangle(\lambda_i)\\ &\ge \widetilde{\bigodot_{i\in K}} \mathscr{B}(\lambda_i) \ge \widetilde{\bigodot_{i\in K}} \mathscr{H}(\lambda_i),\\ \langle\mathscr{B}^*\rangle(\lambda) &\le \langle\mathscr{B}^*\rangle\left(\bigodot_{i\in K}\lambda_i\right) \le \widetilde{\bigoplus_{i\in K}} \langle\mathscr{B}^*\rangle(\lambda_i)\\ &\le \widetilde{\bigoplus_{i\in K}} \mathscr{B}^*(\lambda_i) \le \widetilde{\bigoplus_{i\in K}} \mathscr{H}^*(\lambda_i).\end{aligned} \tag{31}$$

Then, $\langle\mathscr{B}_{\mathscr{H}}\rangle \le \langle\mathscr{B}\rangle$ and $\langle\mathscr{B}^*_{\mathscr{H}^*}\rangle \ge \langle\mathscr{B}^*\rangle$.

Theorem 23. *Let $(\mathscr{B}_1, \mathscr{B}_1^*)$ and $(\mathscr{B}_2, \mathscr{B}_2^*)$ be two (L, M)-dffb's on X and Y, respectively, and $\varphi : X \to Y$ be a function. Then, one has the following properties:*

(i) $\varphi : (X, \langle\mathscr{B}_1\rangle, \langle\mathscr{B}_1^*\rangle) \to (Y, \langle\mathscr{B}_2\rangle, \langle\mathscr{B}_2^*\rangle)$ is a filter map if and only if $\mathscr{B}_2 \le \langle\mathscr{B}_1\rangle \circ \varphi_L^{\leftarrow}$ and $\mathscr{B}_2^* \ge \langle\mathscr{B}_1^*\rangle \circ \varphi_L^{\leftarrow}$.

(ii) $\varphi : (X, \langle\mathscr{B}_1\rangle, \langle\mathscr{B}_1^*\rangle) \to (Y, \langle\mathscr{B}_2\rangle, \langle\mathscr{B}_2^*\rangle)$ is a filter preserving map if and only if $\mathscr{B}_1 \le \langle\mathscr{B}_2\rangle \circ \varphi_L^{\rightarrow}$ and $\mathscr{B}_1^* \ge \langle\mathscr{B}_2^*\rangle \circ \varphi_L^{\rightarrow}$.

(iii) If $\mathscr{B}_2 \le \mathscr{B}_1 \circ \varphi_L^{\leftarrow}$ and $\mathscr{B}_2^* \ge \mathscr{B}_1^* \circ \varphi_L^{\leftarrow}$, then $\varphi : (X, \langle\mathscr{B}_1\rangle, \langle\mathscr{B}_1^*\rangle) \to (Y, \langle\mathscr{B}_2\rangle, \langle\mathscr{B}_2^*\rangle)$ is a filter map.

(iv) If $\mathscr{B}_1 \le \mathscr{B}_2 \circ \varphi_L^{\rightarrow}$ and $\mathscr{B}_1^* \ge \mathscr{B}_2^* \circ \varphi_L^{\rightarrow}$, then $\varphi : (X, \langle\mathscr{B}_1\rangle, \langle\mathscr{B}_1^*\rangle) \to (Y, \langle\mathscr{B}_2\rangle, \langle\mathscr{B}_2^*\rangle)$ is a filter preserving map.

Proof. Proving condition (i) is enough since the other conditions are similarly proved.

(i) (⇒:) Since $\mathscr{B}_2(\mu) \leq \langle \mathscr{B}_2 \rangle(\mu)$ and $\mathscr{B}_2^*(\mu) \geq \langle \mathscr{B}_2^* \rangle(\mu)$, for each $\mu \in L^Y$, it is trivial.

(⇐:) Let $\mathscr{B}_2(\mu) \leq \langle \mathscr{B}_1 \rangle(\varphi_L^{\leftarrow}(\mu))$ and $\mathscr{B}_2^*(\mu) \geq \langle \mathscr{B}_1^* \rangle(\varphi_L^{\leftarrow}(\mu))$, for each $\mu \in L^Y$. We will show that φ is a filter map. For arbitrary $\mu \in L^Y$, we have

$$\begin{aligned}\langle \mathscr{B}_2 \rangle (\mu) &= \bigvee_{\nu \leq \mu} \mathscr{B}_2 (\nu) \leq \bigvee_{\varphi_L^{\leftarrow}(\nu) \leq \varphi_L^{\leftarrow}(\mu)} \langle \mathscr{B}_1 \rangle (\varphi_L^{\leftarrow} (\nu)) \\ &\leq \langle \mathscr{B}_1 \rangle (\varphi_L^{\leftarrow} (\mu)), \\ \langle \mathscr{B}_2^* \rangle (\mu) &= \bigwedge_{\nu \leq \mu} \mathscr{B}_2^* (\nu) \geq \bigwedge_{\varphi_L^{\leftarrow}(\nu) \leq \varphi_L^{\leftarrow}(\mu)} \langle \mathscr{B}_1^* \rangle (\varphi_L^{\leftarrow} (\nu)) \\ &\geq \langle \mathscr{B}_1^* \rangle (\varphi_L^{\leftarrow} (\mu)).\end{aligned} \tag{32}$$

Thus, φ is a filter map.

Example 24. Let $X = \{x, y\}$ be a set, $L = M = [0, 1]$ be the stsc-quantale with Lukasiewicz t-norm, and $\mu, \nu \in [0, 1]^X$ be defined by $\mu(x) = 0.6$, $\mu(y) = 0.5$, $\nu(x) = 0.1$, $\nu(y) = 0$. Define the maps $\mathscr{B}_i, \mathscr{B}_i^* : L^X \to M$ $i = 1, 2, 3$ as follows:

$$\mathscr{B}_1 (\lambda) = \begin{cases} 1, & \text{if } \lambda = \underline{1}; \\ 0.6, & \text{if } \lambda = \mu; \\ 0.3, & \text{if } \lambda = \nu; \\ 0, & \text{otherwise,} \end{cases}$$

$$\mathscr{B}_1^* (\lambda) = \begin{cases} 0, & \text{if } \lambda = \underline{1}; \\ 0.4, & \text{if } \lambda = \mu; \\ 0.7, & \text{if } \lambda = \nu; \\ 1, & \text{otherwise,} \end{cases}$$

$$\mathscr{B}_2 (\lambda) = \begin{cases} 1, & \text{if } \lambda = \underline{1}; \\ 0.6, & \text{if } \lambda = \mu; \\ 0.5, & \text{if } \lambda = \mu \odot \mu; \\ 0, & \text{otherwise,} \end{cases}$$

$$\mathscr{B}_2^* (\lambda) = \begin{cases} 0, & \text{if } \lambda = \underline{1}; \\ 0.4, & \text{if } \lambda = \mu; \\ 0.5, & \text{if } \lambda = \mu \odot \mu; \\ 1, & \text{otherwise,} \end{cases}$$

$$\mathscr{B}_3 (\lambda) = \begin{cases} 1, & \text{if } \lambda = \underline{1}; \\ 0.4, & \text{if } \lambda = \mu; \\ 0.3, & \text{if } \lambda = \mu \odot \mu; \\ 0, & \text{otherwise,} \end{cases}$$

$$\mathscr{B}_3^* (\lambda) = \begin{cases} 0, & \text{if } \lambda = \underline{1}; \\ 0.6, & \text{if } \lambda = \mu; \\ 0.7, & \text{if } \lambda = \mu \odot \mu; \\ 1, & \text{otherwise.} \end{cases} \tag{33}$$

It can be seen by easy computation that

(1) $(\mathscr{B}_1, \mathscr{B}_1^*)$ and $(\mathscr{B}_3, \mathscr{B}_3^*)$ are not (L, M)-double fuzzy filters but they are (L, M)-double fuzzy filter bases, so they generate (L, M)-double fuzzy filters $(\langle \mathscr{B}_1 \rangle, \langle \mathscr{B}_1^* \rangle)$ and $(\langle \mathscr{B}_3 \rangle, \langle \mathscr{B}_3^* \rangle)$.

(2) $(\mathscr{B}_2, \mathscr{B}_2^*)$ is not an (L, M)-double fuzzy filter base and it does not satisfy condition (C2) of Theorem 22.

(3) Since $(\langle \mathscr{B}_3 \rangle, \langle \mathscr{B}_3^* \rangle) \leq (\langle \mathscr{B}_1 \rangle, \langle \mathscr{B}_1^* \rangle)$, $id_X : (X, \langle \mathscr{B}_1 \rangle, \langle \mathscr{B}_1^* \rangle) \to (X, \langle \mathscr{B}_3 \rangle, \langle \mathscr{B}_3^* \rangle)$ is a filter map and $id_X : (X, \langle \mathscr{B}_3 \rangle, \langle \mathscr{B}_3^* \rangle) \to (X, \langle \mathscr{B}_1 \rangle, \langle \mathscr{B}_1^* \rangle)$ is a filter preserving map though $\mathscr{B}_3 \nleq \mathscr{B}_1, \mathscr{B}_3^* \ngeq \mathscr{B}_1^*$ and $\mathscr{B}_1 \nleq \mathscr{B}_3, \mathscr{B}_1^* \ngeq \mathscr{B}_3^*$.

We also note that if $L = M = [0, 1]$ is considered as a frame, then $(\mathscr{B}_1, \mathscr{B}_1^*)$ is an (L, M)-double fuzzy filter base.

4. The Types $(\varphi_1^{\Leftarrow}, \varphi_1^{*\Leftarrow})$, $(\varphi_2^{\Rightarrow}, \varphi_2^{\Rightarrow})$ of Preimages and Images of (L,M)-Double Fuzzy Filter Bases

Theorem 25. *Let $\varphi : X \to Y$ be a function and $(\mathscr{B}, \mathscr{B}^*)$ be an (L, M)-dffb on Y. Then, the following properties are satisfied.*

(i) If $\varphi_L^{\leftarrow}(\mu) = \underline{0}$ implies $\mathscr{B}(\mu) = 0_M$ and $\mathscr{B}^*(\mu) = 1_M$, then $(\varphi_1^{\Leftarrow}(\mathscr{B}), \varphi_1^{*\Leftarrow}(\mathscr{B}^*))$ is an (L, M)-dffb on X and $(\langle \phi_1^{\Leftarrow}(\mathscr{B}) \rangle, \langle \varphi_1^{*\Leftarrow}(\mathscr{B}^*) \rangle)$ is the coarsest (L, M)-dff on X for which $\varphi : (X, \langle \phi_1^{\Leftarrow}(\mathscr{B}) \rangle, \langle \varphi_1^{*\Leftarrow}(\mathscr{B}^*) \rangle) \to (Y, \langle \mathscr{B} \rangle, \langle \mathscr{B}^* \rangle)$ is a filter map.

(ii) If φ is surjective, then $(\varphi_1^{\Leftarrow}(\mathscr{B}), \varphi_1^{*\Leftarrow}(\mathscr{B}^*))$ is an (L, M)-dffb.

(iii) If $\varphi_L^{\leftarrow}(\mu) = \underline{0}$ implies $\mathscr{B}(\mu) = 0_M$ and $\mathscr{B}^*(\mu) = 1_M$, φ is injective, and $(\mathscr{B}, \mathscr{B}^*)$ is an (L, M)-dff on Y, then $(\varphi_1^{\Leftarrow}(\mathscr{B}), \varphi_1^{*\Leftarrow}(\mathscr{B}^*))$ is an (L, M)-dff on X.

Proof. (i) (DFFB1) For each $\lambda \in L^X$, we have

$$\begin{aligned}\varphi_1^{\Leftarrow} (\mathscr{B}) (\lambda) &= \bigvee \{\mathscr{B} (\mu) : \lambda = \varphi_L^{\leftarrow} (\mu)\} \\ &\leq \bigvee \{(\mathscr{B}^* (\mu))^* : \lambda = \varphi_L^{\leftarrow} (\mu)\} \\ &= \left(\bigwedge \{\mathscr{B}^* (\mu) : \lambda = \varphi_L^{\leftarrow} (\mu)\}\right)^* \\ &= (\varphi_1^{*\Leftarrow} (\mathscr{B}^*) (\lambda))^* .\end{aligned} \tag{34}$$

(DFFB2) Since $\varphi_L^{\leftarrow}(\underline{1}) = \underline{1}$, then $\varphi_1^{\Leftarrow}(\mathscr{B})(\underline{1}) = 1_M$ and $\varphi_1^{*\Leftarrow}(\mathscr{B}^*)(\underline{1}) = 0_M$. By assumption, $\varphi_1^{\Leftarrow}(\mathscr{B})(\underline{0}) = 0_M$ and $\varphi_1^{*\Leftarrow}(\mathscr{B}^*)(\underline{0}) = 1_M$.

(DFFB3) Suppose that there exist $\lambda_1, \lambda_2 \in L^X$ such that

$$\begin{aligned}&\langle \varphi_1^{*\Leftarrow} (\mathscr{B}^*) \rangle (\lambda_1 \odot \lambda_2) \\ &\nleq \varphi_1^{*\Leftarrow} (\mathscr{B}^*) (\lambda_1) \tilde{\oplus} \varphi_1^{*\Leftarrow} (\mathscr{B}^*) (\lambda_2).\end{aligned} \tag{35}$$

By the definition of $\varphi_1^{*\Leftarrow}(\mathscr{B}^*)$ and (L4′), there exist $\mu_1, \mu_2 \in L^Y$ with $\lambda_1 = \varphi_L^{\leftarrow}(\mu_1)$ and $\lambda_2 = \varphi_L^{\leftarrow}(\mu_2)$ such that

$$\langle \varphi_1^{*\Leftarrow}(\mathscr{B}^*)\rangle (\lambda_1 \odot \lambda_2) \nleq \mathscr{B}^*(\mu_1) \widetilde{\oplus} \mathscr{B}^*(\mu_2). \tag{36}$$

Since $(\mathscr{B}, \mathscr{B}^*)$ is an (L, M)-dffb, the following is valid:

$$\langle \mathscr{B}^* \rangle (\mu_1 \odot \mu_2) \le \mathscr{B}^*(\mu_1) \widetilde{\oplus} \mathscr{B}^*(\mu_2). \tag{37}$$

Then,

$$\langle \varphi_1^{*\Leftarrow}(\mathscr{B}^*)\rangle (\lambda_1 \odot \lambda_2) \nleq \langle \mathscr{B}^*\rangle (\mu_1 \odot \mu_2). \tag{38}$$

By the definition of $\langle \mathscr{B}^*\rangle$, there exists $\nu \in L^Y$ with $\nu \le \mu_1 \odot \mu_2$ such that

$$\langle \varphi_1^{*\Leftarrow}(\mathscr{B}^*)\rangle (\lambda_1 \odot \lambda_2) \nleq \mathscr{B}^*(\nu). \tag{39}$$

On the other hand, since

$$\begin{aligned} \lambda_1 \odot \lambda_2 &= \varphi_L^{\leftarrow}(\mu_1) \odot \varphi_L^{\leftarrow}(\mu_2) = \varphi_L^{\leftarrow}(\mu_1 \odot \mu_2) \\ &\ge \varphi_L^{\leftarrow}(\nu), \end{aligned} \tag{40}$$

then $\langle \varphi_1^{*\Leftarrow}(\mathscr{B}^*)\rangle(\lambda_1 \odot \lambda_2) \le \varphi_1^{*\Leftarrow}(\mathscr{B}^*)(\varphi_L^{\leftarrow}(\nu)) \le \mathscr{B}^*(\nu)$.

This contradicts the assumption. Thus,

$$\begin{aligned} &\langle \varphi_1^{*\Leftarrow}(\mathscr{B}^*)\rangle (\lambda_1 \odot \lambda_2) \\ &\quad \le \varphi_1^{*\Leftarrow}(\mathscr{B}^*)(\lambda_1) \widetilde{\oplus} \varphi_1^{*\Leftarrow}(\mathscr{B}^*)(\lambda_2), \\ &\qquad\qquad \text{for each } \lambda_1, \lambda_2 \in L^X. \end{aligned} \tag{41}$$

Similarly, $\langle \varphi_1^{\Leftarrow}(\mathscr{B})\rangle(\lambda_1 \odot \lambda_2) \ge \varphi_1^{\Leftarrow}(\mathscr{B})(\lambda_1) \widetilde{\odot} \varphi_1^{\Leftarrow}(\mathscr{B})(\lambda_2)$, for each $\lambda_1, \lambda_2 \in L^X$.

Hence, $(\varphi_1^{\Leftarrow}(\mathscr{B}), \varphi_1^{*\Leftarrow}(\mathscr{B}^*))$ is an (L, M)-dffb on X.

Let $(\mathscr{F}, \mathscr{F}^*)$ be another (L, M)-dff on X such that $\varphi : (X, \mathscr{F}, \mathscr{F}^*) \to (Y, \langle \mathscr{B}\rangle, \langle \mathscr{B}^*\rangle)$ is a filter map. Then, for each $\lambda \in L^X$, the following inequalities are valid:

$$\begin{aligned} \langle \varphi_1^{\Leftarrow}(\mathscr{B})\rangle (\lambda) &= \bigvee \{\mathscr{B}(\mu) : \varphi_L^{\leftarrow}(\mu) \le \lambda\} \\ &\le \bigvee \{\mathscr{F}(\varphi_L^{\leftarrow}(\mu)) : \varphi_L^{\leftarrow}(\mu) \le \lambda\} \\ &\le \mathscr{F}(\lambda), \\ \langle \varphi_1^{*\Leftarrow}(\mathscr{B}^*)\rangle (\lambda) &= \bigwedge \{\mathscr{B}^*(\mu) : \varphi_L^{\leftarrow}(\mu) \le \lambda\} \\ &\ge \bigwedge \{\mathscr{F}^*(\varphi_L^{\leftarrow}(\mu)) : \varphi_L^{\leftarrow}(\mu) \le \lambda\} \\ &\ge \mathscr{F}^*(\lambda). \end{aligned} \tag{42}$$

(ii) Since φ is surjective, $\varphi_L^{\leftarrow}(\mu) = \underline{0}$ implies $\mu = \underline{0}$. So, $\mathscr{B}(\mu) = 0_M$ and $\mathscr{B}^*(\mu) = 1_M$. Then, by (i), $(\varphi_1^{\Leftarrow}(\mathscr{B}), \varphi_1^{*\Leftarrow}(\mathscr{B}^*))$ is an (L, M)-dffb on X.

(iii) (DFF1)–(DFF3) are obvious.

(DFF4) Let $\lambda_1 \le \lambda_2$, for $\lambda_1, \lambda_2 \in L^X$. Since φ is injective, there exists $\nu \in L^Y$ with $\varphi_L^{\leftarrow}(\nu) = \lambda_1$ and $\lambda_2 = \varphi_L^{\leftarrow}(\nu \vee \varphi_L^{\rightarrow}(\lambda_2))$. It implies

$$\begin{aligned} \varphi_1^{\Leftarrow}(\mathscr{B})(\lambda_2) &\ge \mathscr{B}(\nu \vee \varphi_L^{\rightarrow}(\lambda_2)) \ge \mathscr{B}(\nu), \\ \varphi_1^{*\Leftarrow}(\mathscr{B}^*)(\lambda_2) &\le \mathscr{B}^*(\nu \vee \varphi_L^{\rightarrow}(\lambda_2)) \le \mathscr{B}^*(\nu). \end{aligned} \tag{43}$$

If supremum and infimum are taken over $\{\lambda_1 \mid \varphi_L^{\leftarrow}(\nu) = \lambda_1\}$, respectively, then it is clear that $\varphi_1^{\Leftarrow}(\mathscr{B})(\lambda_2) \ge \varphi_1^{\Leftarrow}(\mathscr{B})(\lambda_1)$ and $\varphi_1^{*\Leftarrow}(\mathscr{B}^*)(\lambda_2) \le \varphi_1^{*\Leftarrow}(\mathscr{B}^*)(\lambda_1)$.

Theorem 26. *Let $\{\varphi_i : X \to X_i\}_{i\in\Gamma}$ be a family of functions and $\{(\mathscr{B}_i, \mathscr{B}_i^*)\}_{i\in\Gamma}$ be a family of (L, M)-dffb's on X_i satisfying the following condition:*

(C) For each finite subset K of Γ, if $\bigodot_{i\in K}(\lambda_i \circ \varphi_i) = \underline{0}$, then $\bigodot_{i\in K}\mathscr{B}_i(\lambda_i) = 0_M$ and $\bigoplus_{i\in K}\mathscr{B}_i^*(\lambda_i) = 1_M$.

We define the maps $\bigsqcup_{i\in\Gamma}(\varphi_i)_1^{\Leftarrow}(\mathscr{B}_i)$, $\bigsqcap_{i\in\Gamma}(\varphi_i^*)_1^{\Leftarrow}(\mathscr{B}_i^*) : L^X \to M$ as

$$\begin{aligned} &\bigsqcup_{i\in\Gamma}(\varphi_i)_1^{\Leftarrow}(\mathscr{B}_i)(\lambda) \\ &\quad = \bigvee \left\{ \widetilde{\bigodot_{i\in K}} \mathscr{B}_i(\lambda_i) : \lambda = \bigodot_{i\in K}(\lambda_i \circ \varphi_i) \right\}, \\ &\bigsqcap_{i\in\Gamma}(\varphi_i^*)_1^{\Leftarrow}(\mathscr{B}_i^*)(\lambda) \\ &\quad = \bigwedge \left\{ \widetilde{\bigoplus_{i\in K}} \mathscr{B}_i^*(\lambda_i) : \lambda = \bigodot_{i\in K}(\lambda_i \circ \varphi_i) \right\}, \end{aligned} \tag{44}$$

where $\bigvee$ and $\bigwedge$ are taken for every finite index subset K of Γ such that $\lambda = \bigodot_{i\in K}(\lambda_i \circ \varphi_i)$.

Let $\mathscr{B} = \bigsqcup_{i\in\Gamma}(\varphi_i)_1^{\Leftarrow}(\mathscr{B}_i)$ and $\mathscr{B}^* = \bigsqcap_{i\in\Gamma}(\varphi_i^*)_1^{\Leftarrow}(\mathscr{B}_i^*)$ be given. Then, the following properties are satisfied.

(i) $(\mathscr{B}, \mathscr{B}^*)$ is an (L, M)-dffb on X and $(\langle \mathscr{B}\rangle, \langle \mathscr{B}^*\rangle)$ is the coarsest (L, M)-dff for which $\varphi_i : (X, \langle \mathscr{B}\rangle, \langle \mathscr{B}^*\rangle) \to (X_i, \langle \mathscr{B}_i\rangle, \langle \mathscr{B}_i^*\rangle)$ is a filter map for each $i \in \Gamma$.

(ii) A function $\varphi : (Y, \mathscr{F}, \mathscr{F}^*) \to (X, \langle \mathscr{B}\rangle, \langle \mathscr{B}^*\rangle)$ is a filter map if and only if, for each $i \in \Gamma$, $\varphi_i \circ \varphi : (Y, \mathscr{F}, \mathscr{F}^*) \to (X_i, \langle \mathscr{B}_i\rangle, \langle \mathscr{B}_i^*\rangle)$ is a filter map.

(iii) $\langle \bigsqcup_{i\in\Gamma}(\varphi_i)_1^{\Leftarrow}(\mathscr{B}_i)\rangle = \langle \bigsqcup_{i\in\Gamma}(\varphi_i)_1^{\Leftarrow}(\langle \mathscr{B}_i\rangle)\rangle$ and $\langle \bigsqcap_{i\in\Gamma}(\varphi_i^*)_1^{\Leftarrow}(\mathscr{B}_i^*)\rangle = \langle \bigsqcap_{i\in\Gamma}(\varphi_i^*)_1^{\Leftarrow}(\langle \mathscr{B}_i^*\rangle)\rangle$.

Proof. (i) (DFFB1) Let $\lambda \in L^X$ with $\lambda = \bigodot_{i\in K}(\lambda_i \circ \varphi_i)$. Then, the following inequality is valid:

$$\begin{aligned} \mathscr{B}(\lambda) &= \bigvee \left\{ \widetilde{\bigodot_{i\in K}} \mathscr{B}_i(\lambda_i) : \lambda = \bigodot_{i\in K}(\lambda_i \circ \varphi_i) \right\} \\ &\le \bigvee \left\{ \widetilde{\bigodot_{i\in K}} (\mathscr{B}_i^*(\lambda_i))^* : \lambda = \bigodot_{i\in K}(\lambda_i \circ \varphi_i) \right\} \\ &= \left(\bigwedge \left\{ \widetilde{\bigoplus_{i\in K}} \mathscr{B}_i^*(\lambda_i) : \lambda = \bigodot_{i\in K}(\lambda_i \circ \varphi_i) \right\} \right)^* \\ &= (\mathscr{B}^*(\lambda))^*. \end{aligned} \tag{45}$$

(DFFB2) By condition (C), $\mathscr{B}(\underline{0}) = 0_M$ and $\mathscr{B}^*(\underline{0}) = 1_M$. Since, $\underline{1} = \underline{1} \circ \varphi_i$, then $\mathscr{B}(\underline{1}) = 1_M$ and $\mathscr{B}^*(\underline{1}) = 0_M$.

(DFFB3) Suppose that there exist $\lambda, \mu \in L^X$ such that $\langle \mathscr{B}^*\rangle(\lambda \odot \mu) \nleq \mathscr{B}^*(\lambda) \widetilde{\oplus} \mathscr{B}^*(\mu)$.

By the definition of $\mathscr{B}^*(\lambda)$, and (L4′), there exists a finite subset K of Γ with $\lambda = \bigodot_{k\in K}(\lambda_k \circ \varphi_k)$ such that $\langle \mathscr{B}^*\rangle(\lambda \odot \mu) \nleq (\widetilde{\bigoplus_{k\in K}} \mathscr{B}_k^*(\lambda_k)) \widetilde{\oplus} \mathscr{B}(\mu)$.

Again, by the definition of $\mathscr{B}^*(\mu)$ and (L4′), there exists a finite subset J of Γ with $\mu = \bigodot_{j\in J}(\mu_j \circ \varphi_j)$ such that

$$\langle\mathscr{B}^*\rangle(\lambda\odot\mu) \nleq \left(\widetilde{\bigoplus_{k\in K}} \mathscr{B}_k^*(\lambda_k)\right) \mathbin{\widetilde{\oplus}} \left(\widetilde{\bigoplus_{j\in J}} \mathscr{B}_j^*(\mu_j)\right). \tag{46}$$

Put $m \in (K\cup J)$ such that

$$\rho_m = \begin{cases} \lambda_m, & \text{if } m \in K \setminus (K\cap J) \\ \mu_m, & \text{if } m \in J \setminus (K\cap J) \\ \lambda_m \odot \mu_m, & \text{if } m \in K\cap J. \end{cases} \tag{47}$$

Since, for each $m \in K\cap J$, $\langle\mathscr{B}_m^*\rangle(\lambda_m\odot\mu_m) \le \mathscr{B}_m^*(\lambda_m) \mathbin{\widetilde{\oplus}} \mathscr{B}_m^*(\mu_m)$, we have

$$\langle\mathscr{B}^*\rangle(\lambda\odot\mu) \nleq \left(\widetilde{\bigoplus_{m\in(K\cup J)-(K\cap J)}} \mathscr{B}_m^*(\rho_m)\right) \mathbin{\widetilde{\oplus}} \left(\widetilde{\bigoplus_{m\in(K\cap J)}} \langle\mathscr{B}_m^*\rangle(\lambda_m\odot\mu_m)\right). \tag{48}$$

From the definition of $\langle\mathscr{B}_m^*\rangle$, there exists $\nu_m \in L^{X_m}$ with $\nu_m \le \lambda_m \odot \mu_m$ such that

$$\langle\mathscr{B}^*\rangle(\lambda\odot\mu) \nleq \left(\widetilde{\bigoplus_{m\in(K\cup J)-(K\cap J)}} \mathscr{B}_m^*(\rho_m)\right) \mathbin{\widetilde{\oplus}} \left(\widetilde{\bigoplus_{m\in(K\cap J)}} \mathscr{B}_m^*(\nu_m)\right). \tag{49}$$

On the other hand, since

$$\begin{aligned}\lambda\odot\mu &= \left(\bigodot_{k\in K}(\lambda_k\circ\varphi_k)\right)\odot\left(\bigodot_{j\in J}(\mu_j\circ\varphi_j)\right) \\ &\ge \left(\bigodot_{m\in(K\cup J)-(K\cap J)}(\rho_m\circ\varphi_m)\right) \\ &\quad\odot\left(\bigodot_{m\in(K\cap J)}(\nu_m\circ\phi_m)\right)\end{aligned} \tag{50}$$

and since $K\cup J$ is finite, we have

$$\langle\mathscr{B}^*\rangle(\lambda\odot\mu) \le \left(\widetilde{\bigoplus_{m\in(K\cup J)-(K\cap J)}} \mathscr{B}_m^*(\rho_m)\right) \mathbin{\widetilde{\oplus}} \left(\widetilde{\bigoplus_{m\in(K\cap J)}} \mathscr{B}_m^*(\nu_m)\right). \tag{51}$$

This contradicts the assumption. Then, $\langle\mathscr{B}^*\rangle(\lambda\odot\mu) \le \mathscr{B}^*(\lambda) \mathbin{\widetilde{\oplus}} \mathscr{B}^*(\mu)$, for each $\lambda,\mu \in L^X$. Similarly, $\langle\mathscr{B}\rangle(\lambda\odot\mu) \ge \mathscr{B}(\lambda) \mathbin{\widetilde{\odot}} \mathscr{B}(\mu)$, for each, $\lambda,\mu \in L^X$. Hence, $(\mathscr{B},\mathscr{B}^*)$ is an (L,M)-dffb on X.

Since $\mathscr{B}(\lambda_i\circ\varphi_i) \ge \mathscr{B}_i(\lambda_i)$ and $\mathscr{B}^*(\lambda_i\circ\varphi_i) \le \mathscr{B}_i^*(\lambda_i)$, for each $i\in\Gamma$, by Theorem 23(iii), $\varphi_i : (X,\langle\mathscr{B}\rangle,\langle\mathscr{B}^*\rangle) \to (X_i,\langle\mathscr{B}_i\rangle,\langle\mathscr{B}_i^*\rangle)$ is a filter map.

Let $(\mathscr{F},\mathscr{F}^*)$ be an (L,M)-dff on X such that, for each $i\in\Gamma$, the map $\varphi_i : (X,\mathscr{F},\mathscr{F}^*) \to (X_i,\langle\mathscr{B}_i\rangle,\langle\mathscr{B}_i^*\rangle)$ is a filter map. Then,

$$\begin{aligned}&\mathscr{F}(\nu_i\circ\varphi_i) \ge \langle B_i\rangle(\nu_i), \\ &\mathscr{F}^*(\nu_i\circ\varphi_i) \le \langle B_i^*\rangle(\nu_i), \\ &\qquad\qquad \text{for each } i\in\Gamma,\ \nu_i\in L^{X_i}.\end{aligned} \tag{52}$$

For any finite subset K of Γ with $\nu \ge \bigodot_{k\in K}(\nu_k\circ\varphi_k)$, since $\mathscr{F}(\nu_k\circ\varphi_k) \ge \langle B_k\rangle(\nu_k)$ and $\mathscr{F}^*(\nu_k\circ\varphi_k) \le \langle B_k^*\rangle(\nu_k)$, for each $k\in K$, we have

$$\begin{aligned}\mathscr{F}(\lambda) &\ge \mathscr{F}\left(\bigodot_{k\in K}(\nu_k\circ\varphi_k)\right) \ge \widetilde{\bigodot_{k\in K}}\mathscr{F}(\nu_k\circ\varphi_k) \\ &\ge \widetilde{\bigodot_{k\in K}}\langle B_k\rangle(\nu_k) \ge \widetilde{\bigodot_{k\in K}}\mathscr{B}_k(\nu_k), \\ \mathscr{F}^*(\lambda) &\le \mathscr{F}^*\left(\bigodot_{k\in K}(\nu_k\circ\varphi_k)\right) \le \widetilde{\bigoplus_{k\in K}}\mathscr{F}^*(\nu_k\circ\varphi_k) \\ &\le \widetilde{\bigoplus_{k\in K}}\langle B_k\rangle(\nu_k) \le \widetilde{\bigoplus_{k\in K}}\mathscr{B}_k(\nu_k).\end{aligned} \tag{53}$$

Hence, by the definition of $\langle\mathscr{B}\rangle$ and $\langle\mathscr{B}^*\rangle$, it is obvious that $(\langle\mathscr{B}\rangle,\langle\mathscr{B}^*\rangle) \le (\mathscr{F},\mathscr{F}^*)$.

(ii) Necessity of the composition condition is obvious.

Conversely, for every finite subset K of Γ with $\nu \ge \bigodot_{k\in K}(\nu_k\circ\varphi_k)$, since, for each $k\in K$, $\varphi_k\circ\varphi : (Y,\mathscr{F},\mathscr{F}^*) \to (X_k,\langle\mathscr{B}_k\rangle,\langle\mathscr{B}_k^*\rangle)$ is a filter map, that is, $\langle\mathscr{B}_k\rangle(\nu_k) \le \mathscr{F}(\nu_k\circ(\varphi_k\circ\varphi))$ and $\langle\mathscr{B}_k^*\rangle(\nu_k) \ge \mathscr{F}^*(\nu_k\circ(\varphi_k\circ\varphi))$. Since $\nu\circ\varphi \ge \bigodot_{k\in K}((\nu_k\circ\varphi_k)\circ\varphi)$, we have

$$\begin{aligned}\mathscr{F}(\nu\circ\varphi) &\ge \widetilde{\bigodot_{k\in K}}\mathscr{F}(\nu_k\circ(\varphi_k\circ\varphi)) \\ &\ge \widetilde{\bigodot_{k\in K}}\langle\mathscr{B}_k\rangle(\nu_k) \ge \widetilde{\bigodot_{k\in K}}\mathscr{B}_k(\nu_k), \\ \mathscr{F}^*(\nu\circ\varphi) &\le \widetilde{\bigoplus_{k\in K}}\mathscr{F}^*(\nu_k\circ(\varphi_k\circ\varphi)) \\ &\le \widetilde{\bigoplus_{k\in K}}\langle\mathscr{B}_k^*\rangle(\nu_k) \le \widetilde{\bigoplus_{k\in K}}\mathscr{B}_k^*(\nu_k).\end{aligned} \tag{54}$$

By the definition of $\langle\mathscr{B}\rangle$ and $\langle\mathscr{B}^*\rangle$, we have $\langle\mathscr{B}\rangle(\nu) \le \mathscr{F}(\nu\circ\varphi)$ and $\langle\mathscr{B}^*\rangle(\nu) \ge \mathscr{F}^*(\nu\circ\varphi)$.

(iii) Put $\mathscr{F} = \bigsqcup_{i\in\Gamma}(\varphi_i)_1^{\Leftarrow}(\langle\mathscr{B}_i\rangle)$ and $\mathscr{F}^* = \bigsqcap_{i\in\Gamma}(\varphi_i^*)_1^{\Leftarrow}(\langle\mathscr{B}_i^*\rangle)$; by applying (i) to both $(\langle\mathscr{B}\rangle,\langle\mathscr{B}^*\rangle)$ and $(\langle\mathscr{F}\rangle,\langle\mathscr{F}^*\rangle)$, the desired equality is obtained.

The following corollaries are the direct results of Theorem 26.

Corollary 27. *Let $\{(\mathscr{B}_i, \mathscr{B}_i^*)\}_{i\in\Gamma}$ be a family of (L, M)-dffb's on X satisfying the following condition:*

(C) For any finite subset K of Γ, if $\bigodot_{i\in K}\lambda_i = \underline{0}$, then $\widetilde{\bigodot}_{i\in K}\mathscr{B}_i(\lambda_i) = 0_M$ and $\widetilde{\bigoplus}_{i\in K}\mathscr{B}_i^*(\lambda_i) = 1_M$.

We define the maps $\bigsqcup_{i\in\Gamma}\mathscr{B}_i, \bigsqcap_{i\in\Gamma}\mathscr{B}_i^* : L^X \to M$ as

$$\bigsqcup_{i\in\Gamma}\mathscr{B}_i(\lambda) = \bigvee\left\{\widetilde{\bigodot_{i\in K}}\mathscr{B}_i(\lambda_i) : \lambda = \bigodot_{i\in K}\lambda_i\right\},$$
$$\bigsqcap_{i\in\Gamma}\mathscr{B}_i^*(\lambda) = \bigwedge\left\{\widetilde{\bigoplus_{i\in K}}\mathscr{B}_i^*(\lambda_i) : \lambda = \bigodot_{i\in K}\lambda_i\right\}, \quad (55)$$

where $\bigvee$ and $\bigwedge$ are taken for every finite index subset K of Γ such that $\lambda = \bigodot_{i\in K}\lambda_i$. Then, $(\bigsqcup_{i\in\Gamma}\mathscr{B}_i, \bigsqcap_{i\in\Gamma}\mathscr{B}_i^*)$ is an (L, M)-dffb on X and $(\langle\bigsqcup_{i\in\Gamma}\mathscr{B}_i\rangle, \langle\bigsqcap_{i\in\Gamma}\mathscr{B}_i^*\rangle)$ is the coarsest (L, M)-dff which is finer than $(\langle\mathscr{B}_i\rangle, \langle\mathscr{B}_i^*\rangle)$ for each $i \in \Gamma$.

Example 28. Let $X = \{x, y\}$ be a set and $L = M = [0, 1]$ be an stsc-quantale with $\odot$ (Lukasiewicz t-norm). We define maps $\mathscr{B}_i, \mathscr{B}_i^* : L^X \to M$ as follows ($i = 1, 2$):

$$\mathscr{B}_1(\lambda) = \begin{cases} 1, & \text{if } \lambda = \underline{1}; \\ 0.4, & \text{if } \lambda = 0.6\cdot\underline{1}; \\ 0.5, & \text{if } \lambda = 0.1\cdot\underline{1}; \\ 0, & \text{otherwise}, \end{cases}$$
$$\mathscr{B}_1^*(\lambda) = \begin{cases} 0, & \text{if } \lambda = \underline{1}; \\ 0.6, & \text{if } \lambda = 0.6\cdot\underline{1}; \\ 0.5, & \text{if } \lambda = 0.1\cdot\underline{1}; \\ 1, & \text{otherwise}, \end{cases}$$
$$\mathscr{B}_2(\lambda) = \begin{cases} 1, & \text{if } \lambda = \underline{1}; \\ 0.6, & \text{if } \lambda = 0.7\cdot\underline{1}; \\ 0.3, & \text{if } \lambda = 0.3\cdot\underline{1}; \\ 0, & \text{otherwise}, \end{cases} \quad (56)$$
$$\mathscr{B}_2^*(\lambda) = \begin{cases} 0, & \text{if } \lambda = \underline{1}; \\ 0.4, & \text{if } \lambda = 0.7\cdot\underline{1}; \\ 0.7, & \text{if } \lambda = 0.3\cdot\underline{1}; \\ 1, & \text{otherwise}. \end{cases}$$

Each $(\mathscr{B}_i, \mathscr{B}_i^*)$ for $i = 1, 2$ is an (L, M)-double fuzzy filter base but $(\mathscr{B}_1 \sqcup \mathscr{B}_2, \mathscr{B}_1^* \sqcap \mathscr{B}_2^*)$ is not.

Corollary 29. *Let $\pi_i : X \to X_i$ be projection maps, for all $i \in \Gamma$, where $X = \Pi_{i\in\Gamma}X_i$ is the product set. Let $\{(\mathscr{B}_i, \mathscr{B}_i^*)\}_{i\in\Gamma}$ be a family of (L, M)-dffb's on X_i satisfying the following condition:*

(C) For any finite subset K of Γ, if $\bigodot_{i\in K}(\lambda_i \circ \pi_i) = \underline{0}$, then $\widetilde{\bigodot}_{i\in K}\mathscr{B}_i(\lambda_i) = 0_M$ and $\widetilde{\bigoplus}_{i\in K}\mathscr{B}_i^*(\lambda_i) = 1_M$.

We define the maps $\bigsqcup_{i\in\Gamma}(\pi_i)_1^{\Leftarrow}(\mathscr{B}_i), \bigsqcap_{i\in\Gamma}(\pi_i^*)_1^{\Leftarrow}(\mathscr{B}_i^*) : L^X \to M$ as

$$\bigsqcup_{i\in\Gamma}(\pi_i)_1^{\Leftarrow}(\mathscr{B}_i)(\lambda) = \bigvee\left\{\widetilde{\bigodot_{i\in K}}\mathscr{B}_i(\lambda_i) : \lambda = \bigodot_{i\in K}(\lambda_i \circ \pi_i)\right\},$$
$$\bigsqcap_{i\in\Gamma}(\pi_i^*)_1^{\Leftarrow}(\mathscr{B}_i^*)(\lambda) = \bigwedge\left\{\widetilde{\bigoplus_{i\in K}}\mathscr{B}_i^*(\lambda_i) : \lambda = \bigodot_{i\in K}(\lambda_i \circ \pi_i)\right\}, \quad (57)$$

where $\bigvee$ and $\bigwedge$ are taken for every finite subset K of Γ such that $\lambda = \bigodot_{i\in K}(\lambda_i \circ \pi_i)$.

Let $\mathscr{B} = \bigsqcup_{i\in\Gamma}(\pi_i)_1^{\Leftarrow}(\mathscr{B}_i)$ and $\mathscr{B}^* = \bigsqcap_{i\in\Gamma}(\pi_i^*)_1^{\Leftarrow}(\mathscr{B}_i^*)$ be given. Then, the following properties are satisfied:

(i) $(\mathscr{B}, \mathscr{B}^*)$ is an (L, M)-dffb on X and $(\langle\mathscr{B}\rangle, \langle\mathscr{B}^*\rangle)$ is the coarsest (L, M)-dff on X for which $\pi_i : (X, \langle\mathscr{B}\rangle, \langle\mathscr{B}^*\rangle) \to (X_i, \langle\mathscr{B}_i\rangle, \langle\mathscr{B}_i^*\rangle)$ is a filter map.

(ii) A map $\varphi : (Y, \mathscr{F}, \mathscr{F}^*) \to (X, \langle\mathscr{B}\rangle, \langle\mathscr{B}^*\rangle)$ is a filter map if and only if, for each $i \in \Gamma$, $\pi_i \circ \varphi : (Y, \mathscr{F}, \mathscr{F}^*) \to (X_i, \langle\mathscr{B}_i\rangle, \langle\mathscr{B}_i^*\rangle)$ is a filter map.

In Corollary 29, the structure $(\langle\bigsqcup_{i\in\Gamma}(\pi_i)_1^{\Leftarrow}(\mathscr{B}_i)\rangle, \langle\bigsqcap_{i\in\Gamma}(\pi_i^*)_1^{\Leftarrow}(\mathscr{B}_i^*)\rangle)$ is called a product of (L, M)-dffs's on X.

Theorem 30. *Let $\varphi : X \to Y$ be an injective function and $(\mathscr{B}, \mathscr{B}^*)$ be an (L, M)-dffb on X. Then, the following properties are satisfied:*

(i) $(\varphi_2^{\Rightarrow}(\mathscr{B}), \varphi_2^{\Rightarrow}(\mathscr{B}^*))$ is an (L, M)-dffb on Y, and $(\langle\varphi_2^{\Rightarrow}(\mathscr{B})\rangle, \langle\varphi_2^{\Rightarrow}(\mathscr{B}^*)\rangle)$ is the coarsest (L, M)-dff, for which the function $\varphi : (X, \langle\mathscr{B}\rangle, \langle\mathscr{B}^*\rangle) \to (Y, \langle\varphi_2^{\Rightarrow}(\mathscr{B})\rangle, \langle\varphi_2^{\Rightarrow}(\mathscr{B}^*)\rangle)$ is a filter preserving map.

(ii) $(\varphi_1^{\Leftarrow}(\varphi_2^{\Rightarrow}(\mathscr{B})), \varphi_1^{*\Leftarrow}(\varphi_2^{\Rightarrow}(\mathscr{B}^*)))$ is an (L, M)-dffb on X with $\varphi_1^{\Leftarrow}(\varphi_2^{\Rightarrow}(\mathscr{B})) = \mathscr{B}$ and $\varphi_1^{*\Leftarrow}(\varphi_2^{\Rightarrow}(\mathscr{B}^*)) = \mathscr{B}^*$.

Proof. (i) (DFFB1) For each $\nu \in L^Y$, we have

$$\begin{aligned}\varphi_2^{\Rightarrow}(\mathscr{B})(\nu) &= \mathscr{B}(\varphi_L^{\leftarrow}(\nu)) \le (\mathscr{B}^*(\varphi_L^{\leftarrow}(\nu)))^* \\ &= (\varphi_2^{\Rightarrow}(\mathscr{B}^*)(\nu))^*.\end{aligned} \quad (58)$$

(DFFB2) It is straightforward from the definition.

(DFFB3) Suppose that there exist $\nu_1, \nu_2 \in L^Y$ such that

$$\begin{aligned}\langle\varphi_2^{\Rightarrow}(\mathscr{B}^*)\rangle(\nu_1 \odot \nu_2) &\nleq \varphi_2^{\Rightarrow}(\mathscr{B}^*)(\nu_1) \\ &\tilde{\oplus}\varphi_2^{\Rightarrow}(\mathscr{B}^*)(\nu_2).\end{aligned} \quad (59)$$

By the definition of $\varphi_2^{\Rightarrow}(\mathscr{B}^*)$, we have

$$\langle\varphi_2^{\Rightarrow}(\mathscr{B}^*)\rangle(\nu_1 \odot \nu_2) \nleq \mathscr{B}^*(\nu_1 \circ \varphi)\,\tilde{\oplus}\,\mathscr{B}^*(\nu_2 \circ \varphi). \quad (60)$$

Since $(\mathscr{B}, \mathscr{B}^*)$ is an (L, M)-dffb, the following is obtained:

$$\begin{aligned}\langle\mathscr{B}^*\rangle((\nu_1 \odot \nu_2) \circ \varphi) &= \langle\mathscr{B}^*\rangle((\nu_1 \circ \varphi) \odot (\nu_2 \circ \varphi)) \\ &\le \mathscr{B}^*(\nu_1 \circ \varphi)\,\tilde{\oplus}\,\mathscr{B}^*(\nu_2 \circ \varphi).\end{aligned} \quad (61)$$

Thus,

$$\langle \varphi_2^{\rightrightarrows}(\mathscr{B}^*)\rangle(\nu_1 \odot \nu_2) \not\leq \langle \mathscr{B}^*\rangle((\nu_1 \odot \nu_2)\circ\varphi). \quad (62)$$

By the definition of $\langle \mathscr{B}^*\rangle$, there exists $\lambda \in L^X$ with $\lambda \leq (\nu_1 \odot \nu_2)\circ\varphi$ such that

$$\langle \varphi_2^{\rightrightarrows}(\mathscr{B}^*)\rangle(\nu_1 \odot \nu_2) \not\leq \mathscr{B}^*(\lambda). \quad (63)$$

Since $\varphi_L^{\rightarrow}(\lambda) \leq \varphi_L^{\rightarrow}(\varphi_L^{\leftarrow}(\nu_1 \odot \nu_2)) \leq \nu_1 \odot \nu_2$ and φ is injective,

$$\begin{aligned}\langle \varphi_2^{\rightrightarrows}(\mathscr{B}^*)\rangle(\nu_1 \odot \nu_2) &\leq \varphi_2^{\rightrightarrows}(\mathscr{B}^*)(\varphi_L^{\rightarrow}(\lambda)) \\ &= \mathscr{B}^*(\varphi_L^{\leftarrow}(\varphi_L^{\rightarrow}(\lambda))) \quad (64) \\ &= \mathscr{B}^*(\lambda).\end{aligned}$$

This contradicts the assumption. Then,

$$\begin{aligned}\langle \varphi_2^{\rightrightarrows}(\mathscr{B}^*)\rangle(\nu_1 \odot \nu_2) &\leq \varphi_2^{\rightrightarrows}(\mathscr{B}^*)(\nu_1) \\ &\quad \tilde{\oplus}\,\varphi_2^{\rightrightarrows}(\mathscr{B}^*(\nu_2)), \quad (65) \\ &\quad \text{for each } \nu_1,\nu_2 \in L^Y.\end{aligned}$$

Similarly, it can be verified that

$$\begin{aligned}\langle \varphi_2^{\rightrightarrows}(\mathscr{B})\rangle(\nu_1 \odot \nu_2) &\geq \varphi_2^{\rightrightarrows}(\mathscr{B})(\nu_1)\,\tilde{\odot}\,\varphi_2^{\rightrightarrows}(\mathscr{B}(\nu_2)), \quad (66) \\ &\quad \text{for each } \nu_1,\nu_2 \in L^Y.\end{aligned}$$

Hence, $(\varphi_2^{\rightrightarrows}(\mathscr{B}), \varphi_2^{\rightrightarrows}(\mathscr{B}^*))$ is an (L,M)-dffb on Y.

For each $\lambda \in L^X$, we have

$$\begin{aligned}\langle \varphi_2^{\rightrightarrows}(\mathscr{B})\rangle(\varphi_L^{\rightarrow}(\lambda)) &\geq \varphi_2^{\rightrightarrows}(\mathscr{B})(\varphi_L^{\rightarrow}(\lambda)) \\ &= \mathscr{B}(\varphi_L^{\leftarrow}(\varphi_L^{\rightarrow}(\lambda))) = \mathscr{B}(\lambda), \\ \langle \varphi_2^{\rightrightarrows}(\mathscr{B}^*)\rangle(\varphi_L^{\rightarrow}(\lambda)) &\leq \varphi_2^{\rightrightarrows}(\mathscr{B}^*)(\varphi_L^{\rightarrow}(\lambda)) \quad (67) \\ &= \mathscr{B}^*(\varphi_L^{\leftarrow}(\varphi_L^{\rightarrow}(\lambda))) \\ &= \mathscr{B}^*(\lambda).\end{aligned}$$

Hence, by Theorem 23(ii), $\varphi : (X, \langle\mathscr{B}\rangle, \langle\mathscr{B}^*\rangle) \rightarrow (Y, \langle\varphi_2^{\rightrightarrows}(\mathscr{B})\rangle, \langle\varphi_2^{\rightrightarrows}(\mathscr{B}^*)\rangle)$ is a filter preserving map.

Let $(\mathscr{F}, \mathscr{F}^*)$ be another (L,M)-dff such that $\varphi : (X, \langle\mathscr{B}\rangle, \langle\mathscr{B}^*\rangle) \rightarrow (Y, \mathscr{F}, \mathscr{F}^*)$ is a filter preserving map. So, for each $\mu \in L^Y$, the following is valid:

$$\begin{aligned}\langle \varphi_2^{\rightrightarrows}(\mathscr{B}^*)\rangle(\mu) &= \bigwedge\{\varphi_2^{\rightrightarrows}(\mathscr{B}^*)(\nu) : \nu \leq \mu\} \\ &= \bigwedge\{\mathscr{B}^*(\varphi_L^{\leftarrow}(\nu)) : \nu \leq \mu\} \\ &\geq \bigwedge\{\mathscr{F}^*(\varphi_L^{\rightarrow}(\varphi_L^{\leftarrow}(\nu))) : \nu \leq \mu\} \quad (68) \\ &\geq \bigwedge\{\mathscr{F}^*(\nu) : \nu \leq \mu\} \geq \mathscr{F}^*(\mu).\end{aligned}$$

Hence, $\langle \varphi_2^{\rightrightarrows}(\mathscr{B}^*)\rangle \geq \mathscr{F}^*$. Similarly, it can be proved that $\langle \varphi_2^{\rightrightarrows}(\mathscr{B})\rangle \leq \mathscr{F}$.

(ii) If $\varphi_L^{\leftarrow}(\mu) = \underline{0}$, then $\varphi_2^{\rightrightarrows}(\mathscr{B})(\mu) = \mathscr{B}(\varphi_L^{\leftarrow}(\mu)) = \mathscr{B}(\underline{0}) = 0_M$, and $\varphi_2^{\rightrightarrows}(\mathscr{B}^*)(\mu) = \mathscr{B}^*(\varphi_L^{\leftarrow}(\mu)) = \mathscr{B}^*(\underline{0}) = 1_M$. By Theorem 25(i), $(\varphi_1^{\Leftarrow}(\varphi_2^{\rightrightarrows}(\mathscr{B})), \varphi_1^{*\Leftarrow}(\varphi_2^{\rightrightarrows}(\mathscr{B}^*)))$ is an (L,M)-dffb on X. For each $\nu \in L^X$, following equalities are obtained:

$$\begin{aligned}\varphi_1^{\Leftarrow}(\varphi_2^{\rightrightarrows}(\mathscr{B}))(\nu) &= \bigvee\{\varphi_2^{\rightrightarrows}(\mathscr{B})(\mu) : \nu = \varphi_L^{\leftarrow}(\mu)\} \\ &= \bigvee\{\mathscr{B}(\varphi_L^{\leftarrow}(\mu)) : \nu = \varphi_L^{\leftarrow}(\mu)\} = \mathscr{B}(\nu), \\ \varphi_1^{*\Leftarrow}(\varphi_2^{\rightrightarrows}(\mathscr{B}^*))(\nu) & \quad (69) \\ &= \bigwedge\{\varphi_2^{\rightrightarrows}(\mathscr{B}^*)(\mu) : \nu = \varphi_L^{\leftarrow}(\mu)\} \\ &= \bigwedge\{\mathscr{B}^*(\varphi_L^{\leftarrow}(\mu)) : \nu = \varphi_L^{\leftarrow}(\mu)\} = \mathscr{B}^*(\nu).\end{aligned}$$

Example 31. Let $X = \{a, b\}, Y = \{x, y\}$ be sets and $L = M = [0,1]$ be the stsc-quantale with Lukasiewicz t-norm $\odot$. Let $\varphi : X \rightarrow Y$ be a function defined by $\varphi(a) = \varphi(b) = x$ and $\mu_1, \mu_2 \in [0,1]^X$ be defined by $\mu_1(a) = \mu_1(b) = 0.6$, $\mu_2(a) = 0.1$, $\mu_2(b) = 0$. We define maps $\mathscr{B}, \mathscr{B}^* : [0,1]^X \rightarrow [0,1]$ as follows:

$$\mathscr{B}(\lambda) = \begin{cases} 1, & \text{if } \lambda = \underline{1}; \\ 0.6, & \text{if } \lambda = \mu_1; \\ 0.3, & \text{if } \lambda = \mu_2; \\ 0, & \text{otherwise,} \end{cases}$$
$$\mathscr{B}^*(\lambda) = \begin{cases} 0, & \text{if } \lambda = \underline{1}; \\ 0.4, & \text{if } \lambda = \mu_1; \\ 0.7, & \text{if } \lambda = \mu_2; \\ 1, & \text{otherwise.} \end{cases} \quad (70)$$

Then, $(\mathscr{B}, \mathscr{B}^*)$ is an (L,M)-double fuzzy filter base but $(\varphi_2^{\rightrightarrows}(\mathscr{B}), \varphi_2^{\rightrightarrows}(\mathscr{B}^*))$ is not an (L,M)-double fuzzy filter base.

Theorem 32. *Let $\{\varphi_i : X_i \rightarrow X\}_{i\in\Gamma}$ be a family of injective functions and $\{(\mathscr{B}_i, \mathscr{B}_i^*)\}_{i\in\Gamma}$ be a family of (L,M)-dffb's on X_i satisfying the following condition:*

(C) For any finite subset K of Γ, if $\bigodot_{i\in K}\lambda_i = \underline{0}$, then $\widetilde{\bigodot}_{i\in K}(\varphi_i)_2^{\rightrightarrows}(\mathscr{B}_i)(\lambda_i) = 0_M$, and $\widetilde{\bigoplus}_{i\in K}(\varphi_i)_2^{\rightrightarrows}(\mathscr{B}_i^*)(\lambda_i) = 1_M$.

We define the maps $\mathscr{B}, \mathscr{B}^* : L^X \rightarrow M$ as

$$\begin{aligned}\mathscr{B}(\lambda) &= \bigvee\left\{\widetilde{\bigodot_{i\in K}}(\varphi_i)_2^{\rightrightarrows}(\mathscr{B}_i)(\lambda_i) : \lambda = \bigodot_{i\in K}\lambda_i\right\}, \\ \mathscr{B}^*(\lambda) &= \bigwedge\left\{\widetilde{\bigoplus_{i\in K}}(\varphi_i)_2^{\rightrightarrows}(\mathscr{B}_i^*)(\lambda_i) : \lambda = \bigodot_{i\in K}\lambda_i\right\},\end{aligned} \quad (71)$$

where $\bigvee$ and $\bigwedge$ are taken for every finite index subset K of Γ. Then, the following properties are satisfied:

(i) $(\mathscr{B}, \mathscr{B}^*)$ is an (L,M)-dffb on X and $(\langle\mathscr{B}\rangle, \langle\mathscr{B}^*\rangle)$ is the coarsest (L,M)-dff for which $\varphi_i : (X_i, \langle\mathscr{B}_i\rangle, \langle\mathscr{B}_i^*\rangle) \rightarrow (X, \langle\mathscr{B}\rangle, \langle\mathscr{B}^*\rangle)$ is a filter preserving map.

(ii) A map $\varphi : (X, \langle\mathscr{B}\rangle, \langle\mathscr{B}^*\rangle) \to (Y, \mathscr{F}, \mathscr{F}^*)$ is a filter preserving map if and only if, for each $i \in \Gamma$, $\varphi \circ \varphi_i : (X_i, \langle\mathscr{B}_i\rangle, \langle\mathscr{B}_i^*\rangle) \to (Y, \mathscr{F}, \mathscr{F}^*)$ is a filter preserving map.

Proof. (i) By Corollary 27 and Theorem 30, $(\mathscr{B}, \mathscr{B}^*)$ is an (L, M)-dffb on X.

Since φ_i is injective, for each $i \in \Gamma$,

$$\begin{aligned}
\mathscr{B}\left((\varphi_i)_L^{\rightarrow}(\lambda_i)\right) &\geq (\varphi_i)_2^{\rightrightarrows}\mathscr{B}_i\left((\varphi_i)_L^{\rightarrow}(\lambda_i)\right) \\
&\geq \mathscr{B}_i\left((\varphi_i)_L^{\leftarrow}\left((\varphi_i)_L^{\rightarrow}(\lambda_i)\right)\right) \\
&= \mathscr{B}_i(\lambda_i), \\
\mathscr{B}^*\left((\varphi_i)_L^{\rightarrow}(\lambda_i)\right) &\leq (\varphi_i)_2^{\rightrightarrows}\mathscr{B}_i^*\left((\varphi_i)_L^{\rightarrow}(\lambda_i)\right) \\
&\leq \mathscr{B}_i^*\left((\varphi_i)_L^{\leftarrow}\left((\varphi_i)_L^{\rightarrow}(\lambda_i)\right)\right) \\
&= \mathscr{B}_i^*(\lambda_i).
\end{aligned} \tag{72}$$

Hence, φ_i is a filter preserving map, for each $i \in \Gamma$.

According to Theorem 26(i), other cases are similarly proved.

(ii) It is proved in the same way as Theorem 26(ii).

Definition 33 (see [43]). (a) Let $(\mathbf{A}, U)$ be a concrete category over $\mathbf{X}$. $(\mathbf{A}, U)$ is said to be amnestic provided that its fibres are partially ordered classes; that is, no two different $\mathbf{A}$-objects are equivalent.

(b) Let $\mathbf{A}$ and $\mathbf{B}$ be categories. A functor $G : \mathbf{A} \to \mathbf{B}$ is called topological provided that every G-structured source $(f_i : B \to GA_i)_{i\in\Gamma}$ has a unique G-initial lift $(\overline{f_i} : A \to A_i)_{i\in\Gamma}$.

Proposition 34 (see [43]). *If $G : \mathbf{A} \to \mathbf{B}$ is a functor such that every G-structured source has a G-initial lift, then the following conditions are equivalent:*

(1) G is topological.

(2) $(\mathbf{A}, G)$ is uniquely transportable.

(3) $(\mathbf{A}, G)$ is amnestic.

Theorem 35. *The forgetful functor $V : (L, M)$-**DFIL** $\to$ **SET** defined by $V(X, \mathscr{F}, \mathscr{F}^*) = X$ and $V(\varphi) = \varphi$ is topological.*

Proof. The proof follows from Definition 33, Proposition 34, and Theorem 26.

5. The Types $(\varphi_1^{\rightrightarrows}, \varphi_1^{*\rightrightarrows})$, $(\varphi_2^{\leftleftarrows}, \varphi_2^{\leftleftarrows})$ of Images and Preimages of (L,M)-Double Fuzzy Filter Bases

Theorem 36. *Let $\varphi : X \to Y$ be a surjective function and $(\mathscr{B}, \mathscr{B}^*)$ be an (L, M)-dffb on X. Then, the following properties are satisfied:*

(i) $(\varphi_1^{\rightrightarrows}(\mathscr{B}), \varphi_1^{*\rightrightarrows}(\mathscr{B}^*))$ is an (L, M)-dffb on Y.

(ii) $(\langle\varphi_1^{\rightrightarrows}(\mathscr{B})\rangle, \langle\varphi_1^{*\rightrightarrows}(\mathscr{B}^*)\rangle)$ is the coarsest (L, M)-dff on Y for which $\varphi : (X, \langle\mathscr{B}\rangle, \langle\mathscr{B}^*\rangle) \to (Y, \langle\varphi_1^{\rightrightarrows}(\mathscr{B})\rangle, \langle\varphi_1^{*\rightrightarrows}(\mathscr{B}^*)\rangle)$ is a filter preserving map.

(iii) If $(\mathscr{B}, \mathscr{B}^*)$ is an (L, M)-dff, then $\langle\varphi_1^{\rightrightarrows}(\mathscr{B})\rangle = \varphi_2^{\rightrightarrows}(\mathscr{B})$ and $\langle\varphi_1^{*\rightrightarrows}(\mathscr{B}^*)\rangle\rangle = \varphi_2^{\rightrightarrows}(\mathscr{B}^*)$.

Proof. (i) and (ii) are proved in the same manner as Theorem 25(i).

(iii) Let $(\mathscr{B}, \mathscr{B}^*)$ be an (L, M)-dff. Since φ is surjective, $\varphi_L^{\rightarrow}(\lambda) \leq \nu$ is equivalent to $\lambda \leq \nu \circ \varphi$. Then, for each $\nu \in L^Y$, it is clear that

$$\begin{aligned}
\langle\varphi_1^{*\rightrightarrows}(\mathscr{B}^*)\rangle(\nu) &= \bigwedge\{\varphi_1^{*\rightrightarrows}(\mathscr{B}^*)(\mu) : \mu \leq \nu\} \\
&= \bigwedge\{\mathscr{B}^*(\lambda) : \varphi_L^{\rightarrow}(\lambda) = \mu \leq \nu\} \\
&= \bigwedge\{\mathscr{B}^*(\lambda) : \lambda \leq \nu \circ \varphi\} \\
&= \mathscr{B}^*(\nu \circ \phi) = \varphi_2^{\rightrightarrows}(\mathscr{B}^*)(\nu).
\end{aligned} \tag{73}$$

Hence, $\langle\varphi_1^{*\rightrightarrows}(\mathscr{B}^*)\rangle = \varphi_2^{\rightrightarrows}(\mathscr{B}^*)$. Similarly, $\langle\varphi_1^{\rightrightarrows}(\mathscr{B})\rangle = \varphi_2^{\rightrightarrows}(\mathscr{B})$ is obtained.

Remark 37. Let $\varphi : X \to Y$ be a bijective function, $(\mathscr{K}, \mathscr{K}^*)$ be an (L, M)-dffb on X, and $(\mathscr{B}, \mathscr{B}^*)$ be an (L, M)-dffb on Y. Then, the following equalities are clear.

(i) $\varphi_1^{\rightrightarrows}(\mathscr{K}) = \varphi_2^{\rightrightarrows}(\mathscr{K})$ and $\varphi_1^{*\rightrightarrows}(\mathscr{K}^*) = \varphi_2^{\rightrightarrows}(\mathscr{K}^*)$.

(ii) $\varphi_1^{\leftleftarrows}(\mathscr{B}) = \varphi_2^{\leftleftarrows}(\mathscr{B})$ and $\varphi_1^{*\leftleftarrows}(\mathscr{B}^*) = \varphi_2^{\leftleftarrows}(\mathscr{B}^*)$.

Remark 38. Let $\varphi : X \to Y$ be a bijective function and $(\mathscr{B}, \mathscr{B}^*)$ be an (L, M)-dffb on Y. Then, it follows from Remark 37(ii) and Theorem 25 that $(\varphi_2^{\leftleftarrows}(\mathscr{B}), \varphi_2^{\leftleftarrows}(\mathscr{B}^*))$ is an (L, M)-dffb on X and $(\langle\varphi_2^{\leftleftarrows}(\mathscr{B})\rangle, \langle\varphi_2^{\leftleftarrows}(\mathscr{B}^*)\rangle)$ is the coarsest (L, M)-dff on X for which $\varphi : (X, \langle\varphi_2^{\leftleftarrows}(\mathscr{B})\rangle, \langle\varphi_2^{\leftleftarrows}(\mathscr{B}^*)\rangle) \to (Y, \mathscr{B}, \mathscr{B}^*)$ is a filter map.

Theorem 39. *Let $\varphi : X \to Y$ be a function and $\{(\mathscr{B}_i, \mathscr{B}_i^*)\}_{i\in\Gamma}$ be a family of (L, M)-dffb's on X satisfying the following condition:*

(C) For every finite subset K of Γ, if $\bigodot_{i\in K}\lambda_i = \underline{0}$, then $\widetilde{\bigodot}_{i\in K}\mathscr{B}_i(\lambda_i) = 0_M$ and $\widetilde{\bigoplus}_{i\in K}\mathscr{B}_i^*(\lambda_i) = 1_M$. Then, the following properties are satisfied:

(i) If $\varphi : X \to Y$ is bijective, then $\varphi_1^{\rightrightarrows}(\bigsqcup_{i\in\Gamma}\mathscr{B}_i) = \bigsqcup_{i\in\Gamma}\varphi_1^{\rightrightarrows}(\mathscr{B}_i)$ and $\varphi_1^{*\rightrightarrows}(\bigsqcap_{i\in\Gamma}\mathscr{B}_i^*) = \bigsqcap_{i\in\Gamma}\varphi_1^{*\rightrightarrows}(\mathscr{B}_i^*)$.

(ii) If $\varphi : X \to Y$ is injective, then $\langle\varphi_1^{\rightrightarrows}(\bigsqcup_{i\in\Gamma}\mathscr{B}_i)\rangle = \bigsqcup_{i\in\Gamma}\langle\varphi_1^{\rightrightarrows}(\mathscr{B}_i)\rangle$ and $\langle\varphi_1^{*\rightrightarrows}(\bigsqcap_{i\in\Gamma}\mathscr{B}_i^*)\rangle = \bigsqcap_{i\in\Gamma}\langle\varphi_1^{*\rightrightarrows}(\mathscr{B}_i^*)\rangle$.

Proof. (i) Let us consider the following condition:

(C1) For every finite subset K of Γ, if $\bigodot_{i\in K}\nu_i = \underline{0}$, then $\widetilde{\bigodot}_{i\in K}\varphi_1^{\rightrightarrows}(\mathscr{B}_i)(\nu_i) = 0_M$ and $\widetilde{\bigoplus}_{i\in K}\varphi_1^{*\rightrightarrows}(\mathscr{B}_i^*)(\nu_i) = 1_M$.

For the proof, it is enough to show that (C1) $\Leftrightarrow$ (C).

(C1) $\Rightarrow$ (C): For any finite subset K of Γ with $\bigodot_{i\in K}\lambda_i = \underline{0}$, since φ is injective, by Lemma 6(!),

$$\varphi_L^{\rightarrow}\left(\bigodot_{i\in K}\lambda_i\right) = \bigodot_{i\in K}\varphi_L^{\rightarrow}(\lambda_i) = \underline{0}. \tag{74}$$

By (C1), we have

$$\begin{aligned}
0_M &= \widetilde{\bigodot_{i\in K}}\varphi_1^{\rightrightarrows}(\mathscr{B}_i)\left(\varphi_L^{\rightarrow}(\lambda_i)\right) \geq \widetilde{\bigodot_{i\in K}}\mathscr{B}_i(\lambda_i), \\
1_M &= \widetilde{\bigoplus_{i\in K}}\varphi_1^{*\rightrightarrows}(\mathscr{B}_i^*)\left(\varphi_L^{\rightarrow}(\lambda_i)\right) \leq \widetilde{\bigoplus_{i\in K}}\mathscr{B}_i^*(\lambda_i),
\end{aligned} \tag{75}$$

and thus $\widetilde{\bigodot}_{i\in K}\mathscr{B}_i(\lambda_i) = 0_M$ and $\widetilde{\bigoplus}_{i\in K}\mathscr{B}_i^*(\lambda_i) = 1_M$ is satisfied.

(C) $\Rightarrow$ (C1): Suppose that, for every finite subset K of Γ with $\bigodot_{i\in K}\nu_i = \underline{0}$, $\widetilde{\bigoplus}_{i\in K}\varphi_1^{*\rightrightarrows}(\mathscr{B}_i^*)(\nu_i) \neq 1_M$. Then, for each $i \in K$, there exists $\lambda_i \in L^X$ with $\nu_i = \varphi_L^{\rightarrow}(\lambda_i)$ such that

$$\widetilde{\bigoplus_{i\in K}} \varphi_1^{*\rightrightarrows}(\mathscr{B}_i^*)(\nu_i) \leq \widetilde{\bigoplus_{i\in K}} \mathscr{B}_i^*(\lambda_i) \neq 1_M. \quad (76)$$

By (C), $\bigodot_{i\in K}\lambda_i \neq \underline{0}$. By Lemma 6(!),

$$\varphi_L^{\rightarrow}\left(\bigodot_{i\in K}\lambda_i\right) = \bigodot_{i\in K}\varphi_L^{\rightarrow}(\lambda_i) = \bigodot_{i\in K}\nu_i \neq \underline{0}. \quad (77)$$

This contradicts the assumption. Thus, $\widetilde{\bigoplus}_{i\in K}\varphi_1^{*\rightrightarrows}(\mathscr{B}_i^*)(\nu_i) = 1_M$. Similarly, for every finite subset K of Γ, if $\bigodot_{i\in K}\nu_i = \underline{0}$, then $\widetilde{\bigodot}_{i\in K}\varphi_1^{\rightrightarrows}(\mathscr{B}_i)(\nu_i) = 0_M$.

Since φ is surjective, by Theorem 36, $(\varphi_1^{\rightrightarrows}(\mathscr{B}_i), \varphi_1^{*\rightrightarrows}(\mathscr{B}_i^*))$ exists for each $i \in \Gamma$. By Corollary 27 and (C1), $(\bigsqcup_{i\in\Gamma}\varphi_1^{\rightrightarrows}(\mathscr{B}_i), \bigsqcap_{i\in\Gamma}\varphi_1^{*\rightrightarrows}(\mathscr{B}_i^*))$ exists.

For each finite subset K of Γ such that $\lambda = \bigodot_{k\in K}\lambda_k$ with $\varphi_L^{\rightarrow}(\lambda) = \nu$, the following inequalities are satisfied:

$$\begin{aligned}\bigsqcup_{i\in\Gamma}\varphi_1^{\rightrightarrows}(\mathscr{B}_i)(\nu) &\geq \widetilde{\bigodot_{k\in K}}\varphi_1^{\rightrightarrows}(\mathscr{B}_k)(\varphi_L^{\rightarrow}(\lambda_k)) \\ &\geq \widetilde{\bigodot_{k\in K}}\mathscr{B}_k(\lambda_k), \\ \bigsqcap_{i\in\Gamma}\varphi_1^{*\rightrightarrows}(\mathscr{B}_i^*)(\nu) &\leq \widetilde{\bigoplus_{k\in K}}\varphi_1^{*\rightrightarrows}(\mathscr{B}_k^*)(\varphi_L^{\rightarrow}(\lambda_k)) \\ &\leq \widetilde{\bigoplus_{k\in K}}\mathscr{B}_k^*(\lambda_k).\end{aligned} \quad (78)$$

This implies that

$$\begin{aligned}\bigsqcup_{i\in\Gamma}\mathscr{B}_i(\lambda) &\leq \bigsqcup_{i\in\Gamma}\varphi_1^{\rightrightarrows}(\mathscr{B}_i)(\nu), \\ \bigsqcap_{i\in\Gamma}\mathscr{B}_i^*(\lambda) &\geq \bigsqcap_{i\in\Gamma}\varphi_1^{*\rightrightarrows}(\mathscr{B}_i^*)(\nu).\end{aligned} \quad (79)$$

So, the following are clear:

$$\begin{aligned}\varphi_1^{\rightrightarrows}\left(\bigsqcup_{i\in\Gamma}\mathscr{B}_i\right) &\leq \bigsqcup_{i\in\Gamma}\varphi_1^{\rightrightarrows}(\mathscr{B}_i), \\ \varphi_1^{*\rightrightarrows}\left(\bigsqcap_{i\in\Gamma}\mathscr{B}_i^*\right) &\geq \bigsqcap_{i\in\Gamma}\varphi_1^{*\rightrightarrows}(\mathscr{B}_i^*).\end{aligned} \quad (80)$$

For any finite subset J of Γ with $\mu = \bigodot_{j\in J}\mu_j$, there exist $\eta_j \in L^X$ with $\varphi_L^{\rightarrow}(\eta_j) = \mu_j$. Thus,

$$\begin{aligned}\varphi_1^{\rightrightarrows}\left(\bigsqcup_{i\in\Gamma}\mathscr{B}_i\right)(\mu) &\geq \left(\bigsqcup_{i\in\Gamma}\mathscr{B}_i\right)\left(\bigodot_{j\in J}\eta_j\right) \\ &\geq \widetilde{\bigodot_{j\in J}}\mathscr{B}_j(\eta_j), \\ \varphi_1^{*\rightrightarrows}\left(\bigsqcap_{i\in\Gamma}\mathscr{B}_i^*\right)(\mu) &\leq \left(\bigsqcap_{i\in\Gamma}\mathscr{B}_i^*\right)\left(\bigodot_{j\in J}\eta_j\right) \\ &\leq \widetilde{\bigoplus_{j\in J}}\mathscr{B}_j^*(\eta_j).\end{aligned} \quad (81)$$

This implies that

$$\begin{aligned}\varphi_1^{\rightrightarrows}\left(\bigsqcup_{i\in\Gamma}\mathscr{B}_i\right) &\geq \bigsqcup_{i\in\Gamma}\varphi_1^{\rightrightarrows}(\mathscr{B}_i), \\ \varphi_1^{*\rightrightarrows}\left(\bigsqcap_{i\in\Gamma}\mathscr{B}_i^*\right) &\leq \bigsqcap_{i\in\Gamma}\varphi_1^{*\rightrightarrows}(\mathscr{B}_i^*).\end{aligned} \quad (82)$$

From the above inequalities, we have

$$\begin{aligned}\varphi_1^{\rightrightarrows}\left(\bigsqcup_{i\in\Gamma}\mathscr{B}_i\right) &= \bigsqcup_{i\in\Gamma}\varphi_1^{\rightrightarrows}(\mathscr{B}_i), \\ \varphi_1^{*\rightrightarrows}\left(\bigsqcap_{i\in\Gamma}\mathscr{B}_i^*\right) &= \bigsqcap_{i\in\Gamma}\varphi_1^{*\rightrightarrows}(\mathscr{B}_i^*).\end{aligned} \quad (83)$$

(ii) It is proved by the same method as in (i) and Theorem 36(ii).

Theorem 40. *Let $\{\varphi_i : X_i \to X : i \in \Gamma\}$ be a family of functions and $\{(\mathscr{B}_i, \mathscr{B}_i^*)\}_{i\in\Gamma}$ be a family of (L, M)-dffbs' on X_i satisfying the following condition:*

(C) For any finite subset K of Γ, if $\bigodot_{i\in K}(\varphi_i)_L^{\rightarrow}(\lambda_i) = \underline{0}$, then $\widetilde{\bigodot}_{i\in K}\mathscr{B}_i(\lambda_i) = 0_M$ and $\widetilde{\bigoplus}_{i\in K}\mathscr{B}_i^*(\lambda_i) = 1_M$.

We define the maps $\bigcup_{i\in\Gamma}(\varphi_i)_1^{\rightrightarrows}(\mathscr{B}_i)$, $\bigcap_{i\in\Gamma}(\varphi_i)_1^{*\rightrightarrows}(\mathscr{B}_i^*) : L^X \to M$ as

$$\begin{aligned}&\bigcup_{i\in\Gamma}(\varphi_i)_1^{\rightrightarrows}(\mathscr{B}_i)(\nu) \\ &= \bigvee\left\{\widetilde{\bigodot_{i\in K}}\mathscr{B}_i(\lambda_i) : \nu = \bigodot_{i\in K}(\varphi_i)_L^{\rightarrow}(\lambda_i)\right\}, \\ &\bigcap_{i\in\Gamma}(\varphi_i)_1^{*\rightrightarrows}(\mathscr{B}_i^*)(\nu) \\ &= \bigwedge\left\{\widetilde{\bigoplus_{i\in K}}\mathscr{B}_i^*(\lambda_i) : \nu = \bigodot_{i\in K}(\varphi_i)_L^{\rightarrow}(\lambda_i)\right\},\end{aligned} \quad (84)$$

where $\bigvee$ and $\bigwedge$ are taken for every finite subset K of Γ. Let $\mathscr{B} = \bigcup_{i\in\Gamma}(\varphi_i)_1^{\rightrightarrows}(\mathscr{B}_i)$ and $\mathscr{B}^* = \bigcap_{i\in\Gamma}(\varphi_i)_1^{*\rightrightarrows}(\mathscr{B}_i^*)$. Then, the following properties are satisfied:

(i) If φ_j is surjective for some $j \in \Gamma$, then $(\mathscr{B}, \mathscr{B}^*)$ is an (L, M)-dffb on X and $(\langle\mathscr{B}\rangle, \langle\mathscr{B}^*\rangle)$ is the coarsest (L, M)-dff for which the map $\varphi_i : (X_i, \langle\mathscr{B}_i\rangle, \langle\mathscr{B}_i^*\rangle) \to (X, \langle\mathscr{B}\rangle, \langle\mathscr{B}^*\rangle)$ is a filter preserving map.

(ii) A function $\varphi : (X, \langle\mathscr{B}\rangle, \langle\mathscr{B}^*\rangle) \to (Y, \mathscr{F}, \mathscr{F}^*)$ is a filter preserving map if and only if, for each $i \in \Gamma$, $\varphi \circ \varphi_i : (X_i, \langle\mathscr{B}_i\rangle, \langle\mathscr{B}_i^*\rangle) \to (Y, \mathscr{F}, \mathscr{F}^*)$ is a filter preserving map.

(iii) If φ_i are surjective for all $i \in \Gamma$, then

$$\left\langle \bigcup_{i\in\Gamma} (\varphi_i)_1^{\rightrightarrows}(\mathscr{B}_i) \right\rangle = \left\langle \bigsqcup_{i\in\Gamma} (\varphi_i)_1^{\rightrightarrows}(\mathscr{B}_i) \right\rangle,$$
$$\left\langle \bigcap_{i\in\Gamma} (\varphi_i)_1^{*\rightrightarrows}(\mathscr{B}_i^*) \right\rangle = \left\langle \bigsqcap_{i\in\Gamma} (\varphi_i)_1^{*\rightrightarrows}(\mathscr{B}_i^*) \right\rangle. \quad (85)$$

Proof. (i) (DFFB2) Since φ_j is surjective for some $j \in \Gamma$ and (C), $\mathscr{B}(\underline{1}) = 1_M$, $\mathscr{B}^*(\underline{1}) = 0_M$ and $\mathscr{B}(\underline{0}) = 0_M$, $\mathscr{B}^*(\underline{0}) = 1_M$.

According to Theorems 26(i) and 32(i), other cases are similarly proved.

(ii) The proof is similar to Theorem 26(ii).

(iii) Let us consider the following condition:

(C1) For any finite subset K of Γ, if $\bigodot_{i\in K}\nu_i = \underline{0}$, then $\widetilde{\bigodot}_{i\in K}(\varphi_i)_1^{\rightrightarrows}(\mathscr{B}_i)(\nu_i) = 0_M$ and $\widetilde{\bigoplus}_{i\in K}(\varphi_i)_1^{*\rightrightarrows}(\mathscr{B}_i^*)(\nu_i) = 1_M$.

For the proof, it is enough to show that (C1) $\Leftrightarrow$ (C).

(C1) $\Rightarrow$ (C): For any finite subset K of Γ, if $\bigodot_{i\in K}(\varphi_i)_L^{\rightarrow}(\lambda_i) = \underline{0}$, by (C1), we have

$$0_M = \widetilde{\bigodot_{i\in K}} (\varphi_i)_1^{\rightrightarrows}(\mathscr{B}_i)\left((\varphi_i)_L^{\rightarrow}(\lambda_i)\right) \geq \widetilde{\bigodot_{i\in K}} \mathscr{B}_i(\lambda_i),$$
$$1_M = \widetilde{\bigoplus_{i\in K}} (\varphi_i)_1^{*\rightrightarrows}(\mathscr{B}_i^*)\left((\varphi_i)_L^{\rightarrow}(\lambda_i)\right) \leq \widetilde{\bigoplus_{i\in K}} \mathscr{B}_i^*(\lambda_i). \quad (86)$$

(C) $\Rightarrow$ (C1): Suppose that, for any finite subset K of Γ with $\bigodot_{i\in K}\nu_i = \underline{0}$, we have $\widetilde{\bigoplus}_{i\in K}(\varphi_i)_1^{*\rightrightarrows}(\mathscr{B}_i^*)(\nu_i) \neq 1_M$. Then, for each $i \in K$, there exists $\lambda_i \in L^{X_i}$ with $\nu_i = (\varphi_i)_L^{\rightarrow}(\lambda_i)$ such that

$$\widetilde{\bigoplus_{i\in K}} (\varphi_i)_1^{*\rightrightarrows}(\mathscr{B}_i^*)(\nu_i) \leq \widetilde{\bigoplus_{i\in K}} \mathscr{B}_i^*(\lambda_i) \neq 1_M. \quad (87)$$

By (C),

$$\bigodot_{i\in K}(\varphi_i)_L^{\rightarrow}(\lambda_i) = \bigodot_{i\in K}\nu_i \neq \underline{0}. \quad (88)$$

This is a contradiction. Thus, $\widetilde{\bigoplus}_{i\in K}(\varphi_i)_1^{*\rightrightarrows}(\mathscr{B}_i^*)(\nu_i) = 1_M$. Similarly, $\widetilde{\bigodot}_{i\in K}(\phi_i)_1^{\rightrightarrows}(\mathscr{B}_i)(\nu_i) = 0_M$.

For any finite index set K with $\{\lambda_i : \bigodot_{i\in K}(\varphi_i)_L^{\rightarrow}(\lambda_i) \leq \nu\}$, by the definition of $\langle\bigsqcup_{i\in\Gamma}(\varphi_i)_1^{\rightrightarrows}(\mathscr{B}_i)\rangle$ and $\langle\bigsqcap_{i\in\Gamma}(\varphi_i)_1^{*\rightrightarrows}(\mathscr{B}_i^*)\rangle$, the following inequalities are obtained:

$$\left\langle \bigsqcup_{i\in\Gamma} (\varphi_i)_1^{\rightrightarrows}(\mathscr{B}_i) \right\rangle(\nu) \geq \bigvee_{i\in\Gamma} (\varphi_i)_1^{\rightrightarrows}(\mathscr{B}_i)\left(\bigodot_{i\in K}(\varphi_i)_L^{\rightarrow}(\lambda_i)\right) \geq \widetilde{\bigodot_{i\in K}} (\varphi_i)_1^{\rightrightarrows}(\mathscr{B}_i)\left((\varphi_i)_L^{\rightarrow}(\lambda_i)\right) \text{ (by Corollary 26)} \geq \widetilde{\bigodot_{i\in K}} \mathscr{B}_i(\lambda_i),$$
$$\left\langle \bigsqcap_{i\in\Gamma} (\varphi_i)_1^{*\rightrightarrows}(\mathscr{B}_i^*) \right\rangle(\nu) \leq \bigwedge_{i\in\Gamma} (\varphi_i)_1^{*\rightrightarrows}(\mathscr{B}_i^*)\left(\bigodot_{i\in K}(\varphi_i)_L^{\rightarrow}(\lambda_i)\right) \leq \widetilde{\bigoplus_{i\in K}} (\varphi_i)_1^{*\rightrightarrows}(\mathscr{B}_i^*)\left((\varphi_i)_L^{\rightarrow}(\lambda_i)\right) \text{ (by Corollary 26)} \leq \widetilde{\bigoplus_{i\in K}} \mathscr{B}_i^*(\lambda_i). \quad (89)$$

Hence,

$$\left\langle \bigcup_{i\in\Gamma} (\varphi_i)_1^{\rightrightarrows}(\mathscr{B}_i) \right\rangle \leq \left\langle \bigsqcup_{i\in\Gamma} (\varphi_i)_1^{\rightrightarrows}(\mathscr{B}_i) \right\rangle,$$
$$\left\langle \bigcap_{i\in\Gamma} (\varphi_i)_1^{*\rightrightarrows}(\mathscr{B}_i^*) \right\rangle \geq \left\langle \bigsqcap_{i\in\Gamma} (\varphi_i)_1^{*\rightrightarrows}(\mathscr{B}_i^*) \right\rangle. \quad (90)$$

For any finite index set J with $\{\nu_i : \bigodot_{i\in J}\nu_i \leq \mu\}$, since φ_i is surjective, for each $i \in J$, there exists $\lambda_i \in L^{X_i}$ with $(\varphi_i)_L^{\rightarrow}(\lambda_i) = \nu_i$ such that $\mu \geq \bigodot_{i\in J}\nu_i = \bigodot_{i\in J}(\varphi_i)_L^{\rightarrow}(\lambda_i)$.

Thus,

$$\left\langle \bigcup_{i\in\Gamma} (\varphi_i)_1^{\rightrightarrows}(\mathscr{B}_i) \right\rangle(\mu) \geq \widetilde{\bigodot_{i\in J}} (\varphi_i)_1^{\rightrightarrows}(\mathscr{B}_i)(\nu_i) \geq \widetilde{\bigodot_{i\in J}} \mathscr{B}_i(\lambda_i),$$
$$\left\langle \bigcap_{i\in\Gamma} (\varphi_i)_1^{*\rightrightarrows}(\mathscr{B}_i^*) \right\rangle(\mu) \leq \widetilde{\bigoplus_{i\in J}} (\varphi_i)_1^{*\rightrightarrows}(\mathscr{B}_i^*)(\nu_i) \leq \widetilde{\bigoplus_{i\in J}} \mathscr{B}_i^*(\lambda_i). \quad (91)$$

Hence,

$$\left\langle \bigcup_{i\in\Gamma} (\varphi_i)_1^{\rightrightarrows}(\mathscr{B}_i) \right\rangle \geq \left\langle \bigsqcup_{i\in\Gamma} (\varphi_i)_1^{\rightrightarrows}(\mathscr{B}_i) \right\rangle,$$
$$\left\langle \bigcap_{i\in\Gamma} (\varphi_i)_1^{*\rightrightarrows}(\mathscr{B}_i^*) \right\rangle \leq \left\langle \bigsqcap_{i\in\Gamma} (\varphi_i)_1^{*\rightrightarrows}(\mathscr{B}_i^*) \right\rangle. \quad (92)$$

6. Conclusion

In this study, we introduced the notions of (L, M)-double fuzzy filter space and (L, M)-double fuzzy filter base where L and M are stsc-quantales as an extension of frames. We showed the existence of initial and also final (L, M)-double fuzzy filter structures. We also proved that the category (L, M)-**DFIL** is a topological category over **SET**. By giving illustrative examples, we considered two types of second-order Zadeh image and preimage operators of (L, M)-double fuzzy filter.

Conflicts of Interest

The authors declare that there are no conflicts of interest regarding the publication of this paper.

References

[1] T. Kubiak, *On fuzzy topologies [Ph.D. thesis]*, Adam Mickiewicz University, Poznań, Poland, 1985.

[2] A. P. Šostak, "On a fuzzy topological structure," *Rendiconti del Circolo Matematico di Palermo, Serie II*, vol. 11, pp. 89–103, 1985.

[3] C. L. Chang, "Fuzzy topological spaces," *Journal of Mathematical Analysis and Applications*, vol. 24, pp. 182–190, 1968.

[4] S. E. Abbas and H. Aygün, "Intuitionistic fuzzy semiregularization spaces," *Information Sciences*, vol. 176, no. 6, pp. 745–757, 2006.

[5] V. Çetkin and H. Aygün, "On double fuzzy preuniformity," *Journal of Nonlinear Sciences and Applications*, vol. 6, no. 4, pp. 263–278, 2013.

[6] U. Höhle, "Upper semicontinuous fuzzy sets and applications," *Journal of Mathematical Analysis and Applications*, vol. 78, no. 2, pp. 659–673, 1980.

[7] U. Höhle and A. P. Šostak, "Axiomatic foundations of fixed-basis fuzzy topology," in *Mathematics of Fuzzy Sets*, vol. 3 of *The Handbooks of Fuzzy Sets Series*, pp. 123–272, Kluwer Academic, Boston, Mass, USA, 1999.

[8] T. Kubiak and A. P. Šostak, "Lower set-valued fuzzy topologies," *Quaestiones Mathematicae*, vol. 20, no. 3, pp. 423–429, 1997.

[9] R. Lowen, "Fuzzy topological spaces and fuzzy compactness," *Journal of Mathematical Analysis and Applications*, vol. 56, no. 3, pp. 621–633, 1976.

[10] X. Luo and J. Fang, "Fuzzifying closure systems and closure operators," *Iranian Journal of Fuzzy Systems*, vol. 8, no. 1, pp. 77–94, 167, 2011.

[11] S. E. Rodabaugh, "Categorical foundations of variable-basis fuzzy topology," in *Mathematics of fuzzy sets*, U. Hohle and S. E. Rodabaugh, Eds., vol. 3 of *The Handbooks of Fuzzy Sets Series*, pp. 273–388, Kluwer Academic, Boston, Mass, USA, 1999.

[12] F.-G. Shi, "Countable Compactness and the Lindelof Property of L-Fuzzy Sets," *Iranian Journal of Fuzzy Systems*, vol. 1, no. 1, pp. 79–88, 2004.

[13] A. P. Shostak, "Two decades of fuzzy topology: Basic ideas, notions, and results," *Russian Mathematical Surveys*, vol. 44, no. 6, pp. 125–186, 1989.

[14] A. P. Šostak, "Basic structures of fuzzy topology," *Journal of Mathematical Sciences*, vol. 78, no. 6, pp. 662–701, 1996.

[15] K. T. Atanassov, "Intuitionistic fuzzy sets," *Fuzzy Sets and Systems*, vol. 20, no. 1, pp. 87–96, 1986.

[16] D. Çoker, "An introduction to intuitionistic fuzzy topological spaces," *Fuzzy Sets and Systems*, vol. 88, no. 1, pp. 81–89, 1997.

[17] D. Çoker and M. Demirci, "An introduction to intuitionistic fuzzy topological spaces in Šostak sense," *Busefal*, vol. 67, pp. 67–76, 1996.

[18] T. K. Mondal and S. K. Samanta, "On intuitionistic gradation of openness," *Fuzzy Sets and Systems*, vol. 131, no. 3, pp. 323–336, 2002.

[19] J. Gutiérrez García and S. E. Rodabaugh, "Order-theoretic, topological, categorical redundancies of interval-valued sets, grey sets, vague sets, interval-valued "intuitionistic" sets, "intuitionistic" fuzzy sets and topologies," *Fuzzy Sets and Systems*, vol. 156, no. 3, pp. 445–484, 2005.

[20] M. H. Burton, M. Muraleetharan, and J. Gutiérrez García, "Generalised filters 1," *Fuzzy Sets and Systems*, vol. 106, no. 2, pp. 275–284, 1999.

[21] M. H. Burton, M. Muraleetharan, and J. Gutierrez Garcia, "Generalized filters 2," *Fuzzy Sets and Systems*, vol. 106, no. 3, pp. 393–400, 1999.

[22] W. Gähler, "The general fuzzy filter approach to fuzzy topology, I," *Fuzzy Sets and Systems*, vol. 76, no. 2, pp. 205–224, 1995.

[23] W. Gähler, "The general fuzzy filter approach to fuzzy topology, II," *Fuzzy Sets and Systems*, vol. 76, no. 2, pp. 225–246, 1995.

[24] U. Höhle and E. P. Klement, *Non-Classical Logic and Their Applications to Fuzzy Subsets*, Kluwer Academic Publisher, Boston, Mass, USA, 1995.

[25] Y. M. Liu and M. K. Luo, *Fuzzy Topology*, Scientific Publishing, Singapore, Singapore, 1997.

[26] J. Fang, "Lattice-valued semiuniform convergence spaces," *Fuzzy Sets and Systems*, vol. 195, pp. 33–57, 2012.

[27] G. Jäger, "Lattice-valued convergence spaces and regularity," *Fuzzy Sets and Systems*, vol. 159, no. 19, pp. 2488–2502, 2008.

[28] L. Li and Q. Jin, "Lattice-valued convergence spaces: weaker regularity and p-regularity," *Abstract and Applied Analysis*, vol. 2014, Article ID 328153, pp. 1–11, 2014.

[29] U. Höhle and A. Šostak, "A general theory of fuzzy topological spaces," *Fuzzy Sets and Systems*, vol. 73, no. 1, pp. 131–149, 1995.

[30] J. Luna-Torres and C. O. Ochoa, "L-filters and LF-topologies," *Fuzzy Sets and Systems*, vol. 140, no. 3, pp. 433–446, 2003.

[31] U. Höhle, *Many Valued Topology and Its Application*, Kluwer Academic, Boston, Mass, USA, 2001.

[32] C. J. Mulvey, "&," *Supplemento ai Rendiconti del Circolo Matematico di Palermo. Serie II*, vol. 12, pp. 99–104, 1986.

[33] S. E. Rodabaugh and E. P. Klement, *Toplogical and Algebraic Structures in Fuzzy Sets, The Handbook of Recent Developments in the Mathematics of Fuzzy Sets*, Kluwer Academic Publishers, Boston, Mass, USA, 2003.

[34] Y. C. Kim and Y. S. Kim, "(L, e)-approximation spaces and (L, e)-fuzzy quasi-uniform spaces," *Information Sciences*, vol. 179, no. 12, pp. 2028–2048, 2009.

[35] E. Turunen, *Mathematics Behind Fuzzy Logic*, Springer, New York, NY, USA, 1999.

[36] U. Höhle, "Monoidal closed categories, weak topoi and generalized logics," *Fuzzy Sets and Systems*, vol. 42, no. 1, pp. 15–35, 1991.

[37] U. Höhle, "M-valued sets and sheaves over integral commutative M-valued sets and sheaves over integral commutative cl-monoids," in *Applications of Category Theory of Fuzzy Subsets*, S. Rodabaugh, E. P. Klement, and U. Höhle, Eds., pp. 33–72, Kluwer Academic, Dordrecht, Netherlands, 1992.

[38] U. Höhle, "Commutative, residuated l-monoids," in *Non-Classical Logics and Their Applications to Fuzzy Subsets Theory*, vol. 32 of *Theory and Decision Library (Series B: Mathematical and Statistical Methods)*, pp. 53–106, Kluwer Academic, Dordrecht, Netherlands, 1995.

[39] S. Jenei, "Structure of Girard monoids on $[0, 1]$," in *Topological And Algebraic Structures in Fuzzy Sets*, S. E. Rodabaugh and E. P. Klement, Eds., vol. 20 of *Trends in Logic (Studia Logica Library)*, pp. 277–308, Kluwer Academic, 2003.

[40] Y. C. Kim and J. M. Ko, "Images and preimages of L-filterbases," *Fuzzy Sets and Systems*, vol. 157, no. 14, pp. 1913–1927, 2006.

[41] V. Çetkin and H. Aygün, "On (L, M)-double fuzzy ideals," *International Journal of Fuzzy Systems*, vol. 14, no. 1, pp. 166–174, 2012.

[42] A. A. Abd El-latif, *On fuzzy topological spaces [Ph.D. thesis]*, Beni-Suef University, 2009.

[43] J. Adamek, H. Herrlich, and G. E. Strecker, *Abstract and Concrete Categories*, John Wiley & Sons, New York, NY, USA, 1990.

6

On Application of Ordered Fuzzy Numbers in Ranking Linguistically Evaluated Negotiation Offers

Krzysztof Piasecki[1] **and Ewa Roszkowska**[2]

[1]*Department of Investment and Real Estate, Poznań University of Economics and Business, 61-875 Poznań, Poland*
[2]*Faculty of Economics and Management, University of Bialystok, 15-062 Bialystok, Poland*

Correspondence should be addressed to Krzysztof Piasecki; krzysztof.piasecki@ue.poznan.pl

Academic Editor: Zeki Ayag

The main purpose of this paper is to investigate the application potential of ordered fuzzy numbers (OFN) to support evaluation of negotiation offers. The Simple Additive Weighting (SAW) and the Technique for Order of Preference by Similarity to Ideal Solution (TOPSIS) methods are extended to the case when linguistic evaluations are represented by OFN. We study the applicability of OFN for linguistic evaluation negotiation options and also provide the theoretical foundations of SAW and TOPSIS for constructing a scoring function for negotiation offers. We show that the proposed framework allows us to represent the negotiation information in a more direct and adequate way, especially in ill-structured negotiation problems, allows for holistic evaluation of negotiation offers, and produces consistent rankings, even though new packages are added or removed. An example is presented in order to demonstrate the usefulness of presented fuzzy numerical approach in evaluation of negotiation offers.

1. Introduction

The process of defining, evaluating, and building a negotiation template is an important part of negotiation analysis, as well as constructing a scoring function, which is realized in the prenegotiation phase [1–3]. The negotiation template specifies a negotiation space by defining negotiation issues and acceptable resolution levels (options). The scoring function helps evaluating negotiation packages which take into account negotiator's preferences with respect to all given issues, as well as their relative importance. Because negotiation packages are often characterized by several contradictory criteria, the multicriteria techniques are useful for evaluating negotiation offers and building negotiation-scoring functions [4]. The most popular techniques used for supporting a negotiation process are

(i) Simple Additive Weighting method (SAW)/The Simple Multi Attribute Rating Technique [5, 6].

(ii) The Analytic Hierarchy Process [7].

(iii) Technique for Order of Preference by Similarity to Ideal Solution (TOPSIS) [8, 9].

There are also studies showing applicability of other techniques for building scoring functions such as

(i) Measuring Attractiveness by a Categorical Based Evaluation Technique [10].

(ii) Utility Additive Method [11].

(iii) Measuring Attractiveness near Reference Situations [10, 11].

(iv) Generalized Regression with Intensities of Preference [11].

Each of those methods has its advantaged and disadvantages; thus selecting the "best" method for a particular problem is a really difficult task. The choice between mentioned techniques depends on the negotiation problem, types of criteria, available information, decision maker's cognitive abilities, and properties of the multicriteria technique. In real negotiation situations, the options cannot be assessed precisely in a quantitative form, but still they may be in a qualitative one. This implies the usability of the linguistic approach for evaluating negotiation offers. The approximate technique may represent qualitative/quantitative options verbally by means of linguistic variable, i.e., variable whose values are not numbers but words or sentences in a natural or artificial language. One possibility of modelling linguistic values is the application of fuzzy sets and fuzzy numbers. The application of fuzzy sets in negotiations was competently discussed in [12].

The linguistic approach in negotiation support was considered in [10]. In this paper, we focus on the problem of extending the scale of values used in evaluating the negotiation options to include the linguistic scale that is, for example, expressions: very bad, bad, average, good, and very good together with intermediate values such as "at least good" or "at most good". To deal with a problem defined in this way and involving uncertain information given in linguistic terms *at least* or *at most* we used ordered fuzzy numbers (OFN). The main contribution of the paper is the discussion about applicability of Oriented Fuzzy SAW (OF-SAW) and Oriented Fuzzy TOPSIS (OF-TOPSIS) procedure based on OFNs in scoring negotiation offers.

The main goal is to investigate the effectivity of ordered fuzzy numbers (OFN) application to support evaluation of negotiation offers. The paper is organized as follows. Section 2 outlines the mathematical formulation OFNs. The general linguistic approach based on OFNs is outlined in Section 3. The OF-SAW and OF-TOPSIS are presented in Section 4. The example of application of the OF-SAW and OF-TOPSIS methods for negotiation-scoring system is presented in the Section 5. The numerical example of using OF-SAW and OF-TOPSIS is discussed in Section 6. Finally, Section 7 concludes the article, summarizes the main findings of this research, and proposes some future research directions.

2. Ordered Fuzzy Number Concept

The ordered fuzzy numbers (OFNs) were intuitively introduced by Kosiński and his cowriters [13] as an extension of the concept of fuzzy numbers introduced by Dubois and Prade [14]. OFNs are also called Kosiński's numbers [15]. The Kosiński's theory was revised in [16]. In this paper, we restrict our considerations to the case of trapezoidal OFN (TrOFN) defined in the following way.

Definition 1. For any monotonic sequence $\{a, b, c, d\} \subset \mathbb{R}$, the trapezoidal ordered fuzzy number (TrOFN) $\overleftrightarrow{Tr}(a, b, c, d)$ is determined explicitly by its membership functions $\mu_{Tr}(\bullet \mid a, b, c, d) \in [0, 1]^{\mathbb{R}}$ as follows:

$$\mu_{Tr}(x \mid a, b, c, d) = \begin{cases} 0, & x \notin [a, d] = [d, a], \\ \dfrac{x-a}{b-a}, & x \in [a, b[=]b, a], \\ 1, & x \in [b, c] = [c, b], \\ \dfrac{x-d}{c-d}, & x \in]c, d] = [d, c[. \end{cases} \quad (1)$$

Let us note that this identity describes additionally extended notation of numerical intervals which are used in this work.

The condition $a < d$ fulfilment determines the positive orientation of TrOFN $\overleftrightarrow{Tr}(a, b, c, d)$. Any positively oriented TrOFN is interpreted as such imprecise number, which may increase. The condition $a > d$ fulfilment determines the negative orientation of TrOFN $\overleftrightarrow{Tr}(a, b, c, d)$. Negatively oriented TrOFN is interpreted as such imprecise number, which may decrease. For the case $a = d$, TrOFN $\overleftrightarrow{Tr}(a, a, a, a)$ represents crisp number $a \in \mathbb{R}$, which is not oriented. The space of all TrOFNs we will denote by the symbol $\mathbb{K}_{Tr}$. The following definition fits well with the colloquial understanding of the concept of fuzziness.

Definition 2. A functional $\Psi : \mathbb{K}_{Tr} \longrightarrow \mathbb{R}$ is called significantly fuzzy iff it is dependent on all parameters of its argument.

Kosiński has introduced the arithmetic operators of dot product $\odot$ of K-sum $\oplus$ for TrOFN in following way:

$$\beta \odot \overleftrightarrow{Tr}(a, b, c, d) = \overleftrightarrow{Tr}(\beta \bullet a, \beta \bullet b, \beta \bullet c, \beta \bullet d), \quad (2)$$

$$\overleftrightarrow{Tr}(a, b, c, d) \bigoplus \overleftrightarrow{Tr}(p-a, q-b, r-c, s-d) = \overleftrightarrow{Tr}(p, q, r, s). \quad (3)$$

Kosiński has shown that there exists TrOFNs that their K-sum is not TrOFN [17]. For this reason, in [16] K-sum $\bigoplus$ is replaced by the sum $\boxplus$ determined as follows:

$$\overleftrightarrow{Tr}(a, b, c, d) \boxplus \overleftrightarrow{Tr}(p-a, q-b, r-c, s-d) = \begin{cases} \overleftrightarrow{Tr}(\min\{p, q\}, q, r, \max\{r, s\}) & (q < r) \vee (q = r \wedge p \leq s) \\ \overleftrightarrow{Tr}(\max\{p, q\}, q, r, \min\{r, s\}) & (q > r) \vee (q = r \wedge p > s) \end{cases} \quad (4)$$

The distance between TrOFNs is calculated in the following way:

$$d\left(\overleftrightarrow{Tr}(a, b, c, d), \overleftrightarrow{Tr}(p, q, r, s)\right) = \sqrt{(a-p)^2 + (b-q)^2 + (c-r)^2 + (d-s)^2} \quad (5)$$

Ranking OFN plays a very important role in fuzzy decision-making. Despite many ranking methods proposed in literature, there is no universal technique. Decision makers have to rank fuzzy numbers with their own intuition, preferences, and consistently with considered problem. In this paper we use concept of defuzzification technique extended for TrOFN.

Definition 3. Defuzzification functional is the map $\phi : \mathbb{K}_{Tr} \longrightarrow \mathbb{R}$ satisfying for any monotonic sequence $\{a, b, c, d\} \subset \mathbb{R}$ the following conditions:

$$\min\{a, b, c, d\} \leq \phi\left(\overleftrightarrow{Tr}(a, b, c, d)\right) \leq \max\{a, b, c, d\} \quad (6)$$

$\forall_{r \in \mathbb{R}}$:

$$\phi\left(\overleftrightarrow{Tr}(a, b, c, d) \boxplus \overleftrightarrow{Tr}(r, r, r, r)\right) = \phi\left(\overleftrightarrow{Tr}(a, b, c, d)\right) + r, \quad (7)$$

$$\forall_{r\in\mathbb{R}}:$$

$$\phi\left(r \odot \overleftrightarrow{Tr}(a,b,c,d)\right) = r \bullet \phi\left(\overleftrightarrow{Tr}(a,b,c,d)\right). \tag{8}$$

In [18] we find following defuzzification methods:

(i) the weighted maximum functional

$$\phi_{WM}\left(\overleftrightarrow{Tr}(a,b,c,d) \mid \lambda\right) = \lambda \bullet b + (1-\lambda) \bullet c, \quad \lambda \in [0;1], \tag{9}$$

(ii) the first maximum functional

$$\phi_{FM}\left(\overleftrightarrow{Tr}(a,b,c,d)\right) = \phi_{WM}\left(\overleftrightarrow{Tr}(a,b,c,d) \mid 1\right) = b, \tag{10}$$

(iii) the last maximum functional

$$\phi_{LM}\left(\overleftrightarrow{Tr}(a,b,c,d)\right) = \phi_{WM}\left(\overleftrightarrow{Tr}(a,b,c,d) \mid 0\right) = c, \tag{11}$$

(iv) the middle maximum functional

$$\phi_{MM}\left(\overleftrightarrow{Tr}(a,b,c,d)\right) = \phi_{WM}\left(\overleftrightarrow{Tr}(a,b,c,d) \mid \frac{1}{2}\right) = \frac{1}{2} \bullet (b+c), \tag{12}$$

(v) the gravity center functional

$$\phi_{CG}\left(\overleftrightarrow{Tr}(a,b,c,d)\right) = \begin{cases} \dfrac{a^2 + a \bullet b + b^2 - c^2 - c \bullet d - d^2}{3(a+b-c-d)} & a \neq d \\ a & a = d \end{cases} \tag{13}$$

(vi) the geometrical mean functional

$$\phi_{GM}\left(\overleftrightarrow{Tr}(a,b,c,d)\right) = \begin{cases} \dfrac{a \bullet b - c \bullet d}{a+b-c-d} & a \neq d \\ a & a = d. \end{cases} \tag{14}$$

We see that only the gravity center $\phi_{CG}(\bullet)$ and the geometrical mean $\phi_{GM}(\bullet)$ functionals are significantly fuzzy ones. Only these functionals will be applied for determining benchmarks for research described in the Section 5. Let us note that all defuzzification formulas do not imply much computational effort and does not require a priori knowledge of the set of all alternatives.

3. The Linguistic Approach in Decision-Making

Information in a quantitative setting is usually expressed by means of numerical values. However, there are situations dealing with uncertainty or vague information in which the use of linguistic assessments instead of numerical values may be more useful. Linguistic decision analysis is based on the use of a linguistic approach and it is applied for solving decision-making problems under linguistic information [19]. Herrera-Viedma pointed out also that "*its application in the development of the theory and methods in decision analysis is very beneficial because it introduces a more flexible framework which allows us to represent the information in a more direct and adequate way when we are unable to express it precisely. In this way, the burden of quantifying a qualitative concept is eliminated*" [19].

In information sciences, natural language word is considered as linguistic variable defined as fuzzy subset in the predefined space $\mathbb{X}$. Then these linguistic variables may be transformed with the use of fuzzy set theory [20–22]. From decision-making point view, the linguistic variable transformation methodologies are reviewed in [1, 13, 19]. In the literature, there are many applications of linguistic decision analysis to solve real-world problems such as, e.g., group decision-making, multicriteria decision-making, consensus, marketing, software development, education, material selection, and personnel management as well as negotiation support. For review variety of application linguistic models in decision-making, see, for example, [19].

In [19] the steps of fuzzy linguistic analysis are systematically described. In general, any linguistic value is characterized by means of a label with semantic value. The label is an expression belonging to given linguistic term set. Finally, a mechanism for generating the linguistic descriptors is provided.

The semantic value meaning may be imprecise. Thus each label from applied linguistic term set is represented by fuzzy subset in the real line $\mathbb{R}$.

An important parameter to be determined at the first step is the granularity of uncertainty, i.e., the cardinality of the linguistic term set used for showing the information. The uncertainty granularity indicates the capacity of distinction that may be expressed. The knowledge value is increasing with the increase in granularity. The typical values of cardinality used in the linguistic models are odd ones, usually between 5 and 13. It is worth noting that the idea of granular computing goes from Zadeh [23] who wrote "*fuzzy information granulation underlies the remarkable human ability to make rational decisions in an environment of imprecision, partial knowledge, partial certainty and partial truth.*" Also Yao [24] pointed up that "*the consideration of granularity is motivated by the practical needs for simplification, clarity, low cost, approximation, and the tolerance of uncertainty*".

In our model all linguistic terms are linked with Tentative Order Scale

$$TOS = \{V_1; V_2; \ldots; V_r\} \tag{15}$$

which is previously determined as discrete set of linear ordered linguistic values. Each element of Tentative Order Scale is called reference point. Reference points are ordered from the worst one to the best one. In our paper, we shall use following Tentative Order Scale:

$$TOS = \begin{Bmatrix} V_1 = \text{Very Bad} = VB \\ V_2 = \text{Bad} = B \\ V_3 = \text{Average} = A \\ V_4 = \text{Good} = G \\ V_5 = \text{Very Good} = VG \end{Bmatrix}. \quad (16)$$

It is obvious that each reference point V_j is equivalent to the numerical diagnosis $j \in \mathbb{N}$.

Now, we focus on the problem of Tentative Order Scale enlargement of intermediate values. For this purpose we use the phrases "at least" described by the symbol $\mathscr{L}$ and "at most" described by the symbol $\mathscr{M}$.

The expression $\mathscr{L}.V_j$ means "no worse than V_j and worse that V_{j+1}". The expression $\mathscr{M}.V_j$ means "no better than V_j and better that V_{j-1}". Moreover, we assume that the $\mathscr{L}.V_j$ "is better than" $\mathscr{M}.V_j$.

In this way we obtain extended Order Scale

$$OS = \{VB; \mathscr{L}.VB; \mathscr{M}.B; B; \mathscr{L}.B; \mathscr{M}.A; A; \mathscr{L}.A; \mathscr{M}.G; G; \mathscr{L}.G; \mathscr{M}.VG; VG\}. \quad (17)$$

Let us note that expressions $\mathscr{M}.VB$ and $\mathscr{L}.VG$ do not belong to above extended order scale.

The numerical representation of the phrases "at least" is the relation "great or equal" denoted by the GE. Thus, expression $\mathscr{L}.V_j$ is equivalent to numerical diagnosis $GE.j$.

The numerical representation of the phrases "at most" is the relation "less or equal" denoted by the LE. Thus, expression $\mathscr{M}.V_j$ is equivalent to numerical diagnosis $LE.j$.

In this way we obtain the Numerical Diagnosis Set

$$ND = \{1; GE.2; LE.2; 2; GE.2; LE.3; 3; GE.3; LE.4; 4; GE.4; LE.5; 5\}. \quad (18)$$

All numerical diagnoses $GE.j$ and $LE.j$ are imprecise ones. Therefore according to suggestion given in [25], numerical diagnosis will be represented for any $a \in \mathbb{N}$ by TrONF in following way:

$$a \longrightarrow \overleftrightarrow{S}_1 = \overleftrightarrow{Tr}(a, a, a, a); \quad (19)$$

$$GE.a \longrightarrow \overleftrightarrow{S}_2 = \overleftrightarrow{Tr}\left(a, a, a + \frac{1}{2}, a + 1\right); \quad (20)$$

$$LE.a \longrightarrow \overleftrightarrow{S}_3 = \overleftrightarrow{Tr}\left(a, a, a - \frac{1}{2}, a - 1\right). \quad (21)$$

Let us note that the sum and dot product of the numbers $\overleftrightarrow{S}_1$, $\overleftrightarrow{S}_2$, and $\overleftrightarrow{S}_3$ are also TrOFN. Taking into account (19) together with (20) and (21), we determine the converting system, which transforms numerical diagnoses into performance ratings described by TrOFN in the following way:

$$j \longrightarrow \overleftrightarrow{X}_j = \overleftrightarrow{Tr}(j, j, j, j) \quad \text{for } j = 1, 2, \ldots, 5 \quad (22)$$

$$LE.j \longrightarrow \overleftrightarrow{X}_{Lj} = \overleftrightarrow{Tr}\left(j, j, j - \frac{1}{2}, j - 1\right) \quad \text{for } j = 2, 3, 4, 5 \quad (23)$$

$$GE.j \longrightarrow \overleftrightarrow{X}_{Gj} = \overleftrightarrow{Tr}\left(j, j, j + \frac{1}{2}, j + 1\right) \quad \text{for } j = 1, 2, 3, 4. \quad (24)$$

It is worth noting that the orientation OFN is used for representing relation LE (negative orientation) and GE (positive orientation).

In this way, we obtain following numerical order scale:

$$NOS = \left\{\overleftrightarrow{X}_j : j = 1, 2, \ldots, 5\right\} \cup \left\{\overleftrightarrow{X}_{Lj} : j = 2, 3, 4, 5\right\} \cup \left\{\overleftrightarrow{X}_{Gj} : j = 1, 2, 3, 4\right\} \quad (25)$$

determined by trapezoidal OFN.

The consolidated mechanism of transformation of considered linguistic values into performance ratings is presented in Table 1.

4. The Oriented Fuzzy SAW and Oriented Fuzzy TOPSIS

In this section we describe the SAW and TOPSIS algorithm based on TrOFN. We consider here a multicriteria decision-making problem with m alternatives $A_1, A_2, \ldots, A_m \in A$ and n decision criteria $C_1, C_2, \ldots, C_n$ with which the alternative performances are measured. For n criteria, we have the weight vector

$$w = (w_1, w_2, \ldots, w_n) \in \left(\mathbb{R}_0^+\right)^n \quad (26)$$

where

$$w_1 + w_2 + \cdots + w_n = 1 \quad (27)$$

and w_j is the weight of j-th criterion denoting the importance of this criterion in the evaluation of the alternatives.

The Simple Additive Weighting (SAW) method is a scoring method based on the concept of a weighted average of performance ratings. This method is also known as the Simple Multi Attribute Rating Technique (SMART) [5, 6]. For the case of performance ratings described by TrOFN, the SAW was first generalized in [25]. In this paper, this SAW generalization is converted in the way that the K-sum (3) is replaced by sum (4). Generalized in this way SAW algorithm we will be called Oriented Fuzzy SAW (OF-SAW). The OF-SAW can be described in the following steps.

Step 1. Define the set of evaluation criteria C_j $(j = 1, \ldots, n)$ and the set of feasible alternatives A_i $(i = 1, \ldots, m)$.

Step 2. Define by (17) the Order Scale OS and by (25) the numerical order scale NOS with performance ratings represented by TrOFN.

Step 3. Define the criterion weights w_j which describe the importance of criterion C_j $(j = 1, \ldots, n)$.

TABLE 1: Transformation of linguistic values into numerical order scale* *NOS*.

Linguistic values	Order Scale	Numerical Diagnosis Set	Numerical Order Scale *NOS*
Very Bad	***VB***	**1**	$\overleftrightarrow{Tr}(1,1,1,1)$
at least Very Bad	$\mathscr{L}.VB$	*GE*.1	$\overleftrightarrow{Tr}(1,1,1.5,2)$
at most Bad	$\mathscr{M}.B$	*LE*.2	$\overleftrightarrow{Tr}(2,2,1.5,1)$
Bad	***B***	**2**	$\overleftrightarrow{Tr}(2,2,2,2)$
at least Bad	$\mathscr{L}.B$	*GE*.2	$\overleftrightarrow{Tr}(2,2,2.5,3)$
at most Average	$\mathscr{M}.AV$	*LE*.3	$\overleftrightarrow{Tr}(3,3,2.5,2)$
Average	***AV***	**3**	$\overleftrightarrow{Tr}(3,3,3,3)$
at least Average	$\mathscr{L}.AV$	*GE*.3	$\overleftrightarrow{Tr}(3,3,3.5,4)$
at most Good	$\mathscr{M}.G$	*LE*.4	$\overleftrightarrow{Tr}(4,4,3.5,3)$
Good	***G***	**4**	$\overleftrightarrow{Tr}(4,4,4,4)$
at least Good	$\mathscr{L}.G$	*GE*.4	$\overleftrightarrow{Tr}(4,4,4.5,5)$
at most Very Good	$\mathscr{M}.VG$	*LE*.5	$\overleftrightarrow{Tr}(5,5,4.5,4)$
Very Good	***VG***	**5**	$\overleftrightarrow{Tr}(5,5,5,5)$

* The Tentative **Order Scale** is written in bold.
Source: own elaboration.

Step 4. Evaluate each alternative $A_i \in A$ by the vector $(\overleftrightarrow{X}_{i,1}, \overleftrightarrow{X}_{i,2}, \dots, \overleftrightarrow{X}_{i,n}) \in (NOS)^n$, where $\overleftrightarrow{X}_{i,j}$ is the performance rating of i -th alternative with respect to j -th criterion.

Step 5a. Aggregate the performance ratings with respect to all the criteria for each alternative using the criterion functional given as aggregated evaluation coefficient

$$\overleftrightarrow{SAW}(A_i) = w_1 \odot \overleftrightarrow{X}_{i,1} \boxplus w_2 \odot \overleftrightarrow{X}_{i,2} \boxplus \dots \boxplus w_n \odot \overleftrightarrow{X}_{i,n} \tag{28}$$

Step 6a. Transform the obtained result $\overleftrightarrow{SAW}(A_i)$ to scoring function $\phi(\overleftrightarrow{SAW}(A_i))$ where the function ϕ is one of defuzzification formulae (9)-(14).

Step 7a. Rank all alternatives A_i according to decreasing values $\phi(\overleftrightarrow{SAW}(A_i))$.

Higher value of $\phi(\overleftrightarrow{SAW}(A_i))$ implies higher rank of alternative A_i. Therefore, higher value $\phi(\overleftrightarrow{SAW}(A_i))$ means that alternative A_i is better. Only scoring functions $\phi_{CG}(\overleftrightarrow{SAW}(\bullet))$ and the geometrical mean $\phi_{GM}(\overleftrightarrow{SAW}(\bullet))$ are significantly fuzzy SAW methods which are considered in this paper.

Hwang and Yoon [26] have developed TOPSIS for solving MCDM problems. The basic TOPSIS principle is that the chosen alternative should have the "shortest distance" to the Positive Ideal Solution (PIS) and the "longest distance" to the Negative Ideal Solution (NIS). The PIS is the solution maximizing the benefit criteria and minimizing the cost ones. The NIS is the solution maximizing the cost criteria and minimizing the benefit ones.

Chen and Hwang [27] have proposed Fuzzy TOPSIS, which is a generalization of TOPSIS to the case when fuzzy number describes imprecise performance ratings. Next, Fuzzy TOPSIS is generalized to the case when performance ratings are represented by TrOFN. Generalized in this way TOPSIS algorithm will be called Oriented Fuzzy TOPSIS (OF-TOPSIS).

The first four steps of the OF-TOPSIS algorithm are similar to the corresponding steps of OF-SAW. The fifth and the next steps are as follows [25].

Step 5b. Identify the PIS and the NIS, which are

$$PIS = \left(\overleftrightarrow{X}_5, \overleftrightarrow{X}_5, \dots, \overleftrightarrow{X}_5\right) \in (NOS)^n \tag{29}$$

$$NIS = \left(\overleftrightarrow{X}_1, \overleftrightarrow{X}_1, \dots, \overleftrightarrow{X}_1\right) \in (NOS)^n \tag{30}$$

where $\overleftrightarrow{X}_1 = \overleftrightarrow{Tr}(1,1,1,1)$ and $\overleftrightarrow{X}_5 = \overleftrightarrow{Tr}(5,5,5,5)$.

Step 6b. For each alternative A_i $(i = 1, 2, \dots, m)$, calculate its distances from PIS and NIS.

$$d(A_i, PIS) = \sum_{j=1}^{n} w_j \bullet d\left(\overleftrightarrow{X}_{i,j}, \overleftrightarrow{X}_5\right) \tag{31}$$

$$d(A_i, NIS) = \sum_{j=1}^{n} w_j \bullet d\left(\overleftrightarrow{X}_{i,j}, \overleftrightarrow{X}_1\right) \tag{32}$$

where the mapping $d(\bullet,\bullet)$ is the distance defined by (5).

Step 7a. For each alternative A_i, compute value of criterion functional given as coefficient of relative nearness to PIS

$$CN(A_i) = \frac{d(A_i, NIS)}{d(A_i, NIS) + d(A_i, PIS)} \quad (33)$$

Step 8a. Rank all alternatives A_i according to decreasing values $CN(A_i)$.

The OF-TOPSIS is significantly fuzzy scoring function. It is worth noting that, in general, the ideal solution consists of the maximum values of benefit criteria and of minimal values of cost criteria. At the same time, in this case, the anti-ideal solution consists of the maximum values of benefit criteria and of minimal values of cost criteria. In the proposed OF-TOPSIS, we identify the NIS as a solution consisting of the maximum values from the scale. Simultaneously, we identify the PIS as a solution consisting of the minimum values from the scale. This implies that when including a new alternative or modifying the data of one of the alternatives, decision maker does not need to reevaluate the previously evaluated alternative and scores of all those alternatives remain stable. This technique also avoids rank reversals [19, 22, 24].

5. Application Ordered Fuzzy Number for Evaluation of Negotiation Offers

The scale of linguistic evaluations depends on decision maker. In general, words are less precise than numbers. In the prenegotiation phase the negotiation template is evaluated. It specifies the negotiation space by defining the negotiation issues with their importance and feasible options of these issues [2]. The template is used next to support the negotiators in evaluating the offers, analyzing the negotiation progress, scale of concessions, among many others [11]. In the case of big negotiation problems or continuous options, the wide ranges of options are reduced to salient ones only to discretize the negotiation problem and make it easier to analyze. Next, the negotiation offer scoring system is built to support negotiator in their decisions.

In this paper we propose linguistic approach in evaluation negotiation offers based on TrOFN. The human understanding and human problem solving involve perception, abstraction, representation, and understanding of real-world problems, as well as their solutions, at a different levels of granularity. The linguistic approach in decision-making can deal with the problem of mentioned granularity of information.

In the proposed model we assume that the problem issues are evaluated by linguistics terms instead of numerical values. This approach can be more appropriate and realistic in many negotiation problems for different reasons. First of all, there are situations in which the information cannot be assessed precisely in a quantitative form but may be in the qualitative one (e.g., when the warranty condition may be evaluated by terms like "good", "average", and "bad"). In some cases, we cannot use precise quantitative information because either it is unavailable or its computation is too expensive. Then an "approximate value" represented by the linguistic term may be used instead of numerical one. Also, taking into account the human perception, people are often led to describe their preferences in natural language instead of numerical values (for instance, evaluate price between 25 and 28 as good, price between 29 and 35 as bad, etc.).

Thus, the linguistic variable concept may be especially useful in describing situations which are too complex or ill-defined for evaluation in numerical way. Moreover, the linguistic variable approach is necessary for holistic evaluation of negotiation offers.

To formalize linguistic decision analysis, we start with the following definitions:

(i) negotiation package is an alternative determined as offer, which negotiators may send to or receive from their opponent,

(ii) negotiation issue is a criterion of evaluation of negotiation package,

(iii) negotiation template is the set of all evaluated options.

Then let us use the following notation for evaluating negotiation offers:

(i) $F = \{f_1, f_2, \dots, f_n\}$ is the set of negotiation issues the negotiator uses to evaluate the offers,

(ii) Y_j is the scope of the issue f_j,

(iii) the Cartesian product $Y = \prod_{i=1}^{n} Y_i$ is the set of all feasible negotiation packages,

(iv) $P = \{P_1, P_2, \dots, P_i, \dots, P_m\} \subseteq Y$ is the negotiation template,

(v) OS is the linguistic order scale defined by (17) for evaluation negotiation issues,

(vi) the functions $Z_j : Y_j \longrightarrow OS$ evaluate options of the issue f_j by linguistic order scale OS,

(vii) NOS is the numerical order scale defined by (25),

(viii) the vector $(\overleftrightarrow{X}_{i,1}, \overleftrightarrow{X}_{i,2}, \dots, \overleftrightarrow{X}_{i,n})$ represents negotiation package P_i, where $\overleftrightarrow{X}_{i,j} \in NOS$ is the performance rating of alternative P_i with respect to the criterion f_j.

The evaluation of negotiation offers can be accomplished in the following steps:

(i) determine the set of negotiation issues F, the scopes Y_j for all issue f_j, and the weight vector w;

(ii) determine the negotiation template P;

(iii) define the linguistic order scale OS for evaluation of negotiation issues and transform it into the numerical order scale NOS represented by trapezoidal OFN;

(iv) for each criterion f_j define function Z_j evaluating options by linguistic terms from scale OS;

(v) provide the linguistic evaluation offers from the set P and find the representation of negotiation packages by trapezoidal OFN;

(vi) decide for the OF-SAW with defuzzification formula or for OF-TOPSIS procedure for evaluating offers;

(vii) obtain the scoring system for packages from P and ranking them.

Table 2: The mechanism for Seller' evaluating negotiation templates by linguistic order scale.

Unit Price	Order scale	Complaint conditions	Order scale	Time of payment	Order scale
20-21-22	VB	E	VB	21-22-23-24	VB
23	$\mathscr{L}.VB$	E+	$\mathscr{L}.VB$	20	$\mathscr{L}.VB$
24	$\mathscr{M}.B$	D-	$\mathscr{M}.B$	19	$\mathscr{M}.B$
25-26-27	B	D	B	16-17-18	B
28	$\mathscr{L}.B$	D+	$\mathscr{L}.B$	15	$\mathscr{L}.B$
29	$\mathscr{M}.AV$	C-	$\mathscr{M}.AV$	14	$\mathscr{M}.AV$
30-31-32	AV	C	AV	11-12-13	AV
33	$\mathscr{L}.AV$	C+	$\mathscr{L}.AV$	10	$\mathscr{L}.AV$
34	$\mathscr{M}.G$	B-	$\mathscr{M}.G$	9	$\mathscr{M}.G$
35-36-37	G	B	G	6-7-8	G
38	$\mathscr{L}.G$	B+	$\mathscr{L}.G$	5	$\mathscr{L}.G$
39	$\mathscr{M}.VG$	A-	$\mathscr{M}.VG$	4	$\mathscr{M}.VG$
40-41-42	VG	A	VG	1-2-3	VG

Source: own elaboration.

6. Illustrative Example and Discussion

In this section, we present the numerical example to illustrate the application trapezoidal OFN for evaluation negotiation packages. We test practically all steps of negotiator linguistic preference analysis to show the usefulness of proposed approach. Let as assume that Seller and Buyer negotiate the conditions of the potential business contract. Then the support for Seller's negotiations may be prepared as follows.

At the beginning negotiator defines the negotiation problem, identifies the objectives, transforms them into the negotiation issues, and defines the negotiation space. All packages are measured with regard to every issue, using a related measurement scale. We assume that the negotiators may choose the way of describing the resolution levels of the issues. These evaluations may be based on different types of data (numerical values; linguistic or mixed values) and subjective judgments. Formally, negotiator determines the set of issue F, set of options for each issue Y_i, and the weight vector w.

In our example the following negotiation issues are discussed:

(i) f_1 is unit price expressed in €,

(ii) f_2 is complaint conditions described verbally,

(iii) f_3 is time of payment determined in days.

Next, negotiator defines the negotiation template by the numerical value for criteria f_1 and f_3 and verbally for f_2.

(i) f_1: unit price: from 20€ to 42€ for both sides;

(ii) f_2: complaint conditions: five options described verbally:

A: "5% defects and 2% penalty",
B: "5% defects and 4% penalty",
C: "7% defects and 4% penalty",
D: "3% defects and no penalty"
E: "4% defects and no penalty".

(iii) f_3: time of payment (days): from 1 to 24 days for both sides.

From the Seller's point view, the criterion f_1 is a benefit issue, and the criterion f_3 is a cost issue. In agree with the Seller's preferences, we order the verbal options of the criterion f_2 alphabetically in the way that the option $A \in F_2$ is the best. Moreover, the Seller can accept some complaint conditions near the verbally options described above, for instance, which differ 0.5% in defects or in penalty. Thus we ought to distinguish phrases and evaluated them as

(i) almost as good as the complaint condition Q denoted by the symbol Q-,

(ii) a bit better than the complaint condition Q denoted by the symbol Q+.

Now we potentially have $23 \bullet 13 \bullet 24 = 7176$ different packages The Seller proposes following weight vector:

$$w = [0.6, 0.2, 0.2]. \qquad (34)$$

The scoring system for all negotiation offers can be obtained out of various combinations of the possible options.

In the next step the Seller decides to evaluate all issues using linguistic order scale OS determined by (17). The mechanisms for linguistic evaluating of a negotiation template are described for each issue f_j in Table 2.

In our example, the Seller considers 15 negotiation packages described in Table 3. These packages are evaluated with use by linguistic order scale in way described in Table 2. Obtained linguistic evaluations are presented in Table 3.

In the next step, the Seller numerically evaluates negotiations packages by means of numerical order scale NOS described in Table 1. Obtained numerical evaluations are

TABLE 3: Negotiations packages and their linguistic evaluation.

Package	Unit price		Complaint conditions		Time of payment	
	Option	Linguistic evaluation	Option	Linguistic evaluation	Option	Linguistic evaluation
P1	21	VB	B+	$\mathscr{L}.B$	9	$\mathscr{M}.G$
P2	27	B	E-	$\mathscr{M}.VG$	7	G
P3	23	$\mathscr{L}.VB$	D	G	10	$\mathscr{L}.AV$
P4	24	$\mathscr{M}.B$	C	AV	10	$\mathscr{L}.AV$
P5	25	B	C+	$\mathscr{L}.AV$	8	G
P6	28	$\mathscr{L}.B$	D-	$\mathscr{M}.G$	4	$\mathscr{M}.VG$
P7	32	AV	D-	$\mathscr{M}.G$	10	$\mathscr{L}.AV$
P8	28	$\mathscr{L}.B$	C	AV	4	$\mathscr{M}.VG$
P9	29	$\mathscr{M}.AV$	A+	$\mathscr{L}.VB$	23	VB
P10	31	AV	B-	$\mathscr{M}.B$	20	$\mathscr{L}.VB$
P11	33	$\mathscr{L}.AV$	B	B	18	B
P12	33	$\mathscr{L}.AV$	B+	$\mathscr{L}.B$	17	B
P13	37	G	C	AV	14	$\mathscr{M}.AV$
P14	33	$\mathscr{L}.AV$	D-	$\mathscr{M}.G$	14	$\mathscr{M}.AV$
P15	41	VG	B	B	14	$\mathscr{M}.AV$

Source: own elaboration.

described in Table 4. These numerical evaluations are applied for determining by (28) the aggregated evaluation coefficients $\overleftrightarrow{SAW}(P_i)$, which are presented in Table 4.

Now we juxtapose defuzzificated values of OF-SAW coefficient with SAW coefficient values determined with the use of numerical order scale represented by crisp real numbers. We will consider the numerical order scales NOS1, NOS2, and NOS3 described in Table 5.

The OF-SAW method determines aggregated evaluation coefficients which are TrOFN. TrOFNs comparison may be ambiguous. Thus we have to use one of the defuzzification formulas to obtain rank ordering of negotiation packages. The results of defuzzification formulas are shown in Table 6. The WM defuzzification formula is used with the coefficient λ = 0.1.

In the next step, for each numerical order scale described in Table 5, all negotiations packages are evaluated by values of SAW coefficients. Evaluations obtained in this way are presented in Table 6. It is obvious that, in this case, defuzzification of obtained results is not needed. Finally, Table 6 lists 9 different evaluating methods with evaluations represented by crisp real numbers. All these methods differ in used defuzzification method or numerical order scale. Evaluations obtained by the various evaluating methods are shown in the successive columns of Table 6.

Any column of Table 6 outlines some raking of negotiation packages. We observe that these rankings determine partial or strict orders. Thus, we describe observed rankings by tired ranks [28]. All ranks are presented in Table 7. Any row of this table lists different ranks of chosen negotiation package. The ranking outlined by supposition $\phi_{GC}(FSAV(\bullet))$ is the unique strict order which is significantly fuzzy. For this reason, we choose this ranking as the benchmark. For better clarity of our considerations, the rows of Table 7 are ordered by means of the above benchmark.

We consider a family of rankings linked to the same criterion functional. Considered rankings differ in the method of determining the criterion functional value. Then we evaluate this family in the way that the best ranking is one, which is the most similar to other evaluated rankings. The use of this criterion stems from the belief that the ranking of one most similar to the other is their best substitute. We will measure the similarity using the Spearman's rho (the Spearman's multiple rank correlation coefficient) determined for the case of tired ranks [28] in following way:

$$\rho_s = \frac{12}{m \bullet (m^2 - 1) \bullet (l - 1)} \sum_{\substack{k=1 \\ k \neq s}}^{l} \sum_{i=1}^{m} r_{k,i} \bullet r_{s,i} - \frac{3 \bullet (m+1)}{m-1}, \quad (35)$$

where the individual variables denote the following:

$r_{k,i}$ is i –th package negotiation rank outlined by k –th method of determining the criterion functional value,

s is index of evaluated ranking,

m is the number of ordered package negotiations,

l is the number of considered rankings.

The values of Spearman's rho computed for each evaluated rankings are presented in the last row of Table 7. We see that the best rankings are such one that it is significantly fuzzy. Moreover, there does not exist such not significantly fuzzy ranking, which is better than any significantly fuzzy one.

Now we will consider the case when the Seller applies the OF-TOPSIS criterion functional for evaluating negotiation packages. The OF-TOPSIS is significantly fuzzy method of

TABLE 4: Numerical evaluation of negotiation packages.

Package	Unit price	Complaint conditions	Time of payment	$\overleftrightarrow{SAW}(P_i)$
P1	$\overleftrightarrow{Tr}(1,1,1,1)$	$\overleftrightarrow{Tr}(2,2,2.5,3)$	$\overleftrightarrow{Tr}(4,4,3.5,3)$	$\overleftrightarrow{Tr}(1.8,1.8,1.8,1.8)$
P2	$\overleftrightarrow{Tr}(2,2,2,2)$	$\overleftrightarrow{Tr}(5,5,4.5,4)$	$\overleftrightarrow{Tr}(4,4,4,4)$	$\overleftrightarrow{Tr}(3.0,3.0,2.9,2.8)$
P3	$\overleftrightarrow{Tr}(1,1,1.5,2)$	$\overleftrightarrow{Tr}(4,4,4,4)$	$\overleftrightarrow{Tr}(3,3,3.5,4)$	$\overleftrightarrow{Tr}(2.0,2.0,2.4,2.8)$
P4	$\overleftrightarrow{Tr}(2,2,1.5,1)$	$\overleftrightarrow{Tr}(3,3,3,3)$	$\overleftrightarrow{Tr}(3,3,3.5,4)$	$\overleftrightarrow{Tr}(2.4,2.4,2.2,2.0)$
P5	$\overleftrightarrow{Tr}(2,2,2,2)$	$\overleftrightarrow{Tr}(3,3,3.5,4)$	$\overleftrightarrow{Tr}(4,4,4,4)$	$\overleftrightarrow{Tr}(2.6,2.6,2.7,2.8)$
P6	$\overleftrightarrow{Tr}(2,2,2.5,3)$	$\overleftrightarrow{Tr}(4,4,3.5,3)$	$\overleftrightarrow{Tr}(5,5,4.5,4)$	$\overleftrightarrow{Tr}(3.0,3.0,3.1,3.2)$
P7	$\overleftrightarrow{Tr}(3,3,3,3)$	$\overleftrightarrow{Tr}(4,4,3.5,3)$	$\overleftrightarrow{Tr}(3,3,3.5,4)$	$\overleftrightarrow{Tr}(3.2,3.2,3.2,3.2)$
P8	$\overleftrightarrow{Tr}(2,2,2.5,3)$	$\overleftrightarrow{Tr}(3,3,3,3)$	$\overleftrightarrow{Tr}(5,5,4.5,4)$	$\overleftrightarrow{Tr}(2.8,2.8,3.0,3.2)$
P9	$\overleftrightarrow{Tr}(3,3,2.5,2)$	$\overleftrightarrow{Tr}(1,1,1.5,2)$	$\overleftrightarrow{Tr}(1,1,1,1)$	$\overleftrightarrow{Tr}(2.2,2.2,2.0,1.8)$
P10	$\overleftrightarrow{Tr}(3,3,3,3)$	$\overleftrightarrow{Tr}(2,2,1.5,1)$	$\overleftrightarrow{Tr}(1,1,1.5,2)$	$\overleftrightarrow{Tr}(2.4,2.4,2.4,2.4)$
P11	$\overleftrightarrow{Tr}(3,3,3.5,4)$	$\overleftrightarrow{Tr}(2,2,2,2)$	$\overleftrightarrow{Tr}(2,2,2,2)$	$\overleftrightarrow{Tr}(2.6,2.6,2.9,3.2)$
P12	$\overleftrightarrow{Tr}(3,3,3.5,4)$	$\overleftrightarrow{Tr}(2,2,2.5,3)$	$\overleftrightarrow{Tr}(2,2,2,2)$	$\overleftrightarrow{Tr}(2.6,2.6,3.0,3.4)$
P13	$\overleftrightarrow{Tr}(4,4,4,4)$	$\overleftrightarrow{Tr}(3,3,3,3)$	$\overleftrightarrow{Tr}(3,3,2.5,2)$	$\overleftrightarrow{Tr}(3.6,3.6,3.5,3.4)$
P14	$\overleftrightarrow{Tr}(3,3,3.5,4)$	$\overleftrightarrow{Tr}(4,4,3.5,3)$	$\overleftrightarrow{Tr}(3,3,2.5,2)$	$\overleftrightarrow{Tr}(3.2,3.2,3.3,3.4)$
P15	$\overleftrightarrow{Tr}(5,5,5,5)$	$\overleftrightarrow{Tr}(2,2,2,2)$	$\overleftrightarrow{Tr}(3,3,2.5,2)$	$\overleftrightarrow{Tr}(4.0,4.0,3.9,3.8)$

Source: own elaboration.

TABLE 5: The mechanisms of transforming an order scale into a numerical order scale.

Order Scale	Numerical Diagnosis Set	Numerical Order Scale		
		$NOS1$	$NOS2$	$NOS3$
VB	1	1	1	1
$\mathcal{L}.VB$	$GE.1$	1	1.5	1.25
$\mathcal{M}.B$	$LE.2$	2	1.5	1.75
B	2	2	2	2
$\mathcal{L}.B$	$GE.2$	2	2.5	2.25
$\mathcal{M}.AV$	$LE.3$	3	2.5	2.75
AV	3	3	3	3
$\mathcal{L}.AV$	$GE.3$	3	3.5	3.25
$\mathcal{M}.G$	$LE.4$	4	3.5	3.75
G	4	4	4	4
$\mathcal{L}.G$	$GE.4$	4	4.5	4.25
$\mathcal{M}.VG$	$LE.5$	5	4.5	4.75
VG	5	5	5	5

Source: own elaboration.

evaluation. The results obtained using OF-TOPSIS will be analyzed in the same manner as the above results obtained using the SAW were analyzed.

At the first, the coefficients of relative nearness to PIS $CN(P_i)$ are computed with use of numerical order scale NOS for the numerical evaluations presented in Table 4 which are applied for determining by (33) coefficients $CN(P_i)$, which are presented in Table 8. In currently considered case, the ranking outlined by values $CN(P_i)$ computed for numerical order scale NOS is a unique ranking determining significantly fuzzy strict order. For this reason, we choose this ranking as a benchmark. For better clarity of our considerations, the rows of Table 8 are ordered by means of the above benchmark. Moreover, we compute the values $CN(P_i)$ for numerical order scales $NOS1$, $NOS2$, and $NOS3$ described in Table 5.

Each considered negotiation package ranking increases with decreasing in values of $CN(\bullet)$ coefficient. We describe observed rankings by tired ranks [28]. All ranks are presented in Table 8. We measure the rankings' similarity using the Spearman's rho determined by (35). The values of Spearman's rho computed for each evaluated rankings are presented in

Table 6: Package evaluations obtained by SAW methods.

Numerical Order Scale	NOS						NOS1	NOS2	NOS3
SAW method	OF-SAW						SAW		
Defuzzification method	FM	LM	MM	WM*	CG	GM	-	-	-
Packages	Package evaluation								
P1	1.80	1.80	1.80	1.80	1.80	1.80	1.80	1.80	1.80
P2	3.00	2.90	2.95	2.91	2.92	2.93	3,00	2.90	2.95
P3	2.00	2.40	2.20	2.36	2.31	2.27	2.00	2.40	2.20
P4	2.40	2.20	2.30	2.22	2.24	2.27	2.40	2.20	2.30
P5	2.60	2.70	2.65	2.69	2.68	2.67	2.60	2.70	2.65
P6	3.00	3.10	3.05	3.09	3.08	3.07	3.00	3.20	3.05
P7	3.20	3.20	3.20	3.20	3.20	3.20	3.20	3.20	3.20
P8	2.80	3.00	2.90	2.98	2.96	2.93	2.80	3.00	2.90
P9	2.20	2.00	2.10	2.02	2.04	2.07	2.20	2.00	2.10
P10	2.40	2.40	2.40	2.40	2.40	2.40	2.40	2.40	2.40
P11	2.60	2.90	2.75	2.87	2.83	2.80	2.60	2.90	2.75
P12	2.60	3.00	2.80	2.96	2.91	2.87	2.60	3.00	2.80
P13	3.60	3.50	3.55	3.51	3.52	3.53	3.60	3.50	3.55
P14	3.20	3.30	3.25	3.29	3.28	3.27	3.20	3.30	3.35
P15	4.00	3.90	3.95	3.91	3.92	3.93	4.00	3.90	3.95

Source: own elaboration.

Table 7: The negotiation packages' rankings determined by SAW methods.

Numerical Order Scale	NOS						NOS1	NOS2	NOS3
SAW method	OF-SAW						Crisp		
Defuzzification method	FM	LM	MM	WM	CG	GM	-	-	-
Packages	Tired ranks								
P15	1	1	1	1	1	1	1	1	1
P13	2	2	2	2	2	2	2	2	2
P14	3.5	3	3	3	3	3	3.5	3	3
P7	3.5	4	4	4	4	4	3.5	4.5	4
P6	5.5	5	5	5	5	5	5.5	4.5	5
P8	7	6.5	7	6	6	6.5	7	6.5	7
P2	5.5	8.5	6	8	7	6.5	5.5	8.5	6
P12	9	6.5	8	7	8	8	9	6.5	8
P11	9	8.5	9	9	9	9	9	8.5	9
P10	11.5	11.5	11	11	11	11	9	10	10
P5	9	10	10	10	10	10	11.5	11.5	11
P3	14	11.5	13	12	12	12.5	14	11.5	13
P4	11.5	13	12	13	13	12.5	11.5	13	12
P9	13	14	14	14	14	14	13	14	14
P1	15	15	15	15	15	15	15	15	15
Spearman's rho	0,962	0,970	0,984	0,980	**0,984**	**0,984**	0,958	0,966	0,982

Source: own elaboration.

Table 8: The negotiation packages' rankings determined by OF-TOPSIS method.

Numerical Order Scale	*NOS*		*NOS1*		*NOS2*		*NOS3*	
TOPSIS method	OF-TOPSIS		TOPSIS					
Packages		Rank		Rank		Rank		Rank
P15	0.731	1	0,750	1	0,725	1	0,738	1
P13	0.631	2	0,650	2	0,625	2	0,638	2
P14	0.566	3	0,550	3.5	0,575	3	0,563	3
P7	0.548	4	0,550	3.5	0,550	4	0,550	4
P6	0.516	5	0,500	5.5	0,521	5	0,513	5
P8	0.486	6	0,450	7	0,500	6.5	0,475	7
P2	0.477	7	0,500	5.5	0,475	8.5	0,488	6
P12	0.475	8	0,400	9	0,500	6.5	0,450	8
P11	0.456	9	0,400	9	0,475	8.5	0,438	9
P5	0.419	10	0,400	9	0,425	10	0,413	10
P10	0.358	11	0,350	11.5	0,350	11.5	0,350	11
P3	0.342	12	0,312	13	0,350	11.5	0,288	13
P4	0.324	13	0,350	11.5	0,300	13	0,325	12
P9	0.273	14	0,267	14	0,250	14	0,275	14
P1	0.203	15	0,200	15	0,200	15	0,200	15
Spearman's rho		**0,985**		0,972		0,971		0,985

Source: own elaboration.

the last row of Table 8. We see that the best ranking is order outlined by significantly fuzzy OF-TOPSIS.

We observe that the OF-SAW and OF-TOPSIS methods are linked to methods different criterion functional. It implies that the ranking obtained by using the OF-TOPSIS may differ from the ranking obtained using OF-SAW. The negotiator can evaluate subjectively which of the rankings fit his or her preferences better.

7. Conclusions

In this paper we propose a linguistic approach based on ordered fuzzy numbers which can be implemented to evaluate the negotiations packages. In our approach the negotiator's preferences over the issues are represented by TrOFN and then the OF-SAW or OF-TOPSIS method is adopted to determine the global rating of each package. We show that ranking outlined by significantly fuzzy methods may be better than ranking outlined by conventional methods.

One of the key advantages of the approach proposed is its usefulness for building a general scoring function in the ill-structured negotiation problem, namely, the situation in which the problem itself as well as the negotiator's preferences cannot be precisely defined or the available information is subjective and imprecise. Secondly, both variants of scoring functions produce consistent rankings, even though the new packages are added (or removed) and do not result in rank reversal. The main key advantages of the proposed approach are the following:

(1) Presented techniques allow for linguistic preference attractiveness elicitation of issue options. The linguistic preference analysis based on OFN is useful for building scoring function in the ill-structured negotiation problem in the context of qualitative issues, in satiation holistic evaluation negotiation offers.

(2) The computation processes of determining the scoring function take into account negotiation space of each issue as well the concepts of reservation and aspiration levels.

(3) The procedure makes it possible to expand the negotiation template by introducing new package after the preference elicitation has been conducted (within the actual negotiation space) without modifying ranking preliminary estimated packages. That means that proposed *scoring function* produces consistent ranking even after new packages are added (or removed) and does not lead to rank reversal.

The future work will focus on verifying the usefulness of a linguistic approach in negotiation experiments taking into account different recommendation linguistic sets and usefulness linguistic approach for holistic evaluation negotiation offers. To achieve these goals we should examine possibilities of effective enlargement of thesaurus of imprecise phrases extending Tentative Order Scale.

Moreover, we should test other multicriteria techniques based on OFN for evaluation negotiation packages. Let us note here that compared values of the scoring function are fuzzy. Therefore, in future we ought to study application of fuzzy order to problems described in this paper.

Conflicts of Interest

The authors declare that they have no conflicts of interest.

Acknowledgments

This research was supported with the grants from Polish National Science Centre (2016/21/B/HS4/01583)

References

[1] F. Herrera, S. Alonso, F. Chiclana, and E. Herrera-Viedma, "Computing with words in decision making: Foundations, trends and prospects," *Fuzzy Optimization and Decision Making*, vol. 8, no. 4, pp. 337–364, 2009.

[2] H. Raiffa, "Decision analysis: a personal account of how it got started and evolved," *Operations Research*, vol. 50, no. 1, pp. 179–185, 2002.

[3] M. Schoop, A. Jertila, and T. List, "egoisst: a negotiation support system for electronic business-to-business negotiations in e-commerce," *Data & Knowledge Engineering*, vol. 47, no. 3, pp. 371–401, 2003.

[4] T. Simons and T. M. Tripp, "The Negotiation Checklist," in *Negotiation Reading, Excersises and Cases*, R. J. Lewicki, D. M. Saunders, and J. W. Minton, Eds., pp. 50–63, McGraw-Hill/Irwin, NY, USA, 2003.

[5] G. E. Kersten and S. J. Noronha, "WWW-based negotiation support: Design, implementation, and use," *Decision Support Systems*, vol. 25, no. 2, pp. 135–154, 1999.

[6] S. Schenkerman, "Avoiding rank reversal in AHP decision-support models," *European Journal of Operational Research*, vol. 74, no. 3, pp. 407–419, 1994.

[7] J. Mustajoki and R. P. Hämäläinen, "Web-Hipre: Global Decision Support By Value Tree And AHP Analysis," *INFOR: Information Systems and Operational Research*, vol. 38, no. 3, pp. 208–220, 2016.

[8] E. Roszkowska and T. Wachowicz, "The multi-criteria negotiation analysis based on the membership function," *Studies in Logic, Grammar and Rhetoric*, vol. 37, no. 50, pp. 195–217, 2014.

[9] T. Wachowicz and P. Błaszczyk, "TOPSIS Based Approach to Scoring Negotiating Offers in Negotiation Support Systems," *Group Decision and Negotiation*, vol. 22, no. 6, pp. 1021–1050, 2013.

[10] D. Górecka, E. Roszkowska, and T. Wachowicz, "MARS – a hybrid of ZAPROS and MACBETH for verbal evaluation of the negotiation template," in *Proceedings of the Joint International Conference of the INFORMS GDN Section and the EURO Working Group on DSS*, P. Zaraté, G. Camilleri, D. Kamissoko, and etal., Eds., vol. 25, pp. 24–31, 2014.

[11] E. Roszkowska and T. Wachowicz, "Holistic evaluation of the negotiation template – comparing MARS and GRIP approaches," in *Proceedings of the The 15th International Conference on Group Decision and Negotiation Letters*, B. Kamiński, B. G.E. Kersten, P. Szufel, and etal., Eds., Warsaw School of Economics Press, pp. 139–148, 2015.

[12] A. Mardani, A. Jusoh, and E. K. Zavadskas, "Fuzzy multiple criteria decision-making techniques and applications—two decades review from 1994 to 2014," *Expert Systems with Applications*, vol. 42, no. 8, pp. 4126–4148, 2015.

[13] L. Martinez, D. Ruan, and F. Herrera, "Computing with Words in Decision support Systems: An overview on Models and Applications," *International Journal of Computational Intelligence Systems*, vol. 3, no. 4, pp. 382–395, 2010.

[14] D. Dubois and H. Prade, "Fuzzy real algebra: some results," *Fuzzy Sets and Systems*, vol. 2, no. 4, pp. 327–348, 1979.

[15] P. Prokopowicz, "The Directed Inference for the Kosinski's Fuzzy Number Model," in *Proceedings of the Second International Afro-European Conference for Industrial Advancement AECIA*, vol. 427 of *Advances in Intelligent Systems and Computing*, pp. 493–503, Springer International Publishing, 2016.

[16] K. Piasecki, "Revision of the Kosiński's Theory of Ordered Fuzzy Numbers," *Axioms*, vol. 7, no. 1, p. 16, 2018.

[17] W. Kosiński, "On fuzzy number calculus," *International Journal of Applied Mathematics and Computer Science*, vol. 16, no. 1, pp. 51–57, 2006.

[18] W. Kosinski and D. Wilczynska-Sztyma, "Defuzzification and implication within ordered fuzzy numbers," in *Proceedings of the 2010 IEEE International Conference on Fuzzy Systems (FUZZ-IEEE)*, pp. 1073–1079, Barcelona, Spain, July 2010.

[19] F. Herrera and E. Herrera-Viedma, "Linguistic decision analysis: steps for solving decision problems under linguistic information," *Fuzzy Sets and Systems*, vol. 115, no. 1, pp. 67–82, 2000.

[20] L. A. Zadeh, "The concept of a linguistic variable and its application to approximate reasoning. Part I. Information linguistic variable," *Expert Systems with Applications*, vol. 36, no. 2, pp. 3483–3488, 1975.

[21] L. A. Zadeh, "The concept of a linguistic variable and its application to approximate reasoning—Part II," *Information Sciences*, vol. 8, no. 4, pp. 301–357, 1975.

[22] L. A. Zadeh, "The concept of a linguistic variable and its application to approximate reasoning—part III," *Information Sciences*, vol. 9, no. 1, pp. 43–80, 1975.

[23] L. A. Zadeh, "Toward a theory of fuzzy information granulation and its centrality in human reasoning and fuzzy logic," *Fuzzy Sets and Systems*, vol. 90, no. 2, pp. 111–127, 1997.

[24] Y. Yao, "Granular computing," in *Computer Science*, vol. 31, pp. 1–5, 2004.

[25] E. Roszkowska and D. Kacprzak, "The fuzzy saw and fuzzy TOPSIS procedures based on ordered fuzzy numbers," *Information Sciences*, vol. 369, pp. 564–584, 2016.

[26] C. L. Hwang and K. Yoon, *Multiple Attribute Decision Making: Methods and Applications*, vol. 186, Springer, Berlin, Germany, 1981.

[27] S. J. Chen and C. L. Hwang, *Fuzzy Multiple Attribute Decision Making: Methods and Applications*, vol. 375 of *Lecture Notes in Economics and Mathematical Systems*, Springer, NY, USA, 1992.

[28] M. G. Kendall, *Rank Correlation Methods*, Charles Griffin & Company Limited London, 1955.

7

Solving Fuzzy Volterra Integrodifferential Equations of Fractional Order by Bernoulli Wavelet Method

R. Mastani Shabestari,[1] R. Ezzati,[2] and T. Allahviranloo[1]

[1] *Department of Mathematics, Science and Research Branch, Islamic Azad University, Tehran, Iran*
[2] *Department of Mathematics, Karaj Branch, Islamic Azad University, Karaj, Iran*

Correspondence should be addressed to R. Ezzati; ezati@kiau.ac.ir

Academic Editor: Ferdinando Di Martino

A matrix method called the Bernoulli wavelet method is presented for numerically solving the fuzzy fractional integrodifferential equations. Using the collocation points, this method transforms the fuzzy fractional integrodifferential equation to a matrix equation which corresponds to a system of nonlinear algebraic equations with unknown coefficients. To illustrate the method, it is applied to certain fuzzy fractional integrodifferential equations, and the results are compared.

1. Introduction

Dynamical systems with fractional order derivatives have found many applications in various problems in science and engineering like viscoelasticity, heat conduction, electrode-electrolyte polarization, electromagnetic waves, diffusion wave, control theory, and so on. In fractional equations, the vagueness may be appearing in each part of the equation like initial condition, boundary condition, and so on. So solving fractional equations in the sense of real conditions leads to the use of interval or fuzzy calculations.

The concept of the fuzzy derivative was first introduced by Chang and Zadeh [1], followed by many authors. The starting point of the topic in the set valued differential equation and also fuzzy differential equation is Hukuhara's paper [2]. The Hukuhara derivative was the starting point for the topic of set differential equations and later also for fuzzy fractional differential equations. By the concept of Hukuhara differentiability, the fuzzy Riemann-Liouville fractional differential equation is introduced by Agarwal et al. in [3], which was the starting point of the topic in fuzzy fractional derivative. They have considered the Riemann-Liouville differentiability concept based on the Hukuhara differentiability to solve uncertain fractional differential equations. The existence and uniqueness of solutions of Riemann-Liouville fuzzy fractional differential equations is proved in [4, 5]. Allahviranloo et al. in [6] presented the explicit solutions of uncertain fractional differential equations under Riemann-Liouville H-differentiability using Mittag-Leffler functions and in [7] introduced the fuzzy fractional differential equations under Riemann-Liouville H-differentiability and obtained the solution of this equation by fuzzy Laplace transforms. They showed two new uniqueness results for fuzzy fractional differential equations involving Riemann-Liouville generalized H-differentiability with fuzzy version of Nagumo and Krasnoselskii-Krein conditions [8]. Consequently, the Caputo generalized Hukuhara derivative is introduced in [9]; the authors introduced an ordinary fractional differential equation under the generalized Hukuhara differentiability and studied the existence and uniqueness of the solution. The nonlinear fuzzy fractional integrodifferential equation under generalized fuzzy Caputo derivative is introduced in [10, 11] and proved the existence and uniqueness of the solutions of this set of equations by considering the type of differentiability. Recently, Sahu and Saha Ray [12] applied the two-dimensional Bernoulli wavelet method to solve the fuzzy integrodifferential equations and they developed the Bernoulli wavelet method to solve the nonlinear fuzzy Hammerstein Volterra integral equations with constant delay [13].

The main aim of the presented paper is concerned with the application of the proposed approach to obtain the numerical solution of fuzzy fractional integrodifferential equations of the form

$$\left({}_{\mathrm{gH}}D_{*}^{q} y\right)(t) = f\left(t, y(t), (\mathcal{S} y)(t)\right),$$
$$t \in J = [0, T], \quad q \in (0, 1], \quad (1)$$
$$y(0) = y_0 \in \mathbb{R}_{\mathcal{F}},$$

where ${}_{\mathrm{gH}}D_{*}^{q}$ is the fuzzy Caputo fractional derivative of order q, $f : J \times \mathbb{R}_{\mathcal{F}} \times \mathbb{R}_{\mathcal{F}} \to \mathbb{R}_{\mathcal{F}}$ is a given function satisfying some assumptions that will be specified later, y_0 is an element of $\mathbb{R}_{\mathcal{F}}$, and $\mathcal{S}$ is a nonlinear integral operator given by

$$(\mathcal{S} y)(t) = \int_0^t k(x, t)\, y(x)\, \mathrm{d}x, \qquad (2)$$

where $k : J \times J \to \mathbb{R}^+$, with $\gamma_0 = \max\{\int_0^t k(x,t)\mathrm{d}x : (x,t) \in J \times J\}$.

The paper is organized as follows. Section 2 collects some definitions of basic notions and notations concerning fuzzy calculus. In Section 3 we discuss the properties of Bernoulli wavelets. To determine the approximate solution for the fuzzy fractional integrodifferential equation, two-dimensional Bernoulli wavelet method has been applied in Section 4. Moreover according to the type of differentiability, solutions of a fuzzy fractional integrodifferential equations are investigated in different scenarios. In Section 6, numerical examples are given to solve the fuzzy fractional integrodifferential equation and show the accuracy of the method. Finally, in Section 7, the report ends with a brief conclusion and some remarks.

2. Preliminaries

In this section, we introduce notation, definitions, and preliminary results, which will be used throughout this paper. Let $\mathbb{R}_{\mathcal{F}}$ denote the set of fuzzy subsets of the real axis, if $u : \mathbb{R} \to [0, 1]$, satisfying the following properties:

(i) u is upper semicontinuous on $\mathbb{R}$.

(ii) u is fuzzy convex.

(iii) u is normal.

(iv) closure of $\{x \in \mathbb{R} \mid u(x) > 0\}$ is compact.

Then $\mathbb{R}_{\mathcal{F}}$ is called the space of fuzzy numbers.

For $0 < r \le 1$, set $[u]_r = \{t \in \mathbb{R}^n \mid u(t) \ge r\}$, and $[u]_0 = \mathrm{cl}\{t \in \mathbb{R}^n \mid u(t) > 0\}$. We represent $[u]_r = [\underline{u}(r), \overline{u}(r)]$, so if $u \in \mathbb{R}_{\mathcal{F}}$, the r-level set $[u]_r$ is a closed interval for all $r \in [0, 1]$. For arbitrary $u, v \in \mathbb{R}_{\mathcal{F}}$ and $k \in \mathbb{R}$, the addition and scalar multiplication are defined by $[u+v]_r = [u]_r + [v]_r$, $[ku]^r = k[u]_r$, respectively.

A triangular fuzzy number is defined as a fuzzy set in $\mathbb{R}_{\mathcal{F}}$, which is specified by an ordered triple $u = (a, b, c) \in \mathbb{R}^3$ with $a \le b \le c$ such that $\underline{u}(r) = a + (b-a)r$ and $\overline{u} = c - (c-b)r$ are the endpoints of r-level sets for all $r \in [0, 1]$.

The Hausdorff distance between fuzzy numbers is given by $D : \mathbb{R}_{\mathcal{F}} \times \mathbb{R}_{\mathcal{F}} \to \mathbb{R}^+ \cup \{0\}$ as in [15]

$$\begin{aligned} D(u, v) &= \sup_{t \in [0,1]} d_H\left([u]_r, [v]_r\right) \\ &= \sup_{r \in [0,1]} \max\left\{\left|\underline{u}(r) - \underline{v}(r)\right|, \left|\overline{u}(r) - \overline{u}(r)\right|\right\}, \end{aligned} \qquad (3)$$

where d_H is the Hausdorff metric. The metric space $(\mathbb{R}_{\mathcal{F}}, D)$ is complete, separable, and locally compact and the following properties from [15] for metric D are valid:

(1) $D(u \oplus w, v \oplus w) = D(u, v)$, $\forall u, v, w \in \mathbb{R}_{\mathcal{F}}$;

(2) $D(\lambda u, \lambda v) = |\lambda| D(u, v)$, $\forall \lambda \in \mathbb{R}$, $u, v \in \mathbb{R}_{\mathcal{F}}$;

(3) $D(u \oplus v, w \oplus z) \le D(u, w) + D(v, z)$, $\forall u, v, w, z \in \mathbb{R}_{\mathcal{F}}$;

(4) $D(u \ominus v, w \ominus z) \le D(u, w) + D(v, z)$, as long as $u \ominus v$ and $w \ominus z$ exist, where $u, v, w, z \in \mathbb{R}_{\mathcal{F}}$.

Here, $\ominus$ is the Hukuhara difference (H-difference); it means that $w \ominus v = u$ if and only if $u \oplus v = w$.

Definition 1 (see [16]). The generalized Hukuhara difference of two fuzzy numbers $u, v \in \mathbb{R}_{\mathcal{F}}$ is defined as follows:

$$u \ominus_{\mathrm{gH}} v = w \iff \begin{cases} \text{(i) } u = v + w; \\ \text{or (ii) } v = u + (-1)\, w. \end{cases} \qquad (4)$$

In terms of r-levels one has $[u \ominus_{\mathrm{gH}} v]_r = [\min\{\underline{u}(r) - \underline{v}(r), \overline{u}(r) - \overline{u}(r)\}, \max\{\underline{u}(r) - \underline{v}(r), \overline{u}(r) - \overline{v}(r)\}]$ and if the H-difference exists, then $u \ominus v = u \ominus_{\mathrm{gH}} v$; the conditions for the existence of $w = u \ominus_{\mathrm{gH}} v \in \mathbb{R}_{\mathcal{F}}$ are as follows.

Case (i)

$$\begin{aligned} &\underline{w}(r) = \underline{u}(r) - v(r), \\ &\overline{w}(r) = \overline{u}(r) - \overline{v}(r), \\ &\qquad \forall r \in [0, 1], \\ &\text{with } \underline{w}(r) \text{ increasing, } \overline{w}(r) \text{ decreasing, } \underline{w}(r) \le \overline{w}(r). \end{aligned} \qquad (5)$$

Case (ii)

$$\begin{aligned} &\underline{w}(r) = \overline{u}(r) - \overline{v}(r), \\ &\overline{w}(r) = \underline{u}(r) - \underline{v}(r), \\ &\qquad \forall r \in [0, 1], \\ &\text{with } \underline{w}(r) \text{ increasing, } \overline{w}(r) \text{ decreasing, } \underline{w}(r) \le \overline{w}(r). \end{aligned} \qquad (6)$$

It is easy to show that (i) and (ii) are both valid if and only if w is a crisp number.

Remark 2. Throughout the rest of this paper, we assume that $u \ominus_{\mathrm{gH}} v \in \mathbb{R}_{\mathcal{F}}$.

Definition 3 (see [17]). A fuzzy-valued function $f : [a, b] \to \mathbb{R}_{\mathcal{F}}$ is said to be continuous at $t_0 \in [a, b]$ if for each $\epsilon > 0$ there is $\delta > 0$ such that $D(f(t), f(t_0)) < \epsilon$, whenever $t \in [a, b]$ and $|t - t_0| < \delta$. One says that f is fuzzy continuous on $[a, b]$ if f is continuous at each $t_0 \in [a, b]$.

Definition 4 (see [16]). The generalized Hukuhara derivative of a fuzzy-valued function $f : (a,b) \to \mathbb{R}_{\mathcal{F}}$ at $x_0 \in (a,b)$ is defined as

$$f'_{gH}(x_0) = \lim_{h \to 0} \frac{f(x_0 + h) \ominus_{gH} f(x_0)}{h}. \quad (7)$$

If $f'_{gH}(x_0) \in \mathbb{R}_{\mathcal{F}}$ satisfying (7) exists, one says that f is generalized Hukuhara differentiable (gH-differentiable for short) at x_0.

Definition 5 (see [18]). Let $f : [a,b] \to \mathbb{R}_{\mathcal{F}}$. One says that $f(t)$ is fuzzy Riemann integrable in $\mathbb{I} \in \mathbb{R}_F$ if, for any $\varepsilon > 0$, there exists $\delta > 0$ such that for any division $P = \{[u,v]; \xi\}$ with the norms $\Delta(P) < \delta$ one has

$$D\left(\sum_{P}^{*} (v - u) \odot f(\xi), I\right) < \varepsilon, \quad (8)$$

where $\sum_p^*$ denotes the fuzzy summation. One chooses to write

$$\mathbb{I} := \int_a^b f(t)\,dt. \quad (9)$$

Note that if the fuzzy-valued function $f(t,r) = [\underline{f}(t,r), \overline{f}(t,r)]$ is continuous in the metric D, the Lebesgue integral and the Riemann integral yield the same value, and also

$$\int_a^b f(t,r)\,dt = \left[\int_a^b \underline{f}(t,r)\,dt, \int_a^b \overline{f}(t,r)\,dt\right], \quad 0 \le r \le 1. \quad (10)$$

Throughout this paper, we consider the notations $A^{\mathbb{F}}[a,b]$ for the space of the fuzzy-valued functions from $[a,b]$ into $\mathbb{R}_{\mathcal{F}}$ that are absolutely continuous on $[a,b]$. Also, $C^{\mathbb{F}}[a,b]$ denote the set of the fuzzy-valued functions which are fuzzy continuous on all of $[a,b]$ such that the continuity is one-sided at endpoints a, b. Also, we denote the space of all Lebesgue integrable fuzzy-valued functions on the bounded interval $[a,b] \subset \mathbb{R}$ by $L^{\mathbb{F}}[a,b]$.

Definition 6 (see [7]). Let $f \in L^{\mathbb{F}}[a,b]$. The fuzzy Riemann-Liouville integral of a fuzzy-valued function f is defined as follows:

$$(\mathcal{J}_a^q f)(t) = \frac{1}{\Gamma(q)} \int_a^t (t-s)^{q-1} f(s)\,ds \quad (11)$$

for $a \le t$, and $0 < q \le 1$.

Theorem 7 (see [7]). *Let $f \in L^{\mathbb{F}}[a,b]$ be a fuzzy-valued function. The fuzzy Riemann-Liouville integral of a fuzzy-valued function f can be expressed as follows:*

$$\mathcal{J}_a^q f(t,r) = \left[I_a^q \underline{f}(t,r), I_a^q \overline{f}(t,r)\right], \quad (12)$$

where

$$I_a^q \underline{f}(t,r) = \frac{1}{\Gamma(q)} \int_a^t (t-s)^{q-1} \underline{f}(s,r)\,ds,$$
$$I_a^q \overline{f}(t,r) = \frac{1}{\Gamma(q)} \int_a^t (t-s)^{q-1} \overline{f}(s,r)\,ds. \quad (13)$$

Definition 8 (see [9]). Let $f \in A^{\mathbb{F}}[a,b]$. The fuzzy gH-fractional Caputo differentiability of the fuzzy-valued function f is defined as follows:

$$\left({}_{gH}D_*^q f\right)(t) = \mathcal{J}_a^{1-q}\left(f'_{gH}\right)(t) = \frac{1}{\Gamma(1-q)} \int_a^t \frac{\left(f'_{gH}\right)(s)\,ds}{(t-s)^q}, \quad (14)$$

where $a < s < t$, $q \in (0,1]$.

Definition 9 (see [9]). Let $f : [a,b] \to \mathbb{R}_{\mathcal{F}}$ be ^{cf}[gH]-differentiable at $t_0 \in (a,b)$. One says that f is ^{cf}[(i)-gH]-differentiable at t_0 if

$$\text{(i)}\ \left({}_{gH}D_*^\alpha f\right)(t_0,r) = \left[\left(D_*^\alpha \underline{f}\right)(t_0,r), \left(D_*^\alpha \overline{f}\right)(t_0,r)\right], \quad 0 \le r \le 1, \quad (15)$$

and that f is ^{cf}[(ii)-gH]-differentiable at t_0 if

$$\text{(ii)}\ \left({}_{gH}D_*^\alpha f\right)(t_0,r) = \left[\left(D_*^\alpha \overline{f}\right)(t_0,r), \left(D_*^\alpha \underline{f}\right)(t_0,r)\right], \quad 0 \le r \le 1. \quad (16)$$

Definition 10 (see [9]). Let $f : [a,b] \to \mathbb{R}_{\mathcal{F}}$ be a fuzzy function. A point $t_0 \in (a,b)$ is said to be a switching point for the ^{cf}[gH]-differentiability of f, if in any neighborhood V of t_0 there exist points $t_1 < t_0 < t_2$ such that one has the following.

Type(I). At t_1 (15) holds while (16) does not hold and at t_2 (16) holds and (15) does not hold.

Type(II). At t_1 (16) holds while (15) does not hold and at t_2 (15) holds and (16) does not hold.

Lemma 11 (see [9]). *Let $f : [0,T] \to \mathbb{R}_{\mathcal{F}}$ be a fuzzy-valued function such that $f \in A^{\mathbb{F}}[0,T]$; then*

$$\mathcal{J}_0^q\left({}_{gH}D_*^q f\right)(t) = f(t) \ominus_{gH} f(0). \quad (17)$$

Lemma 12 (see [14]). *Fuzzy initial value problem (1) is equivalent to one of the following integral equations.*

Case 1. If $y(t)$ is ^{cf}[(i)-gH]-differentiable, then

$$y(t) = y_0 + \frac{1}{\Gamma(q)} \int_0^t (t-s)^{q-1} f(s, y(s), (\mathcal{S}y)(s))\,ds. \quad (18)$$

Case 2. If $y(t)$ is ^{cf}[(ii)-gH]-differentiable, hence

$$y(t) = y_0 \ominus \frac{-1}{\Gamma(q)} \int_0^t (t-s)^{q-1} f(s, y(s), (\mathcal{S}y)(s))\,ds. \quad (19)$$

Theorem 13 (see [14]). *Assume that the following conditions hold:*

(H_1) $f : J \times \mathbb{R}_{\mathcal{F}} \times \mathbb{R}_{\mathcal{F}} \rightarrow \mathbb{R}_{\mathcal{F}}$ *is fuzzy continuous.*

(H_2) *There exists a constant* $q_1 \in (0, q)$ *and real-valued positive functions* $m_1(t), m_2(t) \in L^{1/q_1}(J, \mathbb{R}^+)$ *such that*

$$D(f(t, x(t), (\mathcal{S}x)(t)), f(t, y(t), (\mathcal{S}y)(t))) \leq m_1(t) D(x(t), y(t)) + m_2(t) D((\mathcal{S}x)(t), (\mathcal{S}) y(t)), \quad (20)$$

for each $t \in J$, *and all* $x(t), y(t), (\mathcal{S}x)(t), (\mathcal{S}y)(t) \in C^{\mathbb{F}}[0, T]$.

If

$$\Omega_{q,q_1,T} = \frac{MT^{q-q_1}}{\Gamma(q)((q-q_1)/(1-q_1))^{1-q_1}} < 1, \quad (21)$$

then (1) has a unique solution on J.

3. Bernoulli Wavelets

In this section, first we recall the definitions of wavelets and Bernoulli wavelets. Our aim is to approximate the solution $y(t)$ by the truncated Bernoulli series.

3.1. Wavelets and Bernoulli Wavelets. The Bernoulli polynomials play an important role in different areas of mathematics, including number theory and the theory of finite differences. The classical Bernoulli polynomials $\beta_m(x)$ are usually defined by means of following relations:

$$\frac{d\beta_m(x)}{dx} = m\beta_{m-1}(x), \quad m \geq 1, \qquad \beta_0(x) = 1. \quad (22)$$

Also the Bernoulli polynomials can be represented in the form

$$\beta_m(x) = \sum_{i=0}^{m} \binom{i}{m} \alpha_{m-i} t^i, \quad (23)$$

where α_i, $i = 0, 1, \ldots, m$ are the Bernoulli numbers. Thus, the first four such polynomials, respectively, are

$$\beta_0(t) = 1,$$
$$\beta_1(t) = t - \frac{1}{2},$$
$$\beta_2(t) = t^2 - t + \frac{1}{6}, \quad (24)$$
$$\beta_3(t) = t^3 - \frac{3}{2}t^2 + \frac{1}{2}t.$$

Also, these polynomials satisfy the following formula:

$$\int_0^1 \beta_n(t) \beta_m(t) \, dt = (-1)^{n-1} \frac{m!n!}{(m+n)!} \beta_{m+n}, \quad m, n \geq 1. \quad (25)$$

The properties of Bernoulli polynomials ($\beta_m(t)$) and the sequence of Bernoulli numbers (α_m) are

(1) $\beta_m(1-t) = (-1)^m \beta_m(t)$, $m \in \mathbb{Z}^+$.

(2) $\int_0^1 \beta_m(t)\beta_n(t)dt = (-1)^{m-1}(m!n!/(m+n)!)$, $m, n \geq 1$.

(3) $\int_0^1 |\beta_m(t)|dt < 16m!/(2\pi)^{m+1}$, $m \geq 0$.

(4) $\int_a^x \beta_m(t)dt = |\beta_{m+1}(x) - \beta_{m+1}(a)|/(m+1)$.

(5) $\sup_{t\in[0,1]} |\beta_{2m}(t)| = |\alpha_{2m}|$.

(6) $\sup_{t\in[0,1]} |\beta_{2m+1}(t)| \leq ((2m+1)/4)|\alpha_{2m}|$.

(7) $\alpha_{2m+1} = 0$, $\alpha_{2m} = \beta_{2m}(1)$.

(8) $\beta_m(1/2) = (2^{1-m} - 1)\alpha_m$.

(9) $\alpha_m = -(1/(m+1)) \sum_{k=0}^{m-1} \binom{m+1}{k} \alpha_k$.

Wavelets constitute a family of functions constructed from dilation and translation of single function called the mother wavelet $\psi(t)$. They are defined by

$$\psi_{a,b}(t) = \frac{1}{\sqrt{|a|}} \psi\left(\frac{t-b}{a}\right), \quad a, b \in \mathbb{R}, \quad (26)$$

where a is dilation parameter and b is a translation parameter. Bernoulli wavelets $\mathfrak{B}_{n,m}(t) = \mathfrak{B}(k, n, m, t)$ have four arguments, defined on interval $[0, 1)$ by

$$\mathfrak{B}_{n,m}(t) = \begin{cases} 2^{(k-1)/2} \widehat{\beta}_m \left(2^{k-1}t - n + 1\right), & \frac{n-1}{2^{k-1}} \leq t < \frac{n}{2^{k-1}}, \\ 0, & \text{elsewhere}, \end{cases} \quad (27)$$

with

$$\widehat{\beta}_m(t) = \begin{cases} 1, & m = 0, \\ \frac{1}{\sqrt{((-1)^{m-1}(m!)^2/(2m)!)\alpha_{2m}}} \beta_m(t) & m > 0, \end{cases} \quad (28)$$

where $k \in \mathbb{Z}^+$, $n = 1, 2, 3, \ldots, 2^{k-1}$ and $m = 0, 1, \ldots, M-1$ is the order of the Bernoulli polynomials and M is a fixed positive integer. The coefficient $1/\sqrt{((-1)^{m-1}(m!)^2/(2m)!)\alpha_{2m}}$ is for orthonormality, the dilation parameter is $a = 2^{-(k-1)}$, and translation parameter is $b = (n-1)2^{-(k-1)}$.

The two-dimensional Bernoulli wavelets are defined as

$$\mathfrak{B}_{n,i,l,j}(x,t) = \begin{cases} 2^{(k_1-1)/2}2^{(k_2-1)/2}\widehat{\beta}_i\left(2^{k_1-1}x-n+1\right)\widehat{\beta}_j\left(2^{k_2-1}t-l+1\right), & \dfrac{n-1}{2^{k_1-1}} \le x < \dfrac{n}{2^{k_1-1}},\ \dfrac{l-1}{2^{k_2-1}} \le t < \dfrac{l}{2^{k_2-1}} \\ 0, & \text{elsewhere,} \end{cases} \tag{29}$$

where $n = 1,2,\ldots,2^{k_1-1}$, $l = 1,2,\ldots,2^{k_2-1}$, $i = 0,1,\ldots,M_1-1$ and $j = 0,1,\ldots,M_2-1$ and k_1 and k_2 are any positive integers.

3.2. Function Approximation. A function $y(x,t)$ defined over $[0,1)\times[0,1)$ can be expanded in terms of Bernoulli wavelets as

$$y(x,t) = \sum_{n=1}^{\infty}\sum_{i=0}^{\infty}\sum_{l=1}^{\infty}\sum_{j=0}^{\infty} c_{n,i,l,j}\mathfrak{B}_{n,i,l,j}(x,t). \tag{30}$$

If the infinite series in (30) is truncated, then it can be written as

$$\begin{aligned} y(x,t) &= \sum_{n=1}^{2^{k_1-1}}\sum_{i=0}^{M_1-1}\sum_{l=1}^{2^{k_2-1}}\sum_{j=0}^{M_2-1} c_{n,i,l,j}\mathfrak{B}_{n,i,l,j}(x,t) \\ &= C^T\mathbf{B}(x,t), \end{aligned} \tag{31}$$

where $\mathbf{B}(x,t)$ is $(2^{k_1-1}2^{k_2-1}M_1M_2\times 1)$ matrix, given by

$$\begin{aligned} \mathbf{B}(x,t) = \big[&\mathfrak{B}_{1,0,1,0}(x,t), \mathfrak{B}_{1,0,1,1}(x,t), \ldots, \\ &\mathfrak{B}_{1,0,1,M_2-1}(x,t), \mathfrak{B}_{1,0,2,M_2-1}(x,t), \ldots, \\ &\mathfrak{B}_{1,0,2^{k_2-1},M_2-1}(x,t), \mathfrak{B}_{1,1,2^{k_2-1},M_2-1}(x,t), \ldots, \\ &\mathfrak{B}_{1,M_1-1,2^{k_2-1},M_2-1}(x,t), \mathfrak{B}_{2,M_1-1,2^{k_2-1},M_2-1}(x,t), \ldots, \\ &\mathfrak{B}_{2^{k_1-1},M_1-1,2^{k_2-1},M_2-1}(x,t)\big]. \end{aligned} \tag{32}$$

Also, C is $(2^{k_1-1}2^{k_2-1}M_1M_2\times 1)$ matrix and

$$\begin{aligned} C = \big[&c_{1,0,1,0}, c_{1,0,1,1}, \ldots, c_{1,0,1,M_2-1}, c_{1,0,2,M_2-1}, \ldots, \\ &c_{1,0,2^{k_2-1},M_2-1}, \ldots, c_{1,M_1-1,2^{k_2-1},M_2-1}, \ldots, \\ &c_{2^{k_1-1},M_1-1,2^{k_2-1},M_2-1}\big]^T. \end{aligned} \tag{33}$$

3.3. The Fractional Order Integration of the Bernoulli Wavelet. The fractional order integration of the Bernoulli wavelets is as follows:

$$\begin{aligned} I_a^q\Psi(t) = \big[&I_a^q\psi_{1,0}(t), \ldots, I_a^q\psi_{1,M-1}(t), I_a^q\psi_{2,0}(t), \ldots, \\ &I_a^q\psi_{2,M-1}(t), \ldots, I_a^q\psi_{2^{k-1},0}(t), \ldots, I_a^q\psi_{2^{k-1},M-1}(t)\big]^T, \end{aligned} \tag{34}$$

where

$$\begin{aligned} I_a^q\psi_{n,m}(t) &= \begin{cases} 2^{(k-1)/2}I_a^q\left(\widehat{\beta}_m\left(2^{k-1}t-n+1\right)\right), & \dfrac{n-1}{2^{k-1}} \le t < \dfrac{n}{2^{k-1}} \\ 0, & \text{elsewhere,} \end{cases} \\ I_a^q\left(\widehat{\beta}_m\left(2^{k-1}t-n+1\right)\right) &= \frac{1}{\Gamma(q)\sqrt{\left((-1)^{m-1}(m!)^2/(2m)!\right)\alpha_{2m}}}\left(\sum_{m=0}^{r}\binom{i}{m}\alpha_{r-m}\right. \\ &\quad\cdot\left.\int_a^t (t-s)^{q-1}\left(2^{k-1}s-n+1\right)\mathrm{d}s\right) \end{aligned} \tag{35}$$

for $k \in \mathbb{Z}^+$, $n = 1,2,\ldots,2^{k-1}$, and $m = 0,1,2,\ldots,M-1$ is the order of the Bernoulli polynomial and M is a fixed positive integer.

4. The Numerical Method

In this paper, we focus on the fuzzy fractional integrodifferential equation:

$$\left({}_{\mathrm{gH}}D_*^q y\right)(t) = y(t) + \int_0^t k(x,t)\,y(x)\,\mathrm{d}x + g(t), \quad t \in \mathbb{J} = [0,1), \tag{36}$$

with initial condition

$$y(0) = y_0 \in \mathbb{R}_{\mathcal{F}}, \tag{37}$$

where $y(t)$ and $g(t)$ are fuzzy functions and $k(x,t) : \mathbb{J}\times\mathbb{J} \to \mathbb{R}^+$. Applying $\mathcal{J}_0^q$ on both sides of (36), using Lemmas 11 and 12, if $y(t)$ is ${}^{cf}[(\text{i})\text{-gH}]$-differentiable, then

$$y(t) = y_0 + \mathcal{J}_0^q\left(y(t) + \int_0^t k(x,t)\,y(x)\,\mathrm{d}x + g(t)\right) \tag{38}$$

and $y(t)$ is ${}^{cf}[(\text{ii})\text{-gH}]$-differentiable:

$$\begin{aligned} y(t) &= y_0 \\ &\ominus (-1)\mathcal{J}_0^q\left(y(t) + \int_0^t k(x,t)\,y(x)\,\mathrm{d}x + g(t)\right). \end{aligned} \tag{39}$$

Consider $y(t,r) = [\underline{y}(t,r), \overline{y}(t,r)]$ is the solution of (36) and we approximate the unknown function $y(t,r)$ as given by (30). Assume that $y(t)$ is ${}^{cf}[(\text{i})\text{-gH}]$-differentiable, so by Theorem 7 and (38) we have the following fractional integrodifferential equations system:

$$y(x,r) = y(0,r) + \frac{1}{\Gamma(q)} \int_0^t (t-s)^{q-1} y(s,r)\,ds + \frac{1}{\Gamma(q)} \int_0^t \int_0^s (t-s)^{q-1} k(x,s)\, y(x,r)\,dx\,ds + \int_0^t (t-s)^{q-1} g(s,r)\,ds. \tag{40}$$

We see that

$$\begin{pmatrix} \underline{y}(t,r) \\ \overline{y}(t,r) \end{pmatrix} = \begin{pmatrix} \underline{y}(0,r) \\ \overline{y}(0,r) \end{pmatrix} + \begin{pmatrix} \frac{1}{\Gamma(q)} \int_0^t (t-s)^{q-1} \underline{y}(s,r)\,ds \\ \frac{1}{\Gamma(q)} \int_0^t (t-s)^{q-1} \overline{y}(s,r)\,ds \end{pmatrix} + \begin{pmatrix} \frac{1}{\Gamma(q)} \int_0^t \int_0^s (t-s)^{q-1} k(x,s)\, \underline{y}(x,r)\,dx\,ds \\ \frac{1}{\Gamma(q)} \int_0^t \int_0^s (t-s)^{q-1} k(x,s)\, \overline{y}(x,r)\,dx\,ds \end{pmatrix} + \begin{pmatrix} \frac{1}{\Gamma(q)} \int_0^t (t-s)^{q-1} \underline{g}(s,r)\,ds \\ \frac{1}{\Gamma(q)} \int_0^t (t-s)^{q-1} \overline{g}(s,r)\,ds \end{pmatrix}. \tag{41}$$

Hence we obtain

$$\underline{y}(t,r) = \underline{y}(0,r) + \frac{1}{\Gamma(q)} \left(\int_0^t (t-s)^{q-1} \underline{y}(s,r)\,ds + \int_0^t \int_0^s (t-s)^{q-1} k(x,s)\, \underline{y}(x,r)\,dx\,ds + \int_0^t (t-s)^{q-1} \underline{g}(s,r)\,ds \right), \tag{42}$$

$$\overline{y}(t,r) = \overline{y}(0,r) + \frac{1}{\Gamma(q)} \left(\int_0^t (t-s)^{q-1} \overline{y}(s,r)\,ds + \int_0^t \int_0^s (t-s)^{q-1} k(x,s)\, \overline{y}(x,r)\,dx\,ds + \int_0^t (t-s)^{q-1} \overline{g}(s,r)\,ds \right). \tag{43}$$

In order to apply the Bernoulli wavelets in (42), we first approximate the unknown function $\underline{y}(t,r)$ as

$$\underline{y}(t,r) = \sum_{n=1}^{2^{k_1-1}} \sum_{i=0}^{M_1-1} \sum_{l=1}^{2^{k_2-1}} \sum_{j=0}^{M_2-1} c_{n,i,l,j} \mathfrak{B}_{n,i,l,j}(t,r) = C^T \mathbf{B}(t,r). \tag{44}$$

Putting (44) in (42) we obtain

$$\sum_{n=1}^{2^{k_1-1}} \sum_{i=0}^{M_1-1} \sum_{l=1}^{2^{k_2-1}} \sum_{j=0}^{M_2-1} c_{n,i,l,j} \mathfrak{B}_{n,i,l,j}(t,r) = \underline{y}(0,r) + \frac{1}{\Gamma(q)} \left(\sum_{n=1}^{2^{k_1-1}} \sum_{i=0}^{M_1-1} \sum_{l=1}^{2^{k_2-1}} \sum_{j=0}^{M_2-1} c_{n,i,l,j} \int_0^t (t-s)^{q-1} \mathfrak{B}_{n,i,l,j}(s,r)\,ds + \sum_{n=1}^{2^{k_1-1}} \sum_{i=0}^{M_1-1} \sum_{l=1}^{2^{k_2-1}} \sum_{j=0}^{M_2-1} c_{n,i,l,j} \int_0^t \int_0^s (t-s)^{q-1} k(x,s)\, \mathfrak{B}_{n,i,l,j}(x,r)\,dx\,ds + \int_0^t (t-s)^{q-1} \underline{g}(s,r)\,ds \right). \tag{45}$$

Now we collocate (45) at $(2^{k_1-1}2^{k_2-1}M_1M_2)$ collocation points by $t_i = (2i-1)/2^{k_1}M_1$, $r_j = (2j-2)/2^{k_2}M_2$ for $i = 1,2,\ldots,2^{k_1-1}M_1$, $j = 1,2,\ldots,2^{k_2-1}M_2$ yielding

$$\sum_{n=1}^{2^{k_1-1}} \sum_{i=0}^{M_1-1} \sum_{l=1}^{2^{k_2-1}} \sum_{j=0}^{M_2-1} c_{n,i,l,j} \mathfrak{B}_{n,i,l,j}(t_i,r_j) = \underline{y}(0,r_j) + \frac{1}{\Gamma(q)} \left(\sum_{n=1}^{2^{k_1-1}} \sum_{i=0}^{M_1-1} \sum_{l=1}^{2^{k_2-1}} \sum_{j=0}^{M_2-1} c_{n,i,l,j} \cdot \int_0^{t_i} (t_i-s)^{q-1} \mathfrak{B}_{n,i,l,j}(s,r_j)\,ds + \sum_{n=1}^{2^{k_1-1}} \sum_{i=0}^{M_1-1} \sum_{l=1}^{2^{k_2-1}} \sum_{j=0}^{M_2-1} c_{n,i,l,j} \cdot \int_0^{t_i} \int_0^s (t_i-s)^{q-1} k(x,s)\, \mathfrak{B}_{n,i,l,j}(x,r_j)\,dx\,ds + \int_0^{t_i} (t_i-s)^{q-1} \underline{g}(s,r_j)\,ds \right). \tag{46}$$

Now consider (43); we approximate the unknown function $\overline{y}(t,r)$ by Bernoulli wavelet as

$$\overline{y}(t,r) = \sum_{n=1}^{2^{k_1-1}} \sum_{i=0}^{M_1-1} \sum_{l=1}^{2^{k_2-1}} \sum_{j=0}^{M_2-1} c'_{n,i,l,j} \mathfrak{B}_{n,i,l,j}(t,r). \tag{47}$$

Working the same way we find

$$\sum_{n=1}^{2^{k_1-1}M_1-1}\sum_{i=0}^{2^{k_2-1}M_2-1}\sum_{l=1}\sum_{j=0} c'_{n,i,l,j}\mathfrak{B}_{n,i,l,j}\left(t_i,r_j\right)=\overline{y}\left(0,r_j\right)$$
$$+\frac{1}{\Gamma(q)}\left(\sum_{n=1}^{2^{k_1-1}M_1-1}\sum_{i=0}^{2^{k_2-1}M_2-1}\sum_{l=1}\sum_{j=0} c'_{n,i,l,j}\cdot\int_0^{t_i}\left(t_i-s\right)^{q-1}\mathfrak{B}_{n,i,l,j}\left(s,r_j\right)\mathrm{d}s\right.$$
$$+\sum_{n=1}^{2^{k_1-1}M_1-1}\sum_{i=0}^{2^{k_2-1}M_2-1}\sum_{l=1}\sum_{j=0} c'_{n,i,l,j}\cdot\int_0^{t_i}\int_0^{s}\left(t_i-s\right)^{q-1}k(x,s)\,\mathfrak{B}_{n,i,l,j}\left(x,r_j\right)\mathrm{d}x\,\mathrm{d}s$$
$$\left.+\int_0^{t_i}\left(t_i-s\right)^{q-1}\overline{g}\left(s,r_j\right)\mathrm{d}s\right).\qquad(48)$$

Equations (46) and (48) yield $2(2^{k_1-1}M_1)(2^{k_2-1}M_2)$ equations in $2(2^{k_1-1}M_1)(2^{k_2-1}M_2)$ unknowns in $c_{n,i,l,j}$ and $c'_{n,i,l,j}$. By solving this system of equations using mathematical software, the Bernoulli wavelet coefficients $c_{n,i,l,j}$ and $c'_{n,i,l,j}$ can be obtained and, hence, substituting them in (44) and (47), the approximate solutions can be obtained.

Now, consider $y(t)$ is ^{cf}[(ii)-gH]-differentiable; we have the following fractional integrodifferential equations system:

$$y(x,r)=y(0,r)+\frac{1}{\Gamma(q)}\int_0^t(t-s)^{q-1}y(s,r)\,\mathrm{d}s$$
$$+\frac{1}{\Gamma(q)}\int_0^t\int_0^s(t-s)^{q-1}k(x,s)\,y(x,r)\,\mathrm{d}x\,\mathrm{d}s$$
$$+\int_0^t(t-s)^{q-1}g(s,r)\,\mathrm{d}s.\qquad(49)$$

Using (39) and definition of Hukuhara difference, this system can be written in the form

$$\begin{pmatrix}\underline{y}(t,r)\\ \overline{y}(t,r)\end{pmatrix}=\begin{pmatrix}\underline{y}(0,r)\\ \overline{y}(0,r)\end{pmatrix}+\begin{pmatrix}\frac{1}{\Gamma(q)}\int_0^t(t-s)^{q-1}\overline{y}(s,r)\,\mathrm{d}s\\ \frac{1}{\Gamma(q)}\int_0^t(t-s)^{q-1}\underline{y}(s,r)\,\mathrm{d}s\end{pmatrix}$$
$$+\begin{pmatrix}\frac{1}{\Gamma(q)}\int_0^t\int_0^s(t-s)^{q-1}k(x,s)\,\overline{y}(x,r)\,\mathrm{d}x\,\mathrm{d}s\\ \frac{1}{\Gamma(q)}\int_0^t\int_0^s(t-s)^{q-1}k(x,s)\,\underline{y}(x,r)\,\mathrm{d}x\,\mathrm{d}s\end{pmatrix}$$
$$+\begin{pmatrix}\frac{1}{\Gamma(q)}\int_0^t(t-s)^{q-1}\overline{g}(s,r)\,\mathrm{d}s\\ \frac{1}{\Gamma(q)}\int_0^t(t-s)^{q-1}\underline{g}(s,r)\,\mathrm{d}s\end{pmatrix}.\qquad(50)$$

Then in similar way to previous case, we get the values of unknown vectors $c_{n,i,l,j}$ and $c'_{n,i,l,j}$ and then obtain the solutions $\underline{y}(t,r)$ and $\overline{y}(t,r)$ from (44) and (47), respectively.

5. Error Estimation Algorithm and Convergence Analysis

In this section, we will show an efficient estimation for the Bernoulli wavelets approximation and also a technique to obtain the corrected solution of problem (40) by using the residual method and to describe the convergence behavior of the proposed numerical method.

Let $y^*(t,r)$ be the truncated series which approximate the unknown function of (40), so we observe that

$$y^*(t,r)=y(0,r)+\frac{1}{\Gamma(q)}\int_0^t(t-s)^{q-1}y^*(s,r)\,\mathrm{d}s$$
$$+\frac{1}{\Gamma(q)}\int_0^t\int_0^s(t-s)^{q-1}k(x,s)\,y^*(x,r)\,\mathrm{d}x\,\mathrm{d}s$$
$$+\int_0^t(t-s)^{q-1}g(s,r)\,\mathrm{d}s+R^*(t,r),\qquad(51)$$

where $R^*(t,r)$ is the residual function. Now, let us consider

$$L\left[y^*(t,r)\right]=y^*(t,r)-\frac{1}{\Gamma(q)}\int_0^t(t-s)^{q-1}y^*(s,r)\,\mathrm{d}s$$
$$-\frac{1}{\Gamma(q)}\int_0^t\int_0^s(t-s)^{q-1}k(x,s)\,y^*(x,r)\,\mathrm{d}x\,\mathrm{d}s.\qquad(52)$$

Hence, we have

$$L\left[y^*(t,r)\right]=y(0,r)+\int_0^t(t-s)^{q-1}g(s,r)\,\mathrm{d}s+R^*(t,r)\qquad(53)$$

with initial condition $y^*(0,r)=y(0,r)$ for all $0\le r\le 1$.

Furthermore, the error function $E_N(t,r)$ can be defined as

$$E^*(t,r) = y(t,r) - y^*(t,r), \tag{54}$$

where $y(t,r)$ is the exact solution of problem (40).

By using (53) and (54), we have the error equation

$$L[E^*(t,r)] = L[y(t,r)] - L[y^*(t,r)] = -R^*(t,r) \tag{55}$$

with initial condition

$$E^*(0,r) = 0, \quad 0 \le r \le 1. \tag{56}$$

Subsequently, the error problem by using (52) and (55) can be written as

$$\begin{aligned} E^*(t,r) &- \frac{1}{\Gamma(q)}\int_0^t (t-s)^{q-1} E^*(s,r)\,ds \\ &- \frac{1}{\Gamma(q)}\int_0^t\int_0^s (t-s)^{q-1} k(x,s) E^*(x,r)\,dx\,ds \\ &= -R^*(t,r) \\ E^*(0,r) &= 0, \quad 0 \le r \le 1. \end{aligned} \tag{57}$$

Solving (57) in the way as in Section 4, we get the approximation $E^*(t,r)$ which is the error function based on residual function. We note that if the exact solution of problem (40) is unknown, then the error function can be estimated by $E^*(t,r)$ which is found without the exact solution and also clearly seen from given error estimation algorithm.

Let us consider $y(t,r)$ can be expanded in terms of Bernoulli wavelets as

$$y(t,r) = \sum_{n=0}^{\infty}\sum_{i=0}^{\infty}\sum_{l=1}^{\infty}\sum_{j=0}^{\infty} c_{n,i,l,j}\mathfrak{B}_{n,i,l,j}(t,r). \tag{58}$$

And $y^*(t,r)$ can be the truncated series:

$$y^*(t,r) = \sum_{n=1}^{2^{k_1-1}}\sum_{i=0}^{M_1-1}\sum_{l=1}^{2^{k_2-1}}\sum_{j=0}^{M_2-1} c_{n,i,l,j}\mathfrak{B}_{n,i,l,j}(t,r). \tag{59}$$

Hence, the truncated error term can be calculated as

$$E_{n,i,l,j}(t,r) = y(t,r) - y^*(t,r). \tag{60}$$

Now, we have the following Theorems based on [13].

Theorem 14 (see [13]). *If $y(x,t)$ is defined on $[0,1) \times [0,1)$ and $|y(x,t)| \le K$, then the Bernoulli wavelets expansion of $y(x,t)$ defined in (31) converges uniformly and also*

$$\left|c_{n,i,l,j}\right| < K\frac{A_1A_2}{2^{(k_1-1)/2}2^{(k_2-1)/2}}\frac{16_i!}{(2\pi)^{i+1}}\frac{16_j!}{(2\pi)^{j+1}}, \tag{61}$$

where $A_1 = 1/\sqrt{(-1)^{i-1}(i!)^2/(2i)!}$ and $A_1 = 1/\sqrt{(-1)^{j-1}(j!)^2/(2j)!}$.

Theorem 15 (see [13]). *If a continuous function $y(t,r) \in L^2(\mathbb{R}\times\mathbb{R})$ defined on $[0,1)\times[0,1)$ is bounded, namely, $|y(t,r)| \le K$, then*

$$\left\|E_{n,i,l,j}(t,r)\right\|^2_{L^2([0,1)\times[0,1))} \le \sum_{n=2^{k_1-1}+1}^{\infty}\sum_{i=M_1}^{\infty}\sum_{l=2^{k_2-1}+1}^{\infty}\sum_{j=M_2}^{\infty}\left(K\frac{A_1A_2}{2^{(k_1-1)/2}2^{(k_2-1)/2}}\frac{16_i!}{(2\pi)^{i+1}}\frac{16_j!}{(2\pi)^{j+1}}\right)^2, \tag{62}$$

where $A_1 = 1/\sqrt{(-1)^{i-1}(i!)^2/(2i)!}$ and $A_1 = 1/\sqrt{(-1)^{j-1}(j!)^2/(2j)!}$.

6. Numerical Examples

In order to illustrate the effectiveness of the proposed method, we consider numerical examples of fuzzy fractional differential equation.

Example 1. Consider the following fuzzy fractional integrodifferential equation:

$$\begin{aligned} \left({}_{gH}D_*^{1/2}y\right)(t) &= \frac{[r+1,5-3r]}{15\sqrt{\pi}}t^{5/2}\left(48-\sqrt{\pi}t^{7/2}\right) \\ &+ \int_0^t \frac{xt}{3}y(x)\,dx, \\ y(0) &= 0, \end{aligned} \tag{63}$$

where 0 denotes the crisp set $\{0\}$ and the exact solution of (63) is given by $y(t) = [r+1, 5-3r]t^3$. The exact solution is plotted in Figure 1(a) and its ${}^{gH}D_*^{1/2}y(t)$ is plotted in Figure 1(b). As you see, $y(t)$ is ^{cf}[(i)-gH]-differentiable. So, by applying the method which is discussed in detail, we presented numerical solution of this example for $M_1 = M_2 = 4$, $k_1 = k_2 = 1$. Also, we calculated the absolute errors as $|\underline{E}_r| = |\underline{y}(t,r) - \underline{y}^*(t,r)|$ and $|\overline{E}_r| = |\overline{y}(t,r) - \overline{y}^*(t,r)|$. To compare the absolute errors of presented method and Legendre method proposed in [14], see Table 1.

It is evident from Table 1 that the numerical solutions converge to the exact solution. It is also concluded that the proposed method is very efficient for numerical solutions of these problems.

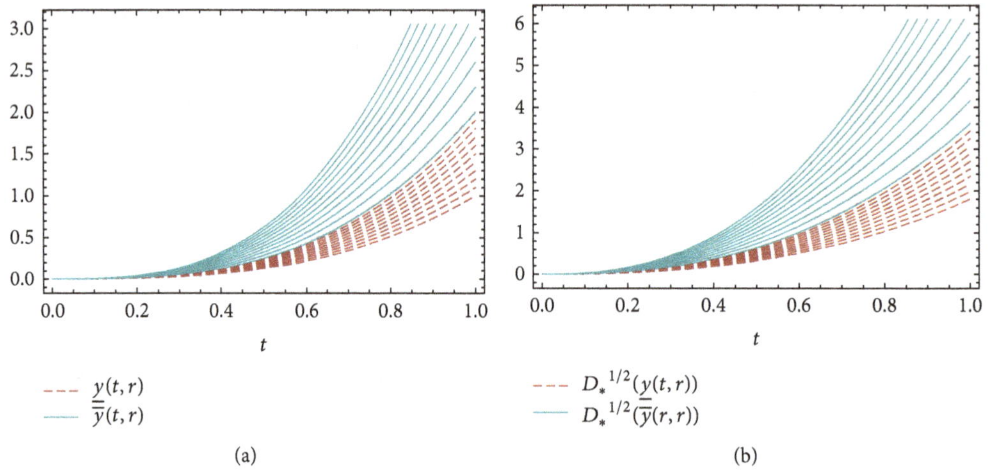

FIGURE 1: The level sets of $y(t)$ (a) and ${}^{\mathrm{gH}}D_*^{1/2}y(t)$ (b) of Example 1.

TABLE 1: Comparing the absolute errors of the presented method and the method of [14] for Example 1.

r	t	$\lvert \underline{E}_r \rvert$ Presented method	Legendre method [14]	$\lvert \overline{E}_r \rvert$ Presented method	Legendre method [14]
0.3	0.3	1.011×10^{-14}	3.81639×10^{-16}	3.497×10^{-15}	6.38378×10^{-16}
	0.6	5.898×10^{-16}	9.71445×10^{-17}	3.413×10^{-15}	7.63278×10^{-16}
	0.9	2.359×10^{-16}	8.18789×10^{-16}	6.453×10^{-16}	1.31839×10^{-16}
0.6	0.3	2.164×10^{-15}	1.66533×10^{-16}	2.664×10^{-15}	9.99201×10^{-16}
	0.6	4.996×10^{-16}	3.33067×10^{-16}	1.110×10^{-16}	4.44089×10^{-16}
	0.9	4.996×10^{-16}	5.55112×10^{-17}	0	4.99693×10^{-16}
0.9	0.3	3.330×10^{-16}	2.22045×10^{-16}	8.881×10^{-16}	1.77636×10^{-15}
	0.6	6.661×10^{-16}	0	1.332×10^{-15}	0
	0.9	8.881×10^{-16}	2.22045×10^{-16}	1.110×10^{-15}	2.22045×10^{-16}

Example 2. Consider the following fuzzy fractional integrodifferential equation:

$$\left({}_{\mathrm{gH}}D_*^{1/2}y\right)(t) = \frac{3s}{140}\left(35\sqrt{\pi} - s^{5/2}\right)[1+2r, 8-5r] + \int_0^t (x+t)\,y(x)\,\mathrm{d}x, \qquad (64)$$

$$y(0) = 0,$$

where 0 denotes the crisp set $\{0\}$. The exact solution of this equation is given by $y(t) = [1+2r, 8-5r]t^{3/2}$. By applying the proposed method to obtain ${}^{cf}[(\mathrm{i})\text{-gH}]$-differentiable solution, we solve this problem numerically and calculate the absolute errors by using $|E_r| = |y(t,r) - y^*(t,r)|$. To see these absolute errors, one can refer to Table 2. Also, in Figure 2, we present the graph of the Bernoulli wavelet approximation error for $r = 1$, and $M_1 = M_2 = 8$, $M_1 = M_2 = 10$.

7. Conclusion

In the present paper, the two-dimensional Bernoulli wavelet method was applied to approximate the solution of fuzzy fractional integrodifferential equation. We transformed our problem to a system of algebraic equations so that by solving this system we obtained the solution of this kind of equation by considering the type of differentiability. Finally, the solution obtained using the suggested method shows that this approach can solve the problem effectively.

TABLE 2: The absolute errors for Example 2.

r	t	$\lvert \underline{E}_r \rvert$ $M_1 = M_2 = 15$	$\lvert \overline{E}_r \rvert$ $M_1 = M_2 = 15$
0.3	0.3	3.69879×10^{-6}	1.39805×10^{-7}
	0.6	3.76203×10^{-6}	7.20378×10^{-6}
	0.9	3.56742×10^{-6}	1.61695×10^{-6}
0.6	0.3	2.42246×10^{-6}	7.26592×10^{-6}
	0.6	3.76347×10^{-6}	1.35717×10^{-6}
	0.9	6.06565×10^{-6}	4.59658×10^{-6}
0.9	0.3	1.09187×10^{-6}	3.75328×10^{-5}
	0.6	8.77415×10^{-6}	1.35462×10^{-5}
	0.9	9.98774×10^{-6}	1.96437×10^{-5}

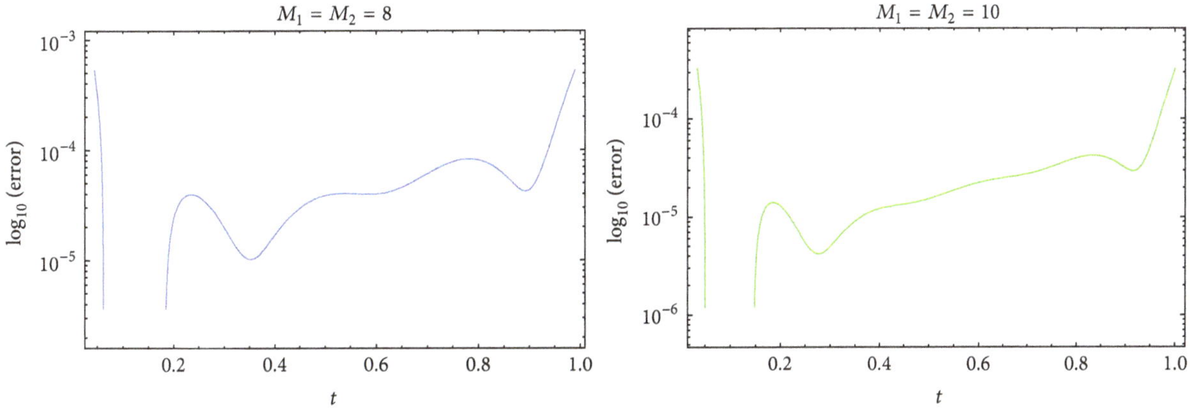

FIGURE 2: Graph of the Bernoulli wavelet approximation error for $r = 1$ and $t \in [0, 1)$ of Example 2.

Conflicts of Interest

The authors declare that they have no conflicts of interest.

References

[1] S. S. L. Chang and L. A. Zadeh, "On fuzzy mapping and control," *IEEE Transactions on Systems, Man, and Cybernetics*, vol. 2, no. 1, pp. 30–34, 1972.

[2] M. Hukuhara, "Intégration des applications mesurables dont la valeur est un compact convex," *Funkcialaj Ekvacioj*, vol. 10, pp. 205–229, 1967.

[3] R. P. Agarwal, V. Lakshmikantham, and J. J. Nieto, "On the concept of solution for fractional differential equations with uncertainty," *Nonlinear Analysis. Theory, Methods & Applications. An International Multidisciplinary Journal*, vol. 72, no. 6, pp. 2859–2862, 2010.

[4] S. Arshad and V. Lupulescu, "On the fractional differential equations with uncertainty," *Nonlinear Analysis. Theory, Methods & Applications. An International Multidisciplinary Journal*, vol. 74, no. 11, pp. 3685–3693, 2011.

[5] S. Arshad and V. Lupulescu, "Fractional differential equation with the fuzzy initial condition," *Electronic Journal of Differential Equations*, vol. 34, pp. 1–8, 2011.

[6] T. Allahviranloo, S. Salahshour, and S. Abbasbandy, "Explicit solutions of fractional differential equations with uncertainty," *Soft Computing*, vol. 16, no. 2, pp. 297–302, 2012.

[7] S. Salahshour, T. Allahviranloo, and S. Abbasbandy, "Solving fuzzy fractional differential equations by fuzzy Laplace transforms," *Communications in Nonlinear Science and Numerical Simulation*, vol. 17, no. 3, pp. 1372–1381, 2012.

[8] T. Allahviranloo, S. Abbasbandy, and S. Salahshour, "Fuzzy fractional differential equations with nagumo and krasnoselskii-krein condition," in *EUSFLAT-LFA 2011*, Aix-les-Bains, France, July 2011.

[9] T. Allahviranloo, A. Armand, and Z. Gouyandeh, "Fuzzy fractional differential equations under generalized fuzzy Caputo derivative," *Journal of Intelligent & Fuzzy Systems*, vol. 26, no. 3, pp. 1481–1490, 2014.

[10] A. Armand and Z. Gouyandeh, "Fuzzy fractional integro-differential equations under generalized Caputo differentiability," *Annals of Fuzzy Mathematics and Informatics*, vol. 10, no. 5, pp. 789–798, 2015.

[11] T. Allahviranloo, A. Armand, Z. Gouyandeh, and H. Ghadiri, "Existence and uniqueness of solutions for fuzzy fractional Volterra-Fredholm integro-differential equations," *Journal of Fuzzy Set Valued Analysis*, vol. 2013, pp. 1–9, 2013.

[12] P. K. Sahu and S. Saha Ray, "Two-dimensional Legendre wavelet method for the numerical solutions of fuzzy integro-differential equations," *Journal of Intelligent & Fuzzy Systems*, vol. 28, pp. 1271–1279, 2015.

[13] P. K. Sahu and S. Saha Ray, "A new Bernoulli wavelet method for accurate solutions of nonlinear fuzzy Hammerstein–Volterra delay integral equations," *Fuzzy Sets and Systems*, vol. 309, pp. 131–144, 2017.

[14] R. Mastani Shabestari, R. Ezzati, and T. Allahviranloo, "Numerical solution of fuzzy fractional integro-differential equation via two-dimensional Legendre wavelet method," *Journal of Intelligent and Fuzzy Systems*, 2017.

[15] V. Lakshmikantham, T. Bhaskar, and J. Devi, *Theory of Set Differential Equations in Metric Spaces*, Cambridge Scientific Publishers, 2006.

[16] B. Bede and L. Stefanini, "Generalized differentiability of fuzzy-valued functions," *Fuzzy Sets and Systems*, vol. 230, pp. 119–141, 2013.

[17] Z. Guang-Quan, "Fuzzy continuous function and its properties," *Fuzzy Sets and Systems*, vol. 43, no. 2, pp. 159–171, 1991.

[18] W. Congxin and M. Ming, "Embedding problem of fuzzy number space: part III," *Fuzzy Sets and Systems*, vol. 46, no. 2, pp. 281–286, 1992.

8

Criterion for Generalized Weakly Fuzzy Invex Monotonocities

Meraj A. Khan[1], **Izhar Ahmad**[2] **and Abdulrahman Aljohani**[1]

[1]*Department of Mathematics, University of Tabuk, Tabuk, Saudi Arabia*
[2]*Department of Mathematics and Statistics, King Fahd University of Petroleum and Minerals, Dhahran 31261, Saudi Arabia*

Correspondence should be addressed to Meraj A. Khan; meraj79@gmail.com

Academic Editor: Pushpinder Singh

The present paper deals with the concepts of generalized fuzzy invex monotonocities and generalized weakly fuzzy invex functions. Some necessary conditions for weakly fuzzy invex monotonocities are presented. Moreover, the concept of fuzzy strong invex monotonocities and fuzzy strong invex functions are also discussed. To strengthen our definitions, we provide nontrivial examples of fuzzy invex monotonocities and weakly fuzzy invex functions.

1. Introduction

In the last few years, the conception of convexity and generalized convexity is well recognized in the optimization theory and accomplished a significant role in the computational economics, management, decision making, and operation research. Consequently, the generalized convexity and generalized monotonocities are fundamental tools in these areas of research.

Hanson [1] introduced the generalized version of convex function namely invex function. Further, generalized invex monotonocities have been explored by Ruiz-Garzon et al. [2] and Yang et al. [3]. A step forward Yang et al. [4] found that there were some errors in [2] and modified the results of Ruiz-Garzon et al. [2] and they also proposed the notion of strong pseudo invex monotonicity and quasi-invex monotonicity. In addition, Antczak [5, 6] and Suneja et al. [7] deliberated the properties and execution of preinvex functions and their generalizations for the nonlinear programming problems. Aghezzaf and Hachimi [8] presented the differentiable type I functions and derived the appropriate duality results for a Mond-Weir type dual. Gulati et al. [9] presented the more generalized class of convex function, called $(F, \alpha, \rho, d) - V$-type I functions, and derived sufficiency and duality results for a multiobjective programming problem. Recently, Ahmad et al. [10] discussed twice weakly differentiable and interval valued bonvex functions. Duality results are also discussed for Mangasarian type dual model.

The notion of fuzzy generalized convex functions has been investigated by several authors and presently, it is a fascinating and exciting area of research. In 1989, Nanda and Kar [11] presented the idea of convex fuzzy mappings and obtained the convex fuzzy mapping under the epigraph of convex function in a convex set. In [12], Furukawa proposed the concept of convexity and Lipschitz continuity for the category of fuzzy valued functions. Yan and Xu [13] investigated the convexity and quasi-convexity of fuzzy mapping involving the concept of ordering [14]. These concepts further generalized for fuzzy functions by Noor [15]. Qiu and Zhang [16] discussed the convexity invariance of fuzzy sets under the extension principles. In [17–21], authors proposed the perception of fuzzy convex mappings and presented the idea of invexity, pseudoconvexity, and pseudoinvexity for fuzzy mappings. Recently, Li et al. [22] introduced the fuzzy generalized convex mappings and discussed the properties. The optimality conditions and duality results are also presented under fuzzy weakly univex functions. Some new types of fuzzy starshapedness and their relationships and basic properties are investigated in [23]. Many authors have studied the applications of generalized fuzzy convex mappings to fuzzy optimization problems (see [24–28]). The purpose of the present paper is to develop the notion of generalized invex monotonocities for fuzzy valued mappings.

2. Notations and Preliminaries

Let U be the universal set whose standard element is denoted by u. A fuzzy set δ in U is a function $\delta : U \to [0,1]$ and support of δ is denoted by $S(\delta)$ and is given by

$$S(\delta) = \{u \in U : \delta(u) > 0\}. \quad (1)$$

Definition 1. If δ be a fuzzy set in U and $\alpha \in [0,1]$, then the $\alpha-$ cut of fuzzy set δ is defined as

$$[\delta]^{\alpha} = \{u \in U : \delta(u) \geq \alpha\}. \quad (2)$$

Definition 2. A fuzzy number δ is a fuzzy set in R^1 and satisfies the following conditions:

(i) δ is normal; that is, there exits $u_0 \in R$ such that $\delta(u_0) = 1$,

(ii) δ is upper semicontinuous,

(iii) δ is fuzzy convex that is, $\delta(\lambda u + (1-\lambda)v) \geq \min\{\delta(u), \delta(v)\}$, $u, v \in R$, $\lambda \in [0,1]$,

(iv) $[\delta]^0$ is compact.

The LR–fuzzy numbers first introduced in [29] are defined as follows:

Definition 3. Let $L, R : [0,1] \to [0,1]$ be two upper semicontinuous and decreasing functions and $L(0) = R(0) = 1$, and $L(1) = R(1) = 0$. Then the fuzzy number δ is given by

$$\delta(u) = \begin{cases} L\left(\dfrac{a-u}{\sigma}\right) & a-\sigma \leq u < a, \\ 1 & a \leq u < b, \\ R\left(\dfrac{u-b}{\beta}\right) & b \leq u < b+\beta, \\ 0 & \text{otherwise}, \end{cases} \quad (3)$$

where $\sigma, \beta > 0$ and $a \leq b$.

Let E be the set of fuzzy numbers and let $\delta \in E$ be a fuzzy number if and only if $[\delta]^{\alpha}$ is a nonempty compact convex subset of R^1. This is denoted by $[\delta_*(\alpha), \delta^*(\alpha)]$ for each $\alpha \in [0,1]$. A fuzzy number is resolved by the end points of the interval $[\delta_*(\alpha), \delta^*(\alpha)]$.

A real number $u \in R^1$ is a particular case of fuzzy number and is specified by

$$\tilde{u}(t) = \begin{cases} 1 & \text{if } t = u, \\ 0 & t \neq u. \end{cases} \quad (4)$$

The parametric form of a fuzzy number δ is presented as

$$\{(\delta_*(\alpha), \delta^*(\alpha), \alpha) : \alpha \in [0,1]\}. \quad (5)$$

The following is the characterization of a fuzzy number in terms of the end point $\delta_*(\alpha)$ and $\delta^*(\alpha)$.

Lemma 4 (see [14]). *Assume that $\delta_* : [0,1] \to R$ and $\delta^* : [0,1] \to R$ satisfy the following conditions:*

(i) $\delta_* : [0,1] \to R$ is bounded increasing function,

(ii) $\delta^* : [0,1] \to R$ is bounded decreasing function,

(iii) $\delta_*(1) \leq \delta^*(1)$,

(iv) For $0 < k \leq 1$, $\lim_{\alpha \to k-} \delta_*(\alpha) = \delta_*(k)$ and $\lim_{\alpha \to k-} \delta^*(\alpha) = \delta^*(k)$,

(v) $\lim_{\alpha \to 0+} \delta_*(\alpha) = \delta_*(0)$ and $\lim_{\alpha \to 0+} \delta^*(\alpha) = \delta^*(0)$.

For $\delta, \gamma \in E$, $\lambda \in R$, the fuzzy addition and scalar multiplication for $u \in R$ are defined as

$$(\delta \tilde{+} \gamma)(u) = \sup_{v+w=u} \min[\delta(v), \gamma(w)],$$
$$(\lambda\delta)(u) = \begin{cases} \delta(\lambda^{-1}u), & \lambda \neq 0, \\ 0, & \lambda = 0. \end{cases} \quad (6)$$

We know that for any $\delta, \gamma \in E$, $\lambda\delta \in E$, $[\delta \tilde{+} \gamma]^{\alpha} = [\delta]^{\alpha} + [\gamma]^{\alpha}$ and $[\lambda\delta]^{\alpha} = \lambda[\delta]^{\alpha}$, that is, for each $\alpha \in [0,1]$,

$$(\delta \tilde{+} \gamma)_*(\alpha) = \delta_*(\alpha) + \gamma_*(\alpha),$$
$$(\delta \tilde{+} \gamma)^*(\alpha) = \delta^*(\alpha) + \gamma^*(\alpha),$$
$$(\lambda\delta)(\alpha) = \begin{cases} \lambda\delta_*(\alpha), & \lambda \geq 0, \\ \lambda\delta^*(\alpha), & \lambda < 0. \end{cases} \quad (7)$$
$$(\lambda\delta)(\alpha) = \begin{cases} \lambda\delta^*(\alpha), & \lambda \geq 0, \\ \lambda\delta_*(\alpha), & \lambda < 0. \end{cases}$$

For any $\delta, \gamma \in E$, $\delta \preceq \gamma$ if for each $\alpha \in [0,1]$, $\delta_*(\alpha) \leq \gamma_*(\alpha)$ and $\delta^*(\alpha) \leq \gamma^*(\alpha)$. If $\delta \preceq \gamma$ and $\gamma \preceq \delta$ then $\delta = \gamma$. We say that $\delta \prec \gamma$, and there exists some $\alpha_0 \in [0,1]$, such that $\delta_*(\alpha_0) < \gamma_*(\alpha_0)$ or $\delta^*(\alpha_0) < \gamma^*(\alpha_0)$. For $\delta, \gamma \in E$, if either $\gamma \preceq \delta$ or $\delta \preceq \gamma$, then δ and γ are comparable, or else they are not comparable, where $\preceq$ is a partial order relation on E.

Definition 5 (triangular fuzzy number). A fuzzy number is called a triangular fuzzy number if $\delta_*(1) = \delta^*(1)$. In addition a fuzzy number $\delta \in E$ is said to be linear, if $\delta_*(\alpha)$ and $\delta^*(\alpha)$ are linear. A fuzzy triangular number is denoted by $\langle \delta_*(0), \delta_*(1), \delta^*(0) \rangle$. For the triangular fuzzy number $\delta = \langle 0, 1, 3 \rangle$, we have $[\delta]^{\alpha} = [\alpha, 3 - 2\alpha]$.

Definition 6 (fuzzy mapping). Let $U \in R^n$ and let $\tilde{F} : U \to E$ be a fuzzy mapping. Then the $\alpha-$ cut of $\tilde{F}(u)$ is given by $\tilde{F}(u)[\alpha] = [F_*(u,\alpha), F^*(u,\alpha)]$, where $F_*(u,\alpha) = \min\{\tilde{F}(u)[\alpha]\}$ and $F^*(u,\alpha) = \max\{\tilde{F}(u)[\alpha]\}$. Thus $\tilde{F}(u)$ is represented by the two functions $F_*(u,\alpha)$ and $F^*(u,\alpha)$; these functions are defined from $U \times [0,1]$ to R. Moreover, $F_*(u,\alpha)$ is bounded increasing function of α and $F^*(u,\alpha)$ is bounded decreasing function of α and $F_*(u,\alpha) \leq F^*(u,\alpha)$ for each $\alpha \in [0,1]$.

Definition 7 (continuity of a fuzzy mapping). If $\tilde{F} : U \subseteq R^n \to E$ is a fuzzy mapping, then $\tilde{F}(u)$ is said to be continuous at $u \in U$, if for each $\alpha \in [0,1]$ both the end point functions $F_*(u,\alpha)$ and $F^*(u,\alpha)$ are continuous at u.

Differentiability of the functions is one of the important tool of the generalized convexity and invex monotonocities. In this paper our aim is to study the concept of fuzzy invex monotone and fuzzy invex function, so we will discuss the notion of differentiability for fuzzy mappings. Puri and Ralescu [30] presented the concept of H-differentiability for fuzzy mappings. Further the concept of S-differentiability, G-differentiability, and weak differentiability was given by Seikkala [31] and Bede and Gal [32], respectively. Rufián-Lizana et al. [33] pointed that G-differentiability is more general than H-differentiability and S-differentiability for fuzzy mappings. In [34], Bede and Stefanini showed that a weak differentiability of fuzzy mappings is different from G-differentiability. Therefore, it will be worthwhile to study the generalized invex monotonocities and invex functions by using weakly differentiable fuzzy mappings.

Definition 8 (weakly differentiable function [26]). Let $\widetilde{F} : U \subseteq R^n \to E$ be a fuzzy mapping. If the derivatives of $F_*(u,\alpha)$, $F^*(u,\alpha)$ with respect to $u \in U$ for each $\alpha \in [0,1]$ exist and are denoted by $F_*'(u,\alpha)$, $F^{*'}(u,\alpha)$, respectively, then $F(u)$ is said to be weakly differentiable.

3. Generalized Convexity and Invex Monotonocities

In this section, we collected some basic definitions of generalized invex functions and generalized invex monotonocities.

Definition 9. A nonempty set $S \subseteq R^n$ is said to be invex if there exists a vector valued function $\eta : R^n \times R^n \to R^n$, such that $v + \lambda\eta(u,v) \in S$ for any $u, v \in S$, $\lambda \in [0,1]$.

Hanson [1] introduced the generalized version of convex function, namely, invex function. A function $f : R^n \to R$ is said to be invex if there exists a vector valued function $\eta : R^n \times R^n \to R$ such that the next inequality

$$f(v + \lambda\eta(u,v)) \le \lambda f(u) + (1-\lambda) f(v) \tag{8}$$

holds, for all $u, v \in R^n$.

Definition 10 (pseudoinvex monotone [2]). Let $S \subseteq R^n$ be an invex set with respect to η. Then the function $\phi : S \to R^n$ is said to be (strictly) pseudoinvex monotone on S if

$$\begin{gathered}\eta(u,v)^T \phi(u) \ge 0 \Longrightarrow \\ \eta(u,v)^T \phi(v)\,(>) \ge 0,\end{gathered} \tag{9}$$

for all $u, v \in S$.

Definition 11 (pseudoinvex function [2]). A differentiable function $\phi : S \subseteq R^n \to R^n$ is said to be (strictly) pseudoinvex function with respect to $\eta : S \times S \to R^n$ if

$$\begin{gathered}\eta(u,v)^T \nabla\phi(u) \ge 0 \Longrightarrow \\ \phi(v)\,(>) \ge \phi(u),\end{gathered} \tag{10}$$

for all $u, v \in S$.

Definition 12 (pseudoinvex monotone [2]). Let $S \subseteq R^n$ be an invex set with respect to η. Then the function $\phi : S \to R^n$ is said to be (strictly) pseudoinvex monotone on S if

$$\begin{gathered}\eta(u,v)^T \phi(u) \ge 0 \Longrightarrow \\ \eta(u,v)^T \phi(v)\,(>) \ge 0,\end{gathered} \tag{11}$$

for all $u, v \in S$.

Definition 13 (quasi-invex monotone [2]). A function $\phi : S \subseteq R^n \to R^n$, defined on an open invex subset S of R^n, is said to be quasi-invex monotone with respect to η on S if

$$\begin{gathered}\eta(u,v)^T \phi(u) > 0 \Longrightarrow \\ \eta(u,v)^T \phi(v) \ge 0,\end{gathered} \tag{12}$$

for all $u, v \in S$.

Definition 14 (quasi-invex function [2]). A function $\phi : S \subseteq R^n \to R^n$ which is differentiable on an open invex subset S of R^n is said to be quasi-invex function with respect to η on S if

$$\begin{gathered}\phi(v) \ge \phi(u) \Longrightarrow \\ \eta(u,v)^T \nabla\phi(v) \le 0,\end{gathered} \tag{13}$$

for all $u, v \in S$.

Now we have the following definitions of strong pseudoinvex monotonicity and strong pseudoinvex functions.

Definition 15 (strong pseudoinvex monotone [3]). A function $\phi : S \subseteq R^n \to R^n$ defined on an open invex subset S of R^n is said to be strong pseudoinvex monotone with respect to η on S, if there exists a scalar $m > 0$, such that

$$\begin{gathered}\eta(u,v)^T \phi(u) \ge 0 \Longrightarrow \\ \eta(u,v)^T \phi(v) \ge m\,\|\eta(u,v)\|,\end{gathered} \tag{14}$$

for all $u, v \in S$.

Definition 16 (strong pseudoinvex function [3]). A function $\phi : S \subseteq R^n \to R^n$ which is differentiable on an open invex subset S of R^n is said to be strong pseudoinvex function with respect to η on S if there exists a scalar $n > 0$ such that

$$\begin{gathered}\eta(u,v)^T \nabla\phi(u) \ge 0 \Longrightarrow \\ \phi(v) \ge \phi(u) + n\,\|\eta(u,v)\|,\end{gathered} \tag{15}$$

for all $u, v \in S$.

4. Weakly Fuzzy Pseudoinvex Monotonicity

In this section, we propose the new concept of weakly fuzzy pseudoinvex monotone and weakly fuzzy pseudoinvex function and we will prove a necessary condition for weakly fuzzy pseudoinvex monotonicity.

Definition 17. A fuzzy mapping $\tilde{\phi} : U \to E$ is said to be (Strictly) weakly fuzzy pseudoinvex monotone with respect to η on $U \in R^n$ if

$$\eta(u,v)^T \phi_*(u,\alpha) \geq 0 \Longrightarrow \eta(u,v)^T \phi_*(\alpha,v)(>) \geq 0, \quad \eta(u,v)^T \phi^*(u,\alpha) \geq 0 \Longrightarrow \eta(u,v)^T \phi^*(\alpha,v)(>) \geq 0, \tag{16}$$

for any $u, v \in U$.

In support of above definition, we have the following example.

Example 18. The triangular fuzzy valued function $\tilde{\phi} = \langle 2,1,3\rangle u$ is a weakly fuzzy pseudoinvex monotone with respect to $\eta(u,v) = u - v$, where $u \geq v$ and $u, v \geq 0$.

The given function can be written as $\tilde{\phi} = [(2-\alpha)u, (3-2\alpha)u]$, where $\phi_*(u,\alpha) = (2-\alpha)u$ and $\phi^*(u,\alpha) = (3-2\alpha)u$ and $\eta(v,u) = u - v$, where $u \geq v$ and $u, v \geq 0$. Then it is easy to see that $\tilde{\phi}(u)$ is a weakly fuzzy monotone.

Definition 19. A fuzzy mapping $\tilde{\psi} : U \to E$, which is weakly differentiable on $U \in R^n$ is said to be (strictly) weakly fuzzy pseudoinvex function with respect to η on U if

$$\eta(u,v)^T \nabla\psi_*(u,\alpha) \geq 0 \Longrightarrow \psi_*(v,\alpha)(>) \geq \psi_*(u,\alpha), \quad \eta(u,v)^T \nabla\psi^*(u,\alpha) \geq 0 \Longrightarrow \psi^*(v,\alpha)(>) \geq \psi^*(u,\alpha) \tag{17}$$

for any $u, v \in U$.

Moreover, we have the following example for weakly fuzzy pseudoinvex function.

Example 20. The triangular fuzzy valued function $\tilde{\psi} = \langle 2,1,3\rangle e^u$ is a weakly fuzzy pseudoinvex function with respect to $\eta(v,u) = v - u$ and $u \leq v$.

The given function can be written as $\tilde{\psi} = [(2-\alpha)e^u, (3-2\alpha)e^u]$; that means $\psi_*(u,\alpha) = (2-\alpha)e^u$, $\psi^*(u,\alpha) = (3-2\alpha)e^u$, and then by some simple computation we can see the following:

$$\eta(u,v)^T \nabla\psi_*(u,\alpha) \geq 0 \Longrightarrow \psi_*(v,\alpha) \geq \psi_*(u,\alpha), \quad \eta(u,v)^T \nabla\psi^*(u,\alpha) \geq 0 \Longrightarrow \psi^*(v,\alpha) \geq \psi^*(u,\alpha). \tag{18}$$

Hence $\tilde{\psi} = \langle 2,1,3\rangle e^u$ is a weakly fuzzy pseudoinvex function.

Now we will define the Condition C as follows.

Condition C. For vector valued function $\eta : U \times U \to R^n$ the following equations are called Condition C:

$$\eta(v, v + \lambda\eta(u,v)) = -\lambda\eta(u,v), \quad \eta(u, v + \lambda\eta(u,v)) = (1-\lambda)\eta(u,v), \tag{19}$$

for any $u, v \in U$ and $\lambda \in [0,1]$. Moreover, we have the following result for subsequent use.

Note 21 (see [2]). From Condition C, we have

$$\eta\left(v + \bar{\lambda}\eta(u,v), v\right) = \bar{\lambda}\eta(u,v). \tag{20}$$

Theorem 22. *Let $\tilde{\phi} : U \subseteq R^n \to E$ be a fuzzy mapping on an open invex set U with respect to η and η satisfies Condition C. Let $\tilde{\phi}$ be weakly differentiable on U with the following assumptions:*

(i) *$\phi_*(v,\alpha) > \phi_*(u,\alpha) \Rightarrow \eta(u,v)^T \nabla\phi_*(v + \bar{\lambda}\eta(u,v),\alpha) < 0$ and $\phi^*(v,\alpha) > \phi^*(u,\alpha) \Rightarrow \eta(u,v)^T \nabla\phi^*(v + \bar{\lambda}\eta(u,v),\alpha) < 0$, for some $\bar{\lambda} \in (0,1)$.*

(ii) *$\nabla\tilde{\phi}$ is weakly fuzzy pseudoinvex monotone with respect to η on U.*

Then $\tilde{\phi}$ is a weakly fuzzy pseudoinvex function with respect to η on U.

Proof. Suppose that

$$\eta(u,v)^T \nabla\phi_*(v,\alpha) \geq 0, \quad \eta(u,v)^T \nabla\phi^*(v,\alpha) \geq 0. \tag{21}$$

Our aim is to show that

$$\phi_*(u,\alpha) \geq \phi_*(v,\alpha), \quad \phi^*(u,\alpha) \geq \phi^*(v,\alpha). \tag{22}$$

Assume to the contrary that

$$\phi_*(u,\alpha) < \phi_*(v,\alpha), \quad \phi^*(u,\alpha) < \phi^*(v,\alpha). \tag{23}$$

By the assumption (i)

$$\eta(u,v)^T \nabla\phi_*\left(v + \bar{\lambda}\eta(u,v),\alpha\right) < 0, \tag{24}$$

$$\eta(u,v)^T \nabla\phi^*\left(v + \bar{\lambda}\eta(u,v),\alpha\right) < 0, \tag{25}$$

for some $\bar{\lambda} \in (0,1)$.

By Note 21 and (25), we have the following inequalities:

$$\eta\left(v + \bar{\lambda}\eta(u,v), v\right)^T \nabla\phi_*\left(v + \bar{\lambda}\eta(u,v),\alpha\right) < 0, \quad \eta\left(v + \bar{\lambda}\eta(u,v), v\right)^T \nabla\phi^*\left(v + \bar{\lambda}\eta(u,v),\alpha\right) < 0. \tag{26}$$

By the assumption that $\nabla\widetilde{\phi}$ is weakly fuzzy pseudo monotone with respect to η, then it follows from (26) that

$$\eta\left(v+\overline{\lambda}\eta(u,v),v\right)^T\nabla\phi_*(v,\alpha)<0,$$
$$\eta\left(v+\overline{\lambda}\eta(u,v),v\right)^T\nabla\phi^*(v,\alpha)<0, \tag{27}$$

where $\overline{\lambda}\in(0,1)$.

Since $\overline{\lambda}\in(0,1)$, then by Note 21, (27) become

$$\eta(u,v)^T\nabla\phi_*(v,\alpha)<0,$$
$$\eta(u,v)^T\nabla\phi^*(v,\alpha)<0. \tag{28}$$

The above inequalities contradict (21). Hence $\widetilde{\phi}$ is a weakly fuzzy pseudoinvex function with respect to η.

Theorem 23. *Let $\widetilde{\phi}:U\subseteq R^n\to E$ be a fuzzy mapping on an open invex set U with respect to η and η satisfies the Condition C. Let $\widetilde{\phi}$ be weakly differentiable on U with the following assumptions:*

(i) $\phi_*(v,\alpha)\geq\phi_*(u,\alpha)\Rightarrow\eta(u,v)^T\nabla\phi_*(v+\overline{\lambda}\eta(u,v),\alpha)\leq 0$ *and* $\phi^*(v,\alpha)\geq\phi^*(u,\alpha)\Rightarrow\eta(u,v)^T\nabla\phi^*(v+\overline{\lambda}\eta(u,v),\alpha)\leq 0$, *for each $u,v\in U$ and $\overline{\lambda}\in(0,1)$,*

(ii) *$\nabla\widetilde{\phi}$ is strictly weakly fuzzy pseudoinvex monotone with respect to η on U.*

Then $\widetilde{\phi}$ is a strictly weakly fuzzy pseudoinvex function with respect to η on U.

Proof. Suppose that

$$\eta(u,v)^T\nabla\phi_*(v,\alpha)\geq 0,$$
$$\eta(u,v)^T\nabla\phi^*(v,\alpha). \tag{29}$$

Our aim is to show that

$$\phi_*(u,\alpha)>\phi_*(v,\alpha),$$
$$\phi^*(u,\alpha)>\phi^*(v,\alpha). \tag{30}$$

Assume to the contrary that

$$\phi_*(u,\alpha)\leq\phi_*(v,\alpha),$$
$$\phi^*(u,\alpha)\leq\phi^*(v,\alpha). \tag{31}$$

By the assumption (i), we have

$$\eta(u,v)^T\nabla\phi_*\left(v+\overline{\lambda}\eta(u,v),\alpha\right)\leq 0,$$
$$\eta(u,v)^T\nabla\phi^*\left(v+\overline{\lambda}\eta(u,v),\alpha\right)\leq 0, \tag{32}$$

for some $\overline{\lambda}\in(0,1)$.

By Condition C, we have

$$\eta\left(v,v+\overline{\lambda}\eta(u,v)\right)=-\overline{\lambda}\eta(u,v). \tag{33}$$

Utilizing (33) in (32), we have

$$\eta\left(v,v+\overline{\lambda}\eta(u,v)\right)^T\nabla\phi_*\left(v+\overline{\lambda}\eta(u,v),\alpha\right)\geq 0,$$
$$\eta\left(v,v+\overline{\lambda}\eta(u,v)\right)^T\nabla\phi^*\left(v+\overline{\lambda}\eta(u,v),\alpha\right)\geq 0, \tag{34}$$

for some $\overline{\lambda}\in(0,1)$.

By the strictly weakly fuzzy pseudo monotonicity of $\nabla\widetilde{\phi}$ with respect to η and (34)

$$\eta\left(v,v+\overline{\lambda}\eta(u,v)\right)^T\nabla\phi_*(v,\alpha)>0,$$
$$\eta\left(v,v+\overline{\lambda}\eta(u,v)\right)^T\nabla\phi^*(v,\alpha)>0. \tag{35}$$

Applying the Condition C and the fact that $\overline{\lambda}\in(0,1)$, then (35) takes the form

$$\eta(u,v)^T\nabla\phi_*(v,\alpha)<0,$$
$$\eta(u,v)^T\nabla\phi^*(v,\alpha)<0. \tag{36}$$

The above inequalities contradict (29). Hence, $\widetilde{\phi}$ is a strictly weakly fuzzy pseudoinvex function with respect to η.

5. Weakly Fuzzy Quasi-invex Monotonicity

In this section, necessary conditions for weakly fuzzy quasi-invex monotone are discussed.

Definition 24. A fuzzy mapping $\widetilde{\phi}:U\subseteq R^n\to E$ is said to be weakly fuzzy quasi-invex monotone with respect to η on $U\subseteq R^n$ if

$$\eta(u,v)^T\phi_*(u,\alpha)>0\Longrightarrow$$
$$\eta(u,v)^T\phi_*(v,\alpha)\geq 0,$$
$$\eta(u,v)^T\phi^*(u,\alpha)>0\Longrightarrow$$
$$\eta(u,v)^T\phi^*(v,\alpha)\geq 0, \tag{37}$$

for any $u,v\in U$.

Example 25. The triangular fuzzy valued function $\widetilde{\phi}(u)=\langle 0,1,2\rangle u^2$ is a weakly fuzzy quasi-invex monotone with respect to $\eta(u,v)=u-v$, where $u>v$ and $u>0$, $v\geq 0$.

$\widetilde{\phi}(u)=\langle 0,1,2\rangle u^2=[\alpha u^2,(2-\alpha)u^2]$; we have

$$\phi^*(u,\alpha)=(2-\alpha)u^2,$$
$$\phi_*(u,\alpha)=\alpha u^2. \tag{38}$$

Now we compute the following

$$\eta(u,v)\phi^*(u,\alpha)=(u-v)(2-\alpha)u^2>0,$$
$$\eta(u,v)\phi^*(v,\alpha)=(u-v)(2-\alpha)v^2\geq 0, \tag{39}$$

for all $u>v$ and $u>0$, $v\geq 0$.

Similarly, we can compute $\eta(u,v)\phi_*(u,\alpha)>0$ and $\eta(u,v)\phi_*(v,\alpha)\geq 0$. It is easy to see that $\widetilde{\phi}(u)$ is not weakly fuzzy pseudoinvex monotone function.

Definition 26. A weakly differentiable fuzzy mapping $\widetilde{\psi} : U \subseteq R^n \rightarrow E$ is said to be weakly fuzzy quasi-invex function with respect to η on $U \subseteq R^n$ if

$$\begin{aligned} &\psi_*(v,\alpha) \geq \psi_*(u,\alpha) \Longrightarrow \\ &\eta(u,v)^T \nabla\psi_*(v,\alpha) \leq 0, \\ &\psi^*(v,\alpha) \geq \psi^*(u,\alpha) \Longrightarrow \\ &\eta(u,v)^T \nabla\psi^*(v,\alpha) \leq 0, \end{aligned} \tag{40}$$

for any $u, v \in U$.

Example 27. The fuzzy valued function $\widetilde{\psi}(u) = \langle 0,1,2\rangle e^u$ is a weakly fuzzy quasi-invex function with respect to $\eta(u,v) = u - v$ and $u \leq v$.

The given function can be written as $\widetilde{\psi}(u) = \langle 0,1,2\rangle e^u = [\alpha e^u, (2-\alpha)e^u]$; we have

$$\begin{aligned} \psi^*(u,\alpha) &= (2-\alpha)e^u, \\ \psi_*(u,\alpha) &= \alpha e^u, \\ \nabla\psi^*(u,\alpha) &= (2-\alpha)e^u, \\ \nabla\psi_*(u,\alpha) &= \alpha e^u. \end{aligned} \tag{41}$$

Now, it is easy to show the following

$$\begin{aligned} &\psi^*(u,\alpha) \geq \psi^*(v,\alpha) \Longrightarrow \\ &\eta(u,v)\nabla\psi^*(v,\alpha) \leq 0, \\ &\psi_*(u,\alpha) \geq \psi_*(v,\alpha) \Longrightarrow \\ &\eta(u,v)\nabla\psi_*(v,\alpha) \leq 0 \end{aligned} \tag{42}$$

for all $u \leq v$. Hence, the function $\widetilde{\psi}(u)$ is a weakly fuzzy quasi-invex function, but not weakly pseudoinvex function.

Theorem 28. *Let $\widetilde{\phi} : U \subseteq R^n \rightarrow E$ be a fuzzy mapping on an open invex set U with respect to η and η satisfying Condition C. Let $\widetilde{\phi}$ be weakly differentiable on U with the following conditions*

(i) $\phi_*(v,\alpha) \geq \phi_*(u,\alpha) \Rightarrow \eta(u,v)^T\nabla\phi_*(v+\overline{\lambda}\eta(u,v),\alpha) < 0$ *and* $\phi^*(v,\alpha) \geq \phi^*(u,\alpha) \Rightarrow \eta(u,v)^T\nabla\phi^*(v+\overline{\lambda}\eta(u,v),\alpha) < 0$, *for some* $\overline{\lambda} \in (0,1)$.

(ii) $\nabla\widetilde{\phi}$ *is weakly fuzzy quasi-invex monotone with respect to* η *on* U.

Then $\widetilde{\phi}$ is a weakly fuzzy quasi-invex function with respect to η on U.

Proof. Suppose that $\widetilde{\phi}$ is not weakly fuzzy quasi-invex function; then

$$\begin{aligned} \phi_*(v,\alpha) &\geq \phi_*(u,\alpha), \\ \phi^*(v,\alpha) &\geq \phi_*(u,\alpha). \end{aligned} \tag{43}$$

But

$$\begin{aligned} \eta(u,v)^T \nabla\phi_*(v,\alpha) &> 0, \\ \eta(u,v)^T \nabla\phi^*(v,\alpha) &> 0. \end{aligned} \tag{44}$$

By the assumption (i) and (43)

$$\begin{aligned} \eta(u,v)^T \nabla\phi_*\left(v+\overline{\lambda}\eta(u,v),\alpha\right) &< 0, \\ \eta(u,v)^T \nabla\phi^*\left(v+\overline{\lambda}\eta(u,v),\alpha\right) &< 0, \end{aligned} \tag{45}$$

for some $\overline{\lambda} \in (0,1)$.

By Condition C, we have

$$\eta\left(v, v+\overline{\lambda}\eta(u,v)\right) = -\overline{\lambda}\eta(u,v). \tag{46}$$

By utilizing (46) in (45), it is easy to see the following

$$\begin{aligned} \eta\left(v, v+\overline{\lambda}\eta(u,v)\right)^T \nabla\phi_*\left(v+\overline{\lambda}\eta(u,v),\alpha\right) &> 0, \\ \eta\left(v, v+\overline{\lambda}\eta(u,v)\right)^T \nabla\phi^*\left(v+\overline{\lambda}\eta(u,v),\alpha\right) &> 0, \end{aligned} \tag{47}$$

for some $\overline{\lambda} \in (0,1)$.

Since $\nabla\widetilde{\phi}$ is weakly fuzzy quasi-invex monotone, then from (47) we have

$$\begin{aligned} \eta\left(v, v+\overline{\lambda}\eta(u,v)\right)^T \nabla\phi_*(v,\alpha) &\geq 0, \\ \eta\left(v, v+\overline{\lambda}\eta(u,v)\right)^T \nabla\phi^*(v,\alpha) &\geq 0. \end{aligned} \tag{48}$$

Again applying the Condition C and the fact $\overline{\lambda} \in (0,1)$, the above equation reduced to

$$\begin{aligned} \eta(u,v)^T \nabla\phi_*(v,\alpha) &\leq 0, \\ \eta(u,v)^T \nabla\phi^*(v,\alpha) &\leq 0, \end{aligned} \tag{49}$$

which contradicts (44). Hence $\widetilde{\phi}$ is a weakly fuzzy quasi-invex function with respect to η.

6. Fuzzy Strong Pseudo Invex Monotonicity

In this section, we propose the idea of fuzzy strong pseudoinvex monotonicity and fuzzy strong pseudo invex functions. Finally, we prove a necessary condition for fuzzy strong pseudoinvex monotone.

Definition 29. A fuzzy mapping $\widetilde{\phi} : U \subseteq R^n \rightarrow E$ is said to be fuzzy strong pseudoinvex monotone with respect to η on an invex set $U \subseteq R^n$ if there exists a scalar m such that

$$\begin{aligned} &\eta(v,u)^T \phi_*(u,\alpha) \geq 0 \Longrightarrow \\ &\eta(u,v)^T \phi^*(v,\alpha) \geq m\left\|\eta(v,u)\right\|, \\ &\eta(v,u)^T \phi^*(u,\alpha) \geq 0 \Longrightarrow \\ &\eta(v,u)^T \phi^*(v,\alpha) \geq m\left\|\eta(v,u)\right\|, \end{aligned} \tag{50}$$

for any $u, v \in U$.

Now we have the following example.

Example 30. The triangular fuzzy valued function $\widetilde{\phi} = \langle 0,1,4\rangle u^2$ is a fuzzy strong pseudoinvex monotone with respect to $\eta(v,u) = u - v$ and $m \le 0$, where $u \ge v$.

For $\widetilde{\phi} = \langle 0,1,4\rangle u^2 = [\alpha u^2, (4-3\alpha)u^2]$, then we have $\phi_*(u,\alpha) = \alpha u^2$ and $\phi^*(u,\alpha) = (4-3\alpha)u^2$. Now we compute the following

$$\eta(v,u)^T \phi_*(u,\alpha) = (u-v)\alpha u^2 \ge 0,$$
$$\eta(v,u)^T \phi_*(v,\alpha) = (u-v)\alpha v^2 \ge m\|u-v\|, \quad (51)$$

for all $m \le 0$.

Similarly, for $\phi^*(u,\alpha) = (4-3\alpha)u^2$, we can show the following

$$\eta(v,u)^T \phi^*(u,\alpha) \ge 0 \Longrightarrow$$
$$\eta(v,u)^T \phi^*(v,\alpha) \ge m\|\eta(v,u)\|, \quad (52)$$

for all $u \ge v$ and $m \le 0$, but $\widetilde{\phi}(u)$ is neither weakly fuzzy pseudoinvex monotone function nor weakly fuzzy quasi-invex monotone function.

Definition 31. A weakly differentiable fuzzy mapping $\widetilde{\phi}: U \subseteq R^n \to E$ is said to be fuzzy strong pseudoinvex function with respect to η on an invex set $U \subseteq R^n$ if there exists a scalar η such that

$$\eta(v,u)^T \nabla\phi_*(u,\alpha) \ge 0 \Longrightarrow$$
$$\phi_*(v,\alpha) \ge \phi_*(u,\alpha) + n\|\eta(v,u)\|,$$
$$\eta(v,u)^T \nabla\phi^*(u,\alpha) \ge 0 \Longrightarrow \quad (53)$$
$$\phi^*(v,\alpha) \ge \phi^*(u,\alpha) + n\|\eta(v,u)\|,$$

for any $u, v \in U$.

Example 32. The triangular fuzzy valued function $\widetilde{\psi}(u) = \langle 0,1,4\rangle u$ is a fuzzy strong pseudoinvex function with respect to $\eta(v,u) = v - u$ and $n \le 0$, where $u \le v$.

We can write $\widetilde{\psi}(u) = \langle 0,1,4\rangle u = [\alpha u, (4-3\alpha)u]$, and then we have $\psi_*(u,\alpha) = \alpha u$ and $\psi^*(u,\alpha) = (4-3\alpha)u$. $\eta(v,u) = v - u$ $\nabla\psi_*(u,\alpha) = \alpha$. Then we have

$$\eta(v,u)\nabla\psi_*(u,\alpha) = (v-u)\alpha \ge 0,$$
$$\alpha v \ge \alpha u + n\|v-u\|, \quad (54)$$

for all $u \le v$ and $n \le 0$; that means

$$\psi_*(v,\alpha) \ge \psi_*(u,\alpha) + n\|\eta(v,u)\|. \quad (55)$$

Similarly, it is easy to show that

$$\eta(v,u)\nabla\psi^*(u,\alpha) \ge 0 \Longrightarrow$$
$$\phi^*(v,\alpha) \ge \phi^*(u,\alpha) + n\|\eta(v,u)\|. \quad (56)$$

Hence the function $\widetilde{\psi}(u) = \langle 0,1,4\rangle u$ is a fuzzy strong pseudoinvex function with respect to $\eta(v,u) = v-u$ and $n \le 0$, where $u \le v$. It is easy see that $\widetilde{\phi}(u)$ is neither weakly fuzzy pseudoinvex function nor weakly fuzzy quasi-invex function.

Theorem 33. *Let $\widetilde{\phi}: U \subseteq R^n \to E$ be a fuzzy mapping on an open invex set with respect to η and η satisfying the Condition C. Let $\widetilde{\phi}$ is weakly differentiable on U with the following conditions*

(i) *$\phi_*(u+\eta(v,u),\alpha) \le \phi_*(v,\alpha)$ and $\phi^*(u+\eta(v,u),\alpha) \le \phi^*(v,\alpha)$, for all $u, v \in U$.*

(ii) *$\nabla\widetilde{\phi}$ is a fuzzy strong pseudoinvex monotone with respect to η on U.*

Then $\widetilde{\phi}$ is a fuzzy pseudoinvex function with respect to η on U.

Proof. Suppose

$$\eta(v,u)^T \nabla\phi_*(u,\alpha) \ge 0,$$
$$\eta(v,u)^T \nabla\phi^*(u,\alpha) \ge 0, \quad (57)$$

for any $u, v \in U$.

From the assumption that U is open invex set with respect to η and η satisfying the Condition C, the above inequalities become

$$\eta(u+\lambda\eta(v,u),u)^T \nabla\phi_*(u,\alpha) \ge 0,$$
$$\eta(u+\lambda\eta(v,u),u)^T \nabla\phi^*(u,\alpha) \ge 0. \quad (58)$$

Since $\nabla\widetilde{\phi}$ is a fuzzy strong pseudoinvex monotone with respect to η, then there exists $m > 0$ such that

$$\eta(u+\lambda\eta(v,u),u)^T \nabla\phi_*(u+\lambda\eta(v,u),\alpha)$$
$$\ge m\|\eta(u+\lambda\eta(v,u),u)\|,$$
$$\eta(u+\lambda\eta(v,u),u)^T \nabla\phi^*(u+\lambda\eta(v,u),\alpha) \quad (59)$$
$$\ge m\|\eta(u+\lambda\eta(v,u),u)\|.$$

Again by Condition C and $\lambda \in (0,1]$ the above inequalities take the form

$$\eta(v,u)^T \nabla\phi_*(u+\lambda\eta(v,u),\alpha) \ge m\|\eta(v,u)\|, \quad (60)$$

$$\eta(v,u)^T \nabla\phi^*(u+\lambda\eta(v,u),\alpha) \ge m\|\eta(v,u)\|, \quad (61)$$

for all $\lambda \in (0,1]$.

Let $\psi_*(\lambda) = \phi_*(u+\lambda\eta(v,u),\alpha)$ and $\psi^*(\lambda) = \phi^*(u+\lambda\eta(v,u),\alpha)$.

Then from (60) and (61) we have,

$$\psi_*'(\lambda) \ge m\|\eta(v,u)\|,$$
$$\psi^{*\prime}(\lambda) \ge m\|\eta(v,u)\|. \quad (62)$$

Integrating from 0 to 1

$$\int_0^1 \psi_*'(\lambda)\,d\lambda \ge m\|\eta(v,u)\| \int_0^1 1\,d\lambda,$$
$$\int_0^1 \psi^{*\prime}(\lambda)\,d\lambda \ge m\|\eta(v,u)\| \int_0^1 1\,d\lambda. \quad (63)$$

After integration, we have the following

$$\begin{aligned} \psi_* (1) - \psi_* (0) &\geq m \left\| \eta (v, u) \right\|, \\ \psi^* (1) - \psi^* (0) &\geq m \left\| \eta (v, u) \right\|, \end{aligned} \tag{64}$$

or

$$\begin{aligned} \phi_* \left(u + \eta (v, u), \alpha \right) - \phi_* (u, \alpha) &\geq m \left\| \eta (v, u) \right\|, \\ \phi^* \left(u + \eta (v, u), \alpha \right) - \phi^* (u, \alpha) &\geq m \left\| \eta (v, u) \right\|. \end{aligned} \tag{65}$$

By the assumption (i), we acquire

$$\begin{aligned} \phi_* (v, \alpha) - \phi_* (u, \alpha) &\geq m \left\| \eta (v, u) \right\|, \\ \phi^* (v, \alpha) - \phi^* (u, \alpha) &\geq m \left\| \eta (v, u) \right\|. \end{aligned} \tag{66}$$

Hence, $\widetilde{\phi}$ is a fuzzy strong pseudo invex function.

7. Conclusion

The concept of fuzzy optimization is well recognized in the literature and many authors are showing their interest in this direction. The idea of fuzzy convexity has been studied by several authors. The motive of the present paper is to project the concept of generalized weakly fuzzy monotonocities and generalized weakly fuzzy invex functions with nontrivial examples. Moreover, we tried to build up the relationship between generalized fuzzy monotones and generalized fuzzy invex functions. The results proved in the present paper generalize the existing results appearing in the literature. Findings of this paper can be used for fuzzy multiobjective programming problems.

Conflicts of Interest

The authors declare that they have no conflicts of interest.

Authors' Contributions

All the authors contributed equally to the writing of this paper and approved the final manuscript.

Acknowledgments

The research of the second author is financially supported by King Fahd University of Petroleum and Minerals, Saudi Arabia, under the Internal Research Project no. IN161058.

References

[1] M. A. Hanson, "On sufficiency of the Kuhn-Tucker conditions," *Journal of Mathematical Analysis and Applications*, vol. 80, no. 2, pp. 545–550, 1981.

[2] G. Ruiz-Garzon, R. Osuna-Gomez, and A. R. Lizana, "Generalized invex monotonicity," *European Journal of Operational Research*, vol. 144, no. 3, pp. 501–512, 2003.

[3] X. M. Yang, X. Q. Yang, and K. L. Teo, "Generalized invexity and generalized invariant monotonicity," *Journal of Optimization Theory and Applications*, vol. 117, no. 3, pp. 607–625, 2003.

[4] X. M. Yang, X. Q. Yang, and K. L. Teo, "Criteria for generalized invex monotonicities," *European Journal of Operational Research*, vol. 164, no. 1, pp. 115–119, 2005.

[5] T. Antczak, "(p, r)-invex sets and functions," *Journal of Mathematical Analysis and Applications*, vol. 263, no. 2, pp. 355–379, 2001.

[6] T. Antczak, "Relationships between pre-invex concepts," *Nonlinear Analysis. Theory, Methods & Applications. An International Multidisciplinary Journal*, vol. 60, no. 2, pp. 349–367, 2005.

[7] S. K. Suneja, C. Singh, and C. R. Bector, "Generalization of preinvex and b-vex functions," *Journal of Optimization Theory and Applications*, vol. 76, no. 3, pp. 577–587, 1993.

[8] B. Aghezzaf and M. Hachimi, "Generalized invexity and duality in multiobjective programming problems," *Journal of Global Optimization*, vol. 18, no. 1, pp. 91–101, 2000.

[9] T. R. Gulati, I. Ahmad, and D. Agarwal, "Sufficiency and duality in multiobjective programming under generalized type I functions," *Journal of Optimization Theory and Applications*, vol. 135, no. 3, pp. 411–427, 2007.

[10] I. Ahmad, D. Singh, B. A. Dar, and S. Al-Homidan, "On interval valued functions and Magasarian type duality involving Hukuhara derivative," *Journal of Computational Analysis and Applications*, vol. 21, no. 5, pp. 881–896, 2016.

[11] S. Nanda and K. Kar, "Convex fuzzy mappings," *Fuzzy Sets and Systems*, vol. 48, no. 1, pp. 129–132, 1992.

[12] N. Furukawa, "Convexity and local Lipschitz continuity of fuzzy-valued mappings," *Fuzzy Sets and Systems*, vol. 93, no. 1, pp. 113–119, 1998.

[13] H. Yan and J. Xu, "A class of convex fuzzy mappings," *Fuzzy Sets and Systems*, vol. 129, no. 1, pp. 47–56, 2002.

[14] R. Goetschel Jr. and W. Voxman, "Elementary fuzzy calculus," *Fuzzy Sets and Systems*, vol. 18, no. 1, pp. 31–43, 1986.

[15] M. A. Noor, "Fuzzy preinvex functions," *Fuzzy Sets and Systems*, vol. 64, no. 1, pp. 95–104, 1994.

[16] D. Qiu and W. Zhang, "Convexity invariance of fuzzy sets under the extension principles," *journal of function spaces and applications*, Article ID 849104, 13 pages, 2012.

[17] J. Li and M. A. Noor, "On characterizations of preinvex fuzzy mappings," *Computers & Mathematics with Applications. An International Journal*, vol. 59, no. 2, pp. 933–940, 2010.

[18] Y.-R. Syau, "On convex and concave fuzzy mappings," *Fuzzy Sets and Systems*, vol. 103, no. 1, pp. 163–168, 1999.

[19] Y.-R. Syau, "Invex and generalized convex fuzzy mappings," *Fuzzy Sets and Systems*, vol. 115, no. 3, pp. 455–461, 2000.

[20] D. Qiu, C. Lu, W. Zhang, and Y. Lan, "Algebraic properties and topological properties of the quotient space of fuzzy numbers based on Mareš equivalence relation," *Fuzzy Sets and Systems*, vol. 245, pp. 63–82, 2014.

[21] D. Qiu, L. Shu, and Z.-W. Mo, "Notes on fuzzy complex analysis," *Fuzzy Sets and Systems*, vol. 160, no. 11, pp. 1578–1589, 2009.

[22] L. Li, S. Liu, and J. Zhang, "On fuzzy generalized convex mappings and optimality conditions for fuzzy weakly univex mappings," *Fuzzy Sets and Systems*, vol. 280, pp. 107–132, 2015.

[23] D. Qiu, L. Shu, and Z.-W. Mo, "On starshaped fuzzy sets," *Fuzzy Sets and Systems*, vol. 160, no. 11, pp. 1563–1577, 2009.

[24] G. Wang and C. Wu, "Directional derivatives and subdifferential of convex fuzzy mappings and application in convex fuzzy programming," *Fuzzy Sets and Systems*, vol. 138, no. 3, pp. 559–591, 2003.

[25] H.-C. Wu, "The Karush-KUHn-Tucker optimality conditions in multiobjective programming problems with interval-valued objective functions," *European Journal of Operational Research*, vol. 196, no. 1, pp. 49–60, 2009.

[26] J. Ramik and M. Vlach, *Generalized Concavity in Fuzzy Optimization and Decision Analysis*, Kluwer Academics Publishers, Dordrecht, Netherlands, 2002.

[27] Y. Chalco-Cano, W. A. Lodwick, and A. Rufian-Lizana, "Optimality conditions of type KKT for optimization problem with interval-valued objective function via generalized derivative," *Fuzzy Optimization and Decision Making. A Journal of Modeling and Computation Under Uncertainty*, vol. 12, no. 3, pp. 305–322, 2013.

[28] D. Qiu, F. Yang, and L. Shu, "On convex fuzzy processes and their generalizations," *International Journal of Fuzzy Systems*, vol. 12, no. 3, pp. 267–272, 2010.

[29] D. Dubois and H. Prade, "Operations on fuzzy numbers," *International Journal of Systems Science*, vol. 9, no. 6, pp. 613–626, 1978.

[30] M. L. Puri and D. A. Ralescu, "Differentials of fuzzy functions," *Journal of Mathematical Analysis and Applications*, vol. 91, no. 2, pp. 552–558, 1983.

[31] S. Seikkala, "On the fuzzy initial value problem," *Fuzzy Sets and Systems*, vol. 24, no. 3, pp. 319–330, 1987.

[32] B. Bede and S. G. Gal, "Generalizations of the differentiability of fuzzy-number-valued functions with applications to fuzzy differential equations," *Fuzzy Sets and Systems*, vol. 151, no. 3, pp. 581–599, 2005.

[33] A. Rufián-Lizana, Y. Chalco-Cano, R. Osuna-Gómez, and G. Ruiz-Garzón, "On invex fuzzy mappings and fuzzy variational-like inequalities," *Fuzzy Sets and Systems*, vol. 200, pp. 84–98, 2012.

[34] B. Bede and L. Stefanini, "Generalized differentiability of fuzzy-valued functions," *Fuzzy Sets and Systems*, vol. 230, pp. 119–141, 2013.

9

σ-Algebra and $\sigma-$Baire in Fuzzy Soft Setting

Shuker Mahmood Khalil, Mayadah Ulrazaq, Samaher Abdul-Ghani, and Abu Firas Al-Musawi

Department of Mathematics, College of Science, University of Basrah, Basrah 61004, Iraq

Correspondence should be addressed to Shuker Mahmood Khalil; shuker.alsalem@gmail.com

Academic Editor: Katsuhiro Honda

We first introduce some new notions of Baireness in fuzzy soft topological space (FSTS). Next, their characterizations and basic properties are investigated in this work. The notions of fuzzy soft dense, fuzzy soft nowhere dense, fuzzy soft meager, fuzzy soft second category, fuzzy soft residual, fuzzy soft Baire, fuzzy soft $\delta-$sets, fuzzy soft $\lambda_\sigma-$sets, fuzzy soft $\sigma-$nowhere dense, fuzzy soft $\sigma-$meager, fuzzy soft $\sigma-$residual, fuzzy soft $\sigma-$Baire, fuzzy soft $\sigma-$second category, fuzzy soft $\sigma-$residual, fuzzy, fuzzy soft submaximal space, fuzzy soft $P-$space, fuzzy soft almost resolvable space, fuzzy soft hyperconnected space, fuzzy soft $A-$embedded, fuzzy soft $D-$Baire, fuzzy soft almost $P-$spaces, fuzzy soft Borel, and fuzzy soft $\sigma-$algebra are introduced. Furthermore, several examples are shown as well.

1. Introduction

The concepts of Baire spaces have been studied and discussed extensively in general topology in [1–4]. Thangaraj and Balasubramanian [5] studied the notion of somewhat fuzzy continuous functions. Next, Thangaraj and Anjalmose investigated and discussed the notion of Baire space in fuzzy topology [6]. After that, they introduced the notion of fuzzy Baire space [7].

Soft sets theory was introduced by Molodtsov [8]. It explains new type of mathematical tool of soft sets and it deals with vagueness when solving problems in practice as in engineering, environment, social science, and economics, which cannot be handled as classical mathematical tools. Also, other authors such as Maji et al. [9–19] have further studied the theory of soft sets and used this theory in pure mathematics to solve some decision making problems. Next, the notion of fuzzy soft set is investigated and discussed [20–22]. Since then much attention has been used to generalize the basic notions of fuzzy topology in soft setting. In other words, the modern theories of fuzzy soft topology have been developed.

In recent years, fuzzy soft topology has been found to be very useful in solving many practical problems [23]. Also, rough fuzzy and other applications are studied [24–26]. The main purpose of this work is to introduce new concepts of fuzzy soft Baireness in fuzzy soft topological spaces. In section three, we introduce fuzzy soft dense sets, fuzzy soft nowhere dense sets, fuzzy soft meager sets, fuzzy soft second category sets, fuzzy soft meager spaces, fuzzy soft second category spaces, fuzzy soft residual sets, fuzzy soft Baire spaces, fuzzy soft $\delta-$sets, fuzzy soft $\lambda_\sigma-$sets, fuzzy soft $\sigma-$nowhere dense, fuzzy soft $\sigma-$meager, fuzzy soft $\sigma-$residual, fuzzy soft $\sigma-$Baire, fuzzy soft $\sigma-$second category, fuzzy soft $\sigma-$residual, fuzzy, fuzzy soft submaximal space, fuzzy soft $P-$space, fuzzy soft almost resolvable space, fuzzy soft hyperconnected space, fuzzy soft $A-$embedded, fuzzy soft $D-$Baire, fuzzy soft almost $P-$spaces, fuzzy soft Borel, and fuzzy soft $\sigma-$algebra. Moreover, several examples are given to illustrate the notions introduced in this work.

2. Preliminaries

In this section, we give few definitions and properties regarding fuzzy soft sets.

Definition 1 ([20]). Assume U is an initial universe set and E is a set of parameters. Let I^U refer to the family of all fuzzy soft sets (FSSs) in U and $A \subseteq E$. The multivalued map $F_A : A \longrightarrow I^U$ defined by $F_A(e) = \mu^e_{F_A}$ is said to be a fuzzy soft set (FSS) over (U, E), where $\mu^e_{F_A} \neq \overline{0}$ if $e \in A$ and $\mu^e_{F_A} = \overline{0}$ if $e \in E \setminus A$. We refer to family of all (FSSs) over (U, E) by $FS(U, E)$.

Definition 2 ([20]). We say $F_\phi \in FS(U, E)$ is null (FSS) and we refer to it by Φ, if $\forall e \in E$, $F(e)$ is the null (FSS) $\overline{0}$ of U, where $\overline{0}(x) = 0\ \forall x \in U$.

Definition 3 ([20]). Assume $F_E \in FS(U, E)$ and $F_E(e) = \overline{1}\ \forall e \in E$, where $\overline{1}(x) = 1\ \forall x \in U$. We say F_E is absolute (FSS) and we refer to it by $\overline{E}$.

Definition 4 ([20]). We say F_A is a fuzzy soft subset of a (FSS) G_B over a common universe U if $A \subseteq B$ and $F_A(e) \subseteq G_B(e)\ \forall e \in A$; i.e., if $\mu^e_{F_A}(x) \leq \mu^e_{G_B}(x)\ \forall x \in U$ and $\forall e \in E$ and denoted by $F_A \tilde{\subseteq} G_B$.

Definition 5 ([20]). Assume F_A and G_B are (FSSs) over a common universe U. We say they are fuzzy soft equal if $F_A \tilde{\subseteq} G_B$ and $G_B \tilde{\subseteq} F_A$.

Definition 6 ([20]). Assume F_A and G_B are (FSSs) over a common universe U. Their union is the (FSS) H_C, defined by $H_C(e) = \mu^e_{H_C} = \mu^e_{F_A} \cup \mu^e_{G_B}\ \forall e \in E$, where $C = A \bigcup B$. For this case, we write $H_C = F_A \tilde{\cup} G_B$.

Definition 7 ([20]). Assume F_A and G_B are (FSSs) over a common universe U. Their intersection is a (FSS) H_C, defined by $H_C(e) = \mu^e_{H_C} = \mu^e_{F_A} \cap \mu^e_{G_B}\ \forall e \in E$, where $C = A \bigcap B$. For this case, we write $H_C = F_A \tilde{\cap} G_B$.

Definition 8 ([27]). Assume $F_A \in FS(U, E)$ is a (FSS). We refer to its complement by $F_A{}^c$, and it is defined as

$$F_A^c = \begin{cases} \overline{1} - \mu^e_{F_A} & \text{if } e \in A, \\ \overline{1} & \text{if } e \notin A. \end{cases} \qquad (1)$$

Remark 9. Let $K = \{F^i_{A_i} \mid i \in I\}$ be a family of fuzzy soft sets. The union (intersection) of any number $(J \subseteq I)$ of family K is defined by $H_C(e) = \mu^e_{H_C} = \bigcup_{j \in J} \{\mu^e_{F^i_{A_i}}\} (H_C(e) = \mu^e_{H_C} = \bigcap_{j \in J} \{\mu^e_{F^i_{A_i}}\})$, where $C = \bigcup_{j \in J} \{A_i\}$ $(C = \bigcap_{j \in J} \{A_i\})$.

Definition 10 ([28]). Assume ψ is the family of (FSSs) over U. We say ψ is a fuzzy soft topology on U if ψ the following axioms hold:

(i) $\Phi, \overline{E}$ belong to ψ.

(ii) The union of any number of (FSSs) in ψ belongs to ψ.

(iii) The intersection of any two (FSSs) in ψ belongs to ψ.

We say (U, E, ψ) is a fuzzy soft topological space (FSTS) over U. Each member in ψ is called (FSOS) (FSOS) in U and its complement is called fuzzy soft closed set (FSCS) in U, where $C = \bigcup_{j \in J} \{A_i\}$ $(C = \bigcap_{j \in J} \{A_i\})$.

Definition 11 ([28]). The union of all fuzzy soft open subsets of F_A over (U, E) is called the interior of F_A and is denoted by $int^{fs}(F_A)$.

Proposition 12 ([28]). $int^{fs}(F_A \tilde{\cap} G_B) = int^{fs}(F_A) \tilde{\cap} int^{fs}(G_B)$.

Definition 13 ([28]). Let $F_A \in FS(U, E)$ be a (FSS). Then the intersection of all closed sets, each containing F_A, is called the closure of F_A and is denoted by $cl^{fs}(F_A)$.

Remark 14. (1) For any (FSS) F_A in a (FSTS) (U, E, ψ), it is easy to see that $(cl^{fs}(F_A))^c = int^{fs}(F_A{}^c)$ and $(int^{fs}(F_A))^c = cl^{fs}(F_A{}^c)$.

(2) For any fuzzy soft F_A subset of a (FSTS) (U, E, ψ), we define the fuzzy soft subspace topology ψ_{F_A} on F_A by $K_D \in \psi_{F_A}$ if $K_D = F_A \tilde{\cap} G_B$ for some $G_B \in \psi$.

(3) For any fuzzy soft H_C in F_A fuzzy soft subspace of a (FSTS), we denote the interior and closure of H_C in F_A by $int^{fs}_{F_A}(H_C)$ and $cl^{fs}_{F_A}(H_C)$, respectively.

3. Fuzzy Soft σ−Baire Spaces

In this section, we introduce new notions of (FSTSs) using new classes of (FSSs) which are introduced in this section and obtained their properties.

Definition 15. A (FSS) F_A in a (FSTS) (U, E, ψ) is called fuzzy soft dense if there exists no (FSCS) G_B in (U, E, ψ) such that $F_A \tilde{\subset} G_B \tilde{\subset} \overline{E}$.

Definition 16. A (FSS) F_A in a (FSTS) (U, E, ψ) is called fuzzy soft nowhere dense if there exists no nonempty (FSOS) G_B in (U, E, ψ) such that $G_B \tilde{\subset} cl^{fs}(F_A)$. That is, $int^{fs}(cl^{fs}(F_A)) = \Phi$.

Definition 17. A (FSS) F_A in a (FSTS) (U, E, ψ) is called fuzzy soft meager if F_A is a countable union of fuzzy soft nowhere dense sets [i.e., if $F_A = \tilde{\bigcup}_{i \in I} \{F^i_{A_i}\}$, where $F^i_{A_i}$'s are fuzzy soft nowhere dense sets in (U, E, ψ), $\forall i \in I \subseteq N$]. Otherwise, F_A will be called a fuzzy soft second category set.

Definition 18. A (FSTS) (U, E, ψ) is called fuzzy soft meager or (fuzzy soft first category) space if the (FSS) $\overline{E}$ is a fuzzy soft meager set in (U, E, ψ). That is, $\overline{E} = \tilde{\bigcup}_{i \in I} \{F^i_{A_i}\}$, where $(F^i_{A_i}{}^c)$'s are fuzzy soft nowhere dense sets in (U, E, ψ). Otherwise, (U, E, ψ) will be called a fuzzy soft second category space.

Definition 19. Assume F_A is a fuzzy soft meager set in (U, E, ψ). We say $F_A{}^c$ is a fuzzy soft residual set in (U, E, ψ).

Definition 20. Assume (U, E, ψ) is a (FSTS). We say (U, E, ψ) is a fuzzy soft Baire space if each sequence $\{F^1_{A_1}, F^2_{A_2}, F^3_{A_3}, \ldots\}$ of fuzzy soft nowhere dense sets in (U, E, ψ) such that $int^{fs}(\tilde{\bigcup}_{i \in I} \{F^i_{A_i}\}) = \Phi$.

Definition 21. A (FSS) F_A in a (FSTS) (U, E, ψ) is called fuzzy soft δ−set in (U, E, ψ) if $F_A = \tilde{\bigcap}_{i \in I} \{F^i_{A_i}\}$, where $F^i_{A_i} \in \psi\ \forall i \in I$.

Remark 22. Definition 20 can be written in other equivalent forms as follows:

(i) We say (U, E, ψ) is a fuzzy soft Baire space if every countable fuzzy soft closed cover $\{F^i_{A_i}; i \in I\}$ of $\overline{E}$, the set $\tilde{\bigcup}_{i \in I} \{int^{fs}(F^i_{A_i})\}$ is fuzzy soft dense in $\overline{E}$.

(ii) We say (U, E, ψ) is a fuzzy soft Baire space if every sequence $\{F^1_{A_1}, F^2_{A_2}, F^3_{A_3}, \ldots\}$ of (FSOSs) with the same closure F_A, we have $F_A = cl^{fs}(\widetilde{\bigcap}_{i\in I}\{F^i_{A_i}\})$.

(iii) We say (U, E, ψ) is a fuzzy soft Baire space if every fuzzy soft meager and fuzzy soft δ−set in $\overline{E}$ is fuzzy soft nowhere dense.

Remark 23. We say $SS(U_E)$ is a family of all soft sets over a universe set U and the parameter set E. Moreover, the cardinality of $SS(U_E)$ is given by $n(SS(U_A)) = 2^{n(U)\times n(E)}$. Therefore, in this paper for each (FSS) F_E over (U, E) we can define F_E by using matrix form as follows:

$$F_E = \begin{pmatrix} a_{11} & a_{12} & a_{13} & \cdots & a_{1k} \\ a_{21} & a_{22} & a_{23} & \cdots & a_{2k} \\ \vdots & \vdots & \vdots & \ddots & \vdots \\ a_{m1} & a_{m2} & a_{m3} & \cdots & a_{mk} \end{pmatrix}_{m\times k} \quad (2)$$

The order of this matrix is given by $m \times k$, where $m = n(E)$, $k = n(U)$, and $a_{ij} = \mu^{e_i}_{F_E}(c_j)$, $\forall 1 \le i \le m$ and $1 \le j \le k$.

Definition 24. A (FSS) F_A in a (FSTS) (U, E, ψ) is called fuzzy soft λ_σ−set in (U, E, ψ) if $F_A = \widetilde{\bigcup}_{i\in I}\{F^i_{A_i}\}$, where $F^{i\ c}_{A_i} \in \psi$ $\forall i \in I$.

Definition 25. A (FSS) F_A in a (FSTS) (U, E, ψ) is called fuzzy soft σ−nowhere dense set if F_A is a fuzzy soft λ_σ−set in (U, E, ψ) such that $int^{fs}(F_A) = \phi$.

Example 26. As an illustration, consider the following example. Suppose the (FSSs) $F_E, G_E, H_E, J_E, K_E, L_E, T_E$ describe attractiveness of the cars with respect to the given parameters, which my friends are going to buy. $U = \{x_1, x_2, x_3\}$ which is the set of all cars under consideration. Let I^U be the collection of all fuzzy subsets of U. Also, let $E = \{e_1, e_2, e_3\}$, where e_1, e_2, e_3 stand for the attributes of cheap, qualification, and colorful, respectively. Let $F_E, G_E, H_E, J_E, K_E, L_E, T_E$ be defined as follows:

$$F_E = \begin{pmatrix} .2 & .3 & .6 \\ .15 & .25 & .10 \\ .3 & .12 & .6 \end{pmatrix},$$

$$G_E = \begin{pmatrix} .4 & .7 & .3 \\ .8 & .75 & .7 \\ .8 & .10 & .9 \end{pmatrix},$$

$$H_E = \begin{pmatrix} .1 & .5 & .3 \\ .5 & .75 & .4 \\ .8 & .4 & .1 \end{pmatrix},$$

$$J_E = \begin{pmatrix} .2 & .5 & .6 \\ .5 & .75 & .4 \\ .8 & .4 & .6 \end{pmatrix},$$

$$K_E = \begin{pmatrix} .2 & .3 & .3 \\ .15 & .25 & .10 \\ .3 & .10 & .6 \end{pmatrix},$$

$$L_E = \begin{pmatrix} .1 & .3 & .3 \\ .15 & .25 & .10 \\ .3 & .12 & .1 \end{pmatrix},$$

$$T_E = \begin{pmatrix} .4 & .7 & .6 \\ .8 & .75 & .7 \\ .8 & .12 & .9 \end{pmatrix}. \quad (3)$$

Then, $\psi = \{\Phi, \overline{E}, F_E, G_E, H_E, J_E, K_E, L_E, T_E\}$ is clearly a fuzzy soft topology on $\overline{E}$. Now consider the (FSS)

$$\alpha = (T_E)^c \,\widetilde{\cup}\, (J_E)^c = \begin{pmatrix} .8 & .5 & .4 \\ .5 & .25 & .6 \\ .2 & .88 & .4 \end{pmatrix} \quad (4)$$

in (U, E, ψ). Then α is a fuzzy soft λ_σ−set in (U, E, ψ) and $int^{fs}(\alpha) = \Phi$ and hence α is a fuzzy soft σ−nowhere dense set in (U, E, ψ). The (FSS)

$$\beta = (F_E)^c \,\widetilde{\cup}\, (H_E)^c = \begin{pmatrix} .9 & .7 & .7 \\ .85 & .75 & .90 \\ .7 & .88 & .9 \end{pmatrix} \quad (5)$$

is a fuzzy soft λ_σ−set in (U, E, ψ) and $int^{fs}(\beta) = F_E \ne \Phi$ and hence β is not a fuzzy soft σ−nowhere dense set in (U, E, ψ).

Definition 27. A (FSS) F_A in a (FSTS) (U, E, ψ) is called fuzzy soft σ−meager if $F_A = \widetilde{\bigcup}_{i\in I}\{F^i_{A_i}\}$, where $(F^i_{A_i})$'s are fuzzy soft σ−nowhere dense sets in (U, E, ψ). Any other (FSS) in (U, E, ψ) is said to be of fuzzy soft σ−second category.

Definition 28. A (FSTS) (U, E, ψ) is called fuzzy soft σ−meager space if the (FSS) $\overline{E}$ is a fuzzy soft σ−meager set in (U, E, ψ). That is, $\overline{E} = \widetilde{\bigcup}_{i\in I}\{F^i_{A_i}\}$, where $(F^i_{A_i})$'s are fuzzy soft σ−nowhere dense sets in (U, E, ψ). Otherwise, (U, E, ψ) will be called a fuzzy soft σ−second category space.

Definition 29. Assume F_A is a fuzzy soft σ−meager set in a (FSTS) (U, E, ψ). We say $F_A{}^c$ is a fuzzy soft σ−residual set in (U, E, ψ).

Proposition 30. *Let (U, E, ψ) be a (FSTS). Then the following are equivalent:*

(1) (U, E, ψ) is a fuzzy soft Baire space.

(2) $int^{fs}(F_A) = \Phi$ for every fuzzy soft meager set F_A in (U, E, ψ).

(3) $cl^{fs}(G_B) = \overline{E}$ for every fuzzy soft residual set G_B in (U, E, ψ).

Proof. (1) $\Longrightarrow$ (2). Let F_A be a fuzzy soft meager set in (U, E, ψ). Then $F_A = (\widetilde{\bigcup}_{i\in I}\{F^i_{A_i}\})$, where $(F^i_{A_i})$'s, $i \in I$ are fuzzy soft nowhere dense sets in (U, E, ψ). Then, we have $int^{fs}(F_A) = int^{fs}(\widetilde{\bigcup}_{i\in I}\{F^i_{A_i}\})$. Since (U, E, ψ) is a fuzzy soft Baire space, $int^{fs}(\widetilde{\bigcup}_{i\in I}\{F^i_{A_i}\}) = \Phi$. Hence, $int^{fs}(F_A) = \Phi$ for any fuzzy soft meager set F_A in (U, E, ψ).

(2) $\Longrightarrow$ (3). Let G_B be a fuzzy soft residual set in (U, E, ψ). Then $G_B{}^c$ is a fuzzy soft σ−meager set in (U, E, ψ). By hypothesis, $int^{fs}(G_B{}^c) = \Phi$. Then $(cl^{fs}(G_B))^c = \Phi$. Hence, $cl^{fs}(G_B) = \overline{E}$ for any fuzzy soft residual set G_B in (U, E, ψ).

(3) $\Longrightarrow$ (1). Let F_A be a fuzzy soft meager set in (U, E, ψ). Then $F_A = (\widetilde{\bigcup}_{i\in I}\{F^i_{A_i}\})$, where $(F^i_{A_i})$'s are fuzzy soft nowhere dense sets in (U, E, ψ). Now F_A is a fuzzy soft meager set in (U, E, ψ) implying that $F_A{}^c$ is a fuzzy soft residual set in (U, E, ψ). By hypothesis, we have $cl^{fs}(F_A)^c = \overline{E}$. Then $(int^{fs}(F_A))^c = \overline{E}$. Hence, $int^{fs}(F_A) = \Phi$. That is, $int^{fs}(\widetilde{\bigcup}_{i\in I}\{F^i_{A_i}\}) = \Phi$, where $(F^i_{A_i})$'s are fuzzy soft nowhere dense sets in (U, E, ψ). Hence, (U, E, ψ) is a fuzzy soft Baire space.

Proposition 31. *If F_A is a fuzzy soft dense and fuzzy soft δ−set in a (FSTS) (U, E, ψ), then $F_A{}^c$ is a fuzzy soft meager set in (U, E, ψ).*

Proof. Since F_A is a fuzzy soft δ−set in (U, E, ψ), $F_A = \widetilde{\bigcap}_{i\in I}\{F^i_{A_i}\}$, where $F^i_{A_i} \in \psi$ and since F_A is a fuzzy soft dense set in (U, E, ψ), $cl^{fs}(F_A) = \overline{E}$. Then $cl^{fs}(\widetilde{\bigcap}_{i\in I}\{F^i_{A_i}\}) = \overline{E}$. But $cl^{fs}(\widetilde{\bigcap}_{i\in I}\{F^i_{A_i}\})\widetilde{\subseteq}\widetilde{\bigcap}_{i\in I}\{cl^{fs}(F^i_{A_i})\}$. Hence, $\overline{E}\widetilde{\subseteq}\widetilde{\bigcap}_{i\in I}\{cl^{fs}(F^i_{A_i})\}$. That is, $\widetilde{\bigcap}_{i\in I}\{cl^{fs}(F^i_{A_i})\} = \overline{E}$. Then we have $cl^{fs}(F^i_{A_i}) = \overline{E}$ for each $F^i_{A_i} \in \psi$ and hence $cl^{fs}(int^{fs}(F^i_{A_i})) = \overline{E}$ which implies that $(cl^{fs}(int^{fs}(F^i_{A_i})))^c = \Phi$ and hence $int^{fs}(cl^{fs}(F^{i\,c}_{A_i})) = \Phi$. Therefore, $(F^{i\,c}_{A_i})$ is a fuzzy soft nowhere dense set in (U, E, ψ). Now $F_A{}^c = (\widetilde{\bigcap}_{i\in I}\{F^i_{A_i}\})^c = \widetilde{\bigcup}_{i\in I}\{F^{i\,c}_{A_i}\}$. Therefore, $F^c_A = \widetilde{\bigcup}_{i\in I}\{F^{i\,c}_{A_i}\}$, where $(F^{i\,c}_{A_i})$'s are fuzzy soft nowhere dense sets in (U, E, ψ). Hence, $F_A{}^c$ is a fuzzy soft meager set in (U, E, ψ).

Lemma 32. *If F_A is a fuzzy soft dense and fuzzy soft δ−set in a (FSTS) (U, E, ψ), then F_A is a fuzzy soft residual set in (U, E, ψ).*

Proof. Since F_A is a fuzzy soft dense and fuzzy soft δ−set in (U, E, ψ), by Proposition 31, we have that $F_A{}^c$ is a fuzzy soft meager set in (U, E, ψ) and hence F_A is a fuzzy soft residual set in (U, E, ψ).

Proposition 33. *In a (FSTS) (U, E, ψ), a (FSS) F_A is a fuzzy soft σ−nowhere dense set in (U, E, ψ) if and only if $F_A{}^c$ is a fuzzy soft dense and fuzzy soft δ−set in (U, E, ψ).*

Proof. Let F_A be a fuzzy soft σ−nowhere dense set in (U, E, ψ). Then $F_A = (\widetilde{\bigcup}_{i\in I}\{F^i_{A_i}\})$, where $F^{i\,c}_{A_i} \in \psi$, $\forall i \in I$ and $int^{fs}(F_A) = \Phi$. Then $(int^{fs}(F_A))^c = \Phi^c = \overline{E}$ implies that $cl^{fs}(F_A{}^c) = \overline{E}$. Also $F_A{}^c = (\widetilde{\bigcup}_{i\in I}\{F^i_{A_i}\})^c = \widetilde{\bigcap}_{i\in I}\{F^{i\,c}_{A_i}\}$, where $F^{i\,c}_{A_i} \in \psi$, for $i \in I$. Hence, we have $F^{i\,c}_{A_i}$ is a fuzzy soft dense and fuzzy soft δ−set in (U, E, ψ).

Conversely, let F_A be a fuzzy soft dense and fuzzy soft δ−set in (U, E, ψ). Then $F_A = \widetilde{\bigcap}_{i\in I}\{F^i_{A_i}\}$, where $F^i_{A_i} \in \psi$ for $i \in I$. Now $F_A{}^c = (\widetilde{\bigcap}_{i\in I}\{F^i_{A_i}\})^c = \widetilde{\bigcup}_{i\in I}\{F^{i\,c}_{A_i}\}$. Hence, $F_A{}^c$ is a λ_σ−set in (U, E, ψ) and $int^{fs}(F_A{}^c) = (cl^{fs}(F_A))^c = (\overline{E})^c = \Phi$ [since F_A is a fuzzy soft dense]. Therefore, $F_A{}^c$ is a fuzzy soft σ−nowhere dense set in (U, E, ψ).

Definition 34. Assume (U, E, ψ) is a (FSTS). We say (U, E, ψ) is a fuzzy soft σ−Baire space if each sequence $\{F^1_{A_1}, F^2_{A_2}, F^3_{A_3}, \ldots\}$ of fuzzy soft σ−nowhere dense sets in (U, E, ψ) such that $int^{fs}(\widetilde{\bigcup}_{i\in I}\{F^i_{A_i}\}) = \Phi$.

Example 35. Let $U = \{c_1, c_2, c_3\}$ be the set of three flats and $E = \{\text{costly } (e_1), \text{ modern } (e_2), \text{ security services } (e_3)\}$ be the set of parameters. Then, we consider that $\psi = \{\Phi, \overline{E}, F_E, G_E, H_E, \}\tilde{\cup}\{T^i_E \mid i = 1, 2, 3, \ldots, 14\}$ is a fuzzy soft topology on $\overline{E}$ defined as follows:

$$F_E = \begin{pmatrix} .7 & .5 & .6 \\ .6 & .4 & .5 \\ .5 & .3 & .4 \end{pmatrix},$$
$$G_E = \begin{pmatrix} .5 & .8 & .7 \\ .4 & .7 & .6 \\ .3 & .6 & .5 \end{pmatrix}, \qquad (6)$$
$$H_E = \begin{pmatrix} .6 & .4 & .8 \\ .5 & .3 & .7 \\ .4 & .2 & .6 \end{pmatrix}.$$

$T^1_E = F_E\tilde{\cap}G_E$, $T^2_E = F_E\tilde{\cap}H_E$, $T^3_E = G_E\tilde{\cap}H_E$, $T^4_E = F_E\tilde{\cup}G_E$, $T^5_E = F_E\tilde{\cup}H_E$, $T^6_E = G_E\tilde{\cup}H_E$, $T^7_E = F_E\tilde{\cap}(G_E\tilde{\cup}H_E)$, $T^8_E = F_E\tilde{\cup}(G_E\tilde{\cap}H_E)$, $T^9_E = G_E\tilde{\cap}(F_E\tilde{\cup}H_E)$, $T^{10}_E = G_E\tilde{\cup}(F_E\tilde{\cap}H_E)$, $T^{11}_E = H_E\tilde{\cap}(F_E\tilde{\cup}G_E)$, $T^{12}_E = H_E\tilde{\cup}(F_E\tilde{\cap}G_E)$, $T^{13}_E = F_E\tilde{\cup}G_E\tilde{\cup}H_E$, and $T^{14}_E = F_E\tilde{\cap}G_E\tilde{\cap}H_E$. Now consider the (FSSs)

$$\alpha = (F_E)^c\,\tilde{\cup}\,(T^8_E)^c\,\tilde{\cup}\,(T^{13}_E)^c = \begin{pmatrix} .3 & .5 & .4 \\ .4 & .6 & .5 \\ .5 & .7 & .6 \end{pmatrix},$$
$$\beta = (G_E)^c\,\tilde{\cup}\,(T^5_E)^c\,\tilde{\cup}\,(T^{10}_E)^c = \begin{pmatrix} .5 & .5 & .3 \\ .6 & .6 & .4 \\ .7 & .7 & .5 \end{pmatrix}, \qquad (7)$$

and

$$\theta = (H_E)^c\,\tilde{\cup}\,(T^1_E)^c\,\tilde{\cup}\,(T^4_E)^c\,\tilde{\cup}\,(T^6_E)^c = \begin{pmatrix} .5 & .6 & .4 \\ .6 & .7 & .5 \\ .7 & .8 & .6 \end{pmatrix}, \qquad (8)$$

in (U, E, ψ). Then α, β and θ are fuzzy soft λ_σ−sets in (U, E, ψ) and $int^{fs}(\alpha) = \Phi$, $int^{fs}(\beta) = \Phi$, and $int^{fs}(\theta) = \Phi$.Then α, β and θ are fuzzy soft σ−nowhere dense sets in (U, E, ψ).

Moreover, $(T_E^2)^c\tilde{\cup}(T_E^3)^c\tilde{\cup}(T_E^7)^c\tilde{\cup}(T_E^9)^c\tilde{\cup}(T_E^{11})^c\tilde{\cup}(T_E^{12})^c\tilde{\cup}(T_E^{14})^c = \theta$ and also $int^{fs}(\alpha\tilde{\cup}\beta\tilde{\cup}\theta) = int^{fs}(\theta) = \Phi$ and therefore (U, E, ψ) is a fuzzy soft $\sigma-$Baire space.

Remark 36. A fuzzy soft $\sigma-$Baire space need not be a fuzzy soft Baire space. For, consider the following example.

Example 37. Assume $X = \{c_1, c_2, c_3\}$ is a set of soldiers under consideration and $E = \{$ courageous (e_1); strong (e_2); smart$(e_3)\}$ is a set of parameters framed to choose the best soldier. Then, we consider that $\psi = \{\Phi, \overline{E}, F_E, G_E, H_E\}\tilde{\cup}\{T_E^i \mid i = 1, 2, 3, \ldots, 9\}$ is a fuzzy soft topology on $\overline{E}$ defined as follows:

$$H_E = \begin{pmatrix} .8 & .5 & .1 \\ .7 & .4 & .1 \\ .9 & .6 & .1 \end{pmatrix},$$

$$F_E = \begin{pmatrix} .1 & .3 & .8 \\ .1 & .2 & .7 \\ .1 & .4 & .9 \end{pmatrix}, \quad (9)$$

$$G_E = \begin{pmatrix} .4 & .1 & .3 \\ .3 & .1 & .2 \\ .5 & .1 & .4 \end{pmatrix}.$$

$T_E^1 = F_E\tilde{\cap}G_E$, $T_E^2 = F_E\tilde{\cap}H_E$, $T_E^3 = G_E\tilde{\cap}H_E$, $T_E^4 = F_E\tilde{\cup}G_E$, $T_E^5 = F_E\tilde{\cup}H_E$, $T_E^6 = G_E\tilde{\cup}H_E$, $T_E^7 = G_E\tilde{\cup}(F_E\tilde{\cap}H_E)$, $T_E^8 = H_E\tilde{\cup}(G_E\tilde{\cap}F_E)$, and $T_E^9 = H_E\tilde{\cap}(F_E\tilde{\cup}G_E)$. Now $\{F_E{}^c, G_E{}^c, H_E{}^c, (T_E^1)^c, (T_E^2)^c, (T_E^3)^c, (T_E^5)^c, (T_E^7)^c, (T_E^8)^c, (T_E^9)^c\}$ are fuzzy soft nowhere dense sets in (U, E, ψ). $(T_E^1)^c = F_E{}^c\tilde{\cup}G_E{}^c\tilde{\cup}H_E{}^c\tilde{\cup}(T_E^2)^c\tilde{\cup}(T_E^3)^c\tilde{\cup}(T_E^5)^c\tilde{\cup}(T_E^7)^c\tilde{\cup}(T_E^8)^c\tilde{\cup}(T_E^9)^c$. Therefore, $(T_E^1)^c$ is a fuzzy soft meager set in (U, E, ψ). $int^{fs}((T_E^1)^c) = T_E^1 \neq \Phi$. Hence, (U, E, ψ) is not a fuzzy soft Baire space. Now consider the (FSSs) $\alpha = (H_E)^c\tilde{\cup}(T_E^5)^c\tilde{\cup}(T_E^6)^c$, and $\beta = (F_E)^c\tilde{\cup}(T_E^2)^c\tilde{\cup}(T_E^4)^c\tilde{\cup}(T_E^7)^c\tilde{\cup}(T_E^8)^c$ in (U, E, ψ) and also $int^{fs}(\alpha) = \Phi$, $int^{fs}(\beta) = \Phi$. Then α and β are fuzzy soft $\sigma-$nowhere dense sets in (U, E, ψ). Now the (FSS) $(\alpha\tilde{\cup}\beta)$ is a fuzzy soft $\sigma-$meager set in (U, E, ψ) and $int^{fs}(\alpha\tilde{\cup}\beta) = \Phi$. Hence, (U, E, ψ) is a fuzzy soft $\sigma-$Baire space.

Proposition 38. *Let (U, E, ψ)be a (FSTS). Then the following are equivalent:*

(1) (U, E, ψ) is a fuzzy soft $\sigma-$Baire space.

(2) $int^{fs}(F_A) = \Phi$ for every fuzzy soft $\sigma-$meager set F_A in (U, E, ψ).

(3) $cl^{fs}(G_B) = \overline{E}$ for every fuzzy soft $\sigma-$residual set G_B in (U, E, ψ).

Proof. (1) $\Longrightarrow$ (2). Let F_A be a fuzzy soft $\sigma-$meager set in (U, E, ψ). Then $F_A = (\tilde{\bigcup}_{i\in I}\{F_{A_i}^i\})$, where $((F_{A_i}^i)$'s, $i \in I$ are fuzzy soft $\sigma-$nowhere dense sets in (U, E, ψ). Then, we have $int^{fs}(F_A) = int^{fs}(\tilde{\bigcup}_{i\in I}\{F_{A_i}^i\})$. Since (U, E, ψ) is a fuzzy soft $\sigma-$Baire space, $int^{fs}(\tilde{\bigcup}_{i\in I}\{F_{A_i}^i\}) = \Phi$. Hence, $int^{fs}(F_A) = \Phi$ for any fuzzy soft $\sigma-$meager set F_A in (U, E, ψ).

(2) $\Longrightarrow$ (3). Let G_B be a fuzzy soft $\sigma-$residual set in (U, E, ψ). Then $G_B{}^c$ is a fuzzy soft $\sigma-$meager set in (U, E, ψ). By hypothesis, $int^{fs}(G_B^c) = \Phi$. Then $(cl^{fs}(G_B^c))^c = \Phi$. Hence, $cl^{fs}(G_B) = \overline{E}$ for any fuzzy soft $\sigma-$residual set G_B in (U, E, ψ).

(3) $\Longrightarrow$ (1). Let F_A be a fuzzy soft $\sigma-$meager set in (U, E, ψ). Then $F_A = (\tilde{\bigcup}_{i\in I}\{F_{A_i}^i\})$, where $(F_{A_i}^i)$'s are fuzzy soft $\sigma-$nowhere dense sets in (U, E, ψ). Now F_A is a fuzzy soft $\sigma-$meager set in (U, E, ψ) implying that $F_A{}^c$ is a fuzzy soft $\sigma-$residual set in (U, E, ψ). By hypothesis, we have $cl^{fs}(F_A)^c = \overline{E}$. Then $(int^{fs}(F_A))^c = \overline{E}$. Hence, $int^{fs}(F_A) = \Phi$. That is, $int^{fs}(\tilde{\bigcup}_{i\in I}\{F_{A_i}^i\}) = \Phi$, where $(F_{A_i}^i)$'s are fuzzy soft $\sigma-$nowhere dense sets in (U, E, ψ). Hence, (U, E, ψ) is a fuzzy soft $\sigma-$Baire space.

Proposition 39. *If the (FSTS) (U, E, ψ) is a fuzzy soft $\sigma-$Baire space, then $cl^{fs}(\tilde{\bigcup}_{i\in I}\{F_{A_i}^i\}) = \overline{E}$, where the (FSSs) $(F_{A_i}^i)$'s $\forall i \in I$ are fuzzy soft dense and fuzzy soft $\delta-$sets in (U, E, ψ).*

Proof. Let $(F_{A_i}^i)$'s, $i \in I$ be fuzzy soft dense and fuzzy soft $\delta-$sets in (U, E, ψ). By Proposition 33, $(F_{A_i}^i)$'s are fuzzy soft $\sigma-$nowhere dense sets in (U, E, ψ). Then the (FSS) $F_A = \tilde{\bigcup}_{i\in I}\{F_{A_i}^i{}^c\}$ is a fuzzy soft $\sigma-$meager set in (U, E, ψ). Now $int^{fs}(F_A) = int^{fs}(\tilde{\bigcup}_{i\in I}\{F_{A_i}^i{}^c\}) = int^{fs}(\tilde{\bigcap}_{i\in I}\{F_{A_i}^i\})^c = (cl^{fs}(\tilde{\bigcap}_{i\in I}\{F_{A_i}^i\}))^c$. Since (U, E, ψ) is a fuzzy soft $\sigma-$Baire space, by Proposition 38, we have $int^{fs}(F_A) = \Phi$. Then $(cl^{fs}(\tilde{\bigcap}_{i\in I}\{F_{A_i}^i\}))^c = \Phi$. This implies that $cl^{fs}(\tilde{\bigcap}_{i\in I}\{F_{A_i}^i\}) = \overline{E}$.

Proposition 40. *If the (FSTS) (U, E, ψ) is a fuzzy soft $\sigma-$Baire space, then $int^{fs}(\tilde{\bigcup}_{i\in I}\{F_{A_i}^i{}^c\}) = \Phi$, where the (FSSs) sets $(F_{A_i}^i)$'s, $i \in I$ are fuzzy soft meager sets formed from the fuzzy soft dense and fuzzy soft $\delta-$sets $F_{A_i}^i$ in (U, E, ψ).*

Proof. Let the (FSTS) (U, E, ψ) be a fuzzy soft $\sigma-$Baire space and the (FSSs) $(F_{A_i}^i)$'s, $i \in I$ be fuzzy soft dense and fuzzy soft $\delta-$sets in (U, E, ψ). By Proposition 39, $cl^{fs}(\tilde{\bigcap}_{i\in I}\{F_{A_i}^i\}) = \overline{E}$. Then $(cl^{fs}(\tilde{\bigcap}_{i\in I}\{F_{A_i}^i\}))^c = \Phi$. This implies that $int^{fs}(\tilde{\bigcup}_{i\in I}\{F_{A_i}^i{}^c\}) = \Phi$. Also by Proposition 31, $(F_{A_i}^i)$'s are fuzzy soft meager sets in (U, E, ψ). Hence, $int^{fs}(\tilde{\bigcup}_{i\in I}\{F_{A_i}^i{}^c\}) = \Phi$, where the (FSSs) $(F_{A_i}^i)$'s, $i \in I$ are fuzzy soft meager sets formed from the fuzzy soft dense and fuzzy soft $\delta-$sets $F_{A_i}^i$ in (U, E, ψ).

Proposition 41. *If the fuzzy soft meager sets are formed from the fuzzy soft dense and fuzzy soft $\delta-$sets in a fuzzy soft $\sigma-$Baire space (U, E, ψ), then (U, E, ψ) is a fuzzy soft Baire space.*

Proof. Let the (FSTS) (U, E, ψ) be a fuzzy soft $\sigma-$Baire space. By Proposition 40, $int^{fs}(\tilde{\bigcup}_{i\in I}\{F_{A_i}^i{}^c\}) = \Phi$, where the (FSSs) $(F_{A_i}^i)$'s, $i \in I$ are fuzzy soft meager sets formed from the fuzzy soft dense and fuzzy soft $\delta-$sets $F_{A_i}^i$ in (U, E, ψ). Now $\tilde{\bigcup}_{i\in I}(int^{fs}\{F_{A_i}^i{}^c\})\tilde{\subseteq}int^{fs}(\tilde{\bigcup}_{i\in I}\{F_{A_i}^i{}^c\})$. Then we have

$\widetilde{\bigcup}_{i\in I}\{int^{fs}({F^i_{A_i}}^c)\} = \Phi$. This implies that $int^{fs}({F^i_{A_i}}^c) = \Phi$, where $({F^i_{A_i}}^c)$ is a fuzzy soft meager set in (U, E, ψ). By Proposition 30, (U, E, ψ) is a fuzzy soft Baire space.

Proposition 42. *If the (FSTS) (U, E, ψ) is a fuzzy soft $\sigma-$meager space, then (U, E, ψ) is not a fuzzy soft $\sigma-$Baire space.*

Proof. Let the (FSTS) (U, E, ψ) be a fuzzy soft $\sigma-$meager space. Then $\widetilde{\bigcup}_{i\in I}\{F^i_{A_i}\} = \overline{E}$, where $(F^i_{A_i})$'s are fuzzy soft $\sigma-$nowhere dense sets in (U, E, ψ). Now $int^{fs}(\widetilde{\bigcup}_{i\in I}\{F^i_{A_i}\}) = int^{fs}(\overline{E}) \neq \Phi$. Hence, by definition, (U, E, ψ) is not a fuzzy soft $\sigma-$Baire space.

Remark 43. By Proposition 42, we consider that if the (FSTS) (U, E, ψ) is a fuzzy soft $\sigma-$Baire space, then (U, E, ψ) is a fuzzy soft $\sigma-$second category space. Moreover, the converse is not true in general. That means a fuzzy soft $\sigma-$second category space need not be a fuzzy soft $\sigma-$Baire space.

Example 44. Let there be three houses in the universe U given by $U = \{s_1, s_2, s_3\}$ and let $E = \{$stone (e_1); steel (e_2); and brick $(e_3)\}$ be the set of parameters framed to choose one house to rent, where (brick) means the brick built houses, (steel) means the steel built houses, and (stone) means the stone built houses. Then, we consider $\psi = \{\Phi, \overline{E}, F_E, G_E, H_E\}\widetilde{\cup}\{T^i_E \mid i = 1, 2, 3, \ldots, 10\}$ is a fuzzy soft topology defined as follows:

$$F_E = \begin{pmatrix} .7 & .3 & .4 \\ .5 & .1 & .2 \\ .8 & .4 & .5 \end{pmatrix},$$
$$G_E = \begin{pmatrix} .4 & .5 & .6 \\ .2 & .3 & .4 \\ .5 & .6 & .7 \end{pmatrix}, \quad (10)$$
$$H_E = \begin{pmatrix} .6 & .4 & .7 \\ .4 & .2 & .5 \\ .7 & .5 & .8 \end{pmatrix}.$$

$T^1_E = F_E\widetilde{\cap}G_E$, $T^2_E = G_E\widetilde{\cap}H_E$, $T^3_E = F_E\widetilde{\cap}H_E$, $T^4_E = F_E\widetilde{\cup}G_E$, $T^5_E = G_E\widetilde{\cup}H_E$, $T^6_E = F_E\widetilde{\cup}H_E$, $T^7_E = G_E\widetilde{\cup}(F_E\widetilde{\cap}H_E)$, $T^8_E = F_E\widetilde{\cup}(G_E\widetilde{\cap}H_E)$, $T^9_E = H_E\widetilde{\cap}(F_E\widetilde{\cup}G_E)$, $T^{10}_E = F_E\widetilde{\cup}G_E\widetilde{\cup}H_E$. Now consider the (FSSs) $\alpha = (G_E)^c\widetilde{\cup}(T^4_E)^c\widetilde{\cup}(T^5_E)^c$, and $\beta = (F_E)^c\widetilde{\cup}(H_E)^c$ in (U, E, ψ). Then α and β are fuzzy soft $\lambda_\sigma-$sets in (U, E, ψ) and $int^{fs}(\alpha) = \Phi$, $int^{fs}(\beta) = \Phi$. Then α and β are fuzzy soft $\sigma-$nowhere dense sets in (U, E, ψ). Now $(\alpha\widetilde{\cup}\beta) \neq \overline{E}$. Therefore, (U, E, ψ) is a fuzzy soft $\sigma-$second category space. But $int^{fs}(\alpha\widetilde{\cup}\beta) \neq \Phi$ and therefore (U, E, ψ) is not a fuzzy soft $\sigma-$Baire space.

Proposition 45. *If $(\widetilde{\bigcap}_{i\in I}\{F^i_{A_i}\}) \neq \Phi$, where the (FSSs) $(F^i_{A_i})$'s are fuzzy soft dense and fuzzy soft $\delta-$sets in a (FSTS) (U, E, ψ), then (U, E, ψ) is a fuzzy soft $\sigma-$second category space.*

Proof. Let $(F^i_{A_i})$'s, $i \in I$ be fuzzy soft dense and fuzzy soft $\delta-$sets in (U, E, ψ). By Proposition 33, $({F^i_{A_i}}^c)$'s are fuzzy soft $\sigma-$nowhere dense sets in (U, E, ψ). Now $(\widetilde{\bigcap}_{i\in I}\{F^i_{A_i}\}) \neq \Phi$ implies that $(\widetilde{\bigcap}_{i\in I}\{F^i_{A_i}\})^c \neq \overline{E}$. Then $\widetilde{\bigcup}_{i\in I}\{(F^i_{A_i})^c\} \neq \overline{E}$. Hence, (U, E, ψ) is not a fuzzy soft $\sigma-$meager space and therefore (U, E, ψ) is a fuzzy soft $\sigma-$second category space.

Proposition 46. *If F_A is a fuzzy soft $\sigma-$meager set in (U, E, ψ), then there is a fuzzy soft $\lambda_\sigma-$set G_B in (U, E, ψ) such that $F_A\widetilde{\subseteq}G_B$.*

Proof. Let F_A be a fuzzy soft $\sigma-$meager set in (U, E, ψ). Thus, $F_A = \widetilde{\bigcup}_{i\in I}\{F^i_{A_i}\}$, where $(F^i_{A_i})$'s are fuzzy soft $\sigma-$nowhere dense sets in (U, E, ψ). Now $[(cl^{fs}(F^i_{A_i}))^c]$'s are (FSOSs) in (U, E, ψ). Then $T = \widetilde{\bigcap}_{i\in I}\{(cl^{fs}(F^i_{A_i}))^c\}$is a fuzzy soft $\delta-$set in (U, E, ψ) and $T^c = (\widetilde{\bigcap}_{i\in I}\{(cl^{fs}(F^i_{A_i}))^c\})^c = \widetilde{\bigcup}_{i\in I}\{cl^{fs}(F^i_{A_i})\}$. Now $F_A = \widetilde{\bigcup}_{i\in I}\{F^i_{A_i}\}\widetilde{\subseteq}\widetilde{\bigcup}_{i\in I}\{cl^{fs}(F^i_{A_i})\} = T^c$. That is, $F_A\widetilde{\subseteq}T^c$ and T^c is a fuzzy soft $\lambda_\sigma-$set in (U, E, ψ). Let ${G_B}^c\widetilde{\subseteq}T$. Hence, if F_A is a fuzzy soft $\sigma-$meager set in (U, E, ψ), then there is a fuzzy soft $\lambda_\sigma-$set G_B in (U, E, ψ) such that $F_A\widetilde{\subseteq}G_B$.

Proposition 47. *If G_B is a fuzzy soft $\sigma-$residual set in a (FSTS) (U, E, ψ) such that G_B in (U, E, ψ) such that $F_A\widetilde{\subseteq}G_B$, where F_A is a fuzzy soft dense and fuzzy soft $\delta-$set in (U, E, ψ), then (U, E, ψ) is a fuzzy soft $\sigma-$Baire space.*

Proof. Let G_B be a fuzzy soft $\sigma-$residual set in a (FSTS) (U, E, ψ). Thus, ${G_B}^c$ is a fuzzy soft $\sigma-$meager set in (U, E, ψ). Now by Proposition 46, there is a fuzzy soft $\lambda_\sigma-$set T in (U, E, ψ) such that ${G_B}^c\widetilde{\subseteq}T$. This implies that $T^c\widetilde{\subseteq}G_B$. Let $F_A = T^c$. Then F_A is a fuzzy soft $\delta-$set in (U, E, ψ) and $F_A\widetilde{\subseteq}G_B$ implies that $cl^{fs}(F_A)\widetilde{\subseteq}cl^{fs}(G_B)$. If $cl^{fs}(F_A) = \overline{E}$, then we have $cl^{fs}(G_B) = \overline{E}$. Hence, by Proposition 30, (U, E, ψ) is a fuzzy soft $\sigma-$Baire space.

Proposition 48. *If the (FSTS) (U, E, ψ) is a fuzzy soft $\sigma-$Baire space and if $\widetilde{\bigcup}_{i\in I}\{F^i_{A_i}\} = \overline{E}$, then there exists at least one $\lambda_\sigma-$set $F^i_{A_i}$ such that $int^{fs}(F^i_{A_i}) \neq \Phi$.*

Proof. Suppose that $int^{fs}(F_A) = \Phi$, $\forall$ $(i \in I)$, where $(F^i_{A_i})$'s are fuzzy soft $\sigma-$nowhere dense sets in (U, E, ψ). Then $\widetilde{\bigcup}_{i\in I}\{F^i_{A_i}\} = \overline{E}$, implying that $int^{fs}(\widetilde{\bigcup}_{i\in I}\{F^i_{A_i}\}) = int^{fs}(\overline{E}) = \overline{E} \neq \Phi$, a contradiction to (U, E, ψ) being a fuzzy soft $\sigma-$Baire space. Hence, $int(F^i_{A_i}) \neq \Phi$, for at least one $\lambda_\sigma-$set $F^i_{A_i}$ in (U, E, ψ).

Proposition 49. *If the (FSTS) (U, E, ψ) is a fuzzy soft $\sigma-$Baire space, then no nonempty (FSOS) is a fuzzy soft $\sigma-$meager set in (U, E, ψ).*

Proof. Let F_A be nonempty (FSOS) in a fuzzy soft $\sigma-$Baire space (U, E, ψ). Suppose that $F_A = \widetilde{\bigcup}_{i\in I}\{F^i_{A_i}\}$, where the (FSSs) $(F^i_{A_i})$'s are fuzzy soft $\sigma-$nowhere dense sets in (U, E, ψ). Then $int^{fs}(F_A) = int^{fs}(\widetilde{\bigcup}_{i\in I}\{F^i_{A_i}\})$. Since (U, E, ψ) is a fuzzy soft σ−Baire space, $int^{fs}(\widetilde{\bigcup}_{i\in I}\{F^i_{A_i}\}) = \Phi$. This implies that $int^{fs}(F_A) = \Phi$. Then we will have $F_A = \Phi$, which is a contradiction, since $F_A \in \psi$ implies that $int^{fs}(F_A) = F_A \neq \Phi$. Hence, no nonempty (FSOS) is a fuzzy soft σ−meager set in (U, E, ψ).

Definition 50. A (FSTS) (U, E, ψ) is called a fuzzy soft submaximal space if for each (FSS) F_A in (U, E, ψ) such that $cl^{fs}(F_A) = \overline{E}$; then $F_A \in \psi$ in (U, E, ψ).

Proposition 51. *If the (FSTS) (U, E, ψ) is a fuzzy soft submaximal space and if F_A is a fuzzy soft σ−meager set in (U, E, ψ), then F_A is a fuzzy soft meager set in (U, E, ψ).*

Proof. Let $F_A = \widetilde{\bigcup}_{i\in I}\{F^i_{A_i}\}$ be a fuzzy soft σ−meager set in (U, E, ψ), where the (FSSs) $(F^i_{A_i})$'s, $i \in I$ are fuzzy soft σ−nowhere dense sets in (U, E, ψ). Then we have $int^{fs}(F^i_{A_i}) = \Phi$ and $(F^i_{A_i})$'s, $i \in I$ are fuzzy soft λ_σ−sets in (U, E, ψ). Now $int^{fs}(F^i_{A_i}) = \Phi$, implying that $(int^{fs}(F^i_{A_i}))^c = \Phi^c = \overline{E}$ and hence $cl^{fs}(F^{i\ c}_{A_i}) = \overline{E}$. Since (U, E, ψ) is a fuzzy soft submaximal space, the fuzzy soft dense sets $(F^{i\ c}_{A_i})$'s are (FSOSs) in (U, E, ψ) and hence $(F^i_{A_i})$'s are (FSCSs) in (U, E, ψ). Then $cl^{fs}(F^i_{A_i}) = (F^i_{A_i})$ and $int^{fs}(F^i_{A_i}) = \Phi$ imply that $int^{fs}(cl^{fs}(F^i_{A_i})) = int^{fs}(F^i_{A_i}) = \Phi$. That is, $(F^i_{A_i})$'s are fuzzy soft nowhere dense sets in (U, E, ψ). Therefore, $F_A = \widetilde{\bigcup}_{i\in I}\{F^i_{A_i}\}$ is a fuzzy soft meager set in (U, E, ψ). □

Proposition 52. *If the (FSTS) (U, E, ψ) is a fuzzy soft σ−Baire space and fuzzy soft submaximal space, then (U, E, ψ) is a fuzzy soft Baire space.*

Proof. Let F_A be a fuzzy soft σ−meager set in (U, E, ψ). Since (U, E, ψ) is a fuzzy soft submaximal space, by Proposition 51, F_A is a fuzzy soft meager set in (U, E, ψ). Since (U, E, ψ) is a fuzzy soft σ−Baire space, by Proposition 38, $int^{fs}(F_A) = \Phi$. Hence, for the fuzzy soft meager set F_A in (U, E, ψ), we have $int^{fs}(F_A) = \Phi$. Therefore, by Proposition 30, (U, E, ψ) is a fuzzy soft Baire space.

Definition 53. A fuzzy soft P−space is a $(FSTS)(U, E, \psi)$ with the property that states that if countable intersection of fuzzy soft open sets in (U, E, ψ) is fuzzy soft open. That is, every non−empty fuzzy soft δ−set in (U, E, ψ) is fuzzy soft open in (U, E, ψ).

Proposition 54. *If the (FSTS) (U, E, ψ) is a fuzzy soft σ−Baire space and fuzzy soft P−space, then (U, E, ψ) is a fuzzy soft Baire space.*

Proof. Let the (FSTS) (U, E, ψ) be a fuzzy soft σ−Baire space. Then, by Proposition 39, $cl^{fs}(\widetilde{\bigcap}_{i\in I}\{F^i_{A_i}\}) = \overline{E}$, where the (FSSs) $(F^i_{A_i})$'s, $i \in I$ are fuzzy soft dense and fuzzy soft δ−sets in (U, E, ψ). Now from $cl^{fs}(\widetilde{\bigcap}_{i\in I}\{F^i_{A_i}\}) = \overline{E}$, we have $(cl^{fs}(\widetilde{\bigcap}_{i\in I}\{F^i_{A_i}\}))^c = \Phi$. This implies that $int^{fs}(\widetilde{\bigcup}_{i\in I}(F^i_{A_i})^c) = \Phi$. Since the (FSSs) $(F^i_{A_i})$'s are fuzzy soft dense in (U, E, ψ), $cl^{fs}(F^i_{A_i}) = \overline{E}$. Then we have $(cl^{fs}(F^i_{A_i}))^c = \Phi$. This implies that $int^{fs}(F^{i\ c}_{A_i}) = \Phi$. Also, since (U, E, ψ) is a fuzzy soft P−space, the non−empty fuzzy soft δ−sets $(F^i_{A_i})$'s in (U, E, ψ) are fuzzy soft open in (U, E, ψ). Then $(F^{i\ c}_{A_i})$'s are (FSCSs) in (U, E, ψ). Then $cl^{fs}(F^{i\ c}_{A_i}) = F^{i\ c}_{A_i}$ and $int^{fs}(F^{i\ c}_{A_i}) = \Phi$ imply that $int^{fs}(cl^{fs}(F^{i\ c}_{A_i})) = int^{fs}(F^{i\ c}_{A_i}) = \Phi$. That is, $(F^{i\ c}_{A_i})$'s are fuzzy soft nowhere dense sets in (U, E, ψ). Therefore, we have $int^{fs}(\widetilde{\bigcup}_{i\in I}\{F^{i\ c}_{A_i}\}) = \Phi$, where $(F^{i\ c}_{A_i})$'s are fuzzy soft nowhere dense sets in (U, E, ψ). Hence, by Proposition 30, (U, E, ψ) is a fuzzy soft Baire space.

Definition 55. A (FSTS) (U, E, ψ) is called a fuzzy soft almost resolvable space if $\widetilde{\bigcup}_{i\in I}\{F^i_{A_i}\} = \overline{E}$, where the (FSSs) $(F^i_{A_i})$'s in (U, E, ψ) are such that $int^{fs}(F^i_{A_i}) = \Phi$. Otherwise, (U, E, ψ) is called a fuzzy soft almost irresolvable space.

Proposition 56. *If the (FSTS) (U, E, ψ) is a fuzzy soft almost irresolvable space, then (U, E, ψ) is a fuzzy soft σ−second category space.*

Proof. Let $(F^i_{A_i})$'s, $i \in I$ be the fuzzy soft dense and fuzzy soft δ−sets in (U, E, ψ). Now $cl^{fs}(F^i_{A_i}) = \overline{E}$ implies that $(cl^{fs}(F^i_{A_i}))^c = \Phi$. That is, $int^{fs}(F^{i\ c}_{A_i}) = \Phi$. Since (U, E, ψ) is a fuzzy soft almost irresolvable space, $\widetilde{\bigcup}_{i\in I}\{F^{i\ c}_{A_i}\} \neq \overline{E}$, where the (FSSs) $(F^i_{A_i})$'s in (U, E, ψ) are such that $int^{fs}(F^{i\ c}_{A_i}) = \Phi$. Now $\widetilde{\bigcup}_{i\in I}\{F^{i\ c}_{A_i}\} \neq \overline{E}$ implies that $(\widetilde{\bigcup}_{i\in I}\{(F^{i\ c}_{A_i}\})^c \neq \Phi$. Hence, we have $\widetilde{\bigcap}_{i\in I}\{F^i_{A_i}\} \neq \Phi$, where the (FSSs) $(F^i_{A_i})$'s are fuzzy soft dense and fuzzy soft δ−sets in a (FSTS) (U, E, ψ). Thus, by Proposition 45, (U, E, ψ) is a fuzzy soft σ−second category space.

Definition 57. A (FSTS) (U, E, ψ) is called a fuzzy soft hyperconnected space if every (FSOS) F_A is fuzzy soft dense in (U, E, ψ). That is, $cl^{fs}(F_A) = \overline{E}\ \forall \Phi \neq F_A \in \psi$.

Proposition 58. *If $cl^{fs}(\widetilde{\bigcap}_{i\in I}\{F^i_{A_i}\}) = \overline{E}$, where $(F^i_{A_i})$'s are fuzzy soft dense and fuzzy soft δ−sets in (U, E, ψ), then (U, E, ψ) is a fuzzy soft σ−Baire space.*

Proof. The proof is obvious.

Proposition 59. *If $cl^{fs}(\widetilde{\bigcap}_{i\in I}\{F^i_{A_i}\}) = \overline{E}$, where the (FSSs) $(F^i_{A_i})$'s are fuzzy soft δ−sets in a fuzzy soft hyperconnected and fuzzy soft P−space (U, E, ψ), then (U, E, ψ) is a fuzzy soft σ−Baire space.*

Proof. Let $(F^i_{A_i})$'s, $i \in I$ be the fuzzy soft δ−sets in (U, E, ψ) such that $cl^{fs}(\widetilde{\bigcap}_{i\in I}\{F^i_{A_i}\}) = \overline{E}$. Since (U, E, ψ) is a fuzzy soft P−space, the fuzzy soft δ−sets $(F^i_{A_i})$'s in (U, E, ψ) are fuzzy soft open in (U, E, ψ). Also since (U, E, ψ) is a fuzzy soft hyperconnected space, the (FSOSs) $(F^i_{A_i})$'s in (U, E, ψ) are fuzzy soft dense sets in (U, E, ψ). Hence, the (FSSs) $(F^i_{A_i})$'s, $i \in I$ are fuzzy soft dense and fuzzy soft δ–sets in (U, E, ψ) and $cl^{fs}(\widetilde{\bigcap}_{i\in I}\{F^i_{A_i}\}) = \overline{E}$. Hence, by Proposition 58, (U, E, ψ) is a fuzzy soft σ–Baire space.

Definition 60. Let F_A be a fuzzy soft subset of a fuzzy soft space $\overline{E}$. Then F_A is said to be fuzzy soft A–embedded in $\overline{E}$ if each fuzzy soft δ–subset G_B of $\overline{E}$ which is contained in F_A is fuzzy soft nowhere dense in F_A (i.e., $int^{fs}_{F_A}(cl^{fs}_{F_A}(H_C)) = \Phi$).

Proposition 61. *Let F_A be a fuzzy soft dense subspace of a fuzzy soft Baire space $\overline{E}$. If $\overline{E} \setminus F_A$ is fuzzy soft A–embedded in $\overline{E}$, then F_A is a fuzzy soft Baire space.*

Proof. Observe that if F_A is not a fuzzy soft Baire space; then there is a sequence $G_B \widetilde{\supseteq} F^1_{A_1} \widetilde{\supseteq} F^2_{A_2} \widetilde{\supseteq} F^3_{A_3} \widetilde{\supseteq} \ldots$ of fuzzy soft open subsets of F_A such that each $F^i_{A_i}$ is fuzzy soft dense in G_B and yet $\widetilde{\bigcap}_{i\in I}\{F^i_{A_i}\} = \Phi$. Then there is a sequence $H_C \widetilde{\supseteq} K^1_{D_1} \widetilde{\supseteq} K^2_{D_2} \widetilde{\supseteq} K^3_{D_3} \widetilde{\supseteq} \ldots$ of fuzzy soft open subsets of $\overline{E}$ such that $G_B = H_C \widetilde{\cap} F_A$ and $F^i_{A_i} = K^i_{D_i} \widetilde{\cap} F_A$. Each $K^i_{D_i}$ is fuzzy soft dense in H_C and H_C is a fuzzy soft Baire space. Hence, $\widetilde{\bigcap}_{i\in I}\{K^i_{D_i}\}$ is fuzzy soft dense in H_C and therefore in $H_C \widetilde{\cap} \overline{E} \setminus F_A$. Since $\widetilde{\bigcap}_{i\in I}\{K^i_{D_i}\} \widetilde{\subseteq} \overline{E} \setminus F_A, \overline{E} \setminus F_A$ is not fuzzy soft A–embedded in $\overline{E}$.

Proposition 62. *Let F_A be a fuzzy soft dense subspace of a fuzzy soft Baire space $\overline{E}$. If $\overline{E} \setminus F_A$ is dense in $\overline{E}$, then F_A is a fuzzy soft Baire space if and only if $\overline{E} \setminus F_A$ is fuzzy soft A–embedded in $\overline{E}$.*

Proof. Assume that $\overline{E} \setminus F_A$ is not fuzzy soft A–embedded in $\overline{E}$. Let G_B be a fuzzy soft δ–subset of $\overline{E}$ which is contained in $\overline{E} \setminus F_A$ and which is fuzzy soft dense in some relatively (FSOS) H_C of $\overline{E} \setminus F_A$. Let K_D be an fuzzy soft open subset of $\overline{E}$ with $K_D \widetilde{\cap} (\overline{E} \setminus F_A) = H_C$. Then $T_M = K_D \widetilde{\cap} G_B$ is a δ–subset of $\overline{E}$ which is fuzzy soft dense in K_D and which is contained in H_C. Let $T_M = \widetilde{\bigcap}_{i\in I}\{T^i_{M_i}\}$, where each $T^i_{M_i}$ is open in $\overline{E}$ and $T^i_{M_i} \widetilde{\subseteq} K_D$. The (FSSs) $\{T^i_{M_i} \widetilde{\cap} F_A\}$ are fuzzy soft open and fuzzy soft dense subsets of $K_D \widetilde{\cap} F_A$ and yet $\widetilde{\bigcap}_{i\in I}\{T^i_{M_i} \widetilde{\cap} F_A\}$. It follows that $K_D \widetilde{\cap} F_A$ is not a fuzzy soft Baire space. Consequently, F_A is not a fuzzy soft Baire space. Conversely, assume that $\overline{E} \setminus F_A$ is fuzzy soft A–embedded in $\overline{E}$ and, by Proposition 61, then F_A is a fuzzy soft Baire space.

Lemma 63. *Let $\overline{E}$ be a fuzzy soft Baire space. If $G_B = \widetilde{\bigcap}_{i\in I}\{T^i_{M_i}\}$ is a nonempty fuzzy soft nowhere dens δ–set of $\overline{E}$, where $T^i_{M_i}$ is a fuzzy soft open subset of $\overline{E}$ $\forall i \in I$, then for every nonempty fuzzy soft open subset T_M of $\overline{E}$ there is $i \in I$ such that $int^{fs}((\overline{E} \setminus T^i_{M_i}) \widetilde{\cap} (T_M \setminus cl^{fs}(G_B))) \neq \Phi$.*

Proof. Let T_M be a nonempty fuzzy soft open subset of $\overline{E}$. Then, $T_M \setminus cl^{fs}(G_B)$ is a nonempty fuzzy soft open subset of $\overline{E}$; hence, $T_M \setminus cl^{fs}(G_B)$ is also fuzzy soft Baire. Since $T_M \setminus cl^{fs}(G_B) \widetilde{\subseteq} \widetilde{\bigcup}_{i\in I}(\overline{E} \setminus T^i_{M_i})$ and each $\overline{E} \setminus T^i_{M_i}$ is a fuzzy soft closed subset of $\overline{E}$, by Remark 22, then there is $i \in I$ such that $int^{fs}((\overline{E} \setminus T^i_{M_i}) \widetilde{\cap} (T_M \setminus cl^{fs}(G_B))) \neq \Phi$.

Proposition 64. *Let $\overline{E}$ be a fuzzy soft Baire space and let $F_A \widetilde{\subseteq} \overline{E}$ be fuzzy soft dense. Then F_A is a fuzzy soft Baire space if and only if every fuzzy soft δ–set in $\overline{E}$ contained in $\overline{E} \setminus F_A$ is fuzzy soft nowhere dense.*

Proof.

Necessity. Let $G_B = \widetilde{\bigcap}_{i\in I}\{T^i_{M_i}\}$, where $T^i_{M_i}$ is a fuzzy soft open subset of $\overline{E}$ for each $i \in I$, which is contained in $\overline{E} \setminus F_A$. Then, $F_A \widetilde{\subseteq} \widetilde{\bigcup}_{i\in I}(\overline{E} \setminus T^i_{M_i})$. In virtue of Remark 22, $\widetilde{\bigcup}_{i\in I} int^{fs}_{F_A}(F_A \widetilde{\cap} (\overline{E} \setminus T^i_{M_i}))$ is fuzzy soft dense in F_A. Suppose that $int_{fs}(cl^{fs}(G_B)) \neq \Phi$. Then, there is $n \in I$ such that $\Phi \neq int^{fs}(cl^{fs}(G_B)) \widetilde{\cap} int^{fs}_{F_A}(F_A \widetilde{\cap} (\overline{E} \setminus T^n_{M_n}))$. On the other hand, we know that $cl^{fs}(G_B) \widetilde{\subseteq} cl^{fs}(T^n_{M_n}) = cl^{fs}(T^n_{M_n} \widetilde{\cap} F_A)$. Hence, $\Phi \neq int^{fs}(cl^{fs}(G_B)) \widetilde{\cap} int^{fs}_{F_A}(F_A \widetilde{\cap} (\overline{E} \setminus T^n_{M_n})) \widetilde{\subseteq} cl^{fs}(T^n_{M_n} \widetilde{\cap} F_A) \widetilde{\cap} F_A = cl^{fs}_{F_A}(T^n_{M_n} \widetilde{\cap} F_A)$ which implies that $int^{fs}(cl^{fs}(G_B)) \widetilde{\cap} int^{fs}_{F_A}(F_A \widetilde{\cap} (\overline{E} \setminus T^n_{M_n})) \widetilde{\cap} cl^{fs}(T^n_{M_n} \widetilde{\cap} F_A) \neq \Phi$, but this is impossible.

Sufficiency. Assume that F_A is no fuzzy soft Baire. According to Remark 22, there is a countable fuzzy soft closed cover $\{F^i_{A_i}; i \in I\}$ of F_A such that $\widetilde{\bigcup}_{i\in I}\{int^{fs}(F^i_{A_i})\}$ is not dense in F_A. For each $i \in I$, choose a fuzzy soft closed subset $T^i_{M_i}$ of $\overline{E}$ such that $F^i_{A_i} = T^i_{M_i} \widetilde{\cap} F_A$, for each $i \in I$. Let $T_M = \widetilde{\bigcap}_{i\in I}(\overline{E} \setminus T^i_{M_i})$ which is a δ–set of $\overline{E}$ contained in $\overline{E} \setminus F_A$. If $T_M = \Phi$, then $\{T^i_{M_i}; i \in I\}$ would be a fuzzy soft closed cover of $\overline{E}$ and, by Remark 22, then $\widetilde{\bigcup}_{i\in I}\{int^{fs}(T^i_{M_i})\}$ would be fuzzy soft dense in $\overline{E}$ which is not possible. So, $T_M \neq \Phi$. Choose a nonempty fuzzy soft open subset K_D of $\overline{E}$ such that $K_D \widetilde{\cap} F_A \widetilde{\cap} int^{fs}_{F_A}(F^i_{A_i}) = \Phi$, $\forall i \in I$. By Lemma 63, we can find $r \in I$ such that $int^{fs}(T^r_{M_r}) \widetilde{\cap} (K_D \setminus int^{fs}(T_M)) \neq \Phi$. Hence, $\Phi \neq int^{fs}(T^r_{M_r}) \widetilde{\cap} (K_D \setminus int^{fs}(T_M)) \widetilde{\cap} F_A \widetilde{\subseteq} int^{fs}_{F_A}(T^r_{M_r} \widetilde{\cap} F_A) \widetilde{\cap} K_D \widetilde{\cap} F_A \widetilde{\subseteq} int^{fs}_{F_A}(F^i_{A_i}) \widetilde{\cap} K_D \widetilde{\cap} F_A$, but this is a contradiction. Thus, F_A is fuzzy soft Baire.

4. Baireness in Fuzzy Soft Setting

In this section, we shall study the new class of fuzzy soft Baire spaces.

Definition 65. We say a space $\overline{E}$ is fuzzy soft D–Baire if every fuzzy soft dense subspace of $\overline{E}$ is fuzzy soft Baire.

An immediate consequence of Proposition 64 is the following.

Corollary 66. *Suppose that $\overline{E}$ is a fuzzy soft Baire space. Then, every fuzzy soft $\delta-$set in $\overline{E}$ with empty interior is fuzzy soft nowhere dense iff $\overline{E}$ is fuzzy soft D–Baire*

Proof. It follows from Proposition 64 and Definition 65.

Definition 67. We say that a (FSTS) $\overline{E}$ is a fuzzy soft almost $P-$space if every non-empty fuzzy soft $\delta-$set in $\overline{E}$ has a nonempty interior.

Corollary 68. *Every fuzzy soft Baire and fuzzy soft almost $P-$space is fuzzy soft D–Baire.*

Proof. This is a consequence of Proposition 64 and Definition 67.

Definition 69. A fuzzy soft Borel set is any (FSS) in a (FSTS) that can be formed from (FSOSs) (or, equivalently, from (FSCSs)) through the operations of countable union, countable intersection, and relative complement.

Definition 70. Let $\overline{E}$ be a (FSTS). Then, the class $FSPB(\overline{E})$ is the fuzzy soft $\sigma-$algebra in $\overline{E}$ generated by all (FSOSs) and all fuzzy soft nowhere dense sets.

Remark 71. (1) For a (FSTS) (U, E, ψ), the collection of all fuzzy soft Borel sets on $\overline{E}$ forms a fuzzy soft $\sigma-$algebra.

(2) The fuzzy soft $\sigma-$algebra of fuzzy soft Borel sets is contained in the class $FSPB(\overline{E})$.

(3) It is clear to show that $F_A \tilde{\subseteq} \overline{E}$ belongs to the class $FSPB(\overline{E})$ if and only if F_A may be expressed in the form $F_A = G_B \tilde{\cup} H_C$, where G_B is a fuzzy soft $\delta-$set and H_C is fuzzy soft meager.

Theorem 72. *The following seven conditions on a space (U, E, ψ) are equivalent.*

(1) (U, E, ψ) is fuzzy soft D–Baire.

(2) (U, E, ψ) is fuzzy soft Baire and every fuzzy soft $\delta-$set with empty interior is fuzzy soft nowhere dense.

(3) Every fuzzy soft meager subset $F_A \tilde{\subseteq} \overline{E}$ is fuzzy soft nowhere dense.

(4) (U, E, ψ) is fuzzy soft Baire and every fuzzy soft dense $\delta-$set has fuzzy soft dense interior.

(5) (U, E, ψ) is fuzzy soft Baire and every set in the class $FSPB(\overline{E})$ with empty interior is fuzzy soft nowhere dense.

(6) (U, E, ψ) is fuzzy soft Baire and every fuzzy soft Borel set with empty interior is fuzzy soft nowhere dense.

(7) (U, E, ψ) is fuzzy soft Baire and the union of a fuzzy soft $\delta-$set with empty interior and a fuzzy soft meager set of $\overline{E}$ is fuzzy soft nowhere dense.

Proof. (1)$\Longleftrightarrow$(2). This is Corollary 66.

(2) $\Longrightarrow$ (3). Let $F_A \tilde{\subseteq} \overline{E}$ be a fuzzy soft meager set. Assume $F_A = \tilde{\bigcup}_{i\in I}\{F^i_{A_i}\}$, where $F^i_{A_i}$ is fuzzy soft nowhere dense $\forall i \in I$. Therefore, $L_M = \overline{E} \setminus \tilde{\bigcup}_{i\in I}\{cl^{fs}(F^i_{A_i})\} = \tilde{\bigcap}_{i\in I}\{\overline{E} \setminus cl^{fs}(F^i_{A_i})\}$ is a $\delta-$set in $\overline{E}$ and L_M is fuzzy soft dense in $\overline{E}$ because its complement is a fuzzy soft meager set and $\overline{E}$ is fuzzy soft Baire. Let $V_D = int^{fs}(L_M)$. The (FSS) $L_M - V_D$ clearly has empty interior. Hence, $L_M - cl^{fs}(V_D)$ is a fuzzy soft $\delta-$set with empty interior; by hypothesis, $L_M - cl^{fs}(V_D)$ is fuzzy soft nowhere dense. Also $L_M \tilde{\cap} Fr^{fs}(V_D)$ is a fuzzy soft nowhere dense set. Therefore, $L_M - V_D = (L_M - cl^{fs}(V_D))\tilde{\cup}(L_M \tilde{\cap} Fr^{fs}(V_D))$ is a fuzzy soft nowhere dense set as well. On the other hand, $\overline{E} \setminus V_D = (L_M \setminus V_D)\tilde{\cup}(\overline{E} \setminus L_M) = (L_M \setminus V_D)\tilde{\cup}(\tilde{\bigcup}_{i\in I}\{F^i_{A_i}\})$ is a fuzzy soft meager set. Since $\overline{E}$ is fuzzy soft Baire, $\Phi = int^{fs}(\overline{E} \setminus V_D) = \overline{E} \setminus cl^{fs}(V_D)$. Therefore, $cl^{fs}(V_D) = \overline{E}$ and $F_A \tilde{\subseteq} \overline{E} \setminus V_D = Fr^{fs}(V_D)$ is fuzzy soft nowhere dense.

(3) $\Longrightarrow$ (4). It follows from Remark 22 that $\overline{E}$ is a fuzzy soft Baire space. Let $L_M \tilde{\subseteq} \overline{E}$ be a fuzzy soft dense $\delta-$set of $\overline{E}$. Since $\overline{E} \setminus L_M$ is a fuzzy soft meager set, the hypothesis implies that $\overline{E} \setminus L_M$ is fuzzy soft nowhere dense; i.e., $cl^{fs}(\overline{E} \setminus L_M)$ has empty interior. Therefore, $V_D = \overline{E} \setminus cl^{fs}(\overline{E} \setminus L_M) = int^{fs}(L_M)$ is a fuzzy soft open dense subspace of $\overline{E}$.

(4) $\Longrightarrow$ (2). Let G_B be a $\delta-$set with empty. First observe that $int^{fs}(cl^{fs}(G_B)) \tilde{\subseteq} cl^{fs}(cl^{fs}(G_B) \setminus G_B)$. Since $cl^{fs}(G_B) \setminus G_B$ is an $\lambda_\sigma-$set with empty interior, $\overline{E} \setminus (cl^{fs}(G_B) \setminus G_B)$ is a fuzzy soft dense $\delta-$set of $\overline{E}$ By assumption, $int^{fs}(\overline{E} \setminus (cl^{fs}(G_B) \setminus G_B))$ is also fuzzy soft dense in $\overline{E}$. That is, $\overline{E} \setminus cl^{fs}(cl^{fs}(G_B) \setminus G_B)$ is fuzzy soft dense in $\overline{E}$. Hence, $int^{fs}(cl^{fs}(cl^{fs}(G_B) \setminus G_B)) = \Phi$ and so $int^{fs}(cl^{fs}(G_B)) = \Phi$.

(4) $\Longrightarrow$ (5). We have already established above the equivalence among clauses (1), (2), (3), and (4). The fifth clause follows directly from the properties of the class $FSPB(\overline{E})$ and clauses (2) and (3).

(5) $\Longrightarrow$ (6). This implication is obvious because the fuzzy soft $\delta-$algebra of fuzzy soft Borel sets is contained in the class $FSPB(\overline{E})$.

(6) $\Longrightarrow$ (1). It is enough to observe that (6) $\Longrightarrow$ (2) $\Longrightarrow$ (1).

(1)$\Longrightarrow$(7). We know the first six statements are equivalent to each other. Thus, clause (7) follows directly from clauses (2) and (3). (7) $\Longrightarrow$ (1). This is a consequence of Corollary 66.

5. Conclusion

In the present paper, we have introduced and discussed new notions of Baireness in fuzzy soft topological spaces. Furthermore, there are many problems and applications in algebra that deal with group theory and spaces. So, future work in this regard would be required to study some applications using the properties of ψ in our new fuzzy soft spaces and new operations depend on fuzzy soft operations $\tilde{\cup}$ and $\tilde{\cap}$ to consider new fuzzy soft groups and fuzzy soft commutative rings. Also, let us say (U, E, ψ) is fuzzy soft $N-$Baire if every fuzzy soft set in (U, E, ψ) with empty interior is fuzzy soft nowhere dense. The question we are concerned with is as follows: what are the possible relationships considered between fuzzy soft $N-$Baire and each concept of our notions that are given in this work?

Conflicts of Interest

The authors declare that they have no conflicts of interest regarding the publication of this paper.

References

[1] R. C. Haworth and R. A. McCoy, "Baire spaces," *Dissertationes Math*, vol. 141, pp. 1–77, 1977.

[2] Z. e. Frolik, "Baire spaces and some generalizations of complete metric spaces," *Czechoslovak Mathematical Journal*, vol. 11 (86), pp. 237–248, 1961.

[3] G. Gruenhage and D. Lutzer, "Baire and Volterra spaces," *Proceedings of the American Mathematical Society*, vol. 128, no. 10, pp. 3115–3124, 2000.

[4] T. Neubrunn, "A note on mappings of Baire spaces," *Math. Slovaca*, vol. 27, no. 2, pp. 173–176, 1977.

[5] G. Thangaraj and G. Balasubramanian, "On somewhat fuzzy continuous functions," *Journal of Fuzzy Mathematics*, vol. 11, no. 3, pp. 725–736, 2003.

[6] G. Thangaraj and S. Anjalmose, "On fuzzy Baire spaces," *Journal of Fuzzy Mathematics*, vol. 21, no. 3, pp. 667–676, 2013.

[7] G. Thangaraj and S. Anjalmose, "On fuzzy D-Baire spaces," *Annals of Fuzzy Mathematics and Informatics*, vol. 7, no. 1, pp. 99–108, 2014.

[8] D. Molodtsov, "Soft set theory-first results," *Computers & Mathematics with Applications*, vol. 37, no. 4-5, pp. 19–31, 1999.

[9] P. K. Maji, A. R. Roy, and R. Biswas, "An application of soft sets in a decision making problem," *Computers & Mathematics with Applications*, vol. 44, no. 8-9, pp. 1077–1083, 2002.

[10] P. K. Maji, R. Biswas, and A. R. Roy, "Soft set theory," *Computers & Mathematics with Applications*, vol. 45, no. 4-5, pp. 555–562, 2003.

[11] S. Mahmood, "Soft Regular Generalized b-Closed Sets in Soft Topological Spaces," *Journal of Linear and Topological Algebra*, vol. 3, no. 4, pp. 195–204, 2014.

[12] S. Mahmood, "Tychonoff Spaces in Soft Setting and their Basic Properties," *International Journal of Applications of Fuzzy Sets and Artificial Intelligence*, vol. 7, pp. 93–112, 2017.

[13] S. Mahmood, "Soft Sequentially Absolutely Closed Spaces," *International Journal of Applications of Fuzzy Sets and Artificial Intelligence*, vol. 7, pp. 73–92, 2017.

[14] S. Mahmood and M. Alradha, "Soft Edge ρ-Algebras of the power sets," *International Journal of Applications of Fuzzy Sets and Artificial Intelligence*, vol. 7, pp. 231–243, 2017.

[15] S. M. Khalil and A. F. al Musawi, "Soft BCL-algebras of the power sets," *International Journal of Algebra*, vol. 11, pp. 329–341, 2017.

[16] S. Mahmood and M. Abd Ulrazaq, "Soft BCH-Algebras of the Power Sets," *American Journal of Mathematics and Statistics*, vol. 8, no. 1, pp. 1–7, 2018.

[17] S. M. Khalil and F. H. Khadhaer, "An algorithm for generating permutation algebras using soft spaces," *Journal of Taibah University for Science*, vol. 12, no. 3, pp. 299–308, 2018.

[18] S. Mahmood and F. Hameed, "An algorithm for generating permutations in symmetric groups using soft spaces with general study and basic properties of permutation spaces," *Journal of Theoretical and Applied Information Technology*, vol. 96, no. 9, pp. 2445–2457, 2018.

[19] S. Mahmood, "Dissimilarity fuzzy soft points and their applications," *Fuzzy Information and Engineering*, vol. 8, no. 3, pp. 281–294, 2016.

[20] P. K. Maji, R. Biswas, and A. R. Roy, "Fuzzy soft sets," *Journal of Fuzzy Mathematics*, vol. 9, no. 3, pp. 589–602, 2001.

[21] P. K. Maji, R. Biswas, and A. R. Roy, "Intuitionistic fuzzy soft sets," *Journal of Fuzzy Mathematics*, vol. 9, no. 3, pp. 677–692, 2001.

[22] S. Mahmood and Z. Al-Batat, "Intuitionistic Fuzzy Soft LA-Semigroups and Intuitionistic Fuzzy Soft Ideals," *International Journal of Applications of Fuzzy Sets and Artificial Intelligence*, vol. 6, pp. 119–132, 2016.

[23] S. Mahmood Khalil, "On intuitionistic fuzzy soft b-closed sets in intuitionistic fuzzy soft topological spaces," *Annals of Fuzzy Mathematics and Informatics*, vol. 10, no. 2, pp. 221–233, 2015.

[24] K. Tyagi and A. Tripathi, "Rough fuzzy automata and rough fuzzy grammar," *International Journal of Fuzzy System Applications*, vol. 6, no. 1, pp. 36–55, 2017.

[25] A. Taeib and A. Chaari, "Optimal tuning strategy for mimo fuzzy predictive controllers," *International Journal of Fuzzy System Applications*, vol. 4, no. 4, pp. 87–99, 2015.

[26] H. Zoulfaghari, J. Nematian, and A. A. K. Nezhad, "A resource-constrained project scheduling problem with fuzzy activity times," *International Journal of Fuzzy System Applications*, vol. 5, no. 4, pp. 1–15, 2016.

[27] N. Cagman, S. Enginoglu, and F. Citak, "Fuzzy soft set theory and its applications," *Iranian Journal of Fuzzy Systems*, vol. 8, no. 3, pp. 137–147, 170, 2011.

[28] S. Roy and T. K. Samanta, "An introduction to open and closed sets on fuzzy soft topological spaces," *Annals of Fuzzy Mathematics and Informatics*, vol. 6, no. 2, pp. 425–431, 2013.

10

KC-Means: A Fast Fuzzy Clustering

Israa Abdzaid Atiyah,[1] Adel Mohammadpour ,[1] and S. Mahmoud Taheri[2]

[1] *Faculty of Mathematics and Computer Science, Amirkabir University of Technology, Tehran, Iran*
[2] *School of Engineering Science, College of Engineering, University of Tehran, Tehran, Iran*

Correspondence should be addressed to Adel Mohammadpour; adel@aut.ac.ir

Academic Editor: Ferdinando Di Martino

A novel hybrid clustering method, named *KC*-Means clustering, is proposed for improving upon the clustering time of the Fuzzy *C*-Means algorithm. The proposed method combines *K*-Means and Fuzzy *C*-Means algorithms into two stages. In the first stage, the *K*-Means algorithm is applied to the dataset to find the centers of a fixed number of groups. In the second stage, the Fuzzy *C*-Means algorithm is applied on the centers obtained in the first stage. Comparisons are then made between the proposed and other algorithms in terms of time processing and accuracy. In addition, the mentioned clustering algorithms are applied to a few benchmark datasets in order to verify their performances. Finally, a class of Minkowski distances is used to determine the influence of distance on the clustering performance.

1. Introduction

Clustering is a method of separating similar data from distinctly different ones into relevant categories or clusters. Being an unsupervised approach, it helps to recognize and extract hidden patterns within the data. The distance, such as Euclidean and Manhattan as a special case of Minkowski, plays an important role in clustering algorithms. Clustering techniques enjoy some advantages as no requirement for domain knowledge or labeled data while they are able to deal with a wide variety of data, including noise and outliers, as well.

Clustering methods may be categorized into two general types: hard and soft. Hard clusters possess well-defined boundaries; examples include *K*-Means (KM) and hierarchical methods [1]. To improve the time processes of fuzzy clustering, we propose a 2-step hybrid method of *K*-Means Fuzzy *C*-Means (KCM) clustering that combines the KM clustering algorithm with that of the Fuzzy *C*-Means (CM).

We begin with a review of the current literature on classical and fuzzy clustering methods. Huang [2] extended the KM algorithm to categorical domains. In order to decrease the computational complexity associated with the conventional CM clustering method, Chang et al. [3] proposed a CM using the cluster center displacement of successive iterative processes clustering method. Volmurgan [4] investigated the performance of two partitions-based clustering methods, i.e., KM and CM algorithms. He made the comparison through clustering randomly distributed data points. Havens et al. [5] compared the efficacy of three different techniques in order to extend the application of CM clustering to very large datasets. Panda et al. [6] implemented clustering techniques in such wide areas as medicine, business, engineering systems, and image processing. Grover [7] studied a wide variety of fuzzy clustering methods such as CM, Possibilistic CM, and Fuzzy Possibilistic CM algorithm and reported their advantages and drawbacks. Bora and Gupta [8] conducted a comparative study of the fuzzy and hard clustering methods. Finally, Fajardo et al. [9] investigated the fuzzy clustering of certain spectra for the objective recognition of soil morphological horizons in soil profiles.

The present study proposes a hybrid clustering algorithm by the name of KCM that combines KM and CM algorithms to achieve its objective by improving the time processing of the CM method. The performances of KM, CM, and KCM techniques are then compared in terms of their accuracy and time processing using simulated data from sub-Gaussian distributions. The methods are also applied to the three standard real datasets, to determine and compare the precision and accuracy of the investigated algorithms.

(1) Let T be the maximum number of iterations allowed, $0 < \varepsilon < 1$, $v_i^{(t)}$, $t = 0$; be the initial centers, $i = 1, \dots, c$, and $V^{(t)}$ be the set of centers in the iteration t;
(2) Compute the value of $\|x_k - v_i\|_p$,
(3) Assign the elements x_k to the clusters, according to

$$A_i(x_k) = \begin{cases} 1, & \|x_k - v_i\|_p = \min_{l=1,\dots,c} \|x_k - v_l\|_p \\ 0, & \text{otherwise;} \end{cases}$$

(4) Update the cluster centers, take $t = t + 1$, $v_i^{(t)} = \sum_{k=1}^{n} A_i(x_k)x_k / \sum_{k=1}^{n} A_i(x_k)$;
(5) If $\|V^{(t-1)} - V^{(t)}\|_p < \varepsilon$; or $T = t$, then stop. Otherwise, go to step (2);
(6) End.

ALGORITHM 1: KM algorithm.

Finally, KM, CM, and KCM are compared using Minkowski distances. The objective is to identify the best combinations of the clustering method and distance measure with higher precision, accuracy measures, and cluster quality in terms of compactness and distinctiveness.

2. Clustering Algorithms

By definition, clustering groups a sample of vectors to c clusters, using an appropriate similarity criterion such as distance from the center of the cluster.

2.1. K-Means Algorithm. KM is one of the most popular clustering algorithms [10, 11]. The clustering results of the KM algorithm are very sensitive to the positions of the initial cluster centers. Being efficient in clustering large data sets, it often terminates at a local optimum and applies only to numeric values [12]. Given a set of n elements $\{x_1, \dots, x_n\}$ and a set of centers $V = \{v_1, \dots, v_c\}$, where $x_k = (x_{k1}, \dots, x_{kd})$, $v_i = (v_{i1}, \dots, v_{id}) \in R^d$, $k = 1, \dots, n$; $i = 1, \dots, c$, We recall that Minkowski distance for two points x_k, v_i is defined as follows:

$$\text{MD}(x_k, v_i) = \|x_k - v_i\|_p = \left(\sum_{j=1}^{d} \left(x_{kj} - v_{ij}\right)^p \right)^{1/p}. \tag{1}$$

Euclidean and Manhattan distances are two special cases of Minkowski distance with $p = 2$ and $p = 1$, respectively. In the rest of the paper, for Minkowski distance, we consider $p = 1.5$. The steps of the KM clustering algorithm are shown in Algorithm 1.

2.2. Fuzzy Clustering Algorithms. In KM clustering, data is divided into disjoint clusters, where each data element belongs to exactly one cluster. In fuzzy clustering, an object can belong to one or more clusters with probabilities [13]. One of the most widely used fuzzy clustering methods is the CM algorithm, originally due to Dunn [14] and later modified by Bezdek [15]. The CM method attempts to partition a finite collection of n elements $X = \{x_1, \dots, x_n\}$ to a collection of c fuzzy clusters with respect to some given criterion, where $x_k \in R^d$ is an observation vector. A fuzzy c-partition of X is a family of fuzzy subsets of X denoted by $U = \{A_1, \dots, A_c\}$, which satisfies $\sum_{i=1}^{c} A_i(x_k) = 1$, $k = 1, \dots, n$, and $0 < \sum_{k=1}^{n} A_i(x_k) < n$, $i = 1, \dots, c$, where $c < n$ is a positive integer. The problem of fuzzy clustering is to find a fuzzy c-partition and the associated cluster centers by which the structure of the data is represented as best as possible. To solve the problem of fuzzy clustering, this criterion needs to be formulated in terms of a performance index. The c cluster centers $v_1, v_2, \dots, v_c$ associated with the partition are calculated as follows:

$$v_i = \frac{\sum_{k=1}^{n} \left[A_i(x_k)\right]^m x_k}{\sum_{k=1}^{n} \left[A_i(x_k)\right]^m}, \quad i = 1, \dots, c, \tag{2}$$

where $m > 1$ is a real number that governs the influence of membership grades, v_i is viewed as the cluster center of the fuzzy class A_i, and the performance index of a fuzzy c-partition U, $J_m(U)$, is then defined in terms of the cluster centers using the formula

$$J_m(U) = \sum_{k=1}^{n} \sum_{i=1}^{c} \left[A_i(x_k)\right]^m \|x_k \quad v_i\|_p^2. \tag{3}$$

This performance index measures the weighted sum of distances between cluster centers and elements in the corresponding fuzzy clusters. The goal of the CM clustering method is to find a fuzzy c-partition that minimizes the performance index $J_m(U)$. In other words, the clustering problem is an optimization problem [16]. The convergence properties of CM algorithms are theoretically important. The optimal cluster centers are the fixed points of CM clustering algorithms. The algorithm is limited by long computational time and sensitivity to noise, outliers, and initial guess [17, 18]. The two steps of the CM clustering algorithm which should be modified in KM algorithm are shown in Algorithm 2.

3. A Hybrid Method: *KC*-Means Algorithm

A novel approach called KCM method is proposed herein that combines the KM and CM methods. The combination is meant to overcome the limitations of both but enjoys their advantages. One of the disadvantages of CM method is long computational time while quick running is one of the advantages of KM method. The goal of the hybrid method is to introduce a fuzzy method faster than CM while its

(3) Compute the membership and assign the elements x_k to the clusters, according to

$$A_i(x_k) = \left(\sum_{j=1}^{c} \left(\frac{\|x_k - v_i\|_p}{\|x_k - v_j\|_p} \right)^{2/(m-1)} \right)^{-1};$$

(4) Compute the new cluster centers $v_i = \sum_{k=1}^{n} [A_i(x_k)]^m x_k / \sum_{k=1}^{n} [A_i(x_k)]^m$;

ALGORITHM 2: CM algorithm.

(1) Apply the KM clustering algorithm on data set X;
(2) Let V be the set of final centers, which obtained from KM algorithm;
(3) Consider V as a new data set;
(4) Apply the CM clustering algorithm on V;
(5) Recover corresponding data set clusters based on CM clustering output.
(6) End.

ALGORITHM 3: KCM algorithm.

accuracy is close to the CM. In the proposed technique, KM is initially applied to individual data objects to generate c clusters, designated as middle-level clusters. Each cluster is then represented by its centroid. The CM clustering is subsequently applied to those centroids in order to structure the final clustering. The distance between two middle-level clusters is measured as the distances between their centroids. The hybrid method considers the final centers produced by KM as the dataset for CM, so that the number of observations in the CM is equal to the number of centers produced by the KM method. Therefore, the KCM time is much less than the time of CM method.

The hybrid method is more suitable for the large dataset, where it has reduced clusters of observations by their centers, eventually computed from the KM. The performance of the proposed approach is evaluated by comparing it with KM and CM algorithms in terms of both accuracy and time processing. It is shown that the proposed technique outperforms CM in time processing; it yields results over shorter times when compared with the CM algorithm. Given a set of n elements $X = \{x_1, \ldots, x_n\}$, where $x_k \in R^d$, $k = 1, \ldots, n$, the steps of the KCM clustering algorithm are shown in Algorithm 3.

4. Evaluation

Simulated datasets are used to evaluate the KM, CM, and KCM clustering methods. We use an external clustering evaluation criterion for comparisons. The Rand index is a criterion used to compare an induced clustering structure (C_1) with a given clustering structure (C_2) defined as follows [13]:

$$\text{RAND} = \frac{a+d}{a+b+c+d}, \qquad (4)$$

where a, b, c, and d are the numbers defined as follows:

(i) a is the number of two points belonging to the same cluster, according to C_1 and C_2.

(ii) b is the number of points belonging to the same cluster according to C_1 but not C_2.

(iii) c is the number of points belonging to the same cluster according to C_2 but not C_1.

(iv) d is the number of points that do not belong to the same cluster, according to C_1 and C_2.

The quantities $\{a, d\}$ can be interpreted as agreements and $\{b, c\}$ as disagreements. The Rand index value lies within the range $[0, 1]$ and the clustering performance is considered to be good if the Rand index value converges to one [4, 13].

We used R 3.3.3 software, on a PC with CPU Core i5-3210 with 4 GB RAM to run all experiments in the next sections. For a fair comparison, termination condition of the algorithms is set as default of R standard codes.

5. Simulation Study

A d dimensional random vector Y has a sub-Gaussian distribution with location vector μ and dispersion matrix Q if its characteristic function is of the form

$$\varphi(u) = \exp\left(i \cdot u^T \mu\right) \exp\left(-\left|u^T Q u\right|^{\alpha/2}\right), \qquad (5)$$

where $u^T = (u_1, \ldots, u_d)$, $\mu \in R^d$, $i = \sqrt{-1}$, $\alpha \in [0, 2]$, and Q is a positive definite matrix. In the case of α equal to 2, we get the multivariate normal distribution that its covariance matrix is $2Q$ [19]. If $\alpha > 1$ then $\mu = E(X)$. However, the expectation of X does not exist for $\alpha \leq 1$.

In this simulation study, a set of real and simulated data generated by the sub-Gaussian and multivariate normal distributions was used. For clustering data using the proposed KCM method, the three Euclidean, Manhattan, and Minkowski distances were used. In addition, the results obtained from the KM, CM, and KCM algorithms were compared in terms of their time processing (in milliseconds) and accuracy. A set of data of 15000 observations having 30

TABLE 1: The time processing of CM versus KCM with Euclidean, Manhattan, and Minkowski ($p = 1.5$) distances when the number of clusters is 40.

α	Euclidean	Manhattan	Minkowski
0.5	KCM = 6.25% CM	KCM = 1.82% CM	KCM = 0.83% CM
1	KCM = 9.09% CM	KCM = 1.15% CM	KCM = 0.79% CM
1.5	KCM = 11.11% CM	KCM = 1.06% CM	KCM = 2.56% CM
2	KCM = 11.76% CM	KCM = 0.59% CM	KCM = 1% CM

attributes and parameter of stability in the range of $\alpha = 0.5, 1, 1.5, 2$ was generated, where if $\alpha = 2$, there will be a multivariate normal distribution. Then, the data were partitioned into 5, 10, 15, 20, 25, 30, and 40 clusters. As previously mentioned, our CM method is based on KM where c-value, which is the cluster number, is to be defined. The simulation results of the test are shown in Figures 1 and 2 showing the accuracy and time of KM, CM, and KCM method with Euclidean, Manhattan, and Minkowski ($p = 1.5$) distances for $\alpha = 0.5, 1, 1.5, 2$.

We have implemented the algorithms 100 times, and the average values of accuracy and time processes were computed. We classify the results as follows.

Time. In general, the time processes of KM were less than the time processes of CM and KCM algorithms for all values of α and the number of clusters, c. CM recorded a long-time process compared with either of the KM or KCM algorithms. The type of distance did not significantly affect the time processing of KM, where the results obtained with the three distances were close to one another. While the time processing of CM and KCM with Minkowski distance is longer than with Euclidean and Manhattan distance.

The increase in values of α does not affect the time processing of KM and KCM, where the values of time are almost close for all α, while the time processing of CM is decreasing with increasing value of α if we used the Euclidean distance, but if we used the Manhattan distance, it is increasing when the value of α increased. The speed of KM and CM is decreased if the number of clusters increases, but it does not affect much the speed of the KCM algorithm with Euclidean and Manhattan distances.

Generally, the processing time of the KCM algorithm is less than the CM algorithm. For example, when the number of clusters is 40, the processing time of CM and KCM is shown in Table 1.

Accuracy. Distance type had no significant effect on the accuracy of the KM and CM algorithms as almost the same results obtained with either. However, accuracy increased with $\alpha > 0.5$. The accuracy of KCM algorithm is increasing with increasing the values of α. In general, the accuracy of KCM and CM with Euclidean and Minkowski distances is better than that with Manhattan.

6. Comparison of Algorithms Using Real Data

In this section, the KM, CM, and KCM algorithms are tested for their performance using Iris (150×4), Wine (178×13), and Lens (24×4) datasets. The three Euclidean (Euc), Manhattan (Man), and Minkowski (Min) distance measures are used to see how they influence the overall clustering performance. The performance of these three techniques has been compared based on the following parameters:

(1) Precision = $T_P/(T_P + F_P)$.

(2) Accuracy = $(T_P + T_N)/(T_P + T_N + F_P + F_N)$.

A true positive (T_P) decision assigns two similar documents in the same cluster; a true negative (T_N) decision assigns two dissimilar documents to different clusters. A (F_P) decision assigns two dissimilar documents to the same cluster. A (F_N) decision assigns two similar documents to different clusters. The experimental results indicating the performance of each technique on the three datasets are reported in Table 2.

Using the Iris dataset led to a greater average precision of clusters formed by KCM-Euc and KCM-Man than those by KM and CM with the three distances. CM-Man recorded a greater accuracy than any of those formed by KM or KCM. Distance and algorithm type had no significant effect on the accuracy. As for the Lens dataset, average precision was generally low with all the algorithms examined. It, however, yielded acceptable accuracy values with the KM, CM, and KCM algorithms, but it does not exceed 0.50. With the Wine dataset, distance and algorithm type had a significant effect on the accuracy and average precision. The average of precision does not exceed 0.70 and the highest average recorded by KCM-Man. The CM-Man recorded a greater accuracy than any of those formed by KM or KCM.

7. Conclusions

In this paper, the two most famous clustering techniques, namely, K-Means and Fuzzy C-Means, were investigated for their performance. To improve the time processes of the fuzzy clustering technique, a hybrid algorithm, named KCM, combining the KM and CM algorithms, was proposed.

It was found that the KM algorithm had shorter time processes than CM and KCM algorithms for all values of α and c. In addition, the speed of CM was observed to be less than those of KM and KCM. However, the time processes of CM with the Euclidean and Manhattan distances were observed to be shorter than that with Minkowski distance. The value of α did not affect time processes under KM and KCM; however, that of CM decreased with increasing values of α with Euclidean but increased with Manhattan distance.

The accuracy of KM, CM, and KCM algorithms was increasing for $\alpha > 0.5$. Distance type had a significant

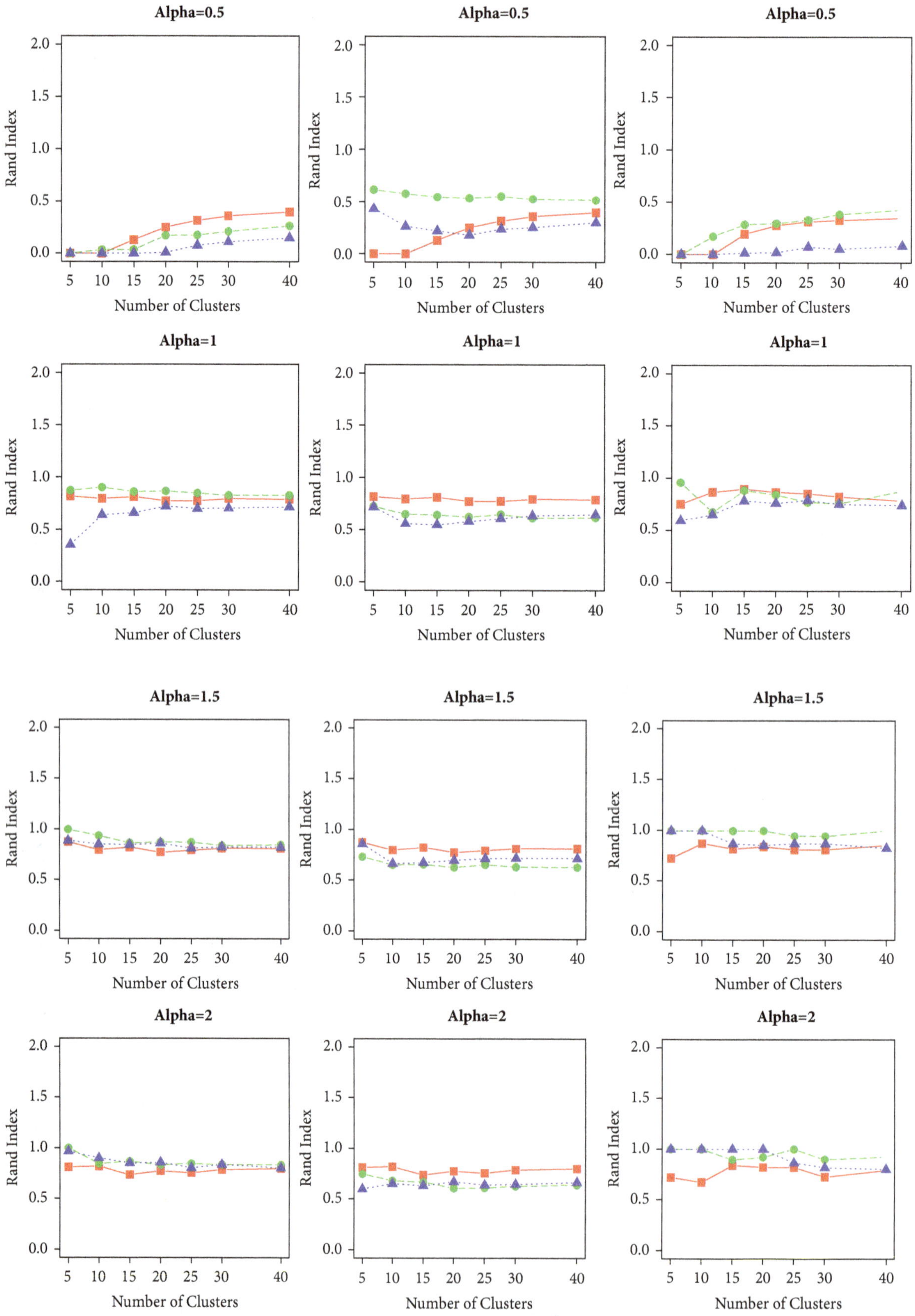

FIGURE 1: Comparison of KM, CM, and KCM algorithms in terms of accuracy based on Euclidean, Manhattan, and Minkowski ($p = 1.5$) distances for $\alpha = 0.5, 1, 1.5, 2$. KM: Red squares, CM: Green circles, KCM: Blue triangles.

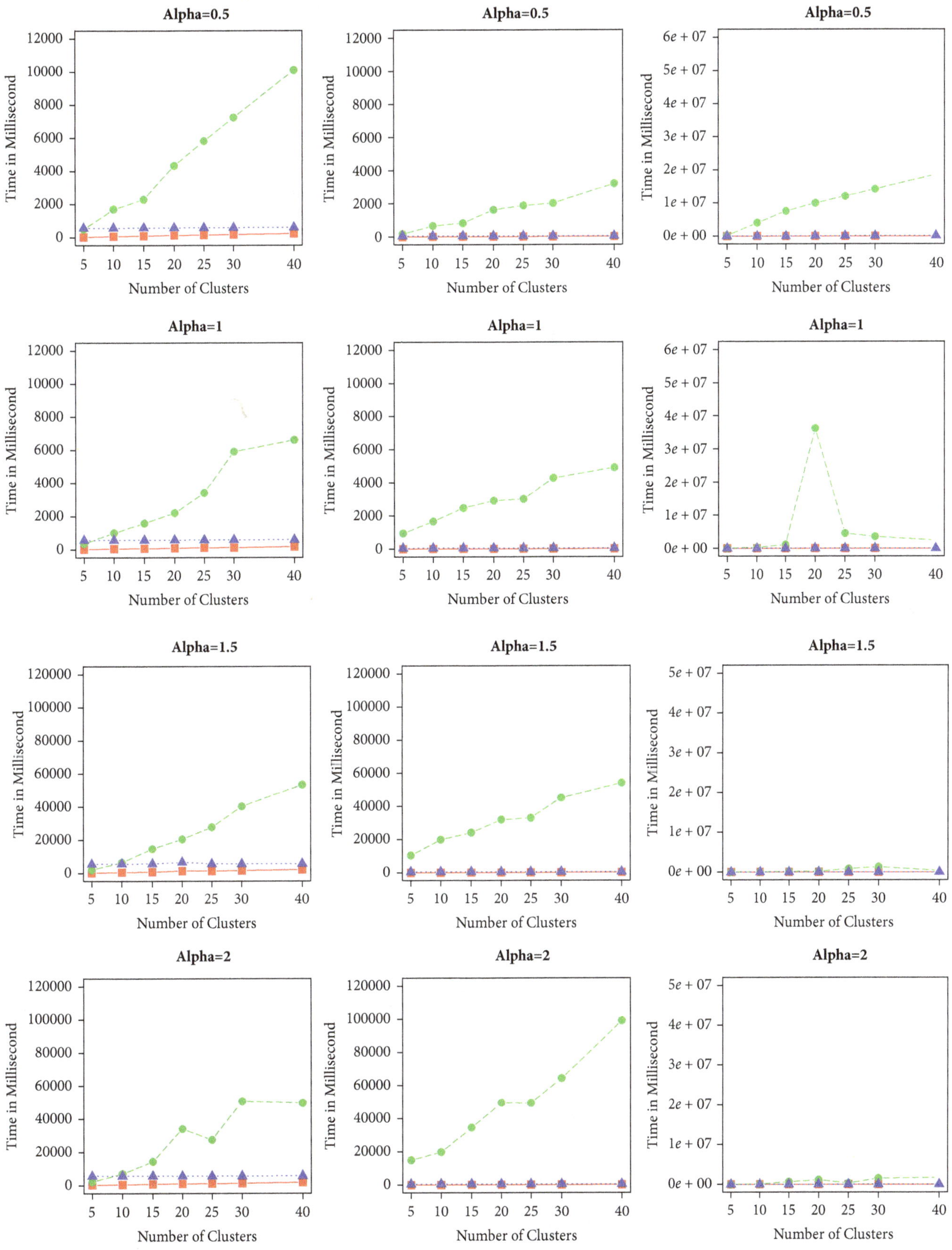

FIGURE 2: Comparison of KM, CM, and KCM algorithms in terms of time processing based on Euclidean, Manhattan, and Minkowski ($p = 1.5$) distances for $\alpha = 0.5, 1, 1.5, 2$. KM: Red squares, CM: Green circles, KCM: Blue triangles.

TABLE 2: Performance analysis of KM, CM, and KCM clustering techniques on data and benchmarks Iris, Lens, and Wine datasets. Note that each data has 3 clusters; the best precision and accuracy for each data set is in bold font.

Dataset	Performance parameters	Clusters	*K*-Means			Fuzzy *C*-Means			*KC*-Means		
			Euc	Man	Min	Euc	Man	Min	Euc	Man	Min
Iris	Precision	1	0.898	0.898	0.898	0.858	0.905	0.861	0.627	0.627	1
		2	1	1	0.645	0.655	0.691	1	1	1	0.895
		3	0.644	0.645	1	1	1	0.670	1	1	0.631
		Average	0.847	0.847	0.848	0.837	0.865	0.844	**0.876**	**0.876**	0.842
		Accuracy	0.880	0.880	0.880	0.880	**0.899**	0.796	0.796	0.880	0.874
Lens	Precision	1	0.167	0.167	0.25	0.286	0.267	0.278	0.200	0.238	0.286
		2	0.273	0.200	0.238	0.200	0.250	0.25	0.167	0.286	0
		3	0.250	0.286	0.278	0.167	0.200	0.238	0.295	0	0.238
		Average	0.230	0.217	**0.255**	0.217	0.239	**0.255**	0.221	0.175	0.175
		Accuracy	0.522	0.507	**0.547**	0.547	0.533	**0.547**	0.504	0.504	0.504
Wine	Precision	1	0.356	0.517	0.957	0.345	0.858	0.957	0.360	0.504	1
		2	0.595	0.430	0.356	0.577	0.605	0.577	0.605	1	0.595
		3	0.957	1	0.595	0.956	0.386	0.345	0.957	0.556	0.338
		Average	0.636	0.649	0.636	0.626	0.616	0.626	0.641	**0.687**	0.644
		Accuracy	0.719	0.685	0.719	0.719	**0.735**	0.711	0.720	0.686	0.695

effect on the accuracy of KM and CM algorithms, but the accuracy of the KCM and CM algorithms with Euclidean and Minkowski distances was better than that with Manhattan distance.

Using the real datasets revealed that the Iris dataset yielded higher precision values for clusters with the three distances. The clusters formed by the combined KCM-Euc were observed to be more distinct. Using the Lens dataset led to poor precision levels but acceptable accuracy values for all the combinations. With the Wine dataset, medium precision levels were achieved with all the combinations. CM-Man and KCM-Euc yielded the most compact clusters, while KCM-Man yielded the most distinct ones. In general, the Iris dataset not only formed the most compact and distinct clusters, but also yielded higher precision and accuracy levels for KM, CM, and KCM clusters with the three distances than did the Lens or Wine datasets.

Finally, we recall that the time computation in a clustering method depends on the algorithm and its implementation, programming language, and hardware. Therefore, based on the complexity of the clustering problem one can consider the best of them.

Conflicts of Interest

The authors declare that they have no conflicts of interest.

References

[1] A. C. Rencher, *Method of Multivariate Analysis*, John Wiley & Sons, 2nd edition, 2002.

[2] Z. Huang, "Extensions to the k-means algorithm for clustering large data sets with categorical values," *Data Mining and Knowledge Discovery*, vol. 2, no. 3, pp. 283–304, 1998.

[3] C.-T. Chang, J. Z. Lai, and M.-D. Jeng, "A fuzzy K-means clustering algorithm using cluster center displacement," *Journal of Information Science and Engineering*, vol. 27, no. 3, pp. 995–1009, 2011.

[4] T. Volmurgan, "Austria performance comparison between K-means and fuzzy C-means," *Wulfenia Journal*, vol. 19, pp. 234–241, 2012.

[5] T. C. Havens, J. C. Bezdek, C. Leckie, L. O. Hall, and M. Palaniswami, "Fuzzy C-means algorithms for very large data," *IEEE Transactions on Fuzzy Systems*, vol. 20, no. 6, pp. 1130–1146, 2012.

[6] B. Panda, S. Sahoo, and S. K. Patnaik, "A comparative study of hard and soft clustering using swarm optimization," *International Journal of Scientific & Engineering Research*, vol. 4, pp. 785–790, 2013.

[7] N. Grover, "A study of various fuzzy clustering algorithms," *International Journal of Engineering Research*, vol. 3, no. 3, pp. 177–181, 2014.

[8] D. J. Bora and D. A. Gupta, "A comparative study between fuzzy clustering algorithm and hard clustering algorithm," *International Journal of Computer Trends and Technology*, vol. 10, no. 2, pp. 108–113, 2014.

[9] M. Fajardo, A. McBratney, and B. Whelan, "Fuzzy clustering of Vis-NIR spectra for the objective recognition of soil morphological horizons in soil profiles," *Geoderma*, vol. 263, pp. 244–253, 2014.

[10] M. R. Anderberg, *Cluster Analysis for Applications*, Academic Press, New York, NY, USA, 1973.

[11] J. MacQueen, "Some methods for classification and analysis of multivariate observations," in *Proceedings of the 5th Berkeley Symposium on Mathematical Statistics and Probability*, vol. 1, pp. 281–297, 1967.

[12] A. Banharnsakun, "A MapReduce-based artificial bee colony for large-scale data clustering," *Pattern Recognition Letters*, vol. 93, pp. 78–84, 2017.

[13] G. Gan, C. Ma, and J. Wu, *Data Clustering Theory: Algorithms and Applications*, SIAM, Virginia, 2007.

[14] J. C. Dunn, "A fuzzy relative of the ISODATA process and its use in detecting compact well-separated clusters," *Journal of Cybernetics*, vol. 3, no. 3, pp. 32–57, 1973.

[15] J. C. Bezdek, *Pattern Recognition with Fuzzy Objective Function Algorithms*, Plenum Press, New York, NY, USA, 1981.

[16] G. J. Klir and B. Yuan, *Fuzzy Sets and Fuzzy Logic: Theory and Applications*, Prentice Hall, New York, NY, USA, 1995.

[17] R. Suganya and R. Shanthi, "Fuzzy C-means algorithm - a review," *International Journal of Scientific and Research Publications*, vol. 2, pp. 440–442, 2012.

[18] M. S. Yang, "Convergence properties of the generalized fuzzy C-means clustering algorithms," *Computers & Mathematics with Applications*, vol. 25, no. 9, pp. 3–11, 1993.

[19] V. Omelchenko, "Parameter estimation of sub-Gaussian stable distributions," *Kybernetika*, vol. 50, no. 6, pp. 929–949, 2014.

11

Fuzzy Sliding Mode Based Series Hybrid Active Power Filter for Power Quality Enhancement

Soumya Ranjan Das,[1] **Prakash K. Ray**,[1] **and Asit Mohanty**[2]

[1]*Department of Electrical Engineering, IIIT Bhubaneswar, Odisha 751003, India*
[2]*Department of Electrical Engineering, CET Bhubaneswar, Odisha 751003, India*

Correspondence should be addressed to Soumya Ranjan Das; srdas1984@gmail.com

Academic Editor: Ying-Yi Hong

This paper contributes an innovative gating signal generation technique based on fuzzy sliding mode pulse width modulation (FSMPWM) with a Series Hybrid Active Power Filter (SEHAPF) for reducing the total harmonic distortion (THD). Hybrid filters are used for compensating reactive power on the load side and mitigate harmonics for the growth of power quality under variable source and load conditions in the utility system. The objective of the paper is to eradicate the various power quality (PQ) problems with development in power factor and reducing distortion in transmission and distribution line due to harmonics. With the implementation of FSMPWM of the proposed filter, the gating pulses are generated by implementing a Mamdani fuzzy rule with sliding surfaces. For producing a fixed pulse the presented method reduces the chattering reaction by controlling the narrow boundary coating on the sliding surface. The results of the projected technique are analyzed and compared with the traditional hysteresis band current controller (HCC). The overall proficiency and results are examined with the help of MATLAB/SIMULINK environment.

1. Introduction

In the present scenario, wide use of power electronic devices generates huge quantities of harmonics in the utility network [1–3]. Due to the presence of harmonics, The load end characteristics behave abruptly. These harmonics are nothing but integer multiples of sinusoidal components of voltage and current. Due to the existence of nonlinear loads [4], different levels of consumers get affected hardly. Current or voltage harmonics and reactive power [5, 6] are the main sources for reducing the quality of power in the distribution system. To overcome these effects, traditionally passive LC filters [7] have been used, but in spite of several advantages still they are ineffectual due to certain conditions like tuning problem, bulky size, and resonance problems. Hence, many researchers conducted analysis with different custom power devices [8] to advance the quality of the power system. Current development in switching appliances has led to focus on a different arrangement of active power filter (APF). This APF is helpful in compensating voltage and current harmonics as well as compensation in unbalanced voltage with other power quality problems. Out of the different configuration of hybrid active filters [9–13], the SEHAPF [14] is the widely used one. This arrangement is linked in series between supply and load. The proposed configuration can mitigate PQ disturbances, like sag and swell insource and load voltage, three-phase imbalance voltage, and many more such that regulation in voltage is maintained perfectly on the load side. SEHAPF is able to supply the voltage harmonics in the utility system with equal magnitude and opposite phase with respect to the nonlinear load. The operation of these custom power devices is determined upon reference current technique and the controlling process used to supply the essential current or voltage compensation into the grid. For better compensation, fast sensing of distorted signal and quick extraction process are highly required.

Several research papers are analyzed based on HCC [15–20] for controlling voltage source inverter (VSI) of active filters. This controller performs a bang-bang type controller, which in turn affects the voltage compensation of APF

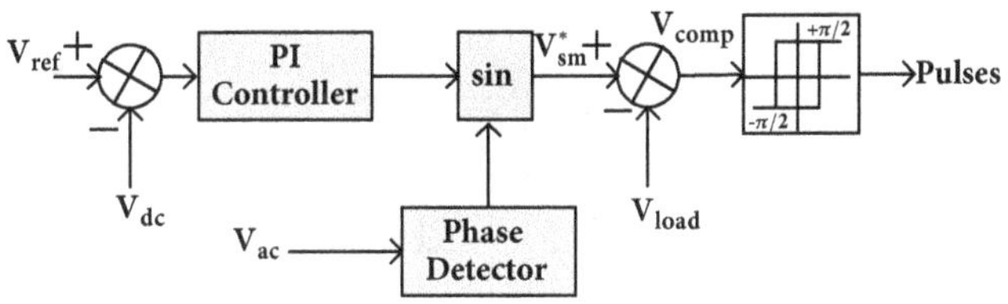

FIGURE 1: Control strategy of HCC.

and follows its estimated reference signal within a specific tolerance limit. The control strategy is depicted in Figure 1. For generating the gating signal to the VSI, the obtained reference voltage signal is related to actual voltage and the error between them is determined to HCC. This controller is employed individually for individual phase and produces the gating signals for the PWM-VSI. But the above controlling techniques are not suitable for tracking the reference signal during distortion in supply and load side.

Sliding mode controllers (SMC) [21, 22] have been utilized widely because of constancy, toughness, and fast switching action with respect to variable load conditions. Despite several merits still this technique suffers from chattering difficulties, in turn, producing a variable switching frequency causing severe losses in power and switching signal.

To overcome this problem, a slight modification has been made with the SMC using fuzzy based method [23] to produce a fixed switching technique. Because of using fuzzy controller the particular mathematical model of the system is not required. Fuzzy controller can perform with indefinite inputs and controls nonlinearity and is dynamic compared to other controllers. This arrangement [24–26] is able to hold power quality issues. A pulse width modulation based fuzzy SMC is presented for better compensation of SEHAPF by tracking accurately the reference signal during uncertainty conditions.

Projected FSMPWM technique eradicates several complications such as voltage harmonics, imbalance in the load, sag and swell in voltage, and distortion in voltage that are found in the utility system. The primary focus of the paper is to lower harmonics by reducing the percentage in THD and managing the reactive power using FSMPWM controlling techniques employed in SEHAPF.

The projected technique produces a constant switching signal in which the loss in power and EMC noise gets reduced. Moreover, it also makes very simple and easy-to-design passive LC filter. Consequently, the applied technique nullifies the mathematical operations and quite effectively to the external disturbances and uncertainties of the concerned system. The proposed scheme is analyzed under different source and load conditions. The result of the projected technique is examined using MATLAB/SIMULINK environment and the results are compared with the conventional HCC. In this paper, the subsequent sections are organized in the following pattern. Section 2 describes the systematic arrangement of the proposed SEHAPF. The control scheme for the SEHAPF using FSMPWM and unit vector is depicted in Section 3. Simulation results analysis is described in Section 4. Section 5 draws the conclusion.

2. System Configuration

The proposed scheme is presented in Figure 2. It comprises active series filter with passive LC filter. The SEHAPF guards the load end from the distorted source voltage, having imbalance characteristics, and delivers cost effective compensation method especially for heavy range of applications. It inserts the compensated voltage to the grid system at the point of common coupling through the coupling transformer. The voltage across the capacitor is maintained at a certain level, which is considered as reference values. The active power difference in the grid system gets affected during deviation of load. The DC link capacitor compensates the power differences. Out of the several configurations of hybrid filters, the series hybrid filter suits the best one which contains a small rating of series active filter with tuned passive LC filters. Series active filters allow the passive filters to tune exactly to reduce harmonics current at load end and to improve the power factor.

3. Control Strategy

3.1. Reference Signal Generation for SEHAPF. The control arrangement for reference voltage generation for SEHAPF is shown in Figure 3. This arrangement provides a fixed frequency pulse width modulation wave for the VSI of SEHAPF. At first a sliding surface is defined and for that a control strategy is designed for compensating the voltage reference. For realizing the value, the required real power can be obtained from the product of three-phase load voltages and three-phase source currents:

$$\begin{aligned} V_{La} &= V_m \sin(\omega_1 t + \varphi) \\ V_{Lb} &= V_m \sin(\omega_1 t - 120^\circ + \varphi) \\ V_{Lc} &= V_m \sin(\omega_1 t + 120^\circ + \varphi) \end{aligned} \quad (1)$$

$$\begin{aligned} I_{Sa} &= I_m \sin(\omega_1 t + \varphi) \\ I_{Sb} &= I_m \sin(\omega_1 t - 120^\circ + \varphi) \\ I_{Sc} &= I_m \sin(\omega_1 t + 120^\circ + \varphi) \end{aligned} \quad (2)$$

where V_m, I_m, and u are, respectively, defined as source magnitude voltage, current, and phase angle.

Now, RMS value of 3-phase load current is shown as

$$I_{Lx} = \sum_{n=1}^{\infty} I^{+}{}_{xn} + \sum_{n=1}^{\infty} I^{-}{}_{xn} + \sum_{n=1}^{\infty} I^{0}{}_{xn} \quad (3)$$

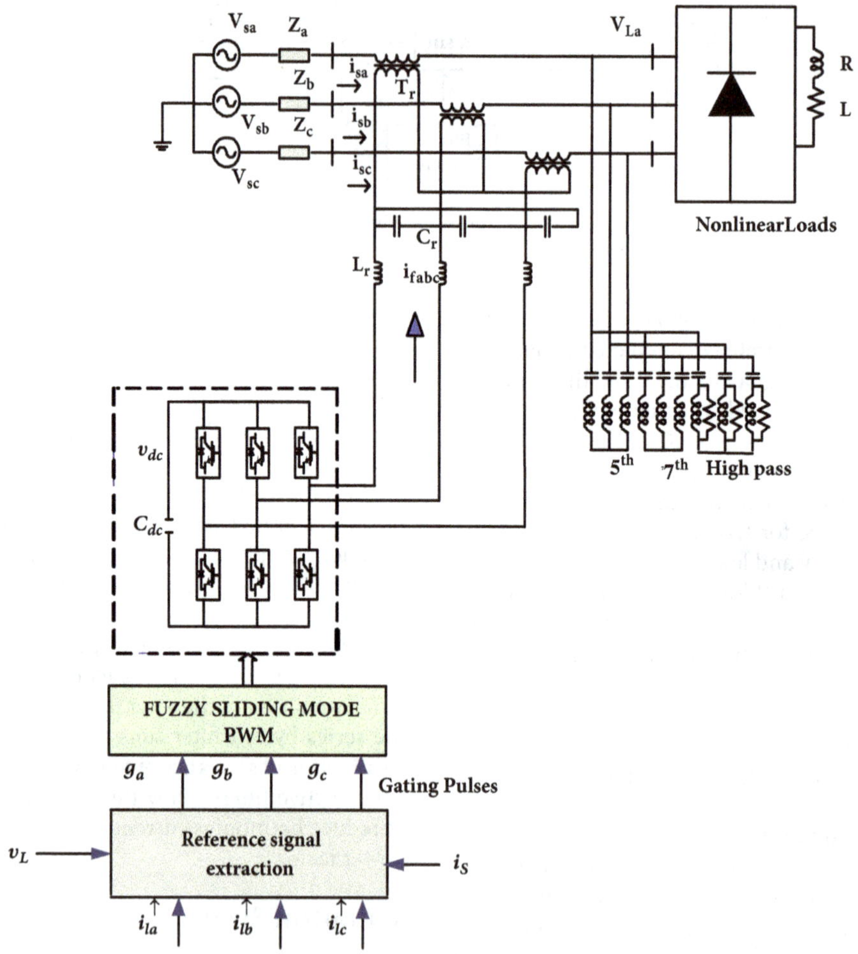

FIGURE 2: Circuit configuration of SEHAPF.

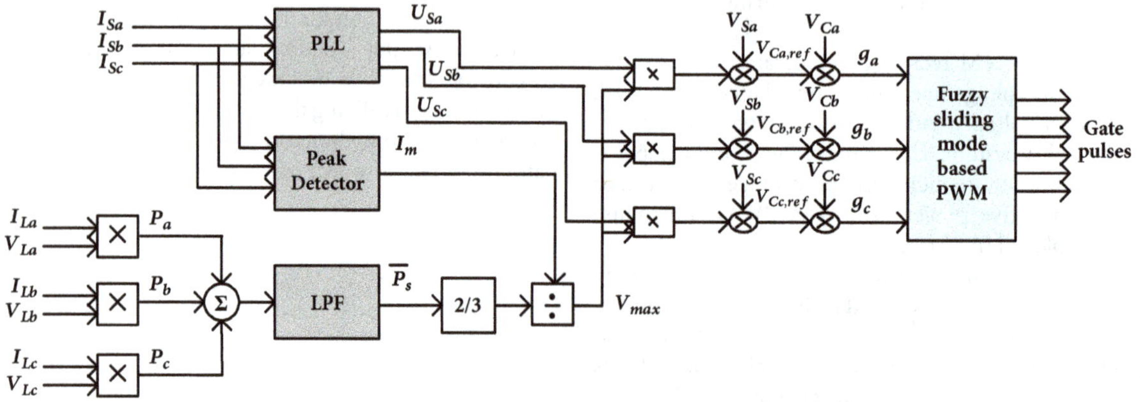

FIGURE 3: Control strategy of series filter using FSMPWM technique.

where $x, +, -, 0$ are, respectively, the sequence in phase, positive, negative, and zero sequence and n denotes number of harmonic components.

$\therefore$ the actual power absorbed by the load in positive sequence is expressed as

$$p^{+}{}_{n} = \frac{1}{T}\int_{0}^{T}\sum_{x\epsilon k} V_{Lx} I_{Lx} dt = \frac{3V_{max} I_{max}}{2} \tag{4}$$

And $p^{0}{}_{l1} = p^{+}{}_{S1}$.

In the aforementioned part, $p^{+}{}_{l1}$ and $p^{+}{}_{S1}$ are, respectively, the actual power absorbed by the load in positive sequence and real power supplied by the grid in positive sequence. Now for obtaining the maximum voltage of the load both the values of low pass filter and maximum value of supply current is used and is shown in (5), and the voltage reference is obtained from the product of

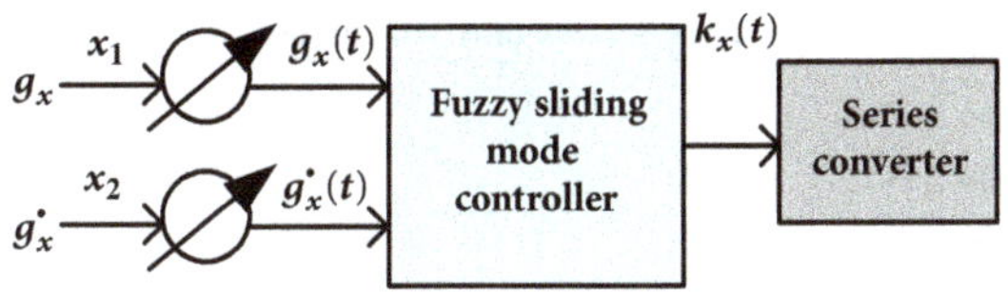

FIGURE 4: Block diagram of FSMC of series converter.

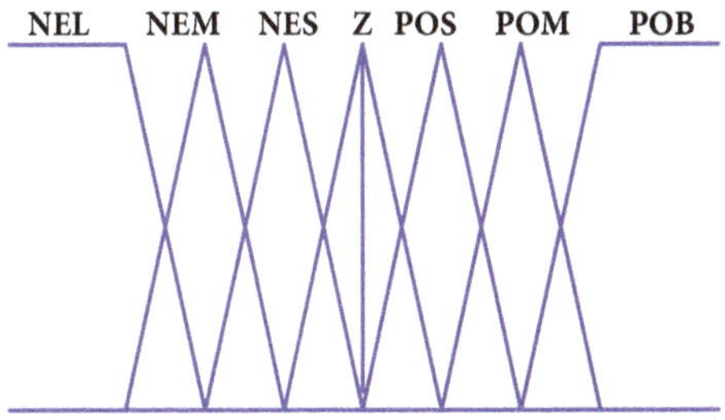

FIGURE 5: Membership function for fuzzy rule base.

maximum load voltage and unit sine vector as shown in (6):

$$V_{max} = \frac{2p^{+}_{l1}}{3I_m} = \frac{2\overline{p_S}}{3I_m} \tag{5}$$

$$V_{xref} = V_{max} * U_{Sx} \tag{6}$$

The compensation value of voltage reference in sliding surface is obtained by removing the supply reference voltage from the supply voltage as shown in

$$V_{Cxref} = V_{Sx} - V_{xref} \tag{7}$$

4. FSMPWM Control of SEHAPF

Block diagram of FSMPWM of series converter is revealed in Figure 4, where $u_c(V_{dc}/2)$ and I_f represent, respectively, the voltage and current along filter side, while, V_{ef}, L_{ef} and V_C, L_{sf}, respectively, represent voltage and current across the capacitor on ac side and at grid side.

The current and voltage at the terminal of the SEHAPF is given as (assume, $V_{ef} = V_C$)

$$\frac{dI_F}{dt} = \frac{V_{dc}}{2L_{sef}}u_c - \frac{R_{sef}}{L_{sef}}I_F - \frac{1}{L_{sef}}V_{ef} \tag{8}$$

$$\frac{dV_{ef}}{dt} = \frac{dV_C}{dt} = \frac{1}{C_{ef}}I_F - \frac{1}{C_{sef}}I_{sf} \tag{9}$$

The path of fuzzy sliding surface for SEHAPF is acquired as

$$S_x(t) = \dot{g}_x \tag{10}$$

[which is obtained from $(V_{Cxref} - V_{Cx})$], where "x" is a sequence of phase.

∴ the Error function is stated as

$$g_x(t) = \left(V_{Cxref} - V_{Cx}\right) \tag{11}$$

For the SEHAPF, $\dot{S}(t)$ is written as

$$\dot{S}(t) = \ddot{g}(t) \tag{12}$$

$$\therefore \dot{s}_x(t) = \left(\ddot{V}_{Cxref} - \ddot{V}_{Cx}\right) \tag{13}$$

Utilizing value of dV_C/dt the value from (13) becomes

$$\begin{aligned}\dot{s}_x(t) &= \left(\ddot{V}_{Cxref} - \frac{1}{C_{ef}}\dot{I}_{Fx} + \frac{1}{C_{ef}}\dot{I}_{sFx}\right)\\ &= \frac{1}{C_{ef}}\left(-\frac{V_{dc}}{2L_{sef}}u_{cx} + \frac{R_{sef}}{L_{sef}}I_{Fx} + \frac{1}{L_{sef}}V_{efx}\right)\\ &\quad + \frac{1}{C_{ef}}\dot{I}_{sFx} + \ddot{V}_{Cxref}\end{aligned} \tag{14}$$

By setting $\dot{s}(t) = 0$ the equation can be written as

$$\begin{aligned}u_{eqcx} = &\left(\frac{2R_{sef}}{V_{dc}}I_{Fx} + \frac{2V_{efx}}{V_{dc}} + \frac{2L_{sef}}{V_{dc}}\dot{I}_{sfx}\right.\\ &\left. + \frac{\ddot{V}_{Cxref}}{V_{dc}}2C_{ef}L_{sef}\right)\end{aligned} \tag{15}$$

$s(\dot{g}_x, t)$ decides the presence of sliding mode.

∴ the switching law is expressed in

$$s\left(\dot{g}_x, t\right)\dot{s}\left(\ddot{g}_x, t\right) < 0 \tag{16}$$

It is evident from the equation that the corresponding equation is linear with regard to u_c subject to

$$\begin{aligned}&u_c < u_{eqsx};\\ &\dot{s}\left(\ddot{g}_x, t\right) > 0\\ &u_c > u_{eqsx};\\ &\dot{s}\left(\ddot{g}_x, t\right) < 0\end{aligned} \tag{17}$$

and u_c, controlling signal of Series APF.

Therefore, corresponding control is reserved by the natural bounds of Series APF, and it is found that

$$\begin{aligned}&u_c = -1;\\ &\dot{s}\left(\ddot{g}_x, t\right) > 0\\ &u_c = 1;\\ &\dot{s}\left(\ddot{g}_x, t\right) < 0\end{aligned} \tag{18}$$

The real error functions are scaled which relates to their time domain variables, implemented to the fuzzy controller, presented in Figure 5, and $k(t)$ fuzzy controller output, which is arranged from -1 to 1.

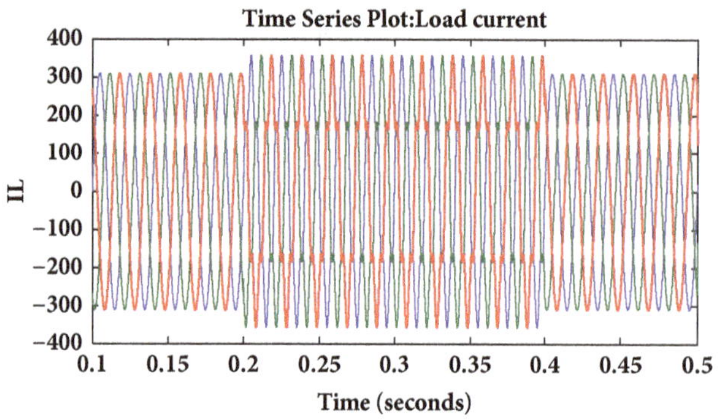

(a) Voltage swell

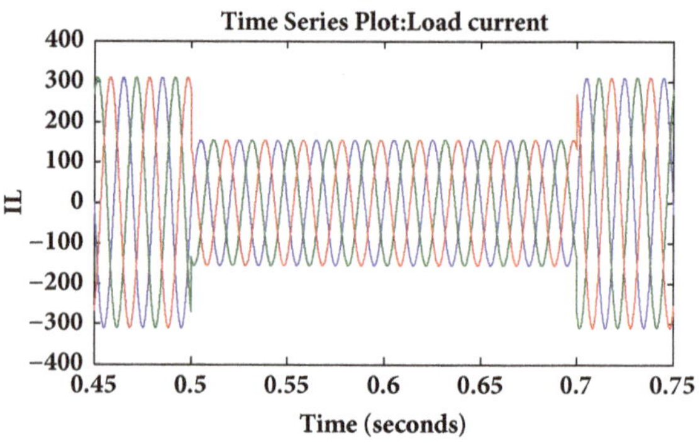

(b) Voltage sag

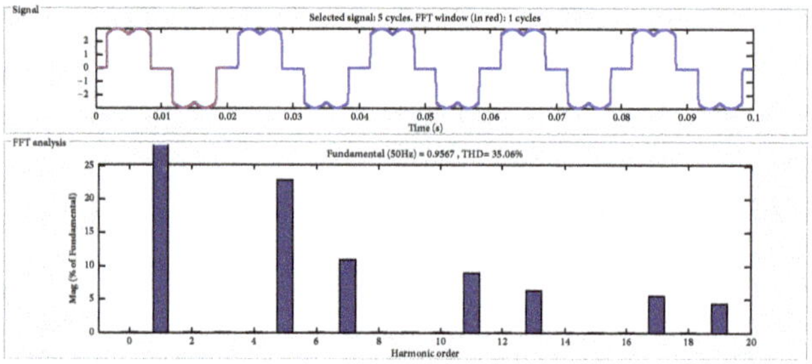

(c) THD of phase a

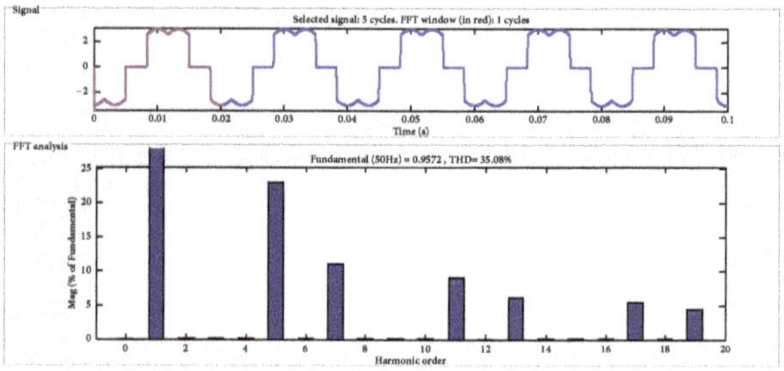

(d) THD of phase b

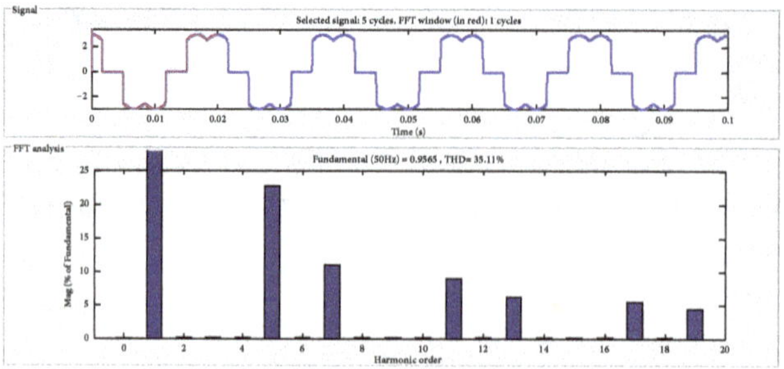

(e) THD of phase c

FIGURE 6: Load voltage and THD during distortions and without connecting SEHAPF.

For controlling the series part of hybrid filters the fuzzy controller obeys the fuzzy set theory as shown in Figure 4, and it is expressed as

$$\begin{aligned} \dot{g}(t) &= \{NEL, NEM, NES, Z, POS, POM, POB\} \\ \ddot{g}(t) &= \{NEL, NEM, NES, ZE, POS, POM, POB\} \\ k(t) &= \{NEL, Z, POB\} \end{aligned} \quad (19)$$

where $NEL, NEM, NES, Z, POS, POM, POB$ are the membership functions in fuzzy set as shown in Table 1.

The Fuzzy rule is designed [13] by using

$$u_{cx} = \begin{cases} 1, & s\left(\dot{g}_x, t\right) > 0 \\ 0, & s\left(\dot{g}_x, t\right) = 0 \\ -1, & s\left(\dot{g}_x, t\right) < 0 \end{cases} \quad (20)$$

Finally, the FSMPWM produces PWM signal depending upon fuzzy rule as shown in Table 1 which is helpful for compensating the power quality problems.

5. Results and Analysis

The modelling and simulation of SEHAPF controlled by a PI controller using HCC and FSMPWM are presented. The system performance is examined using MATLAB/SIMULINK tool. Design of SEHAPF with FSMPWM is employed to remove the voltage harmonics in the power system. The realization of SEHAPF connected through FSMPWM has been examined and the results are compared with the classical HCC using unit vector theory. The data of variable parameters are described in Table 3. The realization of SEHAPF is analyzed by simulating the circuit using the "Power System Block set" simulator.

Initially the system is performed in the absence of SEHAPF and connecting unbalance nonlinear loads. It shows that there is a distortion in voltage signal at load end, which

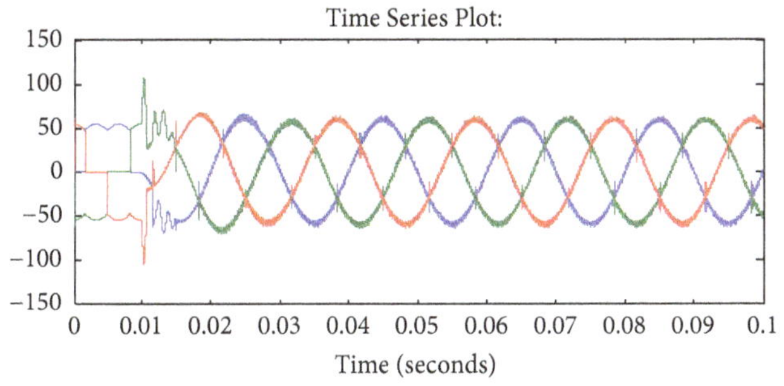

(a) Three-phase load voltage

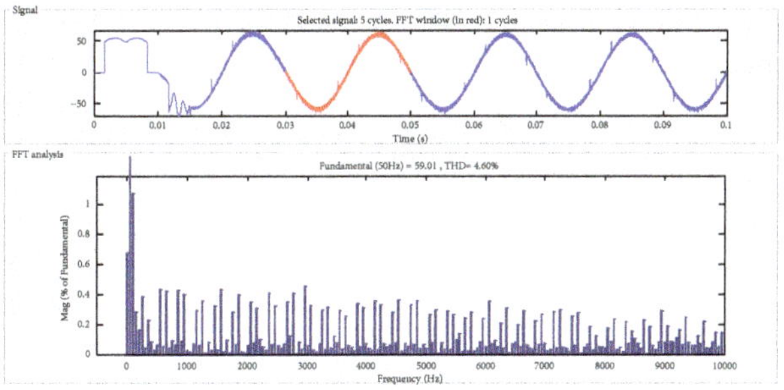

(b) THD for Phase a: 4.60%

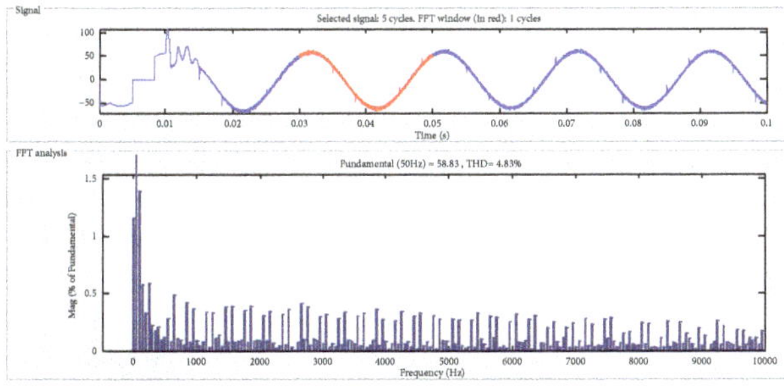

(c) THD for Phase b: 4.83%

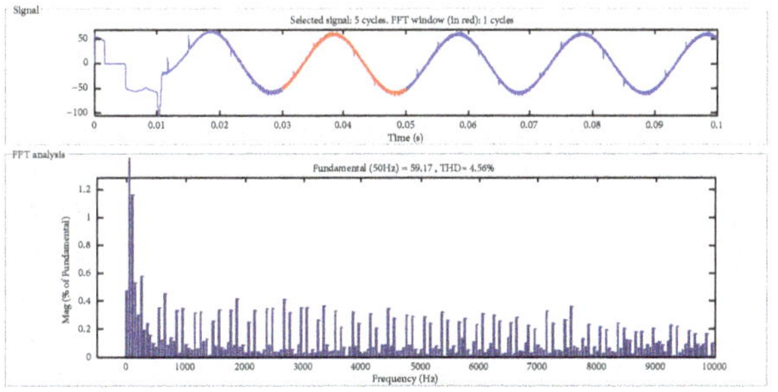

(d) THD for Phase c: 4.56%

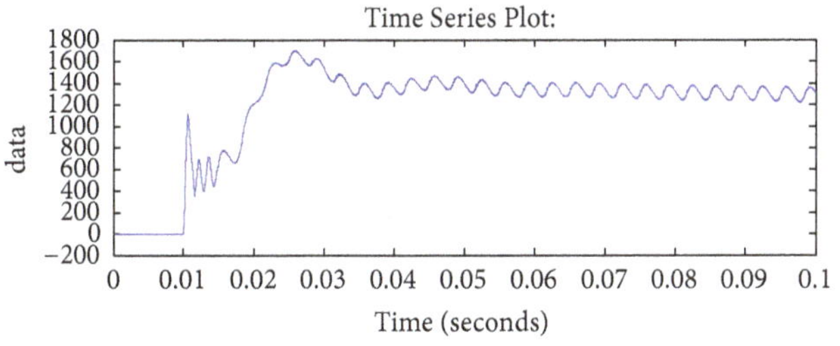

(e) Dc link voltage

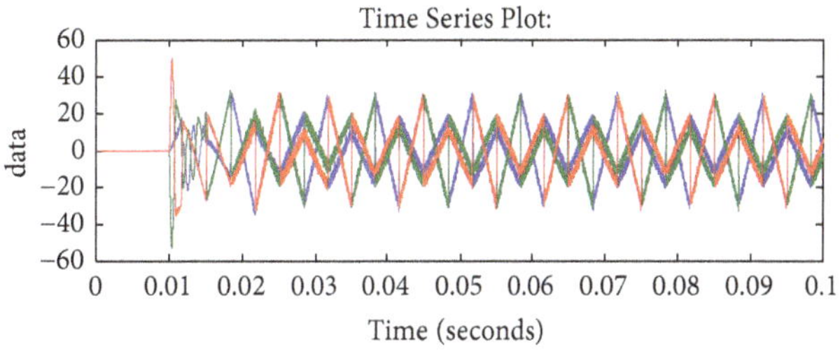

(f) Compensated voltage

FIGURE 7: Performance of load voltage and THD values of three-phase voltages and dc bus voltage and compensation voltage using HCC.

TABLE 1: Fuzzy rule.

	$\dot{g}$						
$\Delta \dot{g}$	NEL	NEM	NES	Z	POS	POM	POB
NEL	NEL	NEL	NEL	Z	POB	POB	POB
NEM	NEL	NEL	NEL	Z	POB	POB	POB
NES	NEL	NEL	NEL	Z	POB	POB	POB
Z	NEL	NEL	NEL	Z	POB	POB	POB
POS	NEL	NEL	Z	Z	POB	POB	POB
POM	NEL	Z	Z	POB	POB	POB	POB
POB	Z	Z	Z	POB	POB	POB	POB

provides sag and swell characteristics of the voltage. From simulations it is observed that voltage swell is found to be present from 0.2 Sec to 0.4 Sec. Similarly, from 0.5 Sec to 0.7 Sec, voltage sag is noticed. The output results are presented in Figures 6(a) and 6(b). It shows that the three-phase load voltages are affected in the sag and swell. THD value in phases a, b, and c is attained to be 35.06%, 35.08%, and 35.11%, respectively, and illustrated in Figures 6(c), 6(d), and 6(e).

5.1. Load Voltage after Compensation Using Hysteresis Current Controller. In this section, SEHAPF configurations based on the hysteresis current controller using unit vector theory are analyzed. The hybrid filters implementing series active filters are performed and simulation results are illustrated in Figure 7(a), and the corresponding THD values of three-phase voltages V_a, V_b, V_c are presented in Figures 7(b), 7(c), 7(d), and 7(e), respectively. The THD value observed from the simulation for three-phase load voltage is found to be 4.60%, 4.83%, and 4.56%, respectively. Figures 7(e) and 7(f) provide the simulation output of the DC bus value and compensation voltage of the SEHAPF using hysteresis controller.

5.2. Load Voltage after Compensation Using FSMPWM. This category represents the performance of the system using the SEHAPF configurations based on FSMPWM. The hybrid filters implementing series filters are performed and simulation results are presented in Figure 8(a), and the corresponding THD values of three-phase voltages V_a, V_b, V_c are illustrated in Figures 8(b), 8(c), and 8(d), respectively. By implementing FSMPWM, a noticeable improvement was found in the THD value 2.36%, 2.46%, and 2.39%, respectively, of phase a, phase b, and phase c. Figures 8(e) and 8(f) provide the simulation

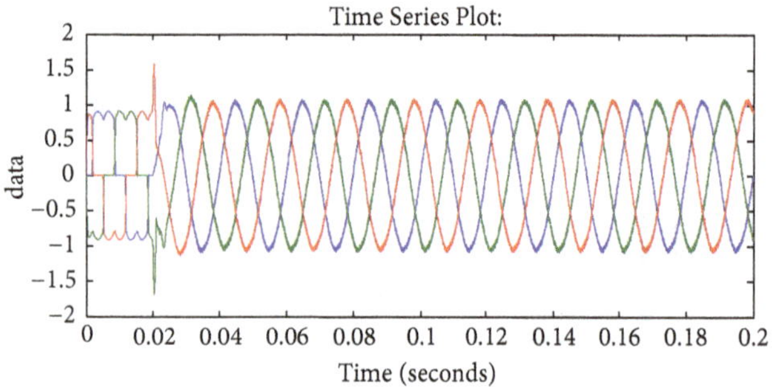

(a) Three-phase load voltage

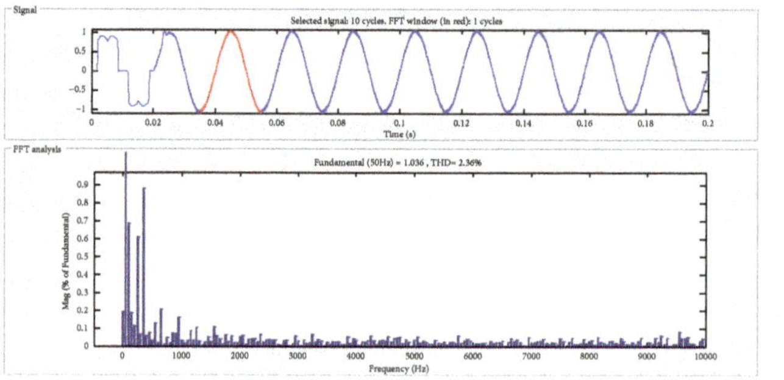

(b) THD for Phase a: 2.36%

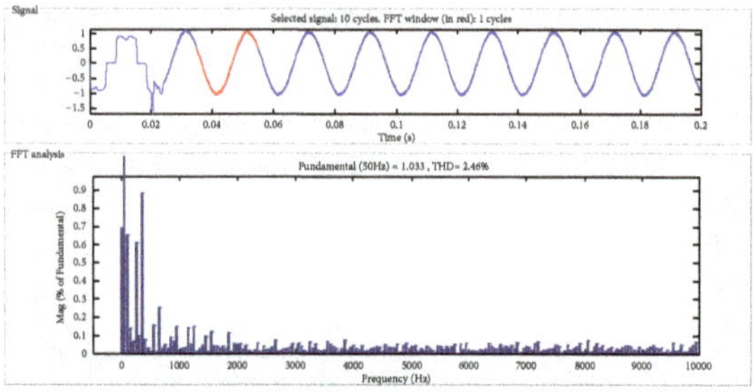

(c) THD for Phase b: 2.46%

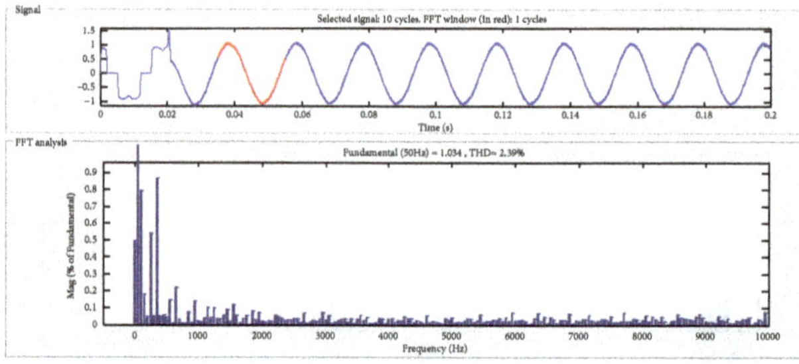

(d) THD for Phase c: 2.39%

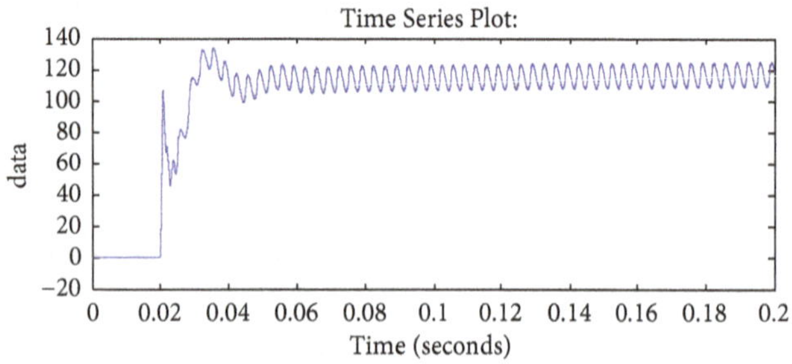

(e) Dc link voltage

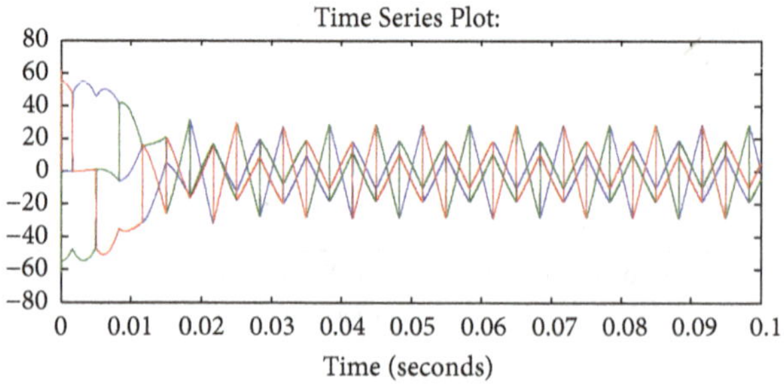

(f) Compensated voltage

FIGURE 8: Performance of load voltage and THD values of three-phase voltages and dc bus voltage and compensation voltage using FSMPWM.

TABLE 2: Comparison of THD % analysis of load voltages between FSMPWM and HCC used in SEHAPF.

Load Voltages	%THD Without SEHAPF	%THD HCC	%THD FSMPWM
Phase a	35.06	4.60	2.36
Phase b	35.08	4.83	2.46
Phase c	35.11	4.56	2.39

output of the dc link value and compensation voltage of SEAPF using FSMPWM.

Table 2 describes a comparative result of the percentage THD between HCC and FSMPWM

TABLE 3: Parameters in the system.

Parameters	Value
Line voltage and Frequency	440V, 50 Hz
Line and Load impedance	Ls= 0.25 mH, Lac= 1.05 mH, C,DC = 2000 μF, RL=30 Ω, L=20 μF (R1 =3 Ω, R2 =6 Ω, R3 =9 Ω)
Passive Filter	Cf5= 50 μF Lf5 =8.10 mH Cf7= 20 μF Lf7 = 8.27 mH Cf11= 20 μF Lf11=8.270 mH
Ripple Filter	CRF = 50 μF, LRF =0.68 mH
Active Filter side	CD = 2000 μF VD =800V
Filter coupling inductance	3 mH
Controller gain	K_P=0.032, K_I=0.00004, for Series PF

6. Conclusions

This paper presents a distinct and dynamic strategy of SEHAPF where the switching signals are controlled by using FSMPWM. The performance of the projected model is analyzed under unbalance load conditions. The simulation results show the improvement in three-phase load voltages with compensating voltage sag and swell. Improved results in THD are also found to be compared with conventional hysteresis controller. The output results with its THD value are shown in Figures 7 and 8, which shows FSMC is the better one and also follows the IEEE-519 standard.

Conflicts of Interest

The authors declare that they have no conflicts of interest.

References

[1] K. H. Kwan, P. L. So, and Y. C. Chu, "An output regulation-based unified power quality conditioner with Kalman filters," *IEEE Transactions on Industrial Electronics*, vol. 59, no. 11, pp. 4248–4262, 2012.

[2] S. Dalai, B. Chatterjee, D. Dey, S. Chakravorti, and K. Bhattacharya, "Rough-set-based feature selection and classification for power quality sensing device employing correlation techniques," *IEEE Sensors Journal*, vol. 13, no. 2, pp. 563–573, 2013.

[3] V. Khadkikar, "Enhancing electric power quality using UPQC: a comprehensive overview," *IEEE Transactions on Power Electronics*, vol. 27, no. 5, pp. 2284–2297, 2012.

[4] M. Hosseini, H. A. Shayanfar, and M. Fotuhi-Firuzabad, "Reliability improvement of distribution systems using SSVR," *ISA Transactions*, vol. 48, no. 1, pp. 98–106, 2009.

[5] J. S. Subjak and J. S. Mcquilkin, "Harmonies—Causes, Effects, Measurements, and Analysis: An Update," *IEEE Transactions on Industry Applications*, vol. 26, no. 6, pp. 1034–1042, 1990.

[6] T. S. Haugan and E. Tedeschi, "Reactive and harmonic compensation using the conservative power theory," in *Proceedings of the 10th International Conference on Ecological Vehicles and Renewable Energies, EVER 2015*, Monaco, April 2015.

[7] O. Prakash Mahela and A. Gafoor Shaik, "Topological aspects of power quality improvement techniques: A comprehensive overview," *Renewable & Sustainable Energy Reviews*, vol. 58, pp. 1129–1142, 2016.

[8] R. Zahira and A. Peer Fathima, "A technical survey on control strategies of active filter for harmonic suppression," in *Proceedings of the International Conference on Communication Technology and System Design 2011, ICCTSD 2011*, pp. 686–693, India, December 2011.

[9] B. Singh, G. Bhuvaneswari, and S. R. Arya, "Review on Power Quality Solution Technology," *Asian Power Electronics Journal*, vol. 6, no. 2, 2012.

[10] C. Kumar and M. K. Mishra, "An improved hybrid DSTATCOM topology to compensate reactive and nonlinear loads," *IEEE Transactions on Industrial Electronics*, vol. 61, no. 12, pp. 6517–6527, 2014.

[11] T.-L. Lee, Y.-C. Wang, J.-C. Li, and J. M. Guerrero, "Hybrid active filter with variable conductance for harmonic resonance suppression in industrial power systems," *IEEE Transactions on Industrial Electronics*, vol. 62, no. 2, pp. 746–756, 2015.

[12] Y. Jiang, J. Chang, and S. Tian, "Multi-objective Optimal Design of Hybrid Active Power Filter," in *Proceedings of the International Conference on Advanced Manufacture Technology and Industrial Application*, 2016.

[13] Z. Shuai, A. Luo, R. Fan, K. Zhou, and J. Tang, "Injection Branch Design of Injection Type Hybrid Active Power Filter," *Automation of Electric Power Systems*, vol. 5, article 012, 2007.

[14] S. Bhattacharya and D. Divan, "Synchronous frame based controller implementation for a hybrid series active filter system," in *Proceedings of the Conference Record of the 1995 IEEE Industry Applications 30th IAS Annual Meeting. Part 3 (of 3)*, pp. 2531–2540, October 1995.

[15] B. Singh, K. Al-Haddad, and A. Chandra, "A new control approach to three-phase active filter for harmonics and reactive power compensation," *IEEE Transactions on Power Systems*, vol. 13, no. 1, pp. 133–138, 1998.

[16] C.-S. Lam, M.-C. Wong, and Y.-D. Han, "Hysteresis current control of hybrid active power filters," *IET Power Electronics*, vol. 5, no. 7, pp. 1175–1187, 2012.

[17] J. Zeng, C. Yu, Q. Qi et al., "A novel hysteresis current control for active power filter with constant frequency," *Electric Power Systems Research*, vol. 68, no. 1, pp. 75–82, 2004.

[18] M. Kale and E. Ozdemir, "An adaptive hysteresis band current controller for shunt active power filter," *Electric Power Systems Research*, vol. 73, no. 2, pp. 113–119, 2005.

[19] S. Buso, S. Fasolo, L. Malesani, and P. Mattavelli, "A dead-beat adaptive hysteresis current control," *IEEE Transactions on Industry Applications*, vol. 36, no. 4, pp. 1174–1180, 2000.

[20] S. Mikkili and A. K. Panda, "Simulation and real-time implementation of shunt active filter id-iq control strategy for mitigation of harmonics with different fuzzy membership functions," *IET Power Electronics*, vol. 5, no. 9, pp. 1856–1872, 2012.

[21] W. Yan, J. Hu, V. Utkin, and L. Xu, "Sliding mode pulsewidth modulation," *IEEE Transactions on Power Electronics*, vol. 23, no. 2, pp. 619–626, 2008.

[22] Y. Fang, J. Fei, and K. Mab, "Model reference adaptive sliding mode control using RBF neural network for active power filter," *Journal of Electrical Power & Energy Systems*, vol. 73, pp. 249–258, 2015.

[23] R. K. Patjoshi, V. R. Kolluru, and K. Mahapatra, "Power quality enhancement using fuzzy sliding mode based pulse width modulation control strategy for unified power quality conditioner," *International Journal of Electrical Power & Energy Systems*, vol. 84, pp. 153–167, 2017.

[24] F. M. Yu, H. Y. Chung, and S. Y. Chen, "Fuzzy sliding mode controller design for uncertain time-delayed systems with nonlinear input," *Fuzzy Sets and Systems*, vol. 140, no. 2, pp. 359–374, 2003.

[25] N. Yagiz, Y. Hacioglu, and Y. Taskin, "Fuzzy sliding-mode control of active suspensions," *IEEE Transactions on Industrial Electronics*, vol. 55, no. 11, pp. 3883–3890, 2008.

[26] S. Morris, P. K. Dash, and K. P. Basu, "A fuzzy variable structure controller for STATCOM," *Electric Power Systems Research*, vol. 65, no. 1, pp. 23–34, 2003.

12

A Multicriteria Decision-Making Approach Based on Fuzzy AHP with Intuitionistic 2-Tuple Linguistic Sets

Shahzad Faizi, Tabasam Rashid, and Sohail Zafar

Department of Mathematics, University of Management and Technology, Lahore 54770, Pakistan

Correspondence should be addressed to Tabasam Rashid; tabasam.rashid@gmail.com

Academic Editor: Ferdinando Di Martino

In the modern literature related to linguistic decision-making, the 2-tuple linguistic representation model and its useful applications in various fields have been extensively studied and used during the last decade. Recently, some useful multicriteria decision-making (MCDM) methods have been introduced based on fuzzy analytic hierarchy process (AHP) for 2-tuple linguistic representation model. By keeping in mind the importance of this linguistic model, in this paper, we introduce a fuzzy AHP methodology for intuitionistic 2-tuple linguistic sets (I2TLSs) which is a useful extension of the 2-tuple linguistic representation model. This study is comprised of four stages. In the first stage, we define some operational laws for I2TL elements (I2TLEs) and prove some related important properties. In the second stage, intuitionistic 2-tuple linguistic preference relation (I2TLPR) and multiplicative I2TLPR are defined using I2TLSs. In the 3rd stage, a transformation mechanism is introduced which can transform an I2TLPR to a corresponding intuitionistic preference relation (IPR) and vice versa. In the fourth stage, an approach is proposed for checking the consistency of an I2TLPR and presented a method to repair the inconsistent one by using the proposed transformation mechanism. Finally, a numerical example is given and comparative analysis is carried out with the TOPSIS method to verify the validity of the proposed method.

1. Introduction

Herrera and Martínez [1, 2] proposed the 2-tuple fuzzy linguistic representation model which can handle linguistic and numerical information in decision-making effectively without loss and distortion of information which formerly occur during the processing of linguistic information. This useful model is basically developed on the basis of symbolic translation of the linguistic variables and has been extensively used in various MCDM problems [3–6] in recent years. The basic shortcoming of this model is that it can only ensure the accuracy in dealing with uniformly distributed linguistic term sets (LTSs). To make up for the above-mentioned shortcoming, Wang and Hao [5] introduced the proportional 2-tuple fuzzy linguistic representation model, which can ensure the accuracy in dealing with the LTSs that are not uniformly distributed. The studies on MCDM problems in the context of 2-tuple fuzzy linguistic models are growing. For example, Beg and Rashid introduced two important extensions of 2-tuple linguistic representation model, namely, the hesitant 2-tuple linguistic information model [7] and the I2TL information model [8], which are very effective in dealing with fuzziness and uncertainty as compared to the ordinary 2-tuple linguistic arguments. Furthermore, Liu and Chen [9] introduced the extended T-norm and T-conorm with the I2TL information and developed a MAGDM method based on the proposed I2TL generalized aggregation operator.

AHP was originally developed by Satty in [10] which is the most powerful technique to solve complex MCDM problems and help the decision-makers (DMs) to set preferences and make the best decision. In addition, to reduce the biasness of the DMs in the decision-making process, the AHP incorporates a useful technique for checking the consistency of the DM's evaluations. Recently, extensive studies have been conducted on AHP in fuzzy context, such as, AHP based on 2-tuple linguistic representation model for supplier segmentation by aggregating quantitative and qualitative criteria [11], a hybrid approach based on 2-tuple fuzzy linguistic method and fuzzy AHP for evaluation in-flight service quality [12], and AHP method based on hesitant fuzzy sets for analyzing the factors affecting the performance of different branches of a cargo company [13]. To collect priorities of the DMs in AHP, different kinds of preference relations are used in the literature, but numerical preference relations [14–16] and linguistic preference relations (LPRs) [17, 18] are the two basic preference relations that are often used in MCDM problems. If DMs cannot guess their preferences of one alternative over the other with actual numerical values [19] and are interested in providing their preferences in linguistic values, then they prefer LPRs which are actually a kind of numerical preference relations. The LPRs have been studied as another important tool to collect preferences and have vast applications in MCDM [20–22].

To identify the inconsistency of preference relations, there is a need of a consistency check to avoid the inconsistent solutions during a decision-making process. Saaty [23] developed an idea of consistency ratio (CR) to measure the inconsistency level of numerical preference relations. He observed that the preference relation is of acceptable consistency if $CR < 0.1$; otherwise, it is inconsistent and it is necessary to return it to the DMs again for the revision of their preferences until acceptable. Extensive studies have been done to measure the degree of inconsistency of numerical preference relations [24–26]. Similar to numerical preference relations, the consistency measure is also a difficult task while using LPRs in various MCDM problems [27]. In order to measure the consistency degree of preference relations, traditional definitions, such as the additive transitivity, the max–min transitivity, and the three-way transitivity, are used. But these definitions are incapable of measuring the consistency degree of LPRs. To make up for the above-mentioned shortcoming, Dong et al. introduced a more flexible method to measure the consistency degree of LPRs in [27]. Xu and Liao [28] proposed a method to check the consistency of an IPR and introduced an interesting procedure to repair the inconsistent IPR without the participation of the DM. Zhu and Xu [29] developed some consistency measures for hesitant fuzzy LPRs and further constructed two optimization methods to improve the consistency of an inconsistent hesitant fuzzy LPR. Zhang and Wu [30] discussed the multiplicative consistency of hesitant fuzzy LPRs and developed a consistency-improving process to adjust hesitant fuzzy LPR with unacceptably multiplicative consistency into an acceptably multiplicative one. Furthermore, Gong et al. [31] introduced the additive consistent conditions of the IPR according to that of intuitionistic fuzzy number preference relation. Wang [32] proved the additive consistency defined in an indirect manner in [31] and proved that the consistency transformation equations' matrix may not always be an IPR.

AHP is a widely used method for solving multicriteria problems in practical situations. The combination of AHP with fuzzy set and 2-tuple representation model can deal with human judgments under fuzzy environment and has no information loss. One of the main strengths of AHP is its ability to deal with subjective opinions of experts and derive a quantitative priority vector that describes the relative importance of each alternative, which makes AHP appealing to a wide variety of MCDM problems [33]. Some authors contend that the applicability of AHP can be attributed to its simplicity, ease of use, and flexibility as well as the possibility of integrating AHP with other techniques such as fuzzy logic and linear programming [34]. Furthermore, the role of AHP is to determine the weights of the criteria in both dimensions. This led to a consistent priority ranking with experts having to make only $(n^2 - n)/2$ pairwise comparisons in a decision problem containing n number of alternatives. The I2TL information model is a more powerful tool in dealing with vagueness and uncertainty that can assign to each element a membership degree as well as a nonmembership degree in the form of 2-tuple linguistic information. Therefore, the aim of this study is to apply AHP method to solve MCDM problems, where the I2TL information should be collected by a tool. First, this paper has developed some operational laws for I2TLEs and proved some of the important properties related to these operational laws. The concepts of I2TLPRs and multiplicative I2TLPR are then developed to collect the preferences of the DMs as an extension of LPRs along with a transformation function that can transform an I2TLPR to a corresponding intuitionistic preference relation (IPR). Finally, an approach is proposed for checking the consistency of an I2TLPR and presented a method to repair the inconsistent one by using the proposed transformation mechanism.

The rest of the paper is organized in the following way. The preliminary concepts related to the study are briefly reviewed in Section 2. Some operational laws for I2TLEs are defined in Section 3 and their important properties are discussed with proofs. In the same section, distance measure between two I2TLEs and comparison method of I2TLEs, I2TLPR, and multiplicative I2TLPR are proposed along with a procedure to get consistent I2TLPR from the inconsistent one. In Section 4, a numerical example is given and comparative analysis is conducted with the TOPSIS method to verify the effectiveness of the proposed method. Finally, the conclusion is presented in the last section.

2. Preliminaries

In this section, we mainly recall some elementary concepts of LTSs and 2-tuple linguistic representation model as well as the I2TL representation model.

2.1. Intuitionistic Fuzzy Set and Intuitionistic Preference Relation

Definition 1 (see [35, 36]). An IFS $\widetilde{A}$ in X is given by $\widetilde{A} = \{(x, \mu_{\widetilde{A}}(x), \nu_{\widetilde{A}}(x)) \mid x \in X\}$, where $\mu_{\widetilde{A}} : X \longrightarrow [0,1]$ and $\nu_{\widetilde{A}} : X \longrightarrow [0,1]$ with the condition that $0 \le \mu_{\widetilde{A}} + \nu_{\widetilde{A}} \le 1$ for every $x \in X$. The numbers $\mu_{\widetilde{A}}(x), \nu_{\widetilde{A}}(x) \in [0,1]$ denote, respectively, the degree of membership and nonmembership of the element $x \in X$ to the set $\widetilde{A}$. For convenience, $A = (\mu_A, \nu_A)$ is called intuitionistic fuzzy element (IFE) and Ω the set of all IFEs. For each IFS A in X, we will call $\pi_A(x) = 1 - \mu_A(x) - \nu_A(x)$ the degree of indeterminacy of x in A.

Definition 2 (see [37]). An IPR R on $X = \{x_1, x_2, \ldots, x_n\}$ is defined by a matrix $R = (r_{ik})_{n\times n}$, where $r_{ik} = (\mu(x_i, x_k), \nu(x_i, x_k))$ for all $1 \le i, k \le n$. For convenience, let r_{ik} be shortly written as (μ_{ik}, ν_{ik}), where μ_{ik} indicates the degree to which x_i is preferred to x_k, and ν_{ik} indicates the degree to which x_i is not preferred to x_k. Furthermore, $\pi(x_i, x_k) = 1 - \mu(x_i, x_k) - \nu(x_i, x_k)$ denotes the indeterminacy degree or a hesitancy degree of the IPR R satisfying the conditions $\mu_{ik}, \nu_{ik} \in [0, 1]$; $\mu_{ii} = \nu_{ii} = 0.5$;$\mu_{ik} + \nu_{ik} \le 1$; $\mu_{ik} = \nu_{ki}$; $\mu_{ki} = \nu_{ik}$; $\pi_{ik} = 1 - \mu_{ik} - \nu_{ik}$ for all $1 \le i, k \le n$.

2.2. Consistency Checking for Multiplicative IPR. A significant property of preference relations is multiplicative consistency. Xu et al. [26] proposed the definition of multiplicative consistent IPR as follows.

Definition 3 (see [26]). An IPR $R = (r_{ik})_{n\times n}$ is multiplicative consistent with $r_{ik} = (\mu_{ik}, \nu_{ik})(i, k = 1, 2, \ldots, n)$, if

$$r_{ik} = \begin{cases} (0,0) & \text{if } (\mu_{it}, \nu_{tk}), (\nu_{it}, \nu_{tk}) \in \{(0,1), (1,0)\} \\ \left(\dfrac{\mu_{it}\mu_{tk}}{\mu_{it}\mu_{tk} + (1-\mu_{it})(1-\mu_{tk})}, \dfrac{\nu_{it}\nu_{tk}}{\nu_{it}\nu_{tk} + (1-\nu_{it})(1-\nu_{tk})}\right) & \text{otherwise, for all } i \le t \le k \end{cases} \quad (1)$$

For IPRs with unacceptable consistency, Xu and Liao [28] proposed a method to measure the consistency of an IPR and then introduced a method to repair the inconsistent IPR. First, they developed an algorithm to build a perfect multiplicative consistent IPR $\overline{R} = (\overline{r}_{ik})_{n\times n}$, where $\overline{r}_{ik} = (\overline{\mu}_{ik}, \overline{\nu}_{ik})$ and

$$\overline{\mu}_{ik} = \frac{\left(\prod_{t=i+1}^{k-1} \mu_{it}\mu_{tk}\right)^{1/(k-i-1)}}{\left(\prod_{t=i+1}^{k-1} \mu_{it}\mu_{tk}\right)^{1/(k-i-1)} + \left(\prod_{t=i+1}^{k-1} (1-\mu_{it})(1-\mu_{tk})\right)^{1/(k-i-1)}} \quad \text{for } k > i+1 \quad (2)$$

$$\overline{\nu}_{ik} = \frac{\left(\prod_{t=i+1}^{k-1} \nu_{it}\nu_{tk}\right)^{1/(k-i-1)}}{\left(\prod_{t=i+1}^{k-1} \nu_{it}\nu_{tk}\right)^{1/(k-i-1)} + \left(\prod_{t=i+1}^{k-1} (1-\nu_{it})(1-\nu_{tk})\right)^{1/(k-i-1)}} \quad \text{for } k > i+1. \quad (3)$$

Definition 4 (see [28]). An IPR R is called an acceptable multiplicative consistent, if the distance measure between R and $\overline{R}$ denoted as $d(R, \overline{R})$ is less than τ, where $\tau = 0.1$ is the consistency threshold. The distance measure $d(R, \overline{R})$ can be determined as follows:

$$d\left(R, \overline{R}\right) = \frac{1}{2(n-1)(n-2)} \cdot \sum_{i=1}^{n}\sum_{k=1}^{n} \left(\left|\overline{\mu}_{ik} - \mu_{ik}\right| + \left|\overline{\nu}_{ik} - \nu_{ik}\right| + \left|\overline{\pi}_{ik} - \pi_{ik}\right|\right) \quad (4)$$

Xu and Liao [28] thought that the transformed IPR $\overline{R}$ cannot represent the initial preferences of the DM for a large value of $d(R, \overline{R})$. Therefore, they fused the IPRs R and $\overline{R}$ into a new IPR $\widetilde{R} = (\widetilde{r}_{ik})_{n\times n}$, where

$$\widetilde{\mu}_{ik} = \frac{(\mu_{ik})^{\sigma-1}(\overline{\mu}_{ik})^{\sigma}}{(1-\mu_{ik})^{\sigma-1}(1-\overline{\mu}_{ik})^{\sigma} + (\mu_{ik})^{\sigma-1}(\overline{\mu}_{ik})^{\sigma}} \quad (5)$$

$$\widetilde{\nu}_{ik} = \frac{(\nu_{ik})^{\sigma-1}(\overline{\nu}_{ik})^{\sigma}}{(1-\nu_{ik})^{\sigma-1}(1-\overline{\nu}_{ik})^{\sigma} + (\nu_{ik})^{\sigma-1}(\overline{\nu}_{ik})^{\sigma}} \quad (6)$$

where σ is called the controlling parameter of the IPR $\widetilde{R}$ that is set by the DM only. If σ is small, then $\widetilde{R}$ is closer to R. For $\sigma = 0$, $\widetilde{R} = R$, and for $\sigma = 1$, $\widetilde{R} = \overline{R}$.

2.3. Basic Concepts of Linguistic Term Set and 2-Tuple Linguistic Information

Definition 5 (see [38, 39]). Let $S = \{s_0, s_1, \ldots, s_g\}$ be a finite LTS with odd cardinality, where each $s_i (0 \le i \le g)$ represents a possible value for a linguistic variable. The following characteristics for S can be defined as follows:

(1) Negation operator: $neg(s_i) = s_j$, such that $i + j = g$;

(2) Ordered set: $s_i \le s_j \Longleftrightarrow i \le j$. Therefore, there exist two operators given as follows:

(a) maximization operator: $\max(s_i, s_j) = s_i$, if $s_j \le s_i$;

(b) minimization operator: $\min(s_i, s_j) = s_i$, if $s_i \le s_j$.

Xu [40, 41] introduced the concept of continuous LTS $\overline{S}$ as an extension of discrete term set S where $\overline{S} = \{s_k | s_0 \le s_k \le s_g\}$. The linguistic term s_k is called the original linguistic term if $s_k \in S$, and is only used by the DMs to evaluate the alternatives during a decision process. If the linguistic term $s_k \notin S$, then s_k is said to be the virtual linguistic term of S and it appears only during the computations.

Herrera and Martínez [2] proposed the 2-tuple linguistic representation model which expresses the linguistic information by a 2-tuple (s_i, α), where $s_i \in S$ and $\alpha \in [-0.5, 0.5)$. The basic purpose of this model is to define a transformation mechanism between linguistic 2-tuples and the numerical values.

Definition 6 (see [2]). Let $S = \{s_0, s_1, \ldots, s_g\}$ be a LTS and $\beta \in [0, g]$ a value representing the result of a symbolic aggregation operation. Then, a function $\Delta : [0, g] \longrightarrow S \times [-0.5, 0.5)$ which provides a linguistic 2-tuple representing the equivalent information to β is defined as follows:

$$\Delta(\beta) = (s_i, \alpha), \quad \text{with } \begin{cases} s_i, & i = \text{round}(\beta), \\ \alpha = \beta - i, & \alpha \in [-0.5, 0.5). \end{cases} \quad (7)$$

Clearly, $\triangle$ is one to one function. The $\triangle$ has an inverse function ∇ with $\nabla((s_i,\alpha)) = i+\alpha$.

2.4. Intuitionistic 2-Tuple Linguistic Information Model. Beg and Rashid [8] proposed the idea of I2TL information model and some operators based on choquet integral to aggregate the I2TL information. They defined I2TL representation model as follows.

Definition 7 (see [8]). For a crisp set X and LTS $S = \{s_0, s_1, \dots, s_g\}$, the set $A = \{(x, h(x), h'(x)) | x \in X\}$ in X where $h, h' : X \longrightarrow S$ is called an intuitionistic LTS if $h(x) = s_i$ and $h'(x) = s_j$ with the condition that $0 \le i+j \le g$, for all $x \in X$. The linguistic values $h(x)$ and $h'(x)$ represent, respectively, the membership and nonmembership degrees of the element x in the set A.

Definition 8 (see [8]). Let $A = \{(x, h(x), h'(x)) | x \in X\}$ be intuitionistic LTS in X and $(s_i, s_j) \in A$; an I2TL model can be defined as $((s_i,\alpha),(s_j,\eta))$, where $s_i, s_j \in S$, and α, η are numeric values in $[-0.5, 0.5]$ denoting the symbolic translation of s_i and s_j, respectively. For our convenience, $((s_i,\alpha),(s_j,\eta))$ is called an I2TLE and $I(S)$ is the set of all I2TLEs.

In order to avoid any loss of information, Beg and Rashid [8] further presented a computational technique to deal with this model as follows.

Definition 9 (see [8]). Let $((s_i,\alpha),(s_j,\eta))$ be an I2TLE for a LTS S. The function $\nabla : (S\times[-0.5,0.5))\times(S\times[-0.5,0.5)) \longrightarrow [-0.5, g+0.5)\times[-0.5, g+0.5)$ from $((s_i,\alpha),(s_j,\eta))$ to an order pair of numerical values $(\beta,\zeta) \in [-0.5, g+0.5) \times [-0.5, g+0.5) \subset \mathbb{R}\times\mathbb{R}$ is defined as $\nabla((s_i,\alpha),(s_j,\eta)) = (i+\alpha, j+\eta) = (\beta,\zeta)$.

It is clear that $\beta, \zeta \in [-0.5, g+0.5)$ with the condition $0 \le \beta+\zeta < g+1$ provided i and j are not simultaneously zero.

The function $\triangle : [-0.5, g+0.5) \times [-0.5, g+0.5) \longrightarrow (S\times[-0.5,0.5))\times(S\times[-0.5,0.5))$ is used to obtain the I2TL information equivalent to the pair (β,ζ). This function $\triangle$ can be defined as $\triangle(\beta,\zeta) = ((s_i,\alpha),(s_j,\eta))$, where $i = \text{round}(\beta)$, $j = \text{round}(\eta)$, $\alpha = \beta - i$, and $\eta = \zeta - j$. The linguistic terms s_i and s_j have the closest index label to β and ζ, respectively. Similarly, the values α, η represent the symbolic translations of s_i and s_j, respectively.

3. Operational Laws of I2TLEs and Consistency Measure

In this section, we define some logical operational laws of I2TLEs and present some properties with proofs. The proposed operational laws for I2TLEs encompass previous operational laws for LTSs and exhibit flexibility. We also define I2TLPR and multiplicative I2TLPR and study a useful method to get a consistent I2TLPR from an inconsistent one. Furthermore, distance measure between two I2TLEs, comparison method of I2TLEs, and a methodology of I2TL AHP method are proposed in the same section to find an optimal alternative in a MCDM problem.

3.1. Some Operational Laws of I2TLEs. Gou and Xu [42] defined some logical operational laws for linguistic variables of a LTS on the basis of two equivalent transformation functions which can avoid the aggregated linguistic values exceeding the bounds of LTSs. They further discussed various related important properties for these operational laws. These operational laws are actually based on a transformation function $f : S \longrightarrow [0,1]$ and inverse transformation function $f^{-1} : [0,1] \longrightarrow S$ which are defined as follows:

$$f(s_i) = \frac{i}{g} = \gamma \in [0,1], \qquad f^{-1}(\gamma) = s_{g\gamma} \qquad \text{for all } s_i \in S \tag{8}$$

Based on these transformation functions, Gou and Xu [42] introduced the following novel operational laws for linguistic values of a LTS as follows:

(1) $s_i \oplus s_j = f^{-1}(f(s_i) + f(s_j) - f(s_i)f(s_j))$

(2) $s_i \otimes s_j = f^{-1}(f(s_i)f(s_j))$

(3) $\lambda s_i = f^{-1}(1-(1-f(s_i))^\lambda)$

(4) $s_i = f^{-1}((f(s_i))^\lambda)$

Gou and Xu [42] also investigated the following important properties for these novel operational laws:

(1) $s_i \oplus s_j = s_j \oplus s_i$

(2) $s_i \otimes s_j = s_j \otimes s_i$

(3) $\lambda(s_i \oplus s_j) = \lambda s_i \oplus \lambda s_j$

(4) $(s_i \otimes s_j)^\lambda = (s_i)^\lambda \otimes (s_j)^\lambda$

(5) $\lambda_1 s_i \oplus \lambda_2 s_i = (\lambda_1 \oplus \lambda_2) s_i$

(6) $(s_i)^{\lambda_1} \otimes (s_i)^{\lambda_2} = (s_i)^{\lambda_1+\lambda_2}$

Motivated by the above operational laws of LTSs, we can also extend these operation laws for I2TLEs as follows.

Definition 10. Let $S = \{s_0, s_1, \dots, s_g\}$ be a LTS and $I = ((s_i,\alpha),(s_j,\beta))$, $I_1 = ((s_{i_1},\alpha_1),(s_{j_1},\beta_1))$, and $I_2 = ((s_{i_2},\alpha_2),(s_{j_2},\beta_2))$ be three I2TLEs in $I(S)$ and $\lambda \in [0,1]$. We define

(1) $I_1 \oplus I_2 = ((s_{i_1} \oplus s_{i_2}, (\alpha_1+\alpha_2)/2), (s_{j_1} \otimes s_{j_2}, (\beta_1+\beta_2)/2))$

(2) $I_1 \otimes I_2 = ((s_{i_1} \otimes s_{i_2}, (\alpha_1+\alpha_2)/2), (s_{j_1} \oplus s_{j_2}, (\beta_1+\beta_2)/2))$

(3) $\lambda I = ((\lambda s_i, \alpha), (s_j^\lambda, \beta))$

(4) $I^\lambda = ((s_i^\lambda, \alpha), (\lambda s_j, \alpha))$

Theorem 11. *Let $S = \{s_0, s_1, \dots, s_g\}$ be a LTS and $I = ((s_i,\alpha),(s_j,\beta))$, $I_1 = ((s_{i_1},\alpha_1),(s_{j_1},\beta_1))$, and $I_2 = ((s_{i_2},\alpha_2),(s_{j_2},\beta_2))$ be three I2TLEs in $I(S)$ and $\lambda, \lambda_1, \lambda_2 \in [0,1]$. Then*

(1) $I_1 \oplus I_2 = I_2 \oplus I_1$

(2) $I_1 \otimes I_2 = I_2 \otimes I_1$

(3) $\lambda(I_1 \oplus I_2) = \lambda I_1 \oplus \lambda I_2$

(4) $(I_1 \otimes I_2)^\lambda = I_1^\lambda \otimes I_2^\lambda$

(5) $\lambda_1 I \oplus \lambda_2 I = (\lambda_1 + \lambda_2) I$

(6) $I^{\lambda_1} \otimes I^{\lambda_2} = I^{\lambda_1+\lambda_2}$

Proof. (1) and (2) are simple, so the proofs of them are omitted here.

$$(3)\ \lambda(I_1 \oplus I_2) = \lambda\left(\left(s_{i_1} \oplus s_{i_2}, \frac{\alpha_1+\alpha_2}{2}\right), \left(s_{j_1} \otimes s_{j_2}, \frac{\beta_1+\beta_2}{2}\right)\right) = \left(\left(\lambda\left(s_{i_1} \oplus s_{i_2}\right), \frac{\alpha_1+\alpha_2}{2}\right), \left(\left(s_{j_1} \otimes s_{j_2}\right)^\lambda, \frac{\beta_1+\beta_2}{2}\right)\right) = \left(\left(\lambda s_{i_1} \oplus \lambda s_{i_2}, \frac{\alpha_1+\alpha_2}{2}\right), \left(s_{j_1}^\lambda \otimes s_{j_2}^\lambda, \frac{\beta_1+\beta_2}{2}\right)\right) = \left(\left(\lambda s_{i_1}, \alpha_1\right), \left(s_{j_1}^\lambda, \beta_1\right)\right) \oplus \left(\left(\lambda s_{i_2}, \alpha_2\right), \left(s_{j_2}^\lambda, \beta_2\right)\right) = \lambda\left(\left(s_{i_1}, \alpha_1\right), \left(s_{j_1}, \beta_1\right)\right) \oplus \lambda\left(\left(s_{i_2}, \alpha_2\right), \left(s_{j_2}, \beta_2\right)\right) = \lambda I_1 \oplus \lambda I_2$$

$$(4)\ (I_1 \otimes I_2)^\lambda = \left(\left(s_{i_1} \otimes s_{i_2}, \frac{\alpha_1+\alpha_2}{2}\right), \left(s_{j_1} \oplus s_{j_2}, \frac{\beta_1+\beta_2}{2}\right)\right)^\lambda = \left(\left(\left(s_{i_1} \otimes s_{i_2}\right)^\lambda, \frac{\alpha_1+\alpha_2}{2}\right), \left(\lambda\left(s_{j_1} \oplus s_{j_2}\right), \frac{\beta_1+\beta_2}{2}\right)\right) = \left(\left(s_{i_1}^\lambda \otimes s_{i_2}^\lambda, \frac{\alpha_1+\alpha_2}{2}\right), \left(\lambda s_{j_1} \oplus \lambda s_{j_2}, \frac{\beta_1+\beta_2}{2}\right)\right) = \left(\left(s_{i_1}^\lambda, \alpha_1\right), \left(\lambda s_{j_1}, \beta_1\right)\right) \otimes \left(\left(s_{i_2}^\lambda, \alpha_2\right), \left(\lambda s_{j_2}, \beta_2\right)\right) = \left(\left(s_{i_1}, \alpha_1\right), \left(s_{j_1}, \beta_1\right)\right)^\lambda \otimes \left(\left(s_{i_2}, \alpha_2\right), \left(s_{j_2}, \beta_2\right)\right)^\lambda = I_1^\lambda \otimes I_2^\lambda$$

$$(5)\ \lambda_1 I \oplus \lambda_2 I = \lambda_1\left(\left(s_i, \alpha\right), \left(s_j, \beta\right)\right) \oplus \lambda_2\left(\left(s_i, \alpha\right), \left(s_j, \beta\right)\right) = \left(\left(\lambda_1 s_i, \alpha\right), \left(s_j^{\lambda_1}, \beta\right)\right) \oplus \left(\left(\lambda_2 s_i, \alpha\right), \left(s_j^{\lambda_2}, \beta\right)\right) = \left(\left(\lambda_1 s_i \oplus \lambda_2 s_i, \frac{\alpha+\alpha}{2}\right), \left(s_j^{\lambda_1} \otimes s_j^{\lambda_2}, \frac{\beta+\beta}{2}\right)\right) = \left(\left(\left(\lambda_1+\lambda_2\right) s_i, \alpha\right), \left(s_j^{\lambda_1+\lambda_2}, \beta\right)\right) = \left(\lambda_1+\lambda_2\right) \cdot \left(\left(s_i, \alpha\right), \left(s_j, \beta\right)\right) = \left(\lambda_1+\lambda_2\right) I$$

$$(6)\ I^{\lambda_1} \otimes I^{\lambda_2} = \left(\left(s_i, \alpha\right), \left(s_j, \beta\right)\right)^{\lambda_1} \otimes \left(\left(s_i, \alpha\right), \left(s_j, \beta\right)\right)^{\lambda_2} = \left(\left(s_i^{\lambda_1}, \alpha\right), \left(\lambda_1 s_j, \beta\right)\right) \otimes \left(\left(s_i^{\lambda_2}, \alpha\right), \left(\lambda_2 s_j, \beta\right)\right) = \left(\left(s_i^{\lambda_1} \otimes s_i^{\lambda_2}, \frac{\alpha+\alpha}{2}\right), \left(\lambda_1 s_j \oplus \lambda_2 s_j, \frac{\beta+\beta}{2}\right)\right) = \left(\left(s_i^{\lambda_1+\lambda_2}, \alpha\right), \left(\left(\lambda_1+\lambda_2\right) s_j, \beta\right)\right) = \left(\left(s_i, \alpha\right), \left(s_j, \beta\right)\right)^{\lambda_1+\lambda_2} = I^{\lambda_1+\lambda_2} \tag{9}$$

In the following, we put forward the axiom of distance measure for I2TLEs, shown as follows:

Definition 12. Let $S = \{s_0, s_1, \ldots, s_g\}$ be a LTS and $I_1 = ((s_{i_1}, \alpha_1), (s_{j_1}, \beta_1))$ and $I_2 = ((s_{i_2}, \alpha_2), (s_{j_2}, \beta_2))$ be two I2TLEs in $I(S)$. The Euclidean distance d_{ed} between I_1 and I_2 can be defined as follows:

$$d_{gd}(I_1, I_2) = \frac{1}{\sqrt{2}g} \cdot \sqrt{\left((i_1+\alpha_1)-(i_2+\alpha_2)\right)^2 + \left((j_1+\beta_1)-(j_2+\beta_2)\right)^2} \tag{10}$$

Definition 13. Let $I_1, I_2 \in I(S)$; then the Euclidean distance d_{ed} between I_1 and I_2 satisfies the following:

(1) $0 \le d_{ed}(I_1, I_2) \le 1$;

(2) $d_{ed}(I_1, I_2) = 0$ if and only if $I_1 = I_2$;

(3) $d_{ed}(I_1, I_2) = d_{ed}(I_2, I_1)$.

Liu and Chen [9] proposed score and accuracy functions for the comparison of two I2TLEs. We now introduce a new comparison method for I2TLEs, which can be seen as follows.

Definition 14. Let $I = ((s_i, \alpha), (s_j, \eta))$ be an I2TLE in $I(S)$ where $\alpha, \eta \notin [-0.5, 0)$ for $i = j = 0$. The score function $S(I)$, the accuracy function $Ac(I)$, and hesitancy degree value $h(I)$ of I can be defined as follows:

$$S(I) = \frac{1}{2g}\left((i+\alpha)-(j+\eta)+g\right) \tag{11}$$

$$Ac(I) = \frac{1}{2g}\left((i+\alpha)+(j+\eta)+g\right) \tag{12}$$

It can be observed that $S(I), Ac(I) \in [0, 1]$. The comparison method of I2TLEs based on score and accuracy functions can be established as follows.

Definition 15. For any two I2FLEs $I_1, I_2 \in I(S)$,

(1) if $S(I_1) > S(I_2)$, then $I_1 \succ I_2$;

(2) if $S(I_1) = S(I_2)$, and

(a) $Ac(I_1) > Ac(I_2)$, then $I_1 \succ I_2$;

(b) $Ac(I_1) = Ac(I_2)$, then $I_1 = I_2$.

Example 16. Let $S = \{s_0, s_1, \ldots, s_6\}$ be a LTS and $I_1 = ((s_2, -0.2), (s_3, 0))$ and $I_2 = ((s_2, 0.2), (s_3, -0.5))$ be two I2TLEs in $I(S)$. Then using (11), $S(I_1) = 0.4000$, $S(I_2) = 0.4750$. This shows that $I_2 \succ I_1$. Similarly, for $I_3 = ((s_4, 0.1), (s_1, -0.3))$, $I_4 = ((s_4, 0.2), (s_1, -0.2))$ in $I(S)$, $S(I_3) = S(I_4) = 0.7833$ but $Ac(I_3) = 0.9$ and $Ac(I_4) = 0.9167$. This implies $I_4 \succ I_3$.

TABLE 1: The $RI(n)$ values for $n \leq 10$.

n	2	3	4	5	6	7	8	9	10
$RI(n)$	0	0.58	0.90	1.12	1.24	1.32	1.41	1.45	1.51

3.2. Consistency Measure of I2TLPR. In preference relations, consistency is an important topic in decision-making and the lack of consistency can lead to inconsistent solutions. Some inconsistencies may typically arise while finding consistent solution to MCDM problems when many pairwise comparisons are performed by the DMs during assessment processes. Saaty [23] proposed a consistency index and a consistency ratio denoted as "CI" and "CR", respectively, in the conventional AHP method to compute the degree or level of consistency for a multiplicative preference relation by using the following formulae:

$$CI = \frac{\lambda_{\max} - n}{n - 1} \tag{13}$$

$$CR = \frac{CI}{RI(n)} \tag{14}$$

where $\lambda_{\max}$ and n are, respectively, the largest eigenvalue and the dimension of the multiplicative preference relation. The term $RI(n)$ is denoted as random index that completely depends on the value of n. The values of $RI(n)$ for $n \leq 10$ are shown in Table 1. The value of CI is always equal to zero for a perfectly consistent DM, but small values of inconsistency may be tolerated during a decision process. However, perfect consistency rarely occurs in practice.

Saaty [23] identified that the multiplicative preference relation is of acceptable level of consistency when $CR < 0.1$; otherwise, it is inconsistent and it is necessary to return it to the DMs again for the revision of their preferences until they are acceptable.

Due to the importance of consistent preference relations, we now focus on the studies of the consistency measures of I2TLPRs. First, we will define I2TLPR and the multiplicative I2TLPR and then propose an intuitionistic 2-tuple transformation function which is useful in obtaining an consistent I2TLPR.

3.2.1. Intuitionistic 2-Tuple Linguistic Preference Relation

Definition 17. Let $X = \{x_1, x_2, \ldots, x_n\}$ be a fixed given set of alternatives and $S = \{s_0, s_1, \ldots, s_g\}$ a LTS. Suppose the DMs provide their pairwise comparison assessments of alternatives by linguistic values based on S and the numeric values representing the symbolic translation are selected from the interval $[-0.5, 0.5)$, and these linguistic values along with symbolic translation are transformed into I2TLSs. The I2TLPR can be defined as follows.

An I2TLPR B is denoted by a matrix $B = (b_{ij})_{n\times n}$, where $b_{ij} = ((x_i, x_j), b_m^{ij}(x_i, x_j), b_n^{ij}(x_i, x_j))$, $b_m^{ij} = (s_p^{ij}, \alpha_p^{ij})$, and $b_n^{ij} = (s_q^{ij}, \eta_q^{ij})$. For convenience, we let $b_{ij} = ((s_p^{ij}, \alpha_p^{ij}), (s_q^{ij}, \eta_q^{ij})) = (b_m^{ij}, b_n^{ij})$, where $0 \leq p + q \leq g$ and $\alpha_p^{ij}, \eta_q^{ij} \in [-0.5, 0.5)$. b_m^{ij} denotes the degree to which x_i is preferred to x_j; b_n^{ij} indicates the degree to which x_i is not preferred to x_j. For all $i, j = 1, 2, \ldots, n, b_{ij} (i < j)$, the I2TLPR should satisfy the following conditions: b_m^{ij}, b_n^{ij} represent, respectively, the linguistic information by 2-tuple $(s_p^{ij}, \alpha_p^{ij})$ and (s_q^{ij}, η_q^{ij}), $b_m^{ij} = b_n^{ji}$, $b_m^{ji} = b_n^{ij}$, $b_m^{ii} = b_n^{ii} = ((s_3, 0.5), (s_3, 0.5))$ for all $i, j = 1, 2, \ldots, n$.

3.2.2. Intuitionistic 2-Tuple Transformation Function. In order to define a multiplicative I2TLPR, we first define a transformation function which can transform an I2TLE to an element of Ω and then the inverse transformation function as follows.

Definition 18. Let $I = ((s_i, \alpha), (s_j, \eta))$ be an intuitionistic 2-tuple. We define a intuitionistic 2-tuple transformation function $h : I(S) \longrightarrow \Omega' \subset \Omega$ as $h(I) = (\beta/(g+1), \zeta/(g+1)) = (a, b)$, where $\beta = i + \alpha$ and $\zeta = j + \eta$ and $\alpha, \eta \notin [-0.5, 0)$ for $i = j = 0$. Clearly $0 \leq a+b = (\beta+\zeta)/(g+1) < (g+1)/(g+1) = 1$ because $0 \leq \beta + \zeta < g + 1$. It should be noted here that, as $\alpha + \eta \longrightarrow 1, a + b \longrightarrow 1$ for $i = g$ and $j = 0$ or $i = 0$ and $j = g$. Similarly, we can define the intuitionistic 2-tuple inverse transformation function $h^{-1} : \Omega' \longrightarrow I(S)$ as $h^{-1}(a, b) = ((s_i, \alpha), (s_j, \eta)) \in I(S)$, where $i =$round$((g + 1)a)$, $j =$round$((g + 1)b)$, $\alpha = (g + 1)a - i$, and $\eta = (g + 1)b - j$ for any $(a, b) \in \Omega'$.

Example 19. Let $S = \{s_0, s_1, \ldots, s_6\}$ be a LTS. Suppose $I = ((s_5, 0.2), (s_1 - 0.5))$ is an I2TLE. Then, $h(I) = (0.74290.0714) \in \Omega'$ is the corresponding IFE. Similarly, by applying intuitionistic 2-tuple inverse transformation function, we get $h^{-1}(0.7429, 0.0714) = I$.

Remark 20. The intuitionistic 2-tuple transformation mechanism can provide a relationship between intuitionistic 2-tuples and IFEs. Obviously, it is convenient to obtain the transformation results according to different situations of decision-making processes.

Remark 21. The intuitionistic 2-tuple transformation mechanism provides a useful relationship between intuitionistic 2-tuples and IFEs. Therefore, the values of parameters τ and σ can be used as discussed in Section 2.2 during the process of obtaining a consistent I2TLPR. Moreover, Table 1 can also be used as it is during the computation process of consistency measure of I2TLPR.

Definition 22. An I2TLPR $B = (b_{ij})_{n\times n}$ with $b_{ij} = (b_m^{ij}, b_n^{ij})$, $b_m^{ij} = (s_p^{ij}, \alpha_p^{ij})$, $b_n^{ij} = (s_q^{ij}, \eta_q^{ij}), 0 \leq p + q \leq g$, and $\alpha_p^{ij}, \eta_q^{ij} \in [-0.5, 0.5)$ is multiplicative consistent if its corresponding IPR $h(B) = (h(b_{ij}))_{n\times n}$ is multiplicative consistent.

3.2.3. How to Find the Consistent Multiplicative Consistent I2TLPR. For I2TLPR $B = (b_{ij})_{n\times n}$, our aim is now to let B approach a consistent one without the interaction of DMs.

TABLE 2: I2TLPR of criteria concerning the overall objective.

B	C_1	C_2	C_3	C_4
C_1	$((s_3, 0.5), (s_3, 0.5))$	$((s_4, 0.3), (s1, 0.4))$	$((s_3, 0.2), (s_2, 0.3))$	$((s_4, -0.45), (s_2, -0.3))$
C_2	$((s_1, 0.4), (s_4, 0.3))$	$((s_3, 0.5), (s_3, 0.5))$	$((s_3, 0.2), (s_1, 0.4))$	$((s_4, -0.45), (s_0, 0.2))$
C_3	$((s_2, 0.3), (s_3, 0.2))$	$((s_1, 0.4), (s_3, 0.2))$	$((s_3, 0.5), (s_3, 0.5))$	$((s_4, 0.2), (s_1, 0.3))$
C_4	$((s_2, -0.3), (s_4, -0.45))$	$((s_0, 0.2), (s_4, -0.45))$	$((s_1, 0.3), (s_4, 0.2))$	$((s_3, 0.5), (s_3, 0.5))$

The following algorithm is developed to obtain a consistent I2TLPR B if B is of unacceptable consistency.

Algorithm 23.

Step 1. Assuming an I2TLPR $B = (b_{ij})_{n\times n}$, and the predefined consistent threshold τ, obtain a corresponding IPR $R = h(B) = (h(b_{ij}))_{n\times n}$ by using the I2T transformation function (see Definition 18).

Step 2. Suppose that y is the number of iterations. Let $y = 1$, and construct a perfect multiplicative consistent IPR $\overline{R}$ from $R_y = R$ using (2)-(4).

Step 3. Construct estimated consistency of R_y by computing $d(\overline{R}, R_y)$ using (4). If $d(\overline{R}, R_y) \le \tau$, then output R_y; otherwise, go to the next step.

Step 4. By using (5) and (6), construct the fused IPR $\overline{R}_y$ by letting a suitable value of the controlling parameter σ.

Step 5. Compute $d(\overline{R}_y, R)$ using (4). If $d(\overline{R}_y, R) \le \tau$, then output $\overline{R}_y$; otherwise, repeat Step 3.

Step 6. Set $R_{y+1} = \overline{R}_y$. Let $y = y + 1$, and, then, go to the next step.

Step 7. Construct the corresponding consistent I2TLPR $B^c = h^{-1}(\overline{R}_y)$ (see Definition 18).

3.3. Intuitionistic 2-Tuple Linguistic AHP Method. Let $A = \{A_1, A_2, \dots, A_m\}$ be a discrete set of m possible alternatives and a set of n objective criteria $C = \{C_1, C_2, \dots, C_n\}$. Now we formulate the I2TL AHP model to solve MCDM problems. The next six steps can sum up the whole procedure of applying the I2TL AHP method.

Step 1. Construct a hierarchical structure for the decision problem to be solved.

Step 2. Establish the I2TLPR $B = (b_{pq})_{n\times n}$ through the pairwise comparison between each criterion. At the same time, further I2TLPRs $B_j = (b^j_{pq})_{m\times m}$ $(j = 1, 2, \dots, n)$ of alternatives concerning the criteria $C_j (j = 1, 2, \dots, n)$ are constructed via the pairwise comparison of alternatives under each criterion.

Step 3. Check the consistency degree of each I2TLPR B and $B_j (j = 1, 2, \dots, n)$ as discussed in Step 4. If all I2TLPRs are already of acceptable consistency, ignore Step 4.

Step 4. Repair the inconsistent I2TLPRs B and $B_j (j = 1, 2, \dots, n)$ by using Algorithm 23 (or return them to the DM for reconsideration until they are acceptable).

Step 5. Calculate the aggregated criteria weights vector $W_C = (w^c_1, w^c_2, \dots, w^c_n)$ of I2TLPR using Definition 10, Part (1). Similarly, also calculate the aggregated weight vector $W_A = (w^A_1, w^A_2, \dots, w^A_m)$ of the alternatives by using the same definition.

Step 6. Rank the overall weights w^A_i $(i = 1, 2, \dots, m)$ of each alternative using Definition 14, and then choose the best alternative.

4. Numerical Example

Based on the availability of information and the scope to get direct, prompt, and appealing information, each student is more willing to select a university option of his/her interest that exactly answers the questions and how the accessibility of this information determine whether one will select one university option over the other. For this, portals of three different universities of Pakistan A_1, A_2, and A_3 are evaluated under the four criteria: C_1 : simple and professional design; C_2 : student services; C_3 : research interface; and C_4 : alumni section.

The three alternatives $A_i (i = 1, 2, 3)$ are evaluated by a DM using the LTS $S = \{s_0$ =Extremely poor, s_1 =Very Poor, s_2 = Poor, s_3 =Medium, s_4 =Good, s_5 =Very Good, and s_6 = Extremely Good$\}$ under the above four criteria. In the following, we use our proposed intuitionistic 2-tuple AHP method to get the best alternative as follows.

The comparison judgments of the priority of one criterion over the other determined by the DM are represented in I2TLPR $B = (b_{pq})_{4\times 4}$ and shown in Table 2.

Similarly, the comparison judgments of the priority of one alternative over the remaining are represented in I2TLPRs $B_j = (b^j_{pq})_{3\times 3}$ $(j = 1, 2, \dots, n)$ and shown in Tables 3–6.

Now, we check the consistency level of each I2TLPR by following the idea presented by Xu and Liao in [28, Algorithm 1].

TABLE 3: I2TLPR of alternatives concerning the criterion C_1.

B_1	A_1	A_2	A_3
A_1	$((s_3, 0.5), (s_3, 0.5))$	$((s_1, 0.4), (s_4, 0.2))$	$((s_3, -0.5), (s_2, -0.2))$
A_2	$((s_4, 0.2), (s_1, 0.4))$	$((s_3, 0.5), (s_3, 0.5))$	$((s_3, 0.1), (s_2, -0.2))$
A_3	$((s_2, -0.2), (s_3, -0.5))$	$((s_2, -0.2), (s_3, 0.1))$	$((s_3, 0.5), (s_3, 0.5))$

TABLE 4: I2TLPR of alternatives concerning the criterion C_2.

B_2	A_1	A_2	A_3
A_1	$((s_3, 0.5), (s_3, 0.5))$	$((s_1, 0.4), (s_4, 0.2))$	$((s_3, 0.45), (s_1, 0.4))$
A_2	$((s_4, 0.2), (s_1, 0.4))$	$((s_3, 0.5), (s_3, 0.5))$	$((s_3, 0.15), (s_2, -0.2))$
A_3	$((s_1, 0.4), (s_3, 0.45))$	$((s_2, -0.2), (s_3, 0.15))$	$((s_3, 0.5), (s_3, 0.5))$

TABLE 5: I2TLPR of alternatives concerning the criterion C_3.

B_3	A_1	A_2	A_3
A_1	$((s_3, 0.5), (s_3, 0.5))$	$((s_1, -0.25), (s_4, -0.15))$	$((s_4, 0.2), (s_1, 0.4))$
A_2	$((s_4, -0.15), (s_1, -0.25))$	$((s_3, 0.5), (s_3, 0.5))$	$((s_3, -0.45), (s_1, -0.3))$
A_3	$((s_1, 0.1), (s_4, 0.2))$	$((s_1, -0.3), (s_3, -0.45))$	$((s_3, 0.5), (s_3, 0.5))$

Suppose we take the I2TLPR B of criteria as an example and discuss the process of checking the consistency.

The perfect multiplicative consistent I2TLPR $\overline{B} = (\overline{b}_{ij})_{4\times4}$ of the I2TLPR B of criteria can be constructed as

$$\overline{B} = \begin{pmatrix} ((s_3, 0.5), & ((s_4, 0.3), & ((s_4, -0.2436), & ((s_4, 0.1297), \\ (s_3, 0.5)) & (s_1, 0.4)) & (s_1, 0.4)) & (s_0, 0.1949)) \\ ((s_1, 0.4), & ((s_3, 0.5), & ((s_3, 0.2), & ((s_4, -0.2958), \\ (s_4, 0.3)) & (s_3, 0.5)) & (s_1, 0.4)) & (s_1, 0.3493)) \\ ((s_1, 0.4), & ((s_1, 0.4), & ((s_3, 0.5), & ((s_4, 0.2), \\ (s_4, -0.2436)) & (s_3, 0.2)) & (s_3, 0.5)) & (s_1, 0.3)) \\ ((s_0, 0.1949), & ((s_1, 0.3493), & ((s_1, 0.3), & ((s_3, 0.5), \\ (s_4, 0.1297)) & (s_4, -0.2958)) & (s_4, 0.2)) & (s_3, 0.5)) \end{pmatrix} \quad (15)$$

By finding the distance between R and $\overline{R}$ corresponding to the I2TLPRs B and $\overline{B}$ with the help of (4), we get $d(R, \overline{R}) = 0.1766$ which is greater than 0.1, which means the I2TLPR $\overline{B}$ is of unacceptable consistency and, therefore, it is necessary to repair it. To improve the consistency, (2.4) and (2.5) in [28, Algorithm 2] are used to get the fused I2TLPR $\overline{B}_1$ by letting $\sigma = 0.8$ as follows:

$$\overline{B}_1 = \begin{pmatrix} ((s_3, 0.5), & ((s_4, 0.3), & ((s_4, -0.3548), & ((s_4, 0.0155), \\ (s_3, 0.5)) & (s_1, 0.4)) & (s_2, -0.4435)) & (s_0, 0.3107)) \\ ((s_1, 0.4), & ((s_3, 0.5), & ((s_3, 0.2), & ((s_4, -0.3266), \\ (s_4, 0.3)) & (s_3, 0.5)) & (s_1, 0.4)) & (s_1, -0.0497)) \\ ((s_2, -0.4435), & ((s_1, 0.4), & ((s_3, 0.5), & ((s_4, 0.2), \\ (s_4, -0.3548)) & (s_3, 0.2)) & (s_3, 0.5)) & (s_1, 0.3)) \\ ((s_0, 0.3107), & ((s_1, -0.0497), & ((s_1, 0.3), & ((s_3, 0.5), \\ (s_4, 0.0155)) & (s_4, -0.3266)) & (s_4, 0.2)) & (s_3, 0.5)) \end{pmatrix} \quad (16)$$

TABLE 6: I2TLPR of alternatives concerning the criterion C_4.

B_4	A_1	A_2	A_3
A_1	$((s_3, 0.5), (s_3, 0.5))$	$((s_1, -0.25), (s_4, -0.15))$	$((s_4, 0.2), (s_1, 0.4))$
A_2	$((s_4, -0.15), (s_1, -0.25))$	$((s_3, 0.5), (s_3, 0.5))$	$((s_3, -0.45), (s_1, -0.3))$
A_3	$((s_1, 0.1), (s_4, 0.2))$	$((s_1, -0.3), (s_3, -0.45))$	$((s_3, 0.5), (s_3, 0.5))$

TABLE 7: Over all aggregated weights of alternatives and criteria.

	w_j^c	$w_j^{A_1}$	$w_j^{A_2}$	$w_j^{A_3}$
C_1	$((s_{5.8889}, 0.0190), (s_0, 0.1570))$	$((s_{4.75}, -0.0250), (s_{0.6667}, 0.0750))$	$((s_{5.5}, 0.2250), (s_{0.1667}, 0.1250))$	$((s_{4.6667}, 0.15), (s_{0.75}, 0.15))$
C_2	$((s_{5.5833}, -0.0008), (s_{0.0556}, 0.1751))$	$((s_{4.75}, -0.0250), (s_{0.6667}, 0.0750))$	$((s_{5.5}, 0.2250), (s_{0.1667}, 0.1250))$	$((s_{4.6667}, 0.15), (s_{0.75}, 0.15))$
C_3	$((s_{5.4444}, 0.2195), (s_{0.1667}, 0.2556))$	$((s_{5.1667}, 0.1625), (s_{0.3333}, 0.1375))$	$((s_{5.5}, -0.1375), (s_{0.0833}, -0.0875))$	$((s_{3.9167}, 0.2), (s_1, 0.1875))$
C_4	$((s_{3.9167}, 0.3576), (s_{0.8889}, 0.2611))$	$((s_{5.5833}, -0.0250), (s_{0.3333}, -0.0750))$	$((s_{5.5}, -0.0250), (s_{0.0833}, 0.25))$	$((s_{3.9167}, 0.2999), (s_{1.25}, 0.1875))$

Corresponding to the I2TLPRs B and $\overline{B}_1$,the distance between R and $\overline{R}_1$ is calculated as $d(R, \overline{R}_1) = 0.0334$ with the help of (4) again, which is now less than 0.1. This means $\overline{B}_1$ is of acceptable multiplicative consistency. For the other intuitionistic preference relations $B_j (j = 1, 2, 3, 4)$ of alternatives concerning the criteria $C_j (j = 1, 2, 3, 4)$, the consistency checking can be done by following the same process. We can see that all the other I2TLPRs are consistent, and we do not need to repair them.

Now, by using Definition 10, Part (1), we can calculate the aggregated criteria weights vector $W_C = (w_1^c, w_2^c, \ldots, w_n^c)$ of the I2TLPR $\overline{B}_1$ as

$$
\begin{aligned}
w_1^c &= ((s_{5.8889}, 0.0190), (s_0, 0.1570)) \\
w_2^c &= ((s_{5.5833}, -0.0008), (s_{0.0556}, 0.1751)) \\
w_3^c &= ((s_{5.4444}, 0.2195), (s_{0.1667}, 0.2556)), \\
w_4^c &= ((s_{3.9167}, 0.3576), (s_{0.8889}, 0.2611))
\end{aligned}
\tag{17}
$$

Similarly, the aggregated weight vectors of the other I2TLPRs $B_j (j = 1, 2, 3, 4)$ of alternatives concerning the criteria $C_j (j = 1, 2, 3, 4)$ are calculated by using again Definition 10, Part 1. The aggregated results are as shown in Table 7.

At the end, we aggregate all the criteria weights and weights of alternatives as computed above by using the operational laws discussed in Definition 10 concerning each alternative as follows:

$$
\begin{aligned}
w_1^A &= \bigoplus_{j=1}^{4} \left(w_j^c \otimes w_j^A\right) = \left(w_1^c \otimes w_1^{A_1}\right) \oplus \left(w_2^c \otimes w_2^{A_1}\right) \\
&\oplus \left(w_3^c \otimes w_3^{A_1}\right) \oplus \left(w_4^c \otimes w_4^{A_1}\right) \\
&= (((s_{5.8889}, 0.0190), (s_0, 0.1570)) \\
&\otimes ((s_{4.75}, -0.0250), (s_{0.6667}, 0.0750))) \\
&\oplus (((s_{5.5833}, -0.0008), (s_{0.0556}, 0.1751)) \\
&\otimes ((s_{4.75}, -0.0250), (s_{0.6667}, 0.0750))) \\
&\oplus (((s_{5.4444}, 0.2195), (s_{0.1667}, 0.2556)) \\
&\otimes ((s_{5.1667}, 0.1625), (s_{0.3333}, 0.1375))) \\
&\oplus (((s_{3.9167}, 0.3576), (s_{0.8889}, 0.2611)) \\
&\otimes ((s_{5.5833}, -0.0250), (s_{0.3333}, -0.0750))) \\
&= ((s_{5.9698}, 0.1586), (s_0, 0.1945))
\end{aligned}
\tag{18}
$$

Similarly, we can determine the overall weights of the remaining alternatives as

$$
\begin{aligned}
w_2^A &= ((s_{5.9940}, 0.1242), (s_0, 0.1752)), \\
w_3^A &= ((s_{5.8911}, 0.2461), (s_0, 0.2226)).
\end{aligned}
\tag{19}
$$

In the following, the score values of the overall weights of alternatives are determined by using (11).

$$
\begin{aligned}
S\left(w_1^A\right) &= 0.9945, \\
S\left(w_2^A\right) &= 0.9952, \\
S\left(w_3^A\right) &= 0.9929
\end{aligned}
\tag{20}
$$

As $S(w_2^A) > S(w_1^A) > S(w_3^A)$, therefore, the ranking order of alternatives is $A_2 > A_1 > A_3$. This implies that A_2 is the most desirable alternative.

4.1. Comparative Analysis. In order to validate the feasibility of our proposed method, TOPSIS method is applied to solve the same problem. The results are shown as follows.

TABLE 8: The aggregated matrix $\widehat{R}$.

	C_1	C_2
A_1	$((s_{4.75}, -0.025), (s_{0.6667}, 0.075))$	$((s_{4.75}, 0.45), (s_{0.3333}, 0.375))$
A_2	$((s_{5.5}, 0.225), (s_{0.1667}, 0.125))$	$((s_{5.5}, 0.25), (s_{0.1667}, 0.125))$
A_3	$((s_{4.6667}, 0.15), (s_{0.75}, 0.15)$	$((s_{4.333}, 0.299), (s_{0.75}, 0.4))$
	C_3	C_4
A_1	$((s_{5.1667}, 0.1625), (s_{0.3333}, 0.1375))$	$((s_{5.5833}, -0.025), (s_{0.3333}, -0.075))$
A_2	$((s_{5.5}, -0.1375), (s_{0.833}, -0.0875))$	$((s_{5.5}, -0.025), (s_{0.833}, 0.25))$
A_3	$((s_{3.9167}, 0.2), (s_1, 0.1875))$	$((s_{3.9167}, 0.299), (s_{1.25}, 0.1875))$

TABLE 9: The result of TOPSIS method.

Alternatives	d_i^+	d_i^-	RC_i	Ranking result
A_1	0.1175	0.8853	0.8828	2
A_2	0.0594	0.9472	0.9410	1
A_3	0.2269	0.7766	0.7739	3

The elements of the aggregated matrix $\widehat{R} = (\widehat{r}_{ij})_{3\times4}$ against criteria C_1 can be computed as in the following:

$$\begin{aligned}\widehat{r}_{11} &= \bigoplus_{q=1}^{3} b_{1q}^1 = b_{11}^1 \oplus b_{12}^1 \oplus b_{13}^1 \\ &= ((s_3, 0.5), (s_3, 0.5)) \oplus ((s_1, 0.4), (s_4, 0.2)) \\ &\quad \oplus ((s_3, -0.5), (s_2, -0.2)) \\ &= ((s_{4.75}, -0.025), (s_{0.6667}, 0.075)) \\ \widehat{r}_{21} &= \bigoplus_{q=1}^{3} b_{2q}^1 = b_{21}^1 \oplus b_{22}^1 \oplus b_{23}^1 \\ &= ((s_4, 0.2), (s_1, 0.4)) \oplus ((s_3, 0.5), (s_3, 0.5)) \\ &\quad \oplus ((s_3, 0.1), (s_2, -0.2)) \\ &= ((s_{5.5}, 0.225), (s_{0.1667}, 0.125) \\ \widehat{r}_{31} &= \bigoplus_{q=1}^{3} b_{3q}^1 = b_{31}^1 \oplus b_{32}^1 \oplus b_{33}^1 \\ &= ((s_2, -0.2), (s_3, -0.5)) \oplus ((s_2, -0.2), (s_3, 0.1)) \\ &\quad \oplus ((s_3, 0.5), (s_3, 0.5)) \\ &= ((s_{4.6667}, 0.15), (s_{0.75}, 0.15) \end{aligned} \tag{21}$$

Similarly, by utilizing the I2TLPRs B_2, B_3, and B_4, we can get the remaining elements of $\widehat{R}$ against criteria C_2, C_3, and C_4, respectively. The final aggregated matrix $\widehat{R} = (\widehat{r}_{ij})_{3\times4}$ can be seen as in Table 8.

The intuitionistic 2-tuple positive ideal and negative ideal solutions are determined. The distances between each alternative to positive ideal alternative $d_i^+(i = 1, 2, 3)$ and negative ideal alternative $d_i^-(i = 1, 2, 3)$ are obtained by using (10). The weight values of the criteria are determined by calculating the score values of each element of $W_C = (w_1^c, w_2^c, \ldots, w_n^c)$ with the help of (11). The relative weight denoted as $w_j(j = 1, 2, 3, 4)$ of each criterion is determined as follows:

$$w_i = \frac{w_j^c}{\sum_{j=1}^{n} w_j^c} \tag{22}$$

It can be observed that $\sum_{j=1}^{n} w_j = 1$. The relative weights are determined as $w_1 = 0.2703$, $w_2 = 0.2612$, $w_3 = 0.2586$, and $w_4 = 0.2099$. The relative closeness coefficients $RC_i(i = 1, 2, 3)$ and ranking result can be seen in Table 9.

Again, the final ranking order is $A_2 > A_1 > A_3$ and the most desirable alternative is A_2.

It is apparent that results of I2TL AHP method and TOPSIS method are identical and the best and worst alternatives have no difference, which can illustrate the validity of our proposed method. As compared to the TOPSIS method, our method is more flexible. In addition, we can find that the proposed method considers bounded rationality of DMs in comparison with TOPSIS method. Obviously, the ranking result obtained by TOPSIS method may conform to the actual decision-making to some extent. As far as the time complexity is concerned, it is in general lower for AHP as compared with the TOPSIS method. The advantage of the AHP method over TOPSIS is that, in the AHP method, decision matrix consistency test is frequently needed. This leads to a consistent priority ranking with pairwise comparisons of the experts. Although both methods are equally adequate to deal with the lack of precision of scores of alternatives as well as the relative importance of different criteria, it is worth noting that the AHP method is more appropriate than the TOPSIS method when the purpose is to avoid the rank reversal phenomenon which lies at the heart of the main MCDM techniques like TOPSIS.

5. Conclusion

In this paper, intuitionistic 2-tuple AHP method has been proposed for solving the MCDM problems based on I2TLSs.

Firstly, we have defined some operational laws for I2TLEs and proved some related important properties. Secondly, by using the idea of I2TLSs, two important preference relations, namely, the I2TLPR and the multiplicative I2TLPR, have been defined along with a transformation mechanism that can transform an I2TLPR to a corresponding IPR and vice versa. Thirdly, we have proposed an approach for checking the consistency of an I2TLPR and presented a method to repair the inconsistent one by using the proposed transformation mechanism. Finally, a comparative example is given to show the effectiveness of the proposed approach and is validated through a comparative analysis. The proposed approach is appropriate for a linguistic preference structure with symbolic translation parameters of linguistic arguments. Furthermore, the DMs remain much easier for collecting pairwise preference information using I2TLSs which are really effective in handling the vagueness and uncertainty in a MCDM problem. Our proposed intuitionistic 2-tuple AHP method is different from all the previous methods of decision-making because the proposed method uses I2TLSs, which always avoid any loss of information in the process. So it is an efficient and most feasible method for real-life applications of decision-making. On the basis of I2TLPRs, more applications should be worked on as our further research, for instance, performance evaluation, emergency management evaluation, and decision support systems, especially expert system.

Conflicts of Interest

The authors declare that they have no conflicts of interest.

References

[1] F. Herrera and L. Martínez, "An approach for combining linguistic and numerical information based on the 2-tuple fuzzy linguistic representation model in decision-making," *International Journal of Uncertainty, Fuzziness and Knowledge-Based Systems*, vol. 8, no. 5, pp. 539–562, 2000.

[2] F. Herrera and L. Martínez, "A 2-tuple fuzzy linguistic representation model for computing with words," *IEEE Transactions on Fuzzy Systems*, vol. 8, no. 6, pp. 746–752, 2000.

[3] F. Herrera, E. Herrera-Viedma, and L. Martínez, "A fusion approach for managing multi-granularity linguistic term sets in decision making," *Fuzzy Sets and Systems*, vol. 114, no. 1, pp. 43–58, 2000.

[4] F. Herrera and L. Martínez, "A model based on linguistic 2-tuples for dealing with multigranular hierarchical linguistic contexts in multi-expert decision-making," *IEEE Transactions on Systems, Man, and Cybernetics, Part B: Cybernetics*, vol. 31, no. 2, pp. 227–234, 2001.

[5] J. Wang and J. Hao, "A new version of 2-tuple fuzzy linguistic representation model for computing with words," *IEEE Transactions on Fuzzy Systems*, vol. 14, no. 3, pp. 435–445, 2006.

[6] G. W. Wei, "Extension of TOPSIS method for 2-tuple linguistic multiple attribute group decision making with incomplete weight information," *Knowledge and Information Systems*, vol. 25, no. 3, pp. 623–634, 2010.

[7] I. Beg and T. Rashid, "Hesitant 2-tuple linguistic information in multiple attributes group decision-making," *Journal of Intelligent and Fuzzy Systems*, vol. 30, pp. 109–116, 2016.

[8] I. Beg and T. Rashid, "An Intuitionistic 2-Tuple Linguistic Information Model and Aggregation Operators," *International Journal of Intelligent Systems*, vol. 31, no. 6, pp. 569–592, 2016.

[9] P. Liu and S.-M. Chen, "Multiattribute group decision making based on intuitionistic 2-tuple linguistic information," *Information Sciences*, vol. 430/431, pp. 599–619, 2018.

[10] T. L. Saaty, *The Analytic Hierarchy Process*, McGraw-Hill, NY, USA, 1980.

[11] L. F. D. O. M. Santos, L. Osiro, and R. H. P. Lima, "A model based on 2-tuple fuzzy linguistic representation and Analytic Hierarchy Process for supplier segmentation using qualitative and quantitative criteria," *Expert Systems with Applications*, vol. 79, pp. 53–64, 2017.

[12] W. Li, S. Yu, H. Pei, C. Zhao, and B. Tian, "A hybrid approach based on fuzzy AHP and 2-tuple fuzzy linguistic method for evaluation in-flight service quality," *Journal of Air Transport Management*, vol. 60, pp. 49–64, 2017.

[13] F. Tüysüz and B. Simsek, "A hesitant fuzzy linguistic term sets-based AHP approach for analyzing the performance evaluation factors: an application to cargo sector," in *Complex & Intelligent Systems DOI*, vol. 3, pp. 167–175, 3 edition, 2017.

[14] F. Chiclana, F. Herrera, and E. Herrera-Viedma, "Integrating three representation models in fuzzy multipurpose decision making based on fuzzy preference relations," *Fuzzy Sets and Systems*, vol. 97, no. 1, pp. 33–48, 1998.

[15] F. Chiclana, E. Herrera-Viedma, F. Alonso, and S. Herrera, "Cardinal consistency of reciprocal preference relations: a characterization of multiplicative transitivity," *IEEE Transactions on Fuzzy Systems*, vol. 17, no. 1, pp. 14–23, 2009.

[16] M. Xia, Z. Xu, and H. Liao, "Preference relations based on intuitionistic multiplicative information," *IEEE Transactions on Fuzzy Systems*, vol. 21, no. 1, pp. 113–133, 2013.

[17] Z. Fan and X. Chen, "Consensus Measures and Adjusting Inconsistency of Linguistic Preference Relations in Group Decision Making," in *Fuzzy Systems and Knowledge Discovery*, vol. 3613 of *Lecture Notes in Computer Science*, pp. 130–139, Springer, Berlin, Germany, 2005.

[18] E. Herrera-Viedma, L. Martínez, F. Mata, and F. Chiclana, "A consensus support system model for group decision-making problems with multigranular linguistic preference relations," *IEEE Transactions on Fuzzy Systems*, vol. 13, no. 5, pp. 644–658, 2005.

[19] M. Delgado, J. L. Verdegay, and M. A. Vila, "On aggregation operations of linguistic labels," *International Journal of Intelligent Systems*, vol. 8, no. 3, pp. 351–370, 1993.

[20] Y. Dong, Y. Xu, and S. Yu, "Linguistic multiperson decision making based on the use of multiple preference relations," *Fuzzy Sets and Systems*, vol. 160, no. 5, pp. 603–623, 2009.

[21] F. Herrera, E. Herrera-Viedma, and J. L. Verdegay, "A sequential selection process in group decision making with a linguistic assessment approach," *Information Sciences*, vol. 85, no. 4, pp. 223–239, 1995.

[22] J. Kacprzyk, "Group decision making with a fuzzy linguistic majority," *Fuzzy Sets and Systems*, vol. 18, no. 2, pp. 105–118, 1986.

[23] T. L. Saaty, "A scaling method for priorities in hierarchical structures," *Journal of Mathematical Psychology*, vol. 15, no. 3, pp. 234–281, 1977.

[24] V. Cutello and J. Montero, "Fuzzy rationality measures," *Fuzzy Sets and Systems*, vol. 62, no. 1, pp. 39–54, 1994.

[25] E. Herrera-Viedma, F. Herrera, F. Chiclana, and M. Luque, "Some issues on consistency of fuzzy preference relations," *European Journal of Operational Research*, vol. 154, no. 1, pp. 98–109, 2004.

[26] Z. Xu, X. Cai, and E. Szmidt, "Algorithms for estimating missing elements of incomplete intuitionistic preference relations," *International Journal of Intelligent Systems*, vol. 26, no. 9, pp. 787–813, 2011.

[27] Y. Dong, Y. Xu, and H. Li, "On consistency measures of linguistic preference relations," *European Journal of Operational Research*, vol. 189, no. 2, pp. 430–444, 2008.

[28] Z. Xu and H. Liao, "Intuitionistic Fuzzy Analytic Hierarchy Process," *IEEE Transactions on Fuzzy Systems*, vol. 22, no. 4, pp. 1–14, 2015.

[29] B. Zhu and Z. Xu, "Consistency measures for hesitant fuzzy linguistic preference relations," *IEEE Transactions on Fuzzy Systems*, vol. 22, no. 1, pp. 35–45, 2014.

[30] Z. Zhang and C. Wu, "On the use of multiplicative consistency in hesitant fuzzy linguistic preference relations," *Knowledge-Based Systems*, vol. 72, pp. 13–27, 2014.

[31] Z.-W. Gong, L.-S. Li, J. Forrest, and Y. Zhao, "The optimal priority models of the intuitionistic fuzzy preference relation and their application in selecting industries with higher meteorological sensitivity," *Expert Systems with Applications*, vol. 38, no. 4, pp. 4394–4402, 2011.

[32] S.-P. Wan, Q.-Y. Wang, and J.-Y. Dong, "The extended VIKOR method for multi-attribute group decision making with triangular intuitionistic fuzzy numbers," *Knowledge-Based Systems*, vol. 52, pp. 65–77, 2013.

[33] N. Subramanian and R. Ramanathan, "A review of applications of Analytic Hierarchy Process in operations management," *International Journal of Production Economics*, vol. 138, no. 2, pp. 215–241, 2012.

[34] W. Ho, "Integrated analytic hierarchy process and its applications—a literature review," *European Journal of Operational Research*, vol. 186, no. 1, pp. 211–228, 2008.

[35] K. T. Atanassov, "Intuitionistic fuzzy sets," *Fuzzy Sets and Systems*, vol. 20, no. 1, pp. 87–96, 1986.

[36] K. Atanassov and G. Gargov, "Interval valued intuitionistic fuzzy sets," *Fuzzy Sets and Systems*, vol. 31, no. 3, pp. 343–349, 1989.

[37] Z. Xu, "Intuitionistic preference relations and their application in group decision making," *Information Sciences*, vol. 177, no. 11, pp. 2363–2379, 2007.

[38] F. Herrera, E. Herrera-Viedma, and J. L. Verdegay, "A model of consensus in group decision making under linguistic assessments," *Fuzzy Sets and Systems*, vol. 78, no. 1, pp. 73–87, 1996.

[39] F. Herrera and E. Herrera-Viedma, "Linguistic decision analysis: steps for solving decision problems under linguistic information," *Fuzzy Sets and Systems*, vol. 115, no. 1, pp. 67–82, 2000.

[40] Z. S. Xu, "A method based on linguistic aggregation operators for group decision making with linguistic preference relations," *Information Sciences*, vol. 166, no. 1–4, pp. 19–30, 2004.

[41] Z. S. Xu, "Group decision making based on multiple types of linguistic preference relations," *Information Sciences*, vol. 178, no. 2, pp. 452–467, 2008.

[42] X. Gou and Z. Xu, "Novel basic operational laws for linguistic terms, hesitant fuzzy linguistic term sets and probabilistic linguistic term sets," *Information Sciences*, vol. 372, pp. 407–427, 2016.

13

Kaizen Selection for Continuous Improvement through VSM-Fuzzy-TOPSIS in Small-Scale Enterprises

Sunil Kumar,[1] **Ashwani Kumar Dhingra**,[1] **and Bhim Singh**[2]

[1]*Mechanical Engineering Department, University Institute of Engineering & Technology, Maharshi Dayanand University, Rohtak, Haryana 124 001, India*
[2]*Mechanical Engineering Department, School of Engineering & Technology, Sharda University, Greater Noida, Uttar Pradesh 201 306, India*

Correspondence should be addressed to Sunil Kumar; sunil.panchal2007@gmail.com

Academic Editor: Zeki Ayag

In the era of cut throat competitive market, Indian industries are under tremendous pressure to continuously reduce the cost and improve product quality. The main objective of this research paper is to provide a road map for investigating the opportunities to reduce cost and improve productivity and quality in the existing production system through the application of Lean-Kaizen concept using value stream mapping (VSM) tool at shop floor of an Indian Small-Scale Enterprise (SSE). On the basis of collected data from the selected industry, a current state map was made. After analysis, the current state map was modified to develop a future state map. By comparing current and future state map, the gap areas were identified and takt time (TT) was calculated considering actual market demand. To overcome the gap between current state map and future state map and to synchronize cycle time of each station with talk time, Kaizen event (KE) was proposed. Fuzzy technology for order preference by similarity to ideal solution (TOPSIS) was applied to prioritize and select the appropriate KE for optimized performance level. After analysis of data obtained after application of the Lean-Kaizen concept, the final improvements in terms of improved productivity (57.15%) and reduced lead time (69.47%), reduced cost (65.61%), and station cycle time (75.25%) were recorded. Through the case study drawn from realistic situations of the industry, the authors highlight that implementation of Lean-Kaizen using VSM and fuzzy TOPSIS provides improvement opportunities across the organization. The findings can inspire the SSEs to adopt the Lean-Kaizen concept for optimizing continuous improvement opportunities in their production industry. This study offers the researchers and practitioners a good example for understanding, selecting, and performing KE program. They can even attain more improvement in various areas by establishing or improving these KE programs.

1. Introduction

In order to meet real world of competitive situations like producing products at the lowest cost, Indian SSEs are now ready to adopt lean tools and techniques for minimizing product cost, eliminating waste, and reducing delivery time [1]. Due to limited budget and financial conditions, some of SSEs are not able to afford any improvement program and lack many lean benefits. Sometimes, the improvement program is not organized in an effective way and ultimately hits the routine activities so badly that the organization does not even think to rearrange it. This study may encourage such SSEs to organize continuous improvement programs through application of "Lean-Kaizen concept" so that optimum benefits of the event can be achieved.

Lean-Kaizen means removal of non-value-added (VA) activities (wastes) by adopting small improvements [2]. The collection of lean tools and techniques, namely, value stream mapping (VSM), poka-yoke, standardization, visual control, and kanban cellular manufacturing, is called lean building block [3, 4]. Henry Ford witnessed that standardization and innovation are equally essential for any organization. He suggested that LM and Kaizen tools are needed to identify opportunity in order to reduce wastes and improve

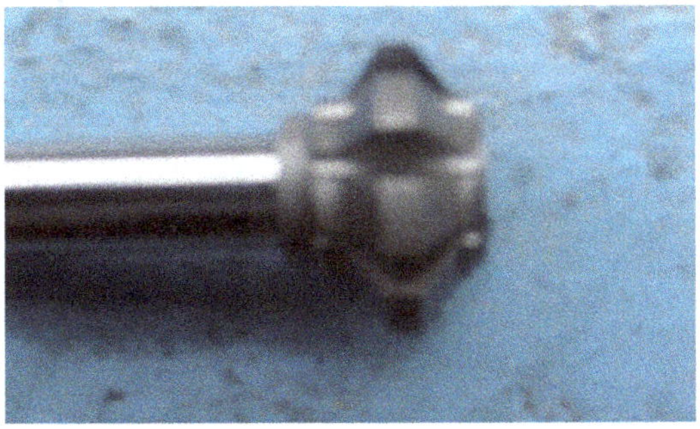

FIGURE 1: Spindle kick disco model.

overall performance. Grunberg [5] notifies that it is most important what to and how to improve in process. Marvel and Standridge [6] suggest that the selection of inappropriate lean tool causes various disruptions in the process of lean implementation, which influence the expected outcomes and reduced employee confidence of organization, which results in increase in various costs, wastes, and production time.

In today's real-world situation, lean tool fitness in particular organization heavily depends on a particular manufacturing process of the organization. The decision making is one of the most critical issue before or while selecting appropriate lean tools and techniques. Many methods have been established and used for solving diverse kinds of multicriteria decision making (MCDM) problems. Chen [7] suggested fuzzy technology for order preference by similarity to ideal solution (TOPSIS), which helps decision makers to know the complete evaluation process and offers a systematic, accurate, and effective decision support tool. It helps to resolve multicriteria decision making difficulties under vagueness such as

$$A = \{x, \mu_A(x)\}, \quad \text{where } x \in X \tag{1}$$

where $\mu_A(x)$ is membership function of A and is defined in universe of course [0, 1].

In general, two types of fuzzy numbers, triangular and trapezoidal, are used, in which triangular fuzzy number (TFN) is commonly used for computation efficient information in fuzzy environment.

$$\mu_A(x) = \begin{cases} 0 & \text{for } x < l \\ \dfrac{x-l}{m-l} & \text{for } l \le x \le m, \\ \dfrac{u-x}{x-m} & \text{for } m \le x \le u \\ 1 & \text{for } x > u \end{cases} \tag{2}$$

where l, m, and u are the parameters showing the least permissible value, the most promising value, and the largest possible value. The fuzzy event is shown by l<m<u (Figure 1), where m shows the maximum relevance degree at m and null outside the internal [l, u].

The algebraic operations of two TFNs, M_1 (l_1, m_1, u_1) and M_2 (l_2, m_2, u_2), as given in Figure 2 are performed as

$$\begin{aligned} M_1 + M_2 &= \{(l_1 + m_1 + u_1) + (l_2 + m_2 + u_2)\} \\ &= \{(l_1 + l_2), (m_1 + m_2), (u_1 + u_2)\} \\ M_1 \times M_2 &= (l_1, m_1, u_1) \times (l_2, m_2, u_2) \\ &= \{(l_1 \times l_2), (m_1 \times m_2), (u_1 \times u_2)\} \end{aligned} \tag{3}$$

Distance calculations for two TFNs are

$$d(M1, M2) = \sqrt{\frac{1}{3\{(l_1 - l_2), (m_1 - m_2) + (u_1 - u_2)\}}} \tag{4}$$

Based on the criteria of benefit and cost categories, the normalization of the fuzzy numbers using (Chen, 2000) is as follows.

For benefit criteria,

$$r_{ij} = \left(\frac{l_{ij}}{u_j^+}, \frac{m_{ij}}{u_j^+}, \frac{u_{ij}}{u_j^+}\right), \quad j \in B, \tag{5}$$

where

$$u_j^+ = \max_i u_{ij} \tag{6}$$

For cost criteria,

$$r_{ij} = \left(\frac{l_j^-}{l_{ij}}, \frac{l_j^-}{m_{ij}}, \frac{l_j^-}{u_{ij}}\right), \quad j \in C, \tag{7}$$

where

$$l_j^- = \min_i l_{ij} \tag{8}$$

Today, many Indian SSEs are still deprived of many lean benefits because Lean-Kaizen is still unfamiliar to employees in most of the Indian SSEs. The goal of this research paper is to deliver a road map for examining the opportunities to reduce cost and improve productivity and quality in the existing production system by implementing Lean-Kaizen concept with VSM tool of an Indian SSE. This paper demonstrates a case study in order to find continuous improvement opportunities in the process and practices of an Indian SSE situated at NCR (noncapital region). Based on the collected information, a current state map (CSM) was made, which stated the current working condition of the organization. The calculated TT pointed out those processes that had station cycle time (CT) more than TT. A future state map (FSM) was prepared by modifying the CSM in which supermarket pull system and continuous flow processing were introduced at workstations to control the production of the process. Gap areas were identified by comparing CSM and FSM of the manufacturing. To synchronize CT of each station with talk time, various lean tools and techniques were used. Fuzzy-TOPSIS approach was used to rank and select the proposed Kaizen event. After implementation, the collected data was investigated and conclusions are made.

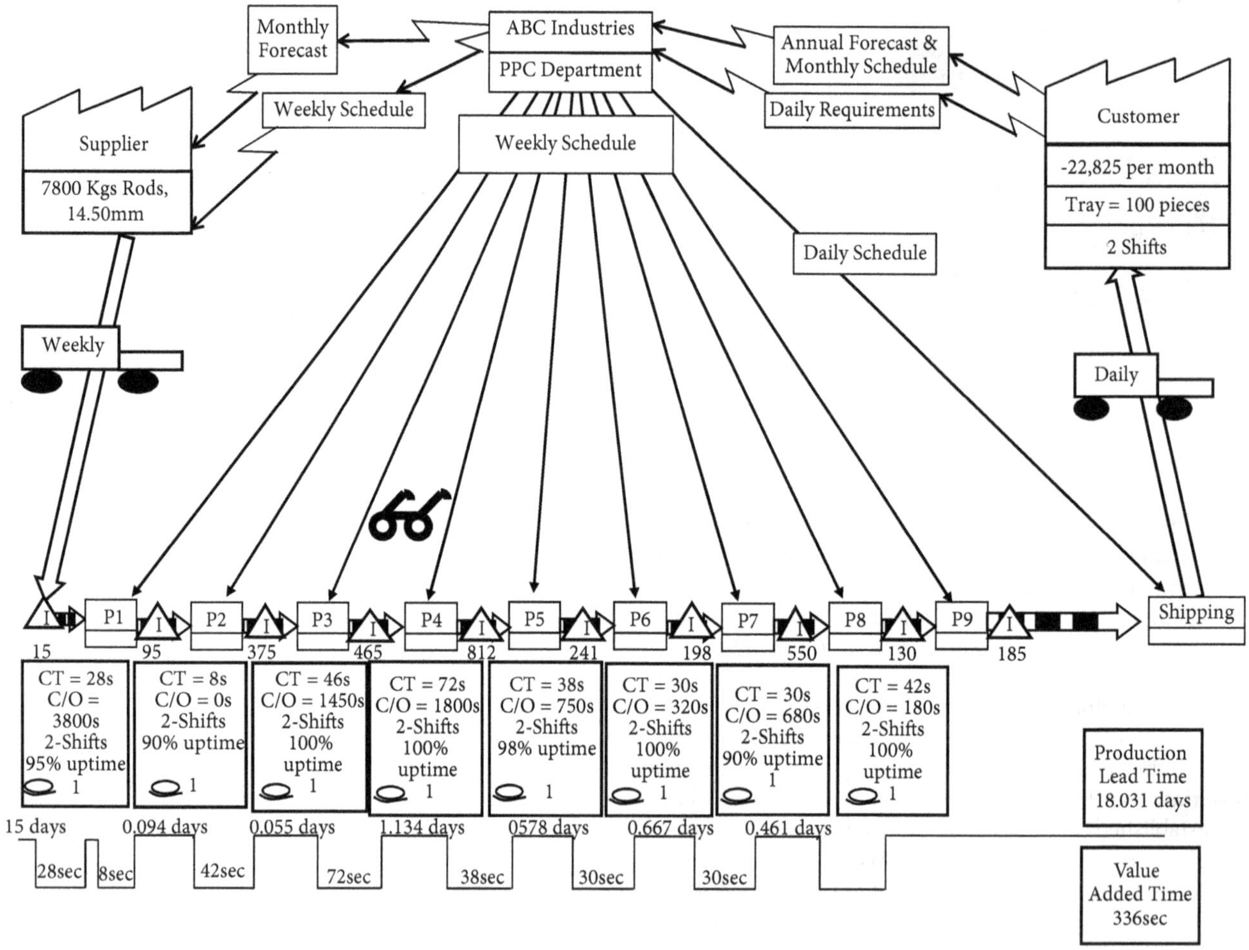

FIGURE 2: Current VSM.

This paper contributes to two dimensions. First, it highlights the restrictions to implement Lean-Kaizen approach in Indian industrial environment. Second, it offers a roadmap for applying lean concept in Indian SSE in order to achieve opportunities of improvement existing in the routine production of products on the shop floor, which keeps the organizations on the right track towards achieving their goals.

2. Literature Review

2.1. Lean Manufacturing, VSM, and Kaizen. Lean manufacturing (LM) is a process or procedure that stresses accumulative work efficiency and minimizing processes expenses [12–16]. The success of LM implementation hinges on proper skills, employee contribution, and training and top management commitment [17, 18]. Kaizen progresses uniquely inside each organization [19]. Bateman [20] notified that lean implementation paves way to progress quality of the processes. Bhuiyan and Baghel [21] demonstrated that continuous improvement program can remove the waste of production and quality of the product. Scyoc [22] assessed various quality tools and techniques for improving process safety and work team. Glover et al. [23] reported that holding the consequences of KEs is challenging over time for many organizations. Suarez-Barraza et al. [24] noticed that three methods (5S, Process mapping and Gemba Kaizen workshops) influenced directly the product, processes, and systems. Lean practices improve the quality in public services [25]. Karim and Arif-Uz-Zaman [26] projected a procedure that helps in systematic improvement of performance of organization. Arya and Choudhary [27] concluded that lean improves product quality and production efficiency. VSM can reduce inventory levels, reduce system waste, minimize resources, and improve the system and performance of the organization [28–37].

2.2. Fuzzy-TOPSIS. Many authors have applied fuzzy techniques for various purposes: to select the location of furniture factories [38], to select robot by seeing both subjective and objective criteria [39, 40], and to choose suitable maintenance strategy [41]. Singh et al. [42] used fuzzy logic to develop efficient measurement lean index to evaluate leanness of manufacturing firm. Kumar et al. [43] used fuzzy-TOPSIS and provided framework to evaluate lean performance of any organization. References [44–46] used fuzzy approach to evaluate leanness of the manufacturing organization. Vidyadhar et al. [47] applied fuzzy logic and developed a conceptual model to evaluate leanness in SMEs. Ravikumar

Table 1: Research contribution for integrated VSM with fuzzy technique.

Research	Tools Used	Contribution
Achanga et al. [8]	Fuzzy-logic advisory system	A decision support tool used for lean manufacturing implementation at the early implementation stage in small-medium-sized enterprises (SMEs)
Wang and Chan [9]	Hierarchical fuzzy-TOPSIS approach	Measure areas of improvements when implementing green supply chain management (GSCM)
Vahdani et al. [10]	Fuzzy-TOPSIS approach with FMEA (failure mode and effect analysis)	Evaluate and overcome the inadequacy of the traditional FMEA technique and rank the group belief structure model
Mohanraj et al. [11]	VSM with fuzzy quality function deployment (QFD)	Evaluate scientific prioritization of improvement proposals to develop leanness

et al. [48] used fuzzy-TOPSIS method to measure the effect of critical success factors of lean in six Indian manufacturing industries. Agrawal et al. [49] identified twelve lean critical success factors and applied fuzzy-TOPSIS technique in five Indian electronics industries to prioritize them. The study concluded that four factors, namely, resource management, top management awareness, contracts terms and conditions, and economic factors, are most important, while process abilities and skilled workers are the smallest important factors in success of lean implementation. The various research contributions are also discussed in Table 1.

2.3. Issues of Lean Manufacturing and Kaizen Implementation. In today's competitive market, SSEs across the globe are seeking continuous improvement opportunities to increase their productivity and quality of product at lowest cost and Indian industries are no exception. In order to meet such requirements, Indian SSEs are adopting lean principles and methodologies in their processes and practices. The benefits of lean adoption for waste elimination are well documented in the literature [50, 51]. Large-Scale Enterprises (LSEs) use lean practices more than SSEs [52–55]. In the study of extensive literature review, it is perceived that, even after a good understanding of lean tools and techniques, the practical application of VSM in Indian SSEs is found feeble. Thus, the adoption of Lean-Kaizen concept can help organizations to find continuous improvement chances for eliminating waste in order to achieve competitive advantage and sustainable growth.

From the extent of literature, it is found that application of VSM to implement Lean-Kaizen concept through Kaizen event in SSEs is found feeble. The selection of Kaizen event using fuzzy-TOPSIS approach is proposed to optimize the lean benefits at all levels of organization.

3. Research Methodology

The case study is conducted in an Indian SSE named ABC Enterprises, situated in NCR (national capital region). The industry is manufacturing many automobile parts such as spindle kick disco model (SKDM, Figure 1), Roller bolt, and Sleeve. The SSE contains a total of 170 machineries generating 1250 crores Indian rupees turnover per annum. A total of 350 employees including supervisors, workers, and managers are working in two shifts of 8 hours. The data like CT, change over time (C/O), shifts, lead time and VA time, customer requirements, product transportation, and work-in-process are collected from production planning and control (PPC) and other departments by taking personal visit to the selected industry. The process symbols are used to prepare the CSM and FSM and gap areas/waste is identified. The fuzzy-TOPSIS approach is used to take decision for selecting suitable Kaizen events that optimize the performance level of the organization. Before and after Kaizen events, the collected data are compared to analyze results.

4. Current State Map

Based on data collected, A CSM (Figure 2) is prepared, in which monthly order for raw material and weekly shipment to customer are drawn. The process stages include nine processes: P1, cutting; P2, deburring; P3, hot forging; P4, CNC; P5, milling; P6, shot blasting; P7, OD grinding; P8, plating, P9, packing and shipping. The PPC department orders supplier for raw material monthly and consignment reaches every week at the industry. The production lead time and VA time are calculated as 18.031 days and 336 seconds, respectively. Process P5 reported high CT (72 seconds) and created bottleneck at P5, milling, due to high work-in-process and rejections.

5. Future State Map

The CSM is analyzed and modified using steps mentioned in Kumar et al. (2017b) for preparing FSM (Figure 3) for identifying waste.

The industry works in two shifts (8 hours each) in order to fulfill average customer demand of 22825 pieces/month within 25 days in a month. From available working time/day (50,400 seconds), the TT will be 55.20 seconds/piece (available working time/customer demand). The calculated TT shows that the industry is fixed to produce each workpiece in every 55.20 seconds to encounter the demand of the customer. As the customer demand varies unsteadily, the selected industry chooses to start a two-day supermarket holding inventory of finished goods for customer. The size of kanban is selected as "100 pieces." From Figure 4, the CT of P5 is 72 seconds (more than TT). The CT of P3 to P7 and P9 are

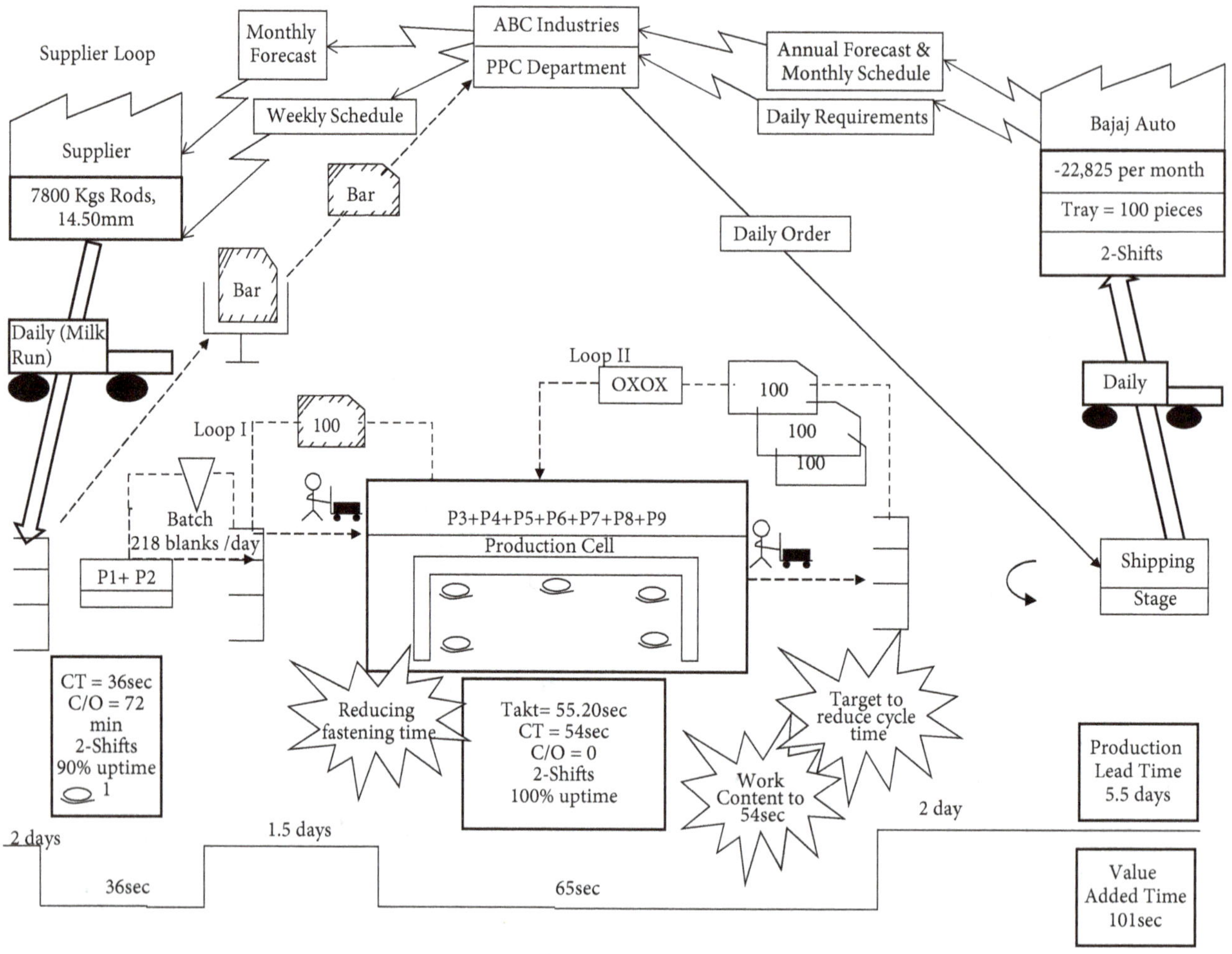

Figure 3: Future VSM.

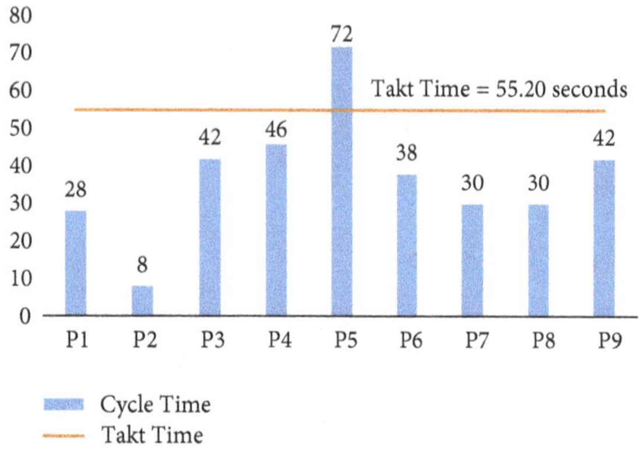

Figure 4: Cycle time and takt time for the processes.

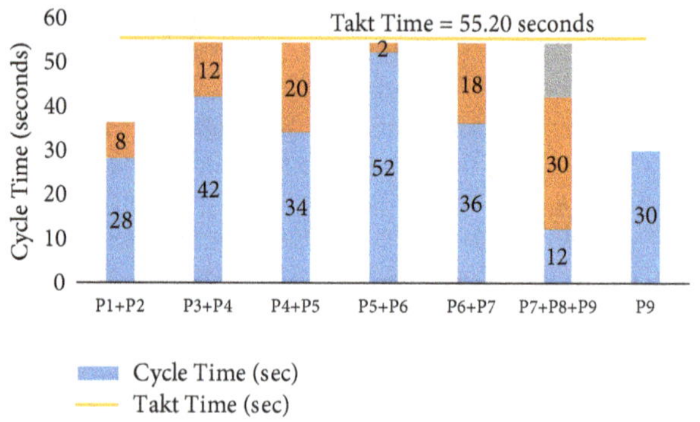

Figure 5: Cycle time with continuous flow.

very close and thus show a continuous flow process, so these processes can be shifted into one cell named "Production Cell" (Figure 5). However, the CT of P2 is small and distant from TT and thus can easily link with P1 to form a batch operation in order to keep the pace of the process.

In order to avoid difficulties in raw material inventories in the current situation, the selected industry generates two-day supermarket for raw material which ensures raw material availability and helps in scheduling the production of blanks. The industry introduces an internal kanban and a milk-run delivery system for raw material to eliminate 90% of raw material inventory. Another supermarket of capacity of 1.5 days is introduced in order to fill requirement of 218 blanks per day.

TABLE 2: Proposed KEs for identified gap areas/waste.

Identified gap area/waste	Proposed KEs
Chattering/tool marks and difficulties in fastening of workpiece	KE-1: training to operator/operator change KE-2: fixture design change with snap fastening KE-3: machine maintenance program for vibration reductions

As batching is difficult in production cell, a load leveling box is introduced (Figure 3). The pitch (5520 seconds/tray) is calculated, which means that each 100-piece tray will be moved on or after every 5520 seconds from one process to another.

The defect and inventory were identified as waste in the production line. The data analysis of machine downtime summary and defect wise rejection report that the main difficulties such as chattering mark and complications in fastening of SKDM are observed, which consequently increases the inventory level at P5. These difficulties of SKDM require to be eliminated.

6. Identifying Possibilities of Continuous Improvement

6.1. Selected Process and Proposed KEs. The milling process (P5) is performed to produce a slot into the shank of the workpiece in order to fulfill the (fitment) requirement of the customer. The screw type fixture is mounted on the milling table that grips the workpiece firmly. The clockwise/anticlockwise rotation of round screw handle is used to move the slide towards rotating cutter, which gradually produces slot on the workpiece.

As the objective of the Lean-Kaizen concept is to eliminate waste continuously through small-small improvements, the comparison of CSM and FSM was made and gap areas were identified. Various lean tools and techniques can be applied to bridge the gap between CSM and FSM. The change from the CSM to FSM may disclose many KEs. In the present study, the P5 process's activities were explored for identifying solution of existing problem by brainstorming process. Thus, a fuzzy condition was recorded as all KEs were based on five criteria: lean practices, customer complaint/communication, in-house quality practices, past experience, and tool room participation. Three KEs (to be implemented at process P5) were proposed in order to eliminate waste/gap areas (Table 2), but due to the limited budget and financial conditions, the industry can bear the expense of any of appropriate KEs, which would give optimum benefits at all levels of organization. In order to select one suitable KE, fuzzy-TOPSIS technique was used for ranking and evaluation of KEs performance.

6.2. Application of Fuzzy-TOPSIS. In order to assess the performance of proposed KEs, three alternatives (KE1, KE2, and KE3) and five criteria (C1, lean practices; C2, customer complaint/communication; C3, in-house quality practices; C4, past experience; and C5, tool room participation) were used to apply fuzzy-TOSIS approach on the basis of steps given in Kumar et al.'s work [43]. The evaluation of each

TABLE 3: Weights and alternatives with fuzzy numbers.

Alternatives	Weight	Fuzzy number
Very low (VL)	Very poor (VP)	(1,1,3)
Low (L)	Poor (P)	(1,3,5)
Medium (M)	Fair (F)	(3,5,7)
High (H)	Good (G)	(5,7,9)
Very high (VH)	Very good (VG)	(7,9,9)

TABLE 4: Weighted criteria for three experts.

Decision makers	C_1	C_2	C_3	C_4	C_5
M1	H(5,7,9)	H(5,7,9)	VH(7,9,9)	H(5,7,9)	H(5,7,9)
M2	M(3,5,7)	M(3,5,7)	H(3,7,9)	L(1,3,5)	M(3,5,7)
M3	L(1,3,5)	H(5,7,9)	L(1,3,5)	L(1,3,9)	M(3,5,7)

Kaizen event to achieve goal of waste elimination/removal is presented in Figure 6.

Three decision makers (experts) M1, M2, and M3 were selected to submit their valuations of criteria importance weights and Kaizen performance rating against the goal of the selected organization. Fuzzy approach of triangular fuzzy numbers (TFN) is used to capture vagueness in evaluation of criteria weights as well as Kaizen performance. The five-point linguistic scale and the corresponding TFNs to evaluate the weight and alternatives are given in Table 3. The importance weight criteria given by the decision makers are provided in Table 4 and rating of Kaizen events based on personal valuation of the three decision makers is presented in Table 5 (average fuzzy rating, normalized fuzzy rating, and weighted normalized fuzzy rating). The linguistic variables, aggregate weight criteria, and the aggregate fuzzy rating of each KE are calculated.

Table 6 shows the FNIS and FPIS values with respect to each criterion. The calculated closeness coefficient (CC_i) for each KE and the comparison of closeness coefficients are given in Table 7. In this way, KE-2 was selected to be performed at P5.

7. Result and Discussion

7.1. Performing KE. The authors visited the workstation (P5) where a screw mounted fixture (design) was observed for performing milling operation of one workpiece at a time and tightening and loosing of SKDM were done by the lever and screw arrangement. This task was found to be time-consuming and tedious for the operators. Moreover, it was also reported for high possibility of accidents during the process and required skilled operator at P5. At the workstation, the authors conducted visual inspection for three

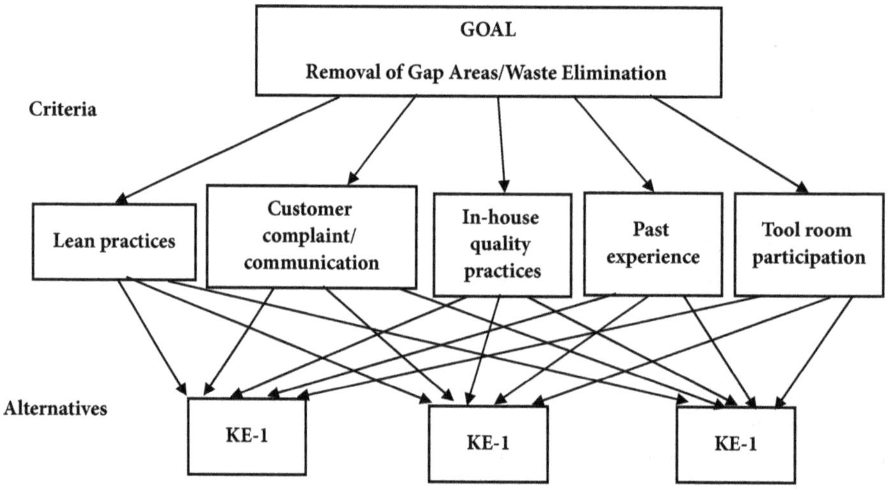

FIGURE 6: Structure for evaluation of Kaizen events.

TABLE 5: Calculations for average fuzzy rating, normalized fuzzy rating, and weighted normalized fuzzy rating.

Criteria	Kaizen events	M1	M2	M3	Aggregate fuzzy rating	Normalized fuzzy rating	Weighted normalized fuzzy rating
C1	KE1	G (5,7,9)	VG(7,9,9)	F(3,5,7)	(3,7,9)	(0.333,0.778,1.000)	(1.000,5.446,9.000)
	KE2	G (5,7,9)	F(3,5,7)	VG(7,7,9)	(3,6.33,9)	(0.333,0.703,1.000)	(1.000,4.449,9.000)
	KE3	F (3,5,7)	P(1,3,5)	G(5,7,9)	(1,5,9)	(0.111,0.556,1.000)	(0.111,2.780,9.000)
C2	KE1	G (5,7,9)	VG(7,9,9)	F(3,5,7)	(3,7,9)	(0.333,0.778,1.000)	(1.000,5.446,9.000)
	KE2	F(3,5,7)	VG(7,9,9)	VG(7,7,9)	(3,7,9)	(0.333,0.778,1.000)	(1.000,5.446,9.000)
	KE3	F(3,5,7)	F(3,5,7)	G(5,7,9)	(3,5.66,7)	(0.333,0.628,0.778)	(1.000,3.554,5.446)
C3	KE1	VG(7,9,9)	VP(1,1,3)	F(3,5,7)	(1,5,9)	(0.111,0.778,1.000)	(0.111,3.890,9.000)
	KE2	F(3,5,7)	F(3,5,7)	VG(7,7,9)	(3,5.66,9)	(0.333,0.628,1.000)	(1.000,3.554,9.000)
	KE3	F (3,5,7)	F(3,5,7)	F(3,5,7)	(3,4.33,9)	(0.333,0.481,1.000)	(1.000,2.082,9.000)
C4	KE1	P (1,3,5)	VG(7,9,9)	F(3,5,7)	(1,5.66,9)	(0.111,0.628,1.000)	(0.111,3.554,9.000)
	KE2	G (5,7,9)	VP (1,3,5)	G(3,5,7)	(1,5,9)	(0.111,0.556,1.000)	(0.111,2.780,9.000)
	KE3	F (3,5,7)	F (3,5,7)	G(5,7,9)	(3,5.66,9)	(0.333,0.628,1.000)	(1.000,3.554,9.000)
C5	KE1	G (5,7,9)	G (5,7,9)	F(3,5,7)	(3,6.33,9)	(0.333,0.703,1.000)	(1.000,4.449,9.000)
	KE2	VG (7,9,9)	G (5,7,9)	F(3,5,7)	(3,7,9)	(0.333,0.778,1.000)	(1.000,5.446,9.000)
	KE3	G (5,7,9)	P(1,3,5)	VG(7,7,9)	(1,5.66,9)	(0.111,0.628,1.000)	(0.111,3.554,9.000)

TABLE 6: Values of FNIS and FPIS.

Criterion	C1	C2	C3	C4	C5
FNIS	0.111	1.000	0.111	0.111	0.111
FPIS	9.000	9.000	9.000	9.000	9.000

TABLE 7: Comparing the closeness coefficient of KEs.

	d_{i+}	d_{i-}	CC_i	Ranking
KE-1	0.527023	0.268328	0.337371	3
KE-2	0.159926	0.496233	0.756269	1
KE-3	0.379473	0.376501	0.498034	2

hours. Three work pieces out of eighty-nine reported "chattering/tool marks" and "chamfer not good" after performing P5 process. In order to eradicate the existing problem, the snap type fixture was recommended to replace the screw mounted fixture for fastening of SKDM (Figure 7).

Implementation of Recommended Snap Type Fixture Design for Reducing Rejection Rate at P5. The problem was resolved by replacing screw type fixture to snap type fixture, in which the fastening activity was completed through the snap. In this fixture, two slots were designed, in which two workpieces were mounted on the same time for milling operation. This newly designed fixture eliminated quality problem along with ease in fastening activities. The production was monitored for two days and no "chattering/tool mark" was reported in the production of 1020

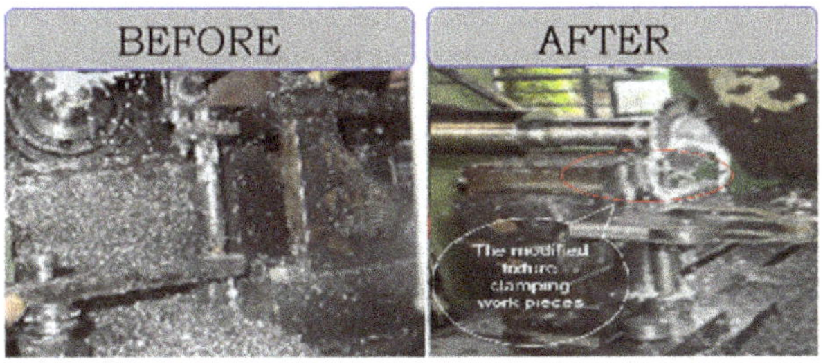

FIGURE 7: Fastening fixture before and after VSM implementation.

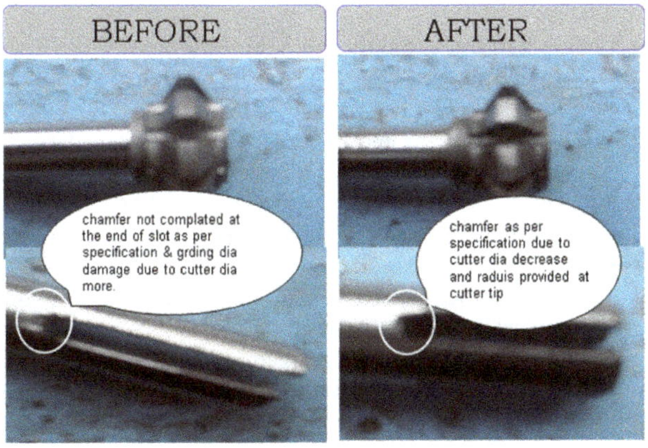

FIGURE 8: The workpiece before and after VSM implementation.

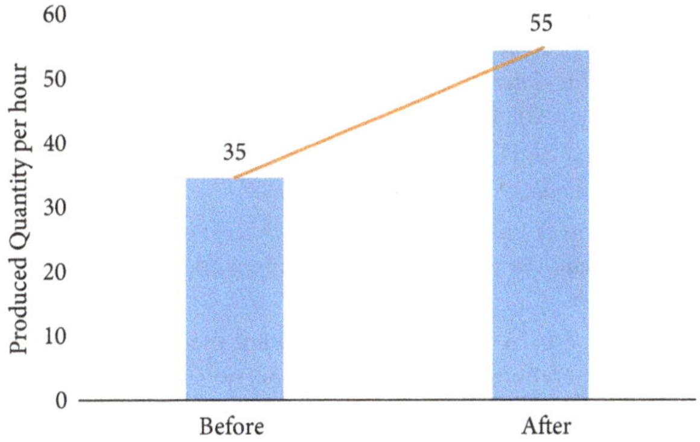

FIGURE 9: Before and after production quantity per hour.

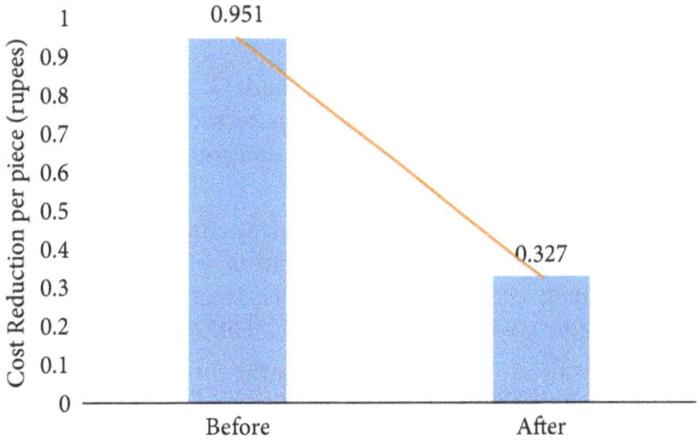

FIGURE 10: Before and after cost per piece.

pieces (Figure 8). This KE was applied within twenty-two days.

After VSM implementation, the result is analyzed, in which lean benefits are observed as reduction in fastening time at P5 by 58.73% (from 63 seconds to 26 seconds), PL time by 69.49% (from 18.031 to 5.5 days), per work piece cost by 60.71% (Figure 10), and total CT by 69.64% (from 336 seconds to 101 seconds) and increase in production per hour by 57.15% at P5 (Figure 9). The new process removed rejection cost of 3.1k USD per year.

8. Conclusion

In order to achieve significant lean benefits, it is important to understand the whole system of any organization. In this study, an effort has been made to apply Lean-Kaizen concept with VSM tool in SKDM line of ABC. The fuzzy-TOPSIS was applied to select the most appropriate KE that optimizes the lean benefits as cost reduction, lead time reduction, inventory level reduction, CT reduction, production improvement, and quality improvement of the product. The Lean-Kaizen with VSM-fuzzy technique is found to be an effective technique that assists in eliminating waste in the organization and motivating individuals to achieve goals of the organization. This study can help managers and practitioners to handle waste removal challenges in real-time working situations at the shop floor. The study highlights the following points:

(i) The endeavor has been made for Lean-Kaizen implementation with VSM-fuzzy technique in order to optimize the benefit of KEs.

(ii) Besides many researches based on the hypothetical concept of Lean-Kaizen and fuzzy-TOPSIS, the efforts have been made to explore the practical validation of combination of the three techniques, Lean, Kaizen, and Fuzzy-TOPSIS, in the context of Indian SSE.

(iii) The study provides a road map for implementing Lean-Kaizen concept in Indian production industry for continuous improvement.

This study provides a wider view of different practical approach of lean manufacturing and thus can help researchers and practitioners to compare their KE performance. A well-established cost-based theoretical framework can also be developed for a wide range of manufacturing industries for getting benefits in product, process, and services.

Conflicts of Interest

The authors declare that they have no conflicts of interest.

References

[1] J. Singh and H. Singh, "Continuous improvement approach: State-of-art review and future implications," *International Journal of Lean Six Sigma*, vol. 3, no. 2, pp. 88–111, 2012.

[2] S. Kumar, A. K. Dhingra, and B. Singh, "Process improvement through Lean-Kaizen using value stream map: a case study in India," *The International Journal of Advanced Manufacturing Technology*, vol. 96, no. 5-8, pp. 2687–2698, 2018.

[3] G. Alukal and A. Manos, *A Simplified Approach to Process Improvements*, ASQ Quality Press, 2007.

[4] J. P. Womack and D. T. Jones, *Lean thinking: Banish waste and create wealth in your organization, Simon and Shuster*, NY, New York, 1996.

[5] T. Grünberg, "A review of improvement methods in manufacturing operations," *Work Study*, vol. 52, no. 2, pp. 89–93, 2003.

[6] J. H. Marvel and C. R. Standridge, "A simulation-enhanced lean design process," *Journal of Industrial Engineering & Management*, vol. 2, no. 1, pp. 90–113, 2009.

[7] C. Chen, "Extensions of the TOPSIS for group decision-making under fuzzy environment," *Fuzzy Sets and Systems*, vol. 114, no. 1, pp. 1–9, 2000.

[8] P. Achanga, E. Shehab, R. Roy, and G. Nelder, "A fuzzy-logic advisory system for lean manufacturing within SMEs," *International Journal of Computer Integrated Manufacturing*, vol. 25, no. 9, pp. 839–852, 2012.

[9] X. Wang and H. K. Chan, "A hierarchical fuzzy TOPSIS approach to assess improvement areas when implementing green supply chain initiatives," *International Journal of Production Research*, vol. 51, no. 10, pp. 3117–3130, 2013.

[10] B. Vahdani, M. Salimi, and M. Charkhchian, "A new FMEA method by integrating fuzzy belief structure and TOPSIS to improve risk evaluation process," *The International Journal of Advanced Manufacturing Technology*, vol. 77, no. 1-4, pp. 357–368, 2015.

[11] R. Mohanraj, M. Sakthivel, S. Vinodh, and K. E. K. Vimal, "A framework for VSM integrated with Fuzzy QFD," *TQM Journal*, vol. 27, no. 5, pp. 616–632, 2015.

[12] M. Hallgren and J. Olhager, "Lean and agile manufacturing: External and internal drivers and performance outcomes," *International Journal of Operations and Production Management*, vol. 29, no. 10, pp. 976–999, 2009.

[13] I. S. Lasa, C. O. Laburu, and R. De Castro Vila, "An evaluation of the value stream mapping tool," *Business Process Management Journal*, vol. 14, no. 1, pp. 39–52, 2008.

[14] M. Dora, D. van Goubergen, M. Kumar, A. Molnar, and X. Gellynck, "Application of lean practices in small and medium-sized food enterprises," *British Food Journal*, vol. 116, no. 1, pp. 125–141, 2014.

[15] S. Kumar, A. K. Dhingra, and B. Singh, "Implementation of the lean-kaizen approach in fastener industries using the data envelopment analysis," *Facta Universitatis, series: Mechanical Engineering*, vol. 15, no. 1, pp. 145–161, 2017.

[16] S. Kumar, A. Dhingra, and B. Singh, "Lean-Kaizen implementation," *Journal of Engineering, Design and Technology*, vol. 16, no. 1, pp. 143–160, 2018.

[17] J. Womack and D. Jones, *Lean Solution*, Free Press Pubs, New York, NY, USA, 2005.

[18] A. Melcher, W. Acar, P. DuMont, and M. Khouja, "Standard-Maintaining and Continuous-Improvement Systems: Experiences and Comparisons," *Interfaces*, vol. 20, no. 3, pp. 24–40, 1990.

[19] A. P. Brunet and S. New, "Kaizen in Japan: An empirical study," *International Journal of Operations and Production Management*, vol. 23, no. 11-12, pp. 1426–1446, 2003.

[20] N. Bateman, "Sustainability: The elusive element of process improvement," *International Journal of Operations and Production Management*, vol. 25, no. 3, pp. 261–276, 2005.

[21] N. Bhuiyan and A. Baghel, "An overview of continuous improvement: From the past to the present," *Management Decision*, vol. 43, no. 5, pp. 761–771, 2005.

[22] K. Van Scyoc, "Process safety improvement-Quality and target zero," *Journal of Hazardous Materials*, vol. 159, no. 1, pp. 42–48, 2008.

[23] W. J. Glover, J. A. Farris, E. M. Van Aken, and T. L. Doolen, "Critical success factors for the sustainability of Kaizen event human resource outcomes: An empirical study," *International Journal of Production Economics*, vol. 132, no. 2, pp. 197–213, 2011.

[24] M. F. S. Barraza, T. Smith, and S. M. Dahlgaard-Park, "Lean-kaizen public service: An empirical approach in Spanish local governments," *TQM Journal*, vol. 21, no. 2, pp. 143–167, 2009.

[25] M. F. Suárez-Barraza and J. Ramis-Pujol, "Implementation of Lean-Kaizen in the human resource service process: A case study in a Mexican public service organisation," *Journal of Manufacturing Technology Management*, vol. 21, no. 3, pp. 388–410, 2010.

[26] A. Karim and K. Arif-Uz-Zaman, "A methodology for effective implementation of lean strategies and its performance evaluation in manufacturing organizations," *Business Process Management Journal*, vol. 19, no. 1, pp. 169–196, 2013.

[27] A. K. Arya and S. Choudhary, "Assessing the application of Kaizen principles in Indian small-scale industry," *International Journal of Lean Six Sigma*, vol. 6, no. 4, pp. 369–396, 2015.

[28] B. Singh, S. Garg, and S. Sharma, "Scope for lean implementation: a survey of 127 Indian industries," *International Journal of Rapid Manufacturing*, vol. 1, no. 3, p. 323, 2010.

[29] J. L. García, A. A. Maldonado, A. Alvarado, and D. G. Rivera, "Human critical success factors for kaizen and its impacts in industrial performance," *The International Journal of Advanced Manufacturing Technology*, vol. 70, no. 9-12, pp. 2187–2198, 2014.

[30] B. Singh, S. K. Garg, S. K. Sharma, and C. Grewal, "Lean implementation and its benefits to production industry," *International Journal of Lean Six Sigma*, vol. 1, no. 2, pp. 157–168, 2010.

[31] B. Das, U. Venkatadri, and P. Pandey, "Applying lean manufacturing system to improving productivity of airconditioning coil manufacturing," *The International Journal of Advanced Manufacturing Technology*, vol. 71, no. 1-4, pp. 307–323, 2014.

[32] S. Gupta and S. K. Jain, "A literature review of lean manufacturing," *International Journal of Management Science and Engineering Management*, vol. 8, no. 4, pp. 241–249, 2013.

[33] J. Bhamu, J. V. S. Kumar, and K. S. Sangwan, "Productivity and quality improvement through value stream mapping: A case study of Indian automotive industry," *International Journal of Productivity and Quality Management*, vol. 10, no. 3, pp. 288–306, 2012.

[34] A. Esfandyari, M. R. Osman, N. Ismail, and F. Tahriri, "Application of value stream mapping using simulation to decrease production lead time: A Malaysian manufacturing case," *International Journal of Industrial and Systems Engineering*, vol. 8, no. 2, pp. 230–250, 2011.

[35] P. Hines and N. Rich, "The seven value stream mapping tools," *International Journal of Operations and Production Management*, vol. 17, no. 1, pp. 46–64, 1997.

[36] A. Prashar, "Redesigning an assembly line through Lean-Kaizen: An Indian case," *TQM Journal*, vol. 26, no. 5, pp. 475–498, 2014.

[37] F. E. Ciarapica, M. Bevilacqua, and G. Mazzuto, "Performance analysis of new product development projects: An approach based on value stream mapping," *International Journal of Productivity and Performance Management*, vol. 65, no. 2, pp. 177–206, 2016.

[38] M. Azizi, N. Mohebbi, R. M. Gargari, and M. Ziaie, "A strategic model for selecting the location of furniture factories: A case of the study of furniture," *International Journal of Multicriteria Decision Making*, vol. 5, no. 1-2, pp. 87–108, 2015.

[39] R. Parameshwaran, S. Praveen Kumar, and K. Saravanakumar, "An integrated fuzzy MCDM based approach for robot selection considering objective and subjective criteria," *Applied Soft Computing*, vol. 26, pp. 31–41, 2014.

[40] T. Rashid, I. Beg, and S. M. Husnine, "Robot selection by using generalized interval-valued fuzzy numbers with TOPSIS," *Applied Soft Computing*, vol. 21, pp. 462–468, 2014.

[41] E. Pourjavad, H. Shirouyehzad, and A. Shahin, "Selecting maintenance strategy in mining industry by analytic network process and TOPSIS," *International Journal of Industrial and Systems Engineering*, vol. 15, no. 2, pp. 171–192, 2013.

[42] B. Singh, S. K. Garg, and S. K. Sharma, "Development of index for measuring leanness: Study of an Indian auto component industry," *Measuring Business Excellence*, vol. 14, no. 2, pp. 46–53, 2010.

[43] S. Kumar, B. Singh, M. A. Qadri, Y. V. S. Kumar, and A. Haleem, "A framework for comparative evaluation of lean performance of firms using fuzzy TOPSIS," *International Journal of Productivity and Quality Management*, vol. 11, no. 4, pp. 371–392, 2013.

[44] S. Vinodh and S. R. Balaji, "Fuzzy logic based leanness assessment and its decision support system," *International Journal of Production Research*, vol. 49, no. 13, pp. 4027–4041, 2011.

[45] S. Vinodh and K. E. K. Vimal, "Thirty criteria based leanness assessment using fuzzy logic approach," *The International Journal of Advanced Manufacturing Technology*, vol. 60, no. 9-12, pp. 1185–1195, 2012.

[46] S. Vinodh and S. K. Chintha, "Leanness assessment using multi-grade fuzzy approach," *International Journal of Production Research*, vol. 49, no. 2, pp. 431–445, 2011.

[47] R. Vidyadhar, R. Sudeep Kumar, S. Vinodh, and J. Antony, "Application of fuzzy logic for leanness assessment in SMEs: a case study," *Journal of Engineering, Design and Technology*, vol. 14, no. 1, pp. 78–103, 2016.

[48] M. M. Ravikumar, K. Marimuthu, P. Parthiban, and H. A. Zubar, "Evaluating lean execution performance in Indian MSMEs using SEM and TOPSIS models," *International Journal of Operational Research*, vol. 26, no. 1, pp. 104–125, 2016.

[49] S. Agrawal, R. K. Singh, and Q. Murtaza, "Prioritizing critical success factors for reverse logistics implementation using fuzzy-TOPSIS methodology," *Journal of Industrial Engineering International*, vol. 12, no. 1, pp. 15–27, 2016.

[50] G. A. Marodin, A. G. Frank, G. L. Tortorella, and T. A. Saurin, "Contextual factors and lean production implementation in the Brazilian automotive supply chain," *Supply Chain Management: An International Journal*, vol. 21, no. 4, pp. 417–432, 2016.

[51] D. Seth and V. Gupta, "Application of value stream mapping for lean operations and cycle time reduction: An Indian case study," *Production Planning and Control*, vol. 16, no. 1, pp. 44–59, 2005.

[52] G. Anand and R. Kodali, "Development of a framework for implementation of lean manufacturing systems," *International Journal of Management Practice*, vol. 4, no. 1, pp. 95–116, 2010.

[53] B. Zhou, "Lean principles, practices, and impacts: a study on small and medium-sized enterprises (SMEs)," *Annals of Operations Research*, vol. 241, no. 1-2, pp. 457–474, 2016.

[54] N. V. K. Jastia and R. Kodali, "Lean production: Literature review and trends," *International Journal of Production Research*, vol. 53, no. 3, pp. 867–885, 2015.

[55] J. R. Jadhav, S. S. Mantha, and S. B. Rane, "Exploring barriers in lean implementation," *International Journal of Lean Six Sigma*, vol. 5, no. 2, pp. 122–148, 2014.

14

Two Approximation Models of Fuzzy Weight Vector from a Comparison Matrix

Tomoe Entani

Graduate School of Applied Informatics, University of Hyogo, Kobe, Hyogo 650-0047, Japan

Correspondence should be addressed to Tomoe Entani; entani@ai.u-hyogo.ac.jp

Academic Editor: Katsuhiro Honda

In this study, our uncertain judgment on multiple items is denoted as a fuzzy weight vector. Its membership function is estimated from more than one interval weight vector. The interval weight vector is obtained from a crisp/interval comparison matrix by Interval Analytic Hierarchy Process (AHP). We redefine it as a closure of the crisp weight vectors which approximate the comparison matrix. The intuitively given comparison matrix is often imperfect so that there could be various approaches to approximate it. We propose two of them: upper and lower approximation models. The former is based on weight possibility and the weight vector with it includes the comparison matrix. The latter is based on comparison possibility and the comparison matrix with it includes the weight vector.

1. Introduction

AHP (Analytic Hierarchy Process) [1] is one of the well-known multicriteria decision-making methods and applied to various decision situations. The decision problem in AHP is structured hierarchically with the criteria and alternatives. The final decision is a priority weight vector of the alternatives reflecting the importance of the criteria. The priority vector of the alternatives represents the decision-makers preference and is obtained by synthesizing two kinds of weight vectors. They are the importance weight vector of the criteria and the priority weight vector of the alternatives under each criterion. Each weight vector is obtained from a pairwise comparison matrix given by a decision-maker intuitively. The core technique in AHP is to derive a weight vector consisting of multiple items, such as alternatives and criteria, from a pairwise comparison matrix. We focus on this technique and derive a fuzzy weight vector from any given comparison matrix.

The often-used techniques are the eigenvector, geometric mean, and logarithm least square methods, all of which basically derive a crisp weight vector from a crisp comparison matrix. Instead of a crisp weight vector, an interval/fuzzy vector could be obtained from a crisp comparison matrix reflecting the inconsistency among the crisp comparisons. On the other hand, the comparisons could be given as interval/fuzzy since a decision-maker's thinking on the items is not always accurate and precise. In order to represent our vague judgment, interval/fuzzy numbers may be more preferable and suitable than crisp numbers. Moreover, when linguistic patterns among the items are given, the comparison matrix is constructed based on transitivity so as to be robust for rank reverse [2]. In case of an interval/fuzzy comparison matrix, a crisp or interval/fuzzy weight vector could be obtained. In this way, there are four cases: a crisp or interval/fuzzy weight vector from a crisp or interval/fuzzy comparison matrix. We review some previously proposed ideas to distinguish ours from them.

A crisp weight vector is useful for ranking the items linearly. It is obtained not only from a crisp comparison matrix but also from an interval comparison matrix. Using the upper or lower bound of each interval comparison, multiple crisp weight vectors are derived [3] and the crisp weight vector is determined by taking a Euclidean center of them in [4]. The other method to derive a crisp weight vector from an interval comparison matrix is based on fuzzy programming approach where a deviation parameter of each comparison is introduced to measure the satisfaction [5]. Moreover, some methods to derive a crisp weight vector from a fuzzy comparison matrix have been proposed [6–8] and the applications have been shown and compared in [9]. These obtained crisp weight vectors are understood as a summary of the given comparison matrix, although a part of its information may be curtailed. Then, in order to reflect the uncertainty of each interval comparison, an interval comparison matrix is summarized in an interval weight vector. The interval weight vector is obtained from the viewpoint of the stability of the rank order of items in [10], where well-known eigenvector method is applied. The upper and lower bounds of each interval weight are obtained from the devised interval comparison matrix to be less inconsistent with logarithmic goal programming in [11]. Furthermore, the crisp comparison matrix is processed into an interval comparison matrix based on transitivity and the interval weight vector is obtained from the upper and lower bounds of the interval comparisons [12]. Each interval comparison depends on two crisp comparisons and does not reflect the other comparisons so that all the given comparisons are not considered equally. In case of a fuzzy comparison matrix, the fuzzy weight vector is derived so as to be close to it in various ways, and they are reviewed in [13]. For instance, the degree of consistency of the obtained weight vector is minimized in [14], the deviations based on geometric mean are measured in [15], and the deviations by logarithm are considered in [16]. As an application, Fuzzy AHP was used for a selection problem in [17, 18]. The fuzzy comparison has been processed from a crisp comparison by assuming its uncertainty in a certainty degree [19]. The primal reason why a weight vector is an interval/fuzzy seems to be the fuzziness of the given comparisons so that these interval/fuzzy weight vectors reflect the uncertainty of each comparison.

It is a human nature to make a decision and we often prefer to do it by ourselves. We do not prefer a system to do it. Since his/her final decision does not always equal the derived result which s/he refers to in making a decision, there is no need to summarize it as a rigid value, crisp weight vector. Instead, it is more important to represent a weight vector of the items in a decision-maker's mind as it is. For a decision-maker, the decision aiding system to derive what s/he cannot easily notice but reflects his/her uncertain judgment would be helpful. The given comparison matrix does not represent the whole decision-maker's thinking, and its imperfectness is embodied in inconsistency among the intuitively given comparisons, as well as fuzziness of each comparison. In Interval AHP [20, 21], the interval weight vector is obtained by reflecting inconsistency among the crisp comparisons. Because of its formulation, the upper and lower bounds of the interval weights are emphasized more than the values within an interval weight. This paper derives a fuzzy weight vector whose membership function is estimated by the interval weight vectors from an interval/crisp comparison matrix. The goal is neither to summarize an interval comparison matrix in a crisp weight vector nor to reflect the fuzziness of each interval comparison into an interval weight vector. We redefine the interval weight vector in Interval AHP based on the relations between the comparison matrix and the weight vector from possibilistic view. The interval weight vector reflects uncertainty in a decision-maker's thinking by inconsistency among the comparisons and fuzziness of the comparisons.

This paper is organized as follows. Sections 2 and 3 explain a comparison matrix and a fuzzy weight vector, respectively. In Section 4, two approaches to derive interval weight vector from an interval/crisp comparison matrix are proposed based on the idea that the given comparisons are not perfect. Then, in Section 5, two example comparison matrices are given to illustrate the proposed approaches to derive the fuzzy weight vectors. The last section concludes this paper.

2. Pairwise Comparison Matrix and Weight Vector

A decision-maker compares a pair of items and gives the comparison based on his/her intuitive judgment. A crisp comparison is used to denote a decision-maker's judgment as quantitative data. The interval comparison is more general than the crisp one since an interval becomes crisp when its upper and lower bounds are equal. We start with an interval pairwise comparison matrix of n items.

$$A = \left[A_{ij}\right] = \begin{bmatrix} 1 & \cdots & [\underline{a}_{1n}, \overline{a}_{1n}] \\ \vdots & [\underline{a}_{ij}, \overline{a}_{ij}] & \vdots \\ [\underline{a}_{n1}, \overline{a}_{n1}] & \cdots & 1 \end{bmatrix}, \tag{1}$$

whose element $A_{ij} = [\underline{a}_{ij}, \overline{a}_{ij}]$ is the comparison representing the importance ratio of item i to item j, and intuitively given by a decision-maker. An interval comparison is sometimes more suitable than a crisp comparison since a decision-maker's judgment is often vague. It may be expressed as follows: item 1 is "probably very much more important" than item 2. The interval comparison of item 1 over item 2 is denoted such that $A_{12} = [6, 8]$. In case that item 1 is "very much more important" than item 2, the crisp comparison is such that $a_{12} = 7$. These comparisons are identical: $a_{ii} = 1$ and reciprocal in interval sense: $\underline{a}_{ij} = 1/\overline{a}_{ji}$. A crisp comparison matrix is the case of $\underline{a}_{ij} = \overline{a}_{ij} = a_{ij}, \forall i, j, i > j$. Those who are more familiar with crisp numbers than interval numbers may prefer crisp numbers to represent their judgments. There are well-known methods in conventional AHP, such as eigenvector method, geometric mean method, and the logarithmic least square method, which derive crisp weight vector $\boldsymbol{w} = (w_1, \ldots, w_n)^t$ from crisp comparison matrix $A = [a_{ij}]$.

A decision-maker can focus on the compared items without regarding the other items at the same time. It is easy for him/her to compare a pair of items, although s/he has to repeat comparing for all the pairs. The comparisons are inconsistent with each other since s/he compares an item several times intuitively. The given comparison matrix surely corresponds to a decision-maker's thinking on the items, and it is reasonable to derive the weight vector of the items from it. However, it is not the best to make the comparison matrix and the weight vector mathematically equal because of the imperfectness of the given comparisons.

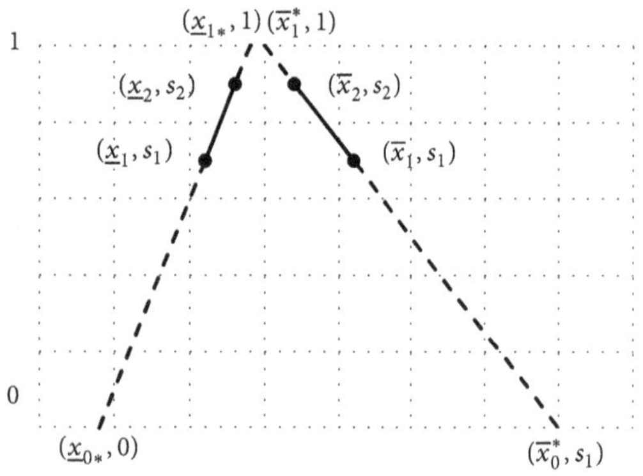

FIGURE 1: Membership function from two α-level sets.

They do not represent the whole decision-maker's thinking, and it is unknown how good and how much reliable they are. Therefore, there could be various weight vectors to approximate a comparison matrix, and this study proposes two approaches to obtain interval weight vector $\boldsymbol{W}$ from interval/crisp comparison matrix A.

3. Membership Function of Fuzzy Weight Vector

The membership function of a triangular/trapezoidal fuzzy number is estimated from two kinds of its α-level sets. Let us assume them as interval $[\underline{x}_1, \overline{x}_1]$ and interval $[\underline{x}_2, \overline{x}_2]$, whose membership values to fuzzy number $\widetilde{X}$ are s_1 and s_2, respectively. The membership function $\mu_{\widetilde{X}}$ is denoted as follows.

$$\mu_{\widetilde{X}}(x) = \begin{cases} s_1 + \dfrac{s_2 - s_1}{\underline{x}_2 - \underline{x}_1}\left(x - \underline{x}_1\right), & x_{0*} \le x \le x_{1*}, \\ 1, & x_{1*} \le x \le x_1^*, \\ s_1 + \dfrac{s_2 - s_1}{\overline{x}_2 - \overline{x}_1}\left(x - \overline{x}_1\right), & x_1^* \le x \le x_0^* \\ 0 & \text{else,} \end{cases} \quad (2)$$

where $x_{0*} = \underline{x}_1 - s_1(\overline{x}_1 - \underline{x}_1)/(s_2 - s_1)$, $x_{1*} = \underline{x}_1 - (1 - s_1)(\overline{x}_1 - \underline{x}_1)/(s_2 - s_1)$, $x_1^* = \overline{x}_1 - (1 - s_1)(\overline{x}_2 - \overline{x}_1)/(s_2 - s_1)$, $x_0^* = \overline{x}_1 - s_1(\overline{x}_2 - \overline{x}_1)/(s_2 - s_1)$. It is illustrated in Figure 1.

A decision-maker's thinking in this study is denoted as a fuzzy weight vector $\widetilde{\boldsymbol{W}} = (\widetilde{W}_1, \ldots, \widetilde{W}_n)^t$, whose α-level set is interval weight vector $\boldsymbol{W} = (W_1, \ldots, W_n)^t$, where $W_i = [\underline{w}_i, \overline{w}_i], \forall i$. We can find the following fact about an interval weight vector [20]. The sum of the lower bounds of the interval weights of all items, $s = \sum_i \underline{w}_i$, is surely assigned to one of the items, while the left, $1 - s$, may be assigned to more than two items. Therefore, s represents a sure degree of $\boldsymbol{W}$ and can be considered as its membership value of interval weight vector $\boldsymbol{W}$ to fuzzy weight vector $\widetilde{\boldsymbol{W}}$. In other words, $\boldsymbol{W}$ is an α-level set of $\widetilde{\boldsymbol{W}}$ with $\alpha = s$. In order to estimate fuzzy weight vector by (2), we need at least two kinds of α-level sets. In the next section, two approaches to derive the representative interval weight vectors from an interval/crisp comparison matrix in Section 2 are proposed.

4. Interval Weight Vectors from Interval Pairwise Comparison Matrix

4.1. Definition of Interval Weight Vector. The interval weight vector $\boldsymbol{W} = ([\underline{w}_i, \overline{w}_i], \forall i)^t$ corresponding to the given interval comparison matrix A in (1) is defined as follows in this study [22].

$$\begin{aligned} \underline{w}_i &= \min\left\{ w_i \middle| A \mathscr{R}_{\boldsymbol{w}}, \sum_i w_i = 1, \varepsilon \le w_i \forall i \right\}, \quad \forall i, \\ \overline{w}_i &= \max\left\{ w_i \middle| A \mathscr{R}_{\boldsymbol{w}}, \sum_i w_i = 1, \varepsilon \le w_i \forall i \right\}, \quad \forall i, \end{aligned} \quad (3)$$

where ε is a small positive number and $\mathscr{R}$ denotes the relation between comparison matrix $A = [A_{ij}]$ and weight vector $\boldsymbol{w} = (w_i, \forall i)^t$. The interval weight vector by (3) is a closure of the normalized crisp weight vectors, and it is also normalized in interval sense: $\boldsymbol{W} \in \mathscr{N}$, where $\mathscr{N}$ denotes a set of the normalized interval weight vectors as follows [23, 24].

$$\mathscr{N} = \left\{ \left([\underline{w}_i, \overline{w}_i], \forall i\right) \middle| \sum_{j \ne i} \underline{w}_j + \overline{w}_i \le 1, \ \forall i, \sum_{j \ne i} \overline{w}_j + \underline{w}_i \ge 1, \ \forall i, \varepsilon \le \underline{w}_i \le \overline{w}_i, \ \forall i \right\}. \quad (4)$$

The first two kinds of inequalities reduce the redundancy to make the sum of the crisp values in the intervals one. A crisp weight vector is an extreme case of interval weight vector with $\underline{w}_i = \overline{w}_i = w_i$, $\forall i$, and the first two kinds of inequalities are replaced into crisp normalization $\sum_i w_i = 1$.

Depending on relation $\mathscr{R}$, various interval weight vectors can be derived from a comparison matrix. Since the comparisons do not represent the decision-maker's thinking perfectly, they are approximated by the weight vector in various ways. This study considers the imperfectness of the comparison matrix with weight or comparison possibility and correspondingly assumes approximation relation $\mathscr{R}$. The upper approximation model in Section 4.2 introduces weight possibility for a weight vector to include the comparison matrix, while the lower approximation model in Section 4.3 introduces comparison possibility for the comparison matrix to include a weight vector.

4.2. Upper Approximation Model. The upper approximation model is based on the idea that the comparison matrix is imperfect because of its scarceness. The given comparison matrix is a part of a decision-maker's thinking so that the weight vector needs to cover it. Interval AHP [21] derives interval weight vector $\boldsymbol{W} = ([\underline{w}_i, \overline{w}_i], \forall i)^t$ from comparison matrix $A = [A_{ij}]$ in (1).

$$\begin{aligned} \min \quad & \sum_i \left(\overline{w}_i - \underline{w}_i\right) \\ \text{s.t.} \quad & \underline{w}_i \le \underline{a}_{ij} \overline{w}_j, \quad \forall i, j, i > j, \\ & \overline{a}_{ij} \underline{w}_j \le \overline{w}_i, \quad \forall i, j, i > j, \end{aligned}$$

$$([\underline{w}_i, \overline{w}_i], \forall i) \in \mathcal{N},$$
$$\varepsilon \le \underline{w}_i \le \overline{w}_i, \quad \forall i, \tag{5}$$

where the variables are interval weights $[\underline{w}_i, \overline{w}_i], \forall i$. It is noted that (5) can derive the interval weight vector from a crisp comparison matrix, $A = [a_{ij}]$, where $a_{ij} = \underline{a}_{ij} = \overline{a}_{ij}$ $\forall i, j$. The third constraint is for the interval normalization in (4). The first two kinds of constraints denote the inclusion relation: $[\underline{a}_{ij}, \overline{a}_{ij}] \subseteq [\underline{w}_i, \overline{w}_i]/[\underline{w}_j, \overline{w}_j] = [\underline{w}_i/\overline{w}_j, \overline{w}_i/\underline{w}_j]$, where a fraction of intervals is defined with the maximum range. The interval weight vector includes the comparison matrix so that (5) is called upper approximation model. By minimizing the widths of the interval weights, the weight vector approximates the comparison matrix as precisely as possible. It synthesizes the fact that the less uncertain interval weight is preferable since the sum of the widths of the interval weights represents how uncertain the interval weight is.

The LP problem (5) is feasible for any comparison matrix, although the uniqueness of the optimal weights $[\underline{w}_i, \overline{w}_i], \forall i$, has not been mentioned in detail [21]. In order to take all possible optimal weights into consideration, this study redefines the interval weight vector as (3) whose ${}_A\mathcal{R}_{\boldsymbol{w}}$ is as follows.

$$\begin{aligned} (w_i - q) &\le \underline{a}_{ij}(w_j + q), \quad \forall i, j, i > j, \\ \overline{a}_{ij}(w_j - q) &\le (w_i + q), \quad \forall i, j, i > j, \end{aligned} \tag{6}$$

where crisp weight w_i is widened into pseudo interval weight $[w_i - q, w_i + q]$ by weight possibility $q \ge 0$ to include comparison matrix A. A pseudo interval weight vector, $([w_i - q, w_i + q], \forall i)^t$, is normalized since it satisfies (4).

The minimum weight possibility q for comparison matrix A is obtained by the following problem.

$$\begin{aligned} q_* = \min \quad & q, \\ \text{s.t.} \quad & (w_i - q) \le \underline{a}_{ij}(w_j + q), \quad \forall i, j, i > j, \\ & \overline{a}_{ij}(w_j + q) \le (w_i - q), \quad \forall i, j, i > j, \\ & \varepsilon \le w_i \\ & \sum_i w_i = 1 \end{aligned} \tag{7}$$

where the variable are q and weight vector $(w_i, \forall i)^t$. The optimal objective function value, q_*, is unique to A, although there may be various optimal weight vectors. In order to include the comparison matrix, crisp weight vector needs to be widened by q_* at least. In other words, no crisp weight vector could be found in the comparison matrix if weight possibility was less than q_*.

The interval weight vector with weight possibility q_* by (7) is obtained by (3) as $\boldsymbol{W}(q_*) = ([\underline{w}_i(q_*), \overline{w}_i(q_*)], \forall i)^t$, where

$$\begin{aligned} \underline{w}_i(q_*) = \min \Big\{ & w_i \mid (w_i - q_*) \\ & \le \underline{a}_{ij}(w_j + q_*), \overline{a}_{ij}(w_j - q_*) \le (w_i + q_*), \ \forall i, j, i \\ & > j, \varepsilon \le w_i, \forall i, \sum_i w_i = 1 \Big\}, \quad \forall i, \\ \overline{w}_i(q_*) = \max \Big\{ & w_i \mid (w_i - q_*) \\ & \le \underline{a}_{ij}(w_j + q_*), \overline{a}_{ij}(w_j - q_*) \le (w_i + q_*), \ \forall i, j, i \\ & > j, \varepsilon \le w_i, \forall i, \sum_i w_i = 1 \Big\}, \quad \forall i. \end{aligned} \tag{8}$$

Interval weight vector $\boldsymbol{W}(q_*)$ includes all the crisp weight vectors which include the comparison matrix by weight possibility q_*. It is noted that $\boldsymbol{W}(q_*)$ is normalized in (4) and the pseudo interval weights, $[w_i - q_*, w_i + q_*]$, $\forall i$, are not always in the range within $[0, 1]$.

Then, the membership value of interval weight vector $\boldsymbol{W}(q_*)$ to fuzzy weight vector $\widetilde{\boldsymbol{W}}$ is $\mu_{\widetilde{W}}(\underline{w}_i(q_*)) = \mu_{\widetilde{W}}(\underline{w}_i(q_*)) = s(q_*) = \sum_i \underline{w}_i(q_*) \le 1$ and $s(q_*)$ is close enough to 1. Interval weight vector $\boldsymbol{W}(q_*)$ is α-level set of fuzzy weight vector $\widetilde{\boldsymbol{W}}$ with $\alpha = s(q_*)$. In order to estimate membership function of fuzzy weight vector $\widetilde{\boldsymbol{W}}$ by (2), we need the other α-level set.

Let us assume maximum weight possibility q^* corresponding to minimum weight possibility q_*. It is reasonable to assume more weight possibility than the minimum one to include the given comparisons since they do not represent the whole decision-maker's thinking on the items. The increase of weight possibility q decreases the lower bound of pseudo interval weight $w_i - q$. Therefore, when $\sum_i(w_i - q) = 0$, q becomes maximum: $q^* = 1/n$. The interval weight vector with weight possibility q^* is obtained by (3) as $\boldsymbol{W}(q^*) = ([\underline{w}_i(q^*), \overline{w}_i(q^*)], \ \forall i)^t$, where

$$\begin{aligned} \underline{w}_i(q^*) = \min \Big\{ & w_i \mid (w_i - q^*) \\ & \le \underline{a}_{ij}(w_j + q^*), \overline{a}_{ij}(w_j - q^*) \le (w_i + q^*), \ \forall i, j, i \\ & > j, \varepsilon \le w_i, \ \forall i, \sum_i w_i = 1 \Big\}, \quad \forall i, \\ \overline{w}_i(q^*) = \max \Big\{ & w_i \mid (w_i - q^*) \\ & \le \underline{a}_{ij}(w_j + q^*), \overline{a}_{ij}(w_j - q^*) \le (w_i + q^*), \ \forall i, j, i \\ & > j, w_i - q^* \ge \varepsilon, \ \forall i, \sum_i w_i = 1 \Big\}, \quad \forall i. \end{aligned} \tag{9}$$

Its membership value to fuzzy weight vector $\widetilde{\boldsymbol{W}}$ is $\mu_{\widetilde{W}}(\underline{w}_i(q^*)) = \mu_{\widetilde{W}}(\overline{w}_i(q^*)) = s(q^*) = \sum_i \underline{w}_i(q^*)$, where

$s(q^*)$ is close to 0, so that interval weight vector $\boldsymbol{W}(q^*)$ is α-level set of fuzzy weight vector $\widetilde{\boldsymbol{W}}$ with $\alpha = s(q^*)$.

Then, the membership function of fuzzy weight vector $\widetilde{\boldsymbol{W}}$ is estimated by (2) with two α-level sets: interval weight vectors $\boldsymbol{W}(q_*)$ and $\boldsymbol{W}(q^*)$. The fuzzy weight vector is estimated more precisely by setting more variety of q between q_* and q^*. The weight possibility, q, is introduced in order not to rely too much on the given imperfect comparisons.

4.3. Lower Approximation Model. The upper approximation model widens a crisp weight vector into the pseudo interval weight vector by weight possibility. This is one of the approaches to handle an imperfect comparison matrix, which do not represent the whole decision-maker's thinking. The other approach is based on comparison possibility and the given comparison matrix is narrowed down or widened into the processed comparison matrix under the condition that it includes a weight vector. Therefore, it is called lower approximation model. In Interval AHP [21], the lower approximation model has been formulated as a counterpart of the upper one (5) as follows.

$$\begin{aligned} \max\quad & \sum_i \left(\overline{w}_i - \underline{w}_i\right) \\ \text{s.t.}\quad & \underline{a}_{ij}\overline{w}_j \le \underline{w}_i, \quad \forall i, j, i > j, \\ & \overline{w}_i \le \overline{a}_{ij}\underline{w}_j, \quad \forall i, j, i > j, \\ & \left(\left[\underline{w}_i, \overline{w}_i\right],\ \forall i\right) \in \mathcal{N}, \\ & \varepsilon \le \underline{w}_i \le \overline{w}_i, \quad \forall i \end{aligned} \tag{10}$$

where the variables are interval weights $[\underline{w}_i, \overline{w}_i], \forall i$. Contrary to (5), the inclusion relation is $[\underline{a}_{ij}, \overline{a}_{ij}] \supseteq [\underline{w}_i, \overline{w}_i]/[\underline{w}_j, \overline{w}_j]$. The widths of the interval weights are maximized to approximate the comparison matrix as closely as possible.

However, this lower approximation model (10) may miss a possible normalized weight vector. A didactic example of missing weight vector is as follows. Let us denote the optimal solutions of (10) as $[\underline{w}_i, \overline{w}_i]$, $i = 1, 2, 3 \in \mathcal{N}$. It can be as follows: if $\overline{w}_2 - \underline{w}_2 < \overline{w}_3 - \underline{w}_3$, then $\overline{w}_1 + \underline{w}_2 + \overline{w}_3 > \overline{w}_1 + \overline{w}_2 + \underline{w}_3 \ge 1$. The former weights $(\overline{w}_1, \underline{w}_2, \overline{w}_3)$, whose sum exceeds 1, are impossible, although the second constraint requires $\overline{w}_1 \le \overline{a}_{12}\underline{w}_2$. This is also shown with a didactic numerical example:

$$A = \begin{bmatrix} 1 & \left[\frac{1}{4}, \frac{3}{2}\right] & \left[\frac{1}{6}, 1\right] \\ - & 1 & \left[\frac{1}{3}, \frac{4}{3}\right] \\ - & - & 1 \end{bmatrix}, \qquad W = \begin{pmatrix} [0.111, 0.333] \\ [0.222, 0.444] \\ [0.333, 0.667] \end{pmatrix}, \tag{11}$$

where A is the given comparison matrix and W is obtained by (10). Crisp weight vector $\boldsymbol{w} = (0.35, 0.3, 0.35)^t$ is not included in $\boldsymbol{W}$, although crisp comparisons by $\boldsymbol{w}$, $a_{12} = 7/6$, $a_{13} = 1$, $a_{23} = 6/7$, are included in the interval comparisons of A.

The conventional lower approximation model (10) has such a drawback as missing a normalized weight vector included in the comparison matrix. In order not to miss any, the interval weight vector for the lower approximation model, $\boldsymbol{W}^*(1) = ([\underline{w}_i(1), \overline{w}_i(1)], \forall i)^t$, is redefined by (3) as follows.

$$\begin{aligned} \underline{w}_i(1) &= \min\left\{ w_i \mid \underline{a}_{ij} w_j \le w_i, w_i \le \overline{a}_{ij} w_j, \forall i, j, i > j, \sum_i w_i = 1, \varepsilon \le w_i, \forall i \right\}, \quad \forall i, \\ \overline{w}_i(1) &= \max\left\{ w_i \mid \underline{a}_{ij} w_j \le w_i, w_i \le \overline{a}_{ij} w_j, \forall i, j, i > j, \sum_i w_i = 1, \varepsilon \le w_i, \forall i \right\}, \quad \forall i, \end{aligned} \tag{12}$$

where ${}_A\mathcal{R}_{\boldsymbol{w}}$ indicates the fact that the comparison matrix includes a weight vector. This problem is not always feasible. We may not find a normalized weight vector $\boldsymbol{w} = (w_i, \forall i)^t$ within comparison matrix A, when the comparisons are very inconsistent with each other or crisp as $\underline{a}_{ij} = \overline{a}_{ij} = a_{ij}$, $\forall i, j, i > j, a_{ij} \neq a_{il}a_{lj}, \exists i, j, l$.

When (12) is feasible, the membership value of interval weight vector $\boldsymbol{W}(1)$ to the fuzzy weight vector is $\mu_{\widetilde{W}}(\underline{w}_i(1)) = \mu_{\widetilde{W}}(\overline{w}_i(1)) = s(1) = \sum_i \underline{w}_i(1)$, $\forall i$.

We introduce comparison possibility p and process the given comparisons, $[\underline{a}_{ij}, \overline{a}_{ij}]$, $\forall i, j, i > j$ into $[\underline{a}_{ij}/p, p\overline{a}_{ij}]$, where $\underline{a}_{ij}/p \le p\overline{a}_{ij}$, $\forall i, j, i > j$. By reducing comparison possibility $p \le 1$ when (12) is feasible, the comparisons are narrowed down, and they are widened by increasing comparison possibility $p \ge 1$. There are the other ways to narrow or to widen the range of the interval comparison such as increasing and decreasing the upper and lower bounds, respectively, by $p > 0$, $[\underline{a}_{ij} + p, \overline{a}_{ij} - p]$, like in [4, 5, 19] and by $0 < p < 1$, $[(1 + p)\underline{a}_{ij}, (1 - p)\overline{a}_{ij}]$. A decision-maker compares a pair of items and gives the importance ratio of one over the other, such as "very" and "extremely." Then, for calculation, it is replaced by a numerical number in such a list as $\{1/9, 1/7, 1/5, 1/3, 1, 3, 5, 7, 9\}$ or that with the even numbers between them in addition. The interval comparison matrix often consists of these numbers. When we compare two numbers, such as 3 and 5, 3 is multiplied by 5/3 into 5 and 5 is multiplied by 3/5. This study is based on this inverse relation, and comparison $[\underline{a}_{ij}, \overline{a}_{ij}]$ is narrowed down or widened into $[\underline{a}_{ij}/p, p\overline{a}_{ij}]$ with $p < 1$ or $p > 1$, respectively.

With comparison possibility p the relation ${}_A\mathcal{R}_{\boldsymbol{w}}$ for the lower approximation model is rewritten as follows.

$$\begin{aligned} &\left(\frac{\underline{a}_{ij}}{p}\right) w_j \le w_i, \quad \forall i, j, i > j, \\ &w_i \le \left(p\overline{a}_{ij}\right) w_j, \quad \forall i, j, i > j, \\ &\frac{\underline{a}_{ij}}{p} \le p\overline{a}_{ij}, \quad \forall i, j, i > j, \end{aligned} \tag{13}$$

where $p = 1$ corresponds to (12) and the last inequality is for the lower bound of p: $\sqrt{\underline{a}_{ij}/\overline{a}_{ij}} \leq p$.

The minimum comparison possibility, p_*, is obtained by the following problem.

$$\begin{aligned} p_* = \min \quad & p \\ \text{s.t.} \quad & \underline{a}_{ij} w_j \leq p w_i, \quad \forall i, j, i > j, \\ & w_i \leq p \overline{a}_{ij} w_j, \quad \forall i, j, i > j, \\ & \sqrt{\frac{\underline{a}_{ij}}{\overline{a}_{ij}}} \leq p, \quad \forall i, j, i > j, \\ & \sum_i w_i = 1, \\ & \varepsilon \leq w_i, \quad \forall i, \end{aligned} \tag{14}$$

where the variables are the crisp weights $w_i, \forall i$, and comparison possibility p. The given comparison matrix could be more focused in case of $p_* \leq 1$. Such a processed comparison would be given if we had asked a decision-maker for more precise comparison. On the other hand, the given comparison matrix could be widened in case of $p_* \geq 1$ to represent the whole decision-maker's thinking, such a processed comparison would be given if s/he had been stuck with something and overlooked the others. The processed comparison matrix with p_* barely includes a normalized crisp weight vector.

Nonlinear programming problem (14) is transformed into the following LP problem by logarithm.

$$\begin{aligned} p'_* = \min \quad & p' \\ \text{s.t.} \quad & \underline{a}'_{ij} + w'_j \leq p' + w'_i, \quad \forall i, j, i > j, \\ & w'_i \leq p' + \overline{a}'_{ij} + w'_j, \quad \forall i, j, i > j, \\ & \underline{a}'_{ij} - \overline{a}'_{ij} \leq 2p'_{ij}, \quad \forall i, j, i > j, \\ & w'_1 = 0, \end{aligned} \tag{15}$$

where $p' = \ln p$, $\underline{a}'_{ij} = \ln \underline{a}_{ij}$, $\overline{a}'_{ij} = \ln \overline{a}_{ij}$, $\forall i, j, i > j$ and $w'_i = \ln w_i$, $\forall i$, and the variables are p' and w'_i, $\forall i$. The last constraint of (15) means $w_1 = 1$, which is different from the corresponding constraint of (14), $\sum_i w_i = 1$. The role of this constraint is to fix the weights whose ratios correspond to the given comparisons, as well as to normalize the weights. Hence, it can be replaced into making one of the weights 1 in (15). The optimal objective function value of (15), p'_*, is transformed into that of (14) as $p_* = e^{p'_*}$. There is no need to transform the optimal weights, w'_i, $\forall i$, since the interval weight vector is obtained by (3).

The interval weights with comparison possibility p_* are obtained by (3) as $\boldsymbol{W}(p_*) = ([\underline{w}_i(p_*), \overline{w}_i(p_*)], \forall i)^t$, where

$$\begin{aligned} \underline{w}_i(p_*) &= \min \Big\{ w_i \mid \underline{a}_{ij} w_j \leq p_* w_i, w_i \\ &\leq (p_* \overline{a}_{ij}) w_j, \forall i, j, i > j, w_i \geq \varepsilon, \forall i, \sum_i w_i = 1 \Big\}, \\ &\forall i, \\ \overline{w}_i(p_*) &= \max \Big\{ w_i \mid \underline{a}_{ij} w_j \leq p_* w_i, w_i \\ &\leq (p_* \overline{a}_{ij}) w_j, \forall i, j, i > j, w_i \geq \varepsilon, \forall i, \sum_i w_i = 1 \Big\}, \\ &\forall i. \end{aligned} \tag{16}$$

All the crisp weight vectors included in the processed comparison matrix with comparison possibility p_* are included in interval weight vector $\boldsymbol{W}(p_*)$. It is noted that the interval ratio of two interval weights by (16) is not always included in the given interval comparison. Assume two normalized crisp weight vectors as $\boldsymbol{w}_1 = (w_{1i}, \forall i)^t$ and $\boldsymbol{w}_2 = (w_{2i}, \forall i)^t$, whose interval comparison matrix includes $w_{1i}/w_{1j} \subseteq A_{ij}$ and $w_{2i}/w_{2j} \subseteq A_{ij}$. The closure of two crisp weight vectors is interval weight vector $\boldsymbol{W} = ([\underline{w}_i, \overline{w}_i], \forall i)^t$, where $\underline{w}_i = \min(w_{1i}, w_{2i})$ and $\overline{w}_i = \max(w_{1i}, w_{2i})$. Since interval weight $[\underline{w}_i, \overline{w}_i]$ includes the values between w_{1i} and w_{2i}, interval comparison matrix does not always include interval ratio $[\underline{w}_i/\overline{w}_j, \overline{w}_i/\underline{w}_j] \not\subseteq A_{ij}$,

The membership value of $\boldsymbol{W}(p_*)$ to fuzzy weight vector $\widetilde{\boldsymbol{W}}$ is $\mu_{\widetilde{W}}(\underline{w}_i(p_*)) = \mu_{\widetilde{W}}(\overline{w}_i(p_*)) = s(p_*) = \sum_i \underline{w}_i(p_*) \leq 1$, $\forall i$, where $s(p_*)$ is not close enough to 1, differently from $s(q_*)$ in the upper approximation model. Therefore, the comparison possibility, p, is reconsidered for each comparison in more detail as follows.

$$\begin{aligned} \min \quad & \sum_{ij} p_{ij} \\ \text{s.t.} \quad & \underline{a}_{ij} w_j \leq p_{ij} w_i, \quad \forall i, j, i > j, \\ & w_i \leq p_{ij} \overline{a}_{ij} w_j, \quad \forall i, j, i > j, \\ & p_* \leq p_{ij}, \quad \forall i, j, i > j, \\ & \sum_i w_i = 1, \\ & \varepsilon \leq w_i, \quad \forall i, \end{aligned} \tag{17}$$

where p_{ij}, $\forall i, j, i > j$ and $w_i, \forall i$, are the variables and p_* is the optimal value of (14). This is a nonlinear programming problem and solved by logarithm in the same way as (14).

When each comparison is processed into $[\underline{a}_{ij}/p_{ij*}, p_{ij*}\overline{a}_{ij}]$, the interval weight vector, $\boldsymbol{W}(p_{ij*})$, is obtained by (16) replacing p_* into p_{ij*}, $\forall i, j$. Its membership value to fuzzy weight vector $\widetilde{\boldsymbol{W}}$ is $\mu_{\widetilde{W}}(\underline{w}_i(p_{ij*})) = \mu_{\widetilde{W}}(\underline{w}_i(p_{ij*})) = s(p_{ij*}) \leq 1$, and $s(p_{ij*})$ is close enough to 1.

Then, the membership function of the fuzzy weight vector is estimated by (2) with two α-level sets, interval weight vectors $\boldsymbol{W}(p_{ij*})$ or $\boldsymbol{W}(p_*)$ and $\boldsymbol{W}(1)$ if $p_*, p_{ij*} \leq 1$. In case of $p_* > 1$, (12) is infeasible so that the interval weight vector with the other comparison possibility, $p \geq p_*$, needs to be obtained. The fuzzy weight vector is estimated more precisely by setting several other p. The comparison possibility is introduced to estimate a decision-maker's thinking precisely from the given imperfect comparison matrix. In other words, each comparison is supplemented by narrowing down or widening.

5. Numerical Example

We consider two types of comparison matrices of three items, A^1 and A^2, whose elements are interval and crisp, respectively, as follows.

$$A^1 = \begin{bmatrix} 1 & [1,2] & [2,6] \\ - & 1 & [2,3] \\ - & - & 1 \end{bmatrix}, \quad A^2 = \begin{bmatrix} 1 & 2 & 3 \\ - & 1 & 3 \\ - & - & 1 \end{bmatrix}. \tag{18}$$

The fuzzy weight vectors are obtained from these comparison matrices, which represent a part of the decision-maker's thinkings. Their imperfectness is taken into consideration by weight possibility or comparison possibility in the upper or lower approximation model, respectively.

5.1. Fuzzy Weight Vector by Upper Approximation Model. The minimum weight possibility by (7) is $q_*^1 = 0.07$ and $q_*^2 = 0.029$, respectively. The maximum weight possibility is $q^{1*} = q^{2*} = 1/3$ because $n = 3$. The corresponding interval weight vectors by (8) and (9), where relation ${}_A\mathscr{R}_w$ in (3) indicates that a pseudo weight vector includes the given comparison matrix as in (6), are obtained as follows.

$$\boldsymbol{W}^1(q_*^1) = \begin{bmatrix} 0.488 \\ 0.349 \\ 0.163 \end{bmatrix},$$

$$\boldsymbol{W}^1(q^{1*}) = \begin{bmatrix} [0.088, 0.819] \\ [0.001, 0.653] \\ [0.001, 0.453] \end{bmatrix},$$

$$\mu_{\widetilde{W}^1}(\boldsymbol{W}^1(q_*^1)) = 1,$$

$$\mu_{\widetilde{W}^1}(\boldsymbol{W}^1(q^{1*})) = 0.09,$$

$$\boldsymbol{W}^2(q_*^2) = \begin{bmatrix} 0.543 \\ 0.314 \\ 0.143 \end{bmatrix},$$

$$\boldsymbol{W}^2(q^{2*}) = \begin{bmatrix} [0038, 0.995] \\ [0.001, 0.662] \\ [0.001, 0.518] \end{bmatrix},$$

$$\mu_{\widetilde{W}^2}(\boldsymbol{W}^2(q_*^2)) = 1,$$

$$\mu_{\widetilde{W}^2}(\boldsymbol{W}^2(q^{2*})) = 0.04 \tag{19}$$

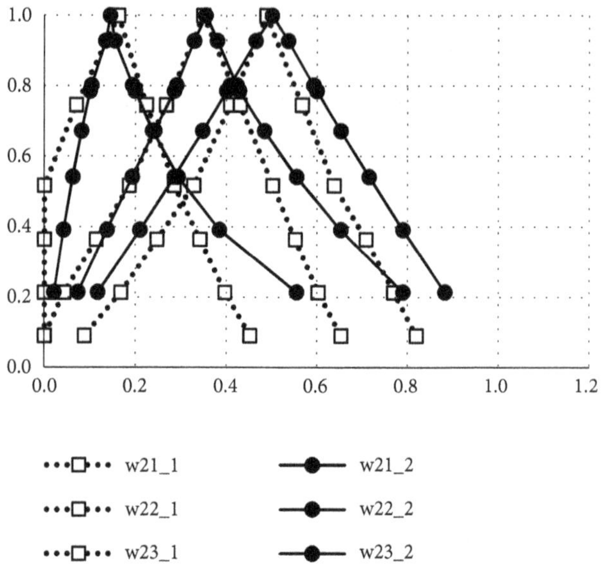

FIGURE 2: Membership functions of fuzzy weight vectors $\widetilde{\boldsymbol{W}}^1$ from A^1.

For instance, $\boldsymbol{W}^1(q_*^1)$ is a closure of 6 kinds of normalized crisp weight vectors. They are obtained by minimizing and maximizing the crisp weight of each item so as to include the comparison matrix with weight possibility q_*^1 as in (8) and are all equal in this example. The membership value of $\boldsymbol{W}^1(q_*^1)$ to fuzzy weight vector $\widetilde{\boldsymbol{W}}^1$ is 1 so that it is α-level set with $\alpha = 1$. In the same way, the interval weight vectors in (19) are considered as α-level sets of the fuzzy weight vectors and their membership values are shown below them. In addition to $q_*^{\cdot}$ and $q^{\cdot *}$, we set several other weight possibilities q between them and obtain the corresponding interval weights by (8) replacing q_* into q. Then, the membership functions of fuzzy weight vectors $\widetilde{\boldsymbol{W}}^1$ and $\widetilde{\boldsymbol{W}}^2$ by the upper approximation models are obtained by (2). They are illustrated as the dotted lines in Figures 2 and 3, respectively. It is noted that very small lower bounds make approximation relation ${}_A\mathscr{R}_w$ by (6) meaningless.

5.2. Fuzzy Weight Vector by Lower Approximation Model. In the lower approximation model, relation ${}_A\mathscr{R}_w$ in (3) indicates that a processed comparison matrix includes a crisp weight vector as in (13). The minimum comparison possibility by (14) is $p_1^* = 0.816$ and $p_*^2 = 1.259$. There is a crisp weight vector included in comparison matrix A^1 because $p_*^1 < 1$. While comparison matrix A^2 needs to be widened because

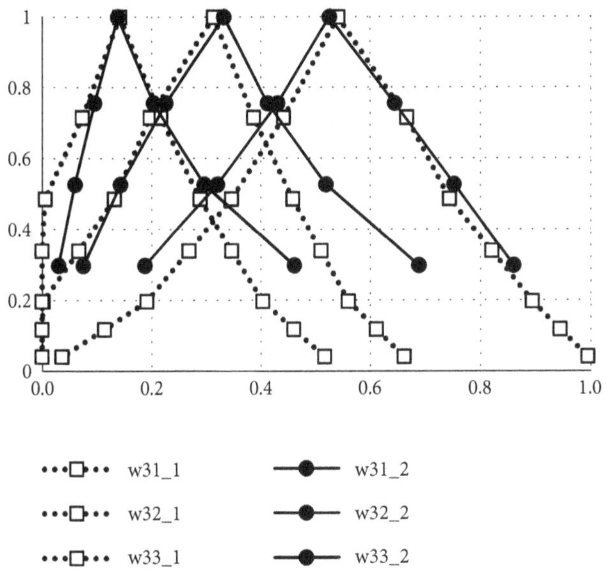

FIGURE 3: Membership functions of fuzzy weight vectors $\widetilde{W}^2$ from A^2.

$p_*^2 > 1$, such that the given comparison, $a_{ij}^2 = 2$, is processed into [1.498, 2.519], therefore, (12) is infeasible with A^2 and feasible with A^1. Furthermore, the minimum comparison possibility for each comparison by (17) is $p_{12*}^1 = 1.414$, $p_{13*}^1 = 1.732$, $p_{23*}^1 = 1.225$, and $p_{12*}^2 = p_{13*}^2 = p_{23*}^2 = 0.794$. The representative interval weight vectors by (16) with $p = p_*^1, p_{ij*}^1, 1$ for A^1 and with $p = p_*^2, p_{ij*}^2$ for A^2 are as follows.

$$W^1(p_*^1) = \begin{bmatrix} [0.465, 0.537] \\ [0.329, 0.380] \\ [0.134, 0.155] \end{bmatrix},$$

$$W^1(p_{ij*}^1) = \begin{bmatrix} 0.501 \\ 0.354 \\ 0.145 \end{bmatrix},$$

$$W^1(1) = \begin{bmatrix} [0.348, 0.652] \\ [0.242, 0.484] \\ [0.082, 0.238] \end{bmatrix},$$

$$\mu_{\widetilde{W}^1}(W^1(p_*^1)) = 0.928,$$

$$\mu_{\widetilde{W}^1}(W^1(p_{ij*}^1)) = 1,$$

$$\mu_{\widetilde{W}^1}(W^1(1)) = 0.672,$$

$$W^2(p_*^2) = \begin{bmatrix} 0.528 \\ 0.333 \\ 0.140 \end{bmatrix},$$

$$W^2(p_{ij*}^2) = \begin{bmatrix} 0.528 \\ 0.333 \\ 0.140 \end{bmatrix},$$

$$\mu_{\widetilde{W}^2}(W^2(p_*^2)) = 1,$$

$$\mu_{\widetilde{W}^2}(W^2(p_{ij*}^2)) = 1. \qquad (20)$$

As for interval comparison matrix A^1, the processed comparison matrix with $p_{ij*}^1, \forall i, j, i > j$ is more focused than that with p_*^1 so that the sure degree of the former weight vector is more than that of the latter one. As for crisp comparison matrix A^2, the other interval weight vectors by setting p are needed to estimate the membership function of the fuzzy weight vector. For both comparison matrices, we set several other comparison possibilities $p \geq p_*$ in approximation relation ${}_A\mathscr{R}_w$ by (13), and from each processed comparison matrix the interval weight vector is obtained by (16) replacing p_* into p. Then, the membership functions fuzzy weight vectors $\widetilde{W}^1$ and $\widetilde{W}^2$ by the lower approximation model are obtained by (2) and illustrated as the bold lines in Figures 2 and 3, respectively.

5.3. Comparing Upper and Lower Approximations. The dotted and bold lines are similar in both figures since they are two kinds of the approximations of the same comparison matrix, A^1 or A^2, respectively. The imperfectness of the given comparison matrix can be well handled by both weight possibility and comparison possibility. We find that the bold lines are wider than the dotted lines. It indicates that the lower approximation model with comparison possibility tends to derive a vague fuzzy weight vector. We can tell neither whether the comparison matrix is given by chance or by well-consideration nor whether a decision-maker is an expert or not. Therefore, unknown uncertainty in the comparison matrix is considered by weight or comparison possibility, and the correspondingly fuzzy weight vector is obtained. The decrease of weight or comparison possibility increases the membership value of the weight vector. When the membership value is more than 0.8, we can find the linear order of three items of both comparison matrices by both models. The interval weight vectors at high membership value could be the results by removing inconsistency of the given comparison matrix. On the contrary, the weight vectors at the lower membership value could be the results by considering the imperfectness more positively.

6. Conclusion

The approaches to obtain a fuzzy weight vector from an interval/crisp comparison matrix have been proposed. We estimated the membership function of a fuzzy vector from more than one interval weight vector. The membership value of the interval weight vector is the sum of the lower bounds of the interval weights, which represents its sure degree. We redefined the interval weight vector as a closure of the crisp weight vectors. The crisp weight vector could approximate the comparison matrix in various ways since the comparison matrix does not represent the whole decision-maker's thinking. We proposed upper and lower approximation models and introduced weight possibility and comparison possibility, respectively. In the upper approximation model, a crisp weight vector with weight possibility should include the comparison matrix. In the lower approximation model, the comparison matrix with comparison possibility should include a crisp weight vector. As the weight or comparison possibility increases, the incompleteness of the comparison matrix is taken into consideration more positively. We have applied the proposed approaches to the crisp and interval comparison matrices. They could be applied to a fuzzy comparison matrix by focusing on its α-level set since it is an interval/crisp comparison matrix. However, it remains an open question how to accumulate the fuzzy weight vectors at various α-level sets.

Conflicts of Interest

The author declares that they have no conflicts of interest.

References

[1] T. L. Saaty, *The Analytic Hierarchy Process*, McGraw-Hill, New York, NY, USA, 1980.

[2] J. Grobelny, "Fuzzy-based linguistic patterns as a tool for the flexible assessment of a priority vector obtained by pairwise comparisons," *Fuzzy Sets and Systems*, vol. 296, pp. 1–20, 2016.

[3] A. Arbel, "Approximate articulation of preference and priority derivation," *European Journal of Operational Research*, vol. 43, no. 3, pp. 317–326, 1989.

[4] A. Arbel and L. Vargas, "Interval judgments and Euclidean centers," *European Journal of Operational Research*, vol. 46, no. 7-8, pp. 976–984, 2007.

[5] L. Mikhailov, "A fuzzy approach to deriving priorities from interval pairwise comparison judgements," *European Journal of Operational Research*, vol. 159, no. 3, pp. 687–704, 2004.

[6] J. F. Chen, H. N. Hsieh, and Q. H. Do, "Evaluating teaching performance based on fuzzy AHP and comprehensive evaluation approach," *Applied Soft Computing*, vol. 28, pp. 100–108, 2015.

[7] Z. Güngör, G. Serhadlıoğlu, and S. E. Kesen, "A fuzzy AHP approach to personnel selection problem," *Applied Soft Computing*, vol. 9, no. 2, pp. 641–646, 2009.

[8] K. Kamvysi, K. Gotzamani, A. Andronikidis, and A. C. Georgiou, "Capturing and prioritizing students' requirements for course design by embedding Fuzzy-AHP and linear programming in QFD," *European Journal of Operational Research*, vol. 237, no. 3, pp. 1083–1094, 2014.

[9] S. Kubler, J. Robert, W. Derigent, A. Voisin, and Y. Le Traon, "A state-of the-art survey & testbed of fuzzy AHP (FAHP) applications," *Expert Systems with Applications*, vol. 65, pp. 398–422, 2016.

[10] T. L. Saaty and L. G. Vargas, "Uncertainty and rank order in the analytic hierarchy process," *European Journal of Operational Research*, vol. 32, no. 1, pp. 107–117, 1987.

[11] Y. M. Wang, J. B. Yang, and D. L. Xu, "A two-stage logarithmic goal programming method for generating weights from interval comparison matrices," *Fuzzy Sets and Systems*, vol. 152, no. 3, pp. 475–498, 2005.

[12] A. A. Salo and R. P. Hämäläinen, "Preference programming through approximate ratio comparisons," *European Journal of Operational Research*, vol. 82, no. 3, pp. 458–475, 1995.

[13] C. Kahraman, S. C. Onar, and B. Oztaysi, "Fuzzy multicriteria decision-making: a literature review," *International Journal of Computational Intelligence Systems*, vol. 8, no. 4, pp. 637–666, 2015.

[14] S. Ohnishi, D. Dubois, and T. Yamanoi, "Uncertainty and Intelligent Information Systems," in *chap. A Fuzzy Constraint-Based Approach to the Analytic Hierarchy Process*, pp. 217–228, World Scientific, Singapore, 2008.

[15] J. Ramík and P. Korviny, "Inconsistency of pair-wise comparison matrix with fuzzy elements based on geometric mean," *Fuzzy Sets and Systems*, vol. 161, no. 11, pp. 1604–1613, 2010.

[16] P. J. M. van Laarhoven and W. Pedrycz, "A fuzzy extension of Saaty's priority theory," *Fuzzy Sets and Systems*, vol. 11, no. 3, pp. 229–241, 1983.

[17] I. Chamodrakas, D. Batis, and D. Martakos, "Supplier selection in electronic marketplaces using satisficing and fuzzy AHP," *Expert Systems with Applications*, vol. 37, no. 1, pp. 490–498, 2010.

[18] O. Kilincci and S. A. Onal, "Fuzzy AHP approach for supplier selection in a washing machine company," *Expert Systems with Applications*, vol. 38, no. 8, pp. 9656–9664, 2011.

[19] K. Zhu, Y. Jing, and D. Chang, "Discussion on extent analysis method and applications of fuzzy AHP," *European Journal of Operational Research*, vol. 116, no. 2, pp. 450–456, 1999.

[20] T. Entani and M. Inuiguchi, "Maximum lower bound estimation of fuzzy priority weights from a crisp comparison matrix," in *Proceedings of the 4th Intentional Symposium on Integrated Uncertainty in Knowledge Modelling and Decision Making*, pp. 65–76, 2015.

[21] K. Sugihara and H. Tanaka, "Interval evaluations in the Analytic Hierarchy Process by possibility analysis," *Computational Intelligence*, vol. 17, no. 3, pp. 567–579, 2001.

[22] T. Entani, "Estimating fuzzy weight vector from interval pairwise comparison matrix with various processed matrices," in *Proceedings of the 2017 IEEE International Conference on Fuzzy Systems*, pp. 1–6, 2017.

[23] T. Entani and M. Inuiguchi, "Pairwise comparison based interval analysis for group decision aiding with multiple criteria," *Fuzzy Sets and Systems*, vol. 274, pp. 79–96, 2015.

[24] H. Tanaka, K. Sugihara, and Y. Maeda, "Non-additive measures by interval probability functions," *Information Sciences*, vol. 164, no. 1-4, pp. 209–227, 2004.

15

A Fuzzy Simulation Model for Military Vehicle Mobility Assessment

Aby K. George,[1] Harpreet Singh,[1] Macam S. Dattathreya,[2] and Thomas J. Meitzler[2]

[1] *Department of Electrical and Computer Engineering, Wayne State University, Detroit, MI 48202, USA*
[2] *Tank Automotive Research, Development and Engineering Center, Warren, MI 48397, USA*

Correspondence should be addressed to Harpreet Singh; hsingh@eng.wayne.edu

Academic Editor: Fanbiao Li

There has been increasing interest in improving the mobility of ground vehicles. The interest is greater in predicting the mobility for military vehicles. In this paper, authors review various definitions of mobility. Based on this review, a new definition of mobility called fuzzy mobility is given. An algorithm for fuzzy mobility assessment is described with the help of fuzzy rules. The simulation is carried out and its implementation, testing, and validation strategies are discussed.

1. Introduction

Recently there has been an increasing interest in various aspects of combat vehicles. Kempinski and Murphy studied the technical challenges of ground combat vehicles in [1]. Dattathreya and Singh discussed energy management strategies of combat vehicles in [2, 3]. Several authors have shown interest in predicting the mobility of military vehicles [4–6]. Mobility in general can be defined as the ability to move or to be moved freely and easily. Mobility in case of a ground combat vehicle is defined as, "vehicle's capability to move over a specified terrain, which is influenced by other environmental conditions such as weather" [4]. The basic function of a military combat vehicle is the transportation of the soldiers and weapons. According to a recent research on the cost/benefit analysis for the military combat vehicle, a 10% weightage is given for mobility by the ground combat vehicle analysis of alternatives [1]. Other main attributes obtained from the analysis are total life cost, lethality, survivability, and so forth, as given in [1] and shown in Figure 1.

Mobility has a different definition when viewed from a military vehicle's perspective. Such a definition in [7] for the mobility of military vehicle is "the ability to move freely and rapidly over the terrain of interest to accomplish varied combat objectives." From this definition, mobility is the freedom of movement in diverse terrains under different environmental conditions. Freedom of movement can be defined as good speed, less vibration, and so forth. A report by Unger discusses several aspects of mobility in [8]. Mobility assessment for the military vehicle is carried out in diverse terrains with different environmental conditions. It is necessary to test the mobility of a vehicles before using it on the field. The conventional methods for defining mobility revolve around pure mathematical modeling. Different mathematical models are available in the literature. Studies by Engineering Research and Development Center (ERDC) treat mobility as a function of trafficability [9]. A modified mathematical model based on trafficability studies for wheeled vehicles was also proposed in [10, 11]. A stochastic approach for predicting mobility is developed by González et al. [12]. Mobility model for ground vehicles based on soil-moisture can be found in [13]. An extended Kalman filter based mobility estimation for unmanned ground vehicles is presented in [14]. The tire-soil interaction simulation based on absolute nodal coordinate formulation (ANCF) is developed by Recuero et al. in [15]. A physics-based simulation model is discussed in [16]. This deals with light tracked vehicles, weigh less than 100 lb, operating on deformable terrains. These complex mathematical models focus on the effect of individual input attributes on the mobility. With the improvements in technology, newer

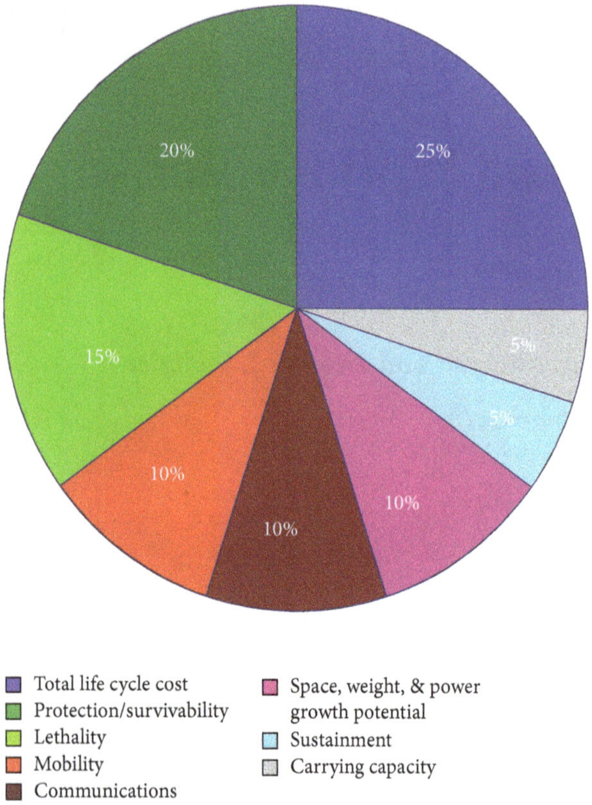

FIGURE 1: Ground combat vehicle (GCV) criteria and weighting from cost/benefit analysis in the Army's analysis of alternatives.

compact vehicles are being developed, and a generic mobility model which can be applicable to all these vehicles is still a challenge.

The military vehicle has to travel in different terrains under different environmental conditions. Thus, mobility is an important attribute for the combat vehicle. The mobility is subjective in nature, and, hence, an accurate measure of mobility is difficult to define. The terrain information which is one of the major attribute in predicting the mobility, is highly associated with uncertainty. It is difficult to define them mathematically and modify them later with the technological advances in the field of military vehicles and sensors. The best way to incorporate such uncertainties is by using linguistic variables. These linguistic variables can be used in the rules. These are developed from the expert knowledge. For example, if the vehicle is traveling in off-road, with slower speed, with higher vibration, then the mobility will be less.

In this paper, the authors are giving a brief review of the existing mathematical models for calculating mobility first. The mobility is then determined from a set of linguistic fuzzy rules. Two of the authors of this article gave an introductory idea of fuzzy logic in mobility [4]. The authors used the parameters such as speed and tire pressure as the input variables. These attributes are depending only on the vehicle. The mobility of a vehicle depends on the vehicle parameters as well as the environmental parameters and terrain of travel. In this paper, we are considering practical and implementable parameters such as type of vehicle, terrain information, speed, and weight of the vehicle as input parameters. Each of these attributes is defined using a set of membership functions and a rule set which relates these attributes to the mobility. Another advantage of using fuzzy based model is that the models can incorporate new findings by simply modifying the fuzzy rules. A procedure for fabricating an integrated chip from the fuzzy model is also discussed in this paper. This can be integrated in the military vehicle.

The organization of the paper is as follows. The attributes to be considered for calculating the mobility of a combat vehicle are discussed in Section 2. Section 3 gives a brief review of the existing mathematical methods for computing mobility. Section 4 gives a detailed explanation of the new fuzzy model proposed for calculating the mobility of a military combat vehicle. The procedure for implementing this fuzzy mobility model in a real-time application is also discussed in this section. The simulation results for hypothetical inputs and an overview of simulation based testing are given in Section 5. Section 6 gives the concluding remarks.

2. Main Attributes of Mobility

The attributes used to determine mobility in a normal vehicle may not be applicable to a military combat vehicle. Normal vehicles are commonly used on roads where they can move fast. Therefore, weight and speed are not prime concerns for them. In case of a military combat vehicle, the terrain of operation is entirely different. The mobility of the military combat vehicle is primarily based on the terrain. For instance, in military operations and peace keeping missions the time frame will not be short. They have to be there for a long time and have to move through narrow bridges, ruined roads, tunnels, and so forth. In such a situation the terrain, weight, and speed will be the key contributors in determining the mobility of the vehicle. It can be noticed that the attributes for mobility in different terrains will be different. We can address this problem by classifying the mobility into on-road and off-road mobility in general [1].

2.1. On-Road Mobility. The on-road mobility primarily depends on the type of the vehicle used. The achievement of required speed will be easy in the case of a wheeled vehicle. While focusing on military combat vehicles, a majority of them are tracked vehicles. The tracked vehicles are particularly designed for off-road operations. It is not always the case; they may have to travel on-road as well. Since the tracked vehicles are primarily designed for off-road operations, the rigorous operation of these kinds of vehicles for on-road operations may cause problems [17].

In on-road operation the wheeled vehicles give better mobility. For wheeled vehicles the dash speed is a function of horsepower and weight. Consequently, mobility of the wheeled vehicle increases with increasing horsepower and decreases with increasing the vehicle weight. Another important factor to be considered for wheeled vehicle is the friction between the tire and the road. This will vary with the environmental conditions such as snowy and rainy. Under these circumstances the mobility of the wheeled vehicles will decrease, and, thus, a good choice will be the tracked vehicles.

2.2. Off-Road Mobility. For military operations, off-road mobility is essential. A military vehicle is said to have good mobility if it is having good off-road as well as on-road mobility. The military vehicles are constrained to travel through off-road in many operations. This may be to avoid the danger of IEDs planted on raod, or sometimes there is no proper road at all. The statistics shows that the casualty is severe by IEDs compared to other kinds of attacks [18]. Thus, a military vehicle has to travel off-road intensively.

Compared to on-road mobility, off-road mobility is very complex and depends upon a number of factors. The measurability of these parameters is also challenging. The weight of the vehicle is taken as an attribute for off-road mobility, but not in a direct form. Here, the resistance of the surface is playing a vital role. This resistance is related to the type of surface and the ground pressure. The ground pressure is calculated as the ratio of gross weight of the vehicle to the surface area of ground contact with the vehicle. Another term used in literature to calculate off-road mobility is vehicle cone index (VCI) [7]. VCI is a function of soil strength and vehicle ground pressure. Priddy and Willoughby [19] define VCI as "the minimum soil strength necessary for a self-propelled vehicle to consistently make a prescribed number of passes in track without becoming immobilized."

The ground pressure and VCI are having an inverse relationship with mobility. Another important factor to be considered for computing the mobility of an off-road combat vehicle is the freedom of movement of the vehicle in that particular terrain. The freedom of movement depends on the minimum acceptable value of ground pressure or threshold ground pressure. This value varies for different combat vehicles. Another terminology given in literature is percentage no-go terrain [7]. It is a measure of immobile terrain and can be directly related to the ground pressure [20].

The tracked vehicles are preferred for off-road movement over wheeled vehicle, as the tracked vehicles provide greater surface area and, consequently, lesser ground pressure [21]. For better off-road mobility, there should be a higher horse power to weight ratio, low VCI, low ground pressure, and advanced suspension system for the vehicle.

3. Review of Mobility Models

There are different models available in the literature for predicting the mobility of a combat vehicle. Some of these are used as simulation models while others are mathematical models with complex equations. Some of the popular models available in the literature are discussed in this section.

3.1. Simulation Models. The simulation model is useful to understand the performance of the vehicle in different terrains and other conditions. It is always a good idea to test the mobility in the simulation model with expected attributes for a particular kind of military vehicle. The simulation models differ from each other on the attributes they are using for computing the mobility.

One of the most popular simulation models for analyzing mobility is NATO Reference Mobility Model (NRMM) [22]. There are three modules associated with NRMM: (1) vehicle dynamic module, (2) obstacle crossing performance module, and (3) primary prediction module. This model can predict the mobility of a combat vehicle for both on-road and off-road operations. The mobility is predicted as the effective maximum speed by analyzing different attributes. The disadvantage of NRMM is its limited range of operation. This model will not fit in complex terrains. Lessem et al. proposed NRMM adaptation to stochastic orientation [5], so that the model can be used for high resolution combat zones. In [15], a mobility simulation model based on tire flexibility and deformation of terrains is presented. This simulation model is limited to wheeled vehicles. A simulation model which focuses on small autonomous vehicles is discussed in [16]. This simulation deals with the obstacles in the path of vehicles and its impact on the mobility of the light autonomous vehicle and can simulate different terrains such as flat rigid terrain and deformable terrain.

The testing of simulation models can be performed by developing a set of virtual operating conditions. Such a framework is called Virtual Evaluation Test (VET) framework [6]. Evaluation suites like VET can be used for many simulation models to study mobility, stability, durability, and so forth. These frameworks evaluate the existing simulation models and provide a report on the progress of the model.

3.2. Mathematical Models. The Engineer Research and Development Center (ERDC) defines vehicle cone index in two different ways. One is one-pass vehicle cone index (VCI_1) and other is fifty-pass vehicle cone index (VCI_{50}) [9]. These values are calculated from the vehicle attributes such as weight and dimensions, by conducting multipass experiment, and are expressed in PSI [19]. As discussed earlier, the soil strength is an important parameter for calculating mobility. The International Society for Terrain Vehicle Systems (ISTVS) defines VCI as the "minimum soil strength in the critical soil layer, in terms of rating cone index for fine grained soils or in cone index for coarse grained soils, required for a specific number of passes of a vehicle, usually one pass (VCI_1) or 50 passes (VCI_{50})" [23]. VCI is a common parameter for both on-road and off-road analysis; hence, it can be used as a common parameter for defining mobility of all ground combat vehicles.

Other important factors used in the calculation of mobility are the mobility index (MI) and deflection correction factor (DCF) [19]. Combat vehicle mobility in soft soil terrain is defined by a parameter called "mean maximum pressure (MMP)." MMP was developed by UK MOD's Defence Science and Technology Laboratory (DSTL) [19]. According to this model, the mobility is calculated from the ground contact pressure. MMP is calculated by taking the average of magnitudes of maximum pressure at each wheel. Therefore, MMP is related to the dimensions of the wheel and also to the weight of the vehicle.

Modifications on original MMP calculations based on terrains and different sensors are discussed in [24–27]. An extensive review of mathematical modeling of mobility is best given by Priddy and Willoughby in [19]. Mobility index (MI) for wheeled vehicle is given in [19] as

$$\mathrm{MI} = \left(\frac{(\mathrm{CPF})(\mathrm{WF})}{(\mathrm{TEF})(\mathrm{GF})} + \mathrm{WLF} - \mathrm{CF}\right)(\mathrm{EF})(\mathrm{TF}), \quad (1)$$

where CPF is the contact pressure factor, WF is the weight factor, TEF is the traction element factor, GF is the grouser factor, WLF is the wheel load factor, CF is the clearance factor, EF is the engine factor, and TF is the transmission factor.

Now CPF can be calculated as follows:

$$\mathrm{CPF} = \frac{w}{0.5ndb}, \quad (2)$$

where n is the average axle loading in lb, n is the average number of tires per axle, d is the average tire outside diameter (inflated; unloaded) in in., and b is the average tire section width (inflated; unloaded) in in. Similarly, other parameters are given by

$$\mathrm{TEF} = \frac{10 + b}{100},$$
$$\mathrm{WLF} = \frac{w}{2000}, \quad (3)$$
$$\mathrm{CF} = \frac{h_c}{10},$$

where h_c is the vehicle minimum clearance height in in.,

$$\mathrm{GF} = 1 + 0.05c_{\mathrm{GF}}, \quad (4)$$

where $c_{\mathrm{GF}} = 1$, if tire chains are used or 0 if not.

$$\mathrm{EF} = 1 + 0.05c_{\mathrm{EF}}, \quad (5)$$

where $c_{\mathrm{EF}} = 1$, if PWR < 10 hp/ton or 0 if not.

$$\mathrm{TF} = 1 + 0.05c_{\mathrm{TF}}, \quad (6)$$

where $c_{\mathrm{TF}} = 1$, if manual transmission or 0 if automatic.

$$\mathrm{WF} = c_{\mathrm{WF1}}\frac{w}{1000} + c_{\mathrm{WF2}}, \quad (7)$$

where $w < 2000$ lb $\Rightarrow c_{\mathrm{WF1}} = 0.553$ and $c_{\mathrm{WF2}} = 0$; $2000 \leq w < 13{,}500$ lb $\Rightarrow c_{\mathrm{WF1}} = 0.033$ and $c_{\mathrm{WF2}} = 1.050$; $13{,}500 \leq w < 20{,}000$ lb $\Rightarrow c_{\mathrm{WF1}} = 0.142$ and $c_{\mathrm{WF2}} = -0.420$; $20{,}000 \leq w < 31{,}500$ lb $\Rightarrow c_{\mathrm{WF1}} = 0.278$ and $c_{\mathrm{WF2}} = -3.115$; $31{,}500 \leq w \Rightarrow c_{\mathrm{WF1}} = 0.836$ and $c_{\mathrm{WF2}} = -20.686$.

A deflection correction factor (DCF) is required to account the effect of tire deflection on VCI performance [19].

$$\mathrm{DCF} = \left(\frac{0.15}{\delta/h}\right)^{0.25}, \quad (8)$$

where δ is the average hard-surface tire deflection expressed in in. and h is the average tire section height (inflated; unloaded) in in.

The analysis and test data of different vehicles under different environments were studied. Based on the past 50 years' data, the researchers came up with an expression in which the VCI_1 is a function of MI and DCF [19].

$$\mathrm{MI} \leq 115 \Longrightarrow$$
$$\mathrm{VCI}_1 = \left(11.48 + 0.2\mathrm{MI} - \frac{39.2}{\mathrm{MI} + 3.74}\right)\mathrm{DCF}, \quad (9)$$
$$\mathrm{MI} > 115 \Longrightarrow$$
$$\mathrm{VCI}_1 = \left(4.1\mathrm{MI}^{0.446}\right)\mathrm{DCF}.$$

A stochastic model which relates the geometry of the surface and soil type to the mobility map was proposed by González et al. in [12]. The mobility map produced in this model shows the surface elevation at each location. In [13], the effect of soil-moisture on the off-road mobility is studied based on satellite soil-moisture data. It is found that the type of vehicle, the environmental conditions, and so forth will significantly vary the soil-moisture and, thus, the off-road mobility. Another approach for mobility estimation of unmanned ground vehicle, which uses a Gauss-Markov state space dynamic model and a first-order semi-Markov model along with an extended Kalman Filter, is discussed in [14]. This approach can give a real-time path planning for the unmanned ground vehicle.

4. Proposed Method

Mobility models explained in Section 3 consist of very complex mathematical equations. The modification of these models is very difficult and cumbersome. Mobility is subjective and depends on the user comfortableness. A fuzzy based model gives the freedom for the designer to improve the model without much effort by taking feedback from the user.

4.1. Fuzzy Logic Outline. Fuzzy logic is a technique which uses the degree of truth instead of discrete values such as 0 or 1. Fuzzy logic also uses "linguistic" variables such as *low, medium,* and *high,* along with numerical variables for the calculations. The relationship between inputs and outputs is given by some simple statements rather than complex mathematical equations [28]. Fuzzy based systems are widely used in many real-time applications [29].

Let A be a fuzzy subset. We can represent A as

$$A = \frac{\mu}{y}, \quad (10)$$

where μ is the degree of membership of y in fuzzy subset A. If A is having more number of membership functions associated with it then A can be represented as

$$A = \sum_{i=1}^{N}\frac{\mu_i}{y_i}, \quad (11)$$

where μ_i is the degree of membership of corresponding y_i in A. Here, the + sign indicates the union of different

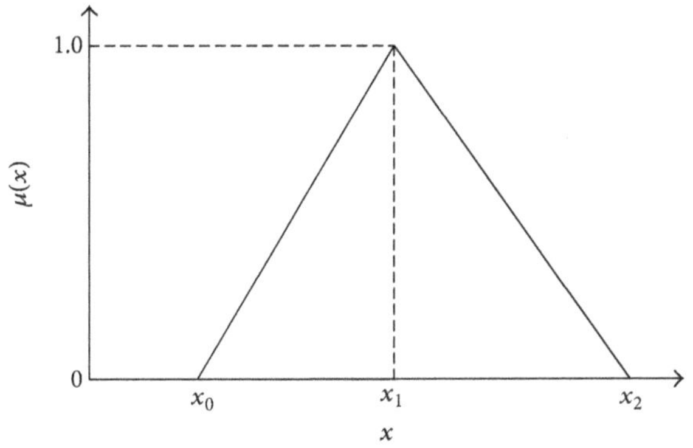

FIGURE 2: Triangular membership function.

memberships. The membership degree is having a value between 0 and 1. For a fuzzy subset A,

$$\begin{aligned} \mu(x) &= 0 \quad \text{if } x \notin A, \\ \mu(x) &= 1 \quad \text{if } x \in A, \\ \mu(x) &\in (0,1) \quad \text{if } x \text{ possibly in } A, \text{ but not sure.} \end{aligned} \tag{12}$$

There are different types of membership functions available in Fuzzy. Triangular function and Gaussian distribution function and so forth are some of the commonly used membership functions. A triangular membership function is shown in Figure 2. The degree of the membership function $\mu(x)$ is given by

$$\mu = \begin{cases} 0, & \text{if } x \le x_0, \\ \dfrac{x - x_0}{x_1 - x_0}, & \text{if } x_0 \le x \le x_1, \\ \dfrac{x_2 - x}{x_2 - x_1}, & \text{if } x_1 \le x \le x_2, \\ 0, & \text{if } x_2 \le x. \end{cases} \tag{13}$$

4.2. Fuzzy Mobility Model. The block diagram of the proposed fuzzy mobility assessment model is shown in Figure 3. The proposed scheme consists of three parts. The first is the input section. The inputs can be analog signals directly from the sensors such as speed and weight of the vehicle in the proposed method. The other inputs are vehicle and the terrain types. The data is either preset or entered manually by the operator.

The second part of the proposed method is the fuzzy logic controller. The fuzzy logic controller consists of four different sections. The first section is a fuzzifier. The fuzzifier converts the numerical value obtained from the input signal into fuzzy sets. That is, the input variables are represented in terms of their degree of membership for different membership functions. The second section is a rule base. The rule base contains all the rules for the fuzzy model. The rules use if-then structure to relate the input fuzzy sets with the output fuzzy sets. The third section is an inference engine. The inference engine takes the input fuzzy sets and decides which rules should be applicable from the rule base and creates the output fuzzy set. The last section is a defuzzifier which takes the fuzzy set output created by the inference engine and produce the corresponding output.

TABLE 1: Linguistic parameters for the weight variable.

Weight in lb.	Linguistic parameter
10,000–30,000	Low
30,000–50,000	Medium
50,000–70,000	High

TABLE 2: Linguistic parameters for the speed variable.

Speed in mph	Linguistic parameter
0–20	Low
20–40	Medium
40–60	High

TABLE 3: Input and output variables for fuzzy model.

Input variables				Output variable
Terrain	Vehicle type	Weight	Speed	Mobility
Dry	Wheel	Low	Low	Low
Wet	Track	Medium	Medium	MediumLow
Snow		High	High	Medium
Sand				MediumHigh
				High

The fuzzifier consists of membership functions. The membership functions are expressed using linguistic parameters. Important attributes considered for calculating mobility of the military vehicle are type of vehicle, terrain information, speed of the vehicle, and the weight of the vehicle. The weight and speed of the vehicle are numerical values. They are converted to linguistic parameters such as *low, medium,* and *high*. In the proposed example, the weight of a vehicle varies between 10,000 and 70,000 lb. This range is divided into three sections. The linguistic parameters for the weight are shown in the Table 1.

Linguistic parameters for the speed are defined in Table 2. In this case the normal range is from 0 to 60 mph. Likewise, the membership functions for the mobility of the military vehicle are also expressed by 5 different linguistic parameters. The input and output linguistic variables used for the proposed fuzzy model can be summarized as shown in Table 3. Different types of membership functions are available in fuzzifier. A trigonometric membership function is used for the proposed model. The arrangement of membership functions for the speed of military vehicle and normalized mobility output is shown in Figures 4 and 5, respectively. The fuzzifier converts the numerical value of input to fuzzy set which is a set of membership function and corresponding degree of membership. For example, if the speed is 26 mph, then, the fuzzifier will represent it by *low* with a degree of 0.2

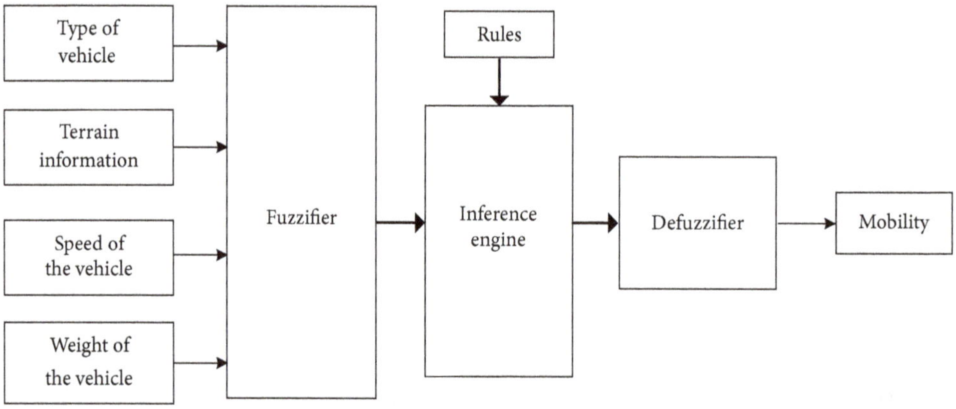

FIGURE 3: Block diagram of the proposed mobility assessment method.

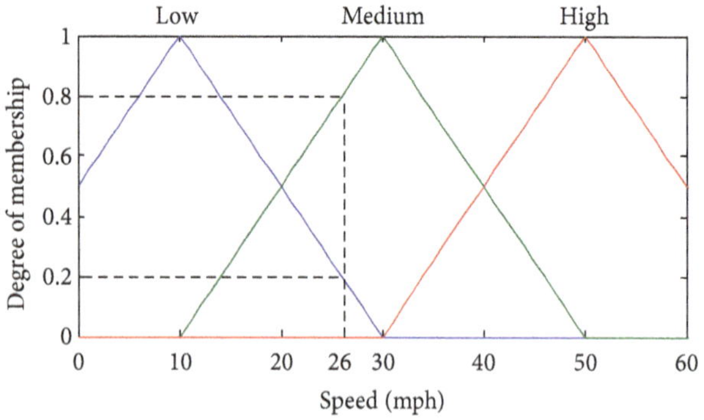

FIGURE 4: Speed input variable partitions using triangular functions.

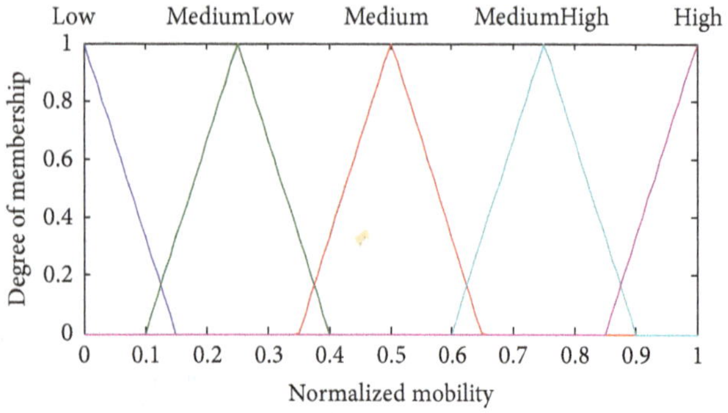

FIGURE 5: Military vehicle mobility variable partitions using triangular functions.

and *medium* with a degree of 0.8 and *high* with a degree of zero.

After studying the characteristics of the military vehicle mobility defined by different mathematical models, it is possible to define a set of approximate rules for the mobility assessment. The linguistic parameters are used in deriving appropriate rules for the proposed fuzzy logic model. These rules are made up of simple logic employing IF, AND, OR, and NOT operators. The mobility output is categorized as *Low, MediumLow, Medium, MediumHigh,* and *High*. In other words, the input and output parameters of the proposed models are fuzzy sets. The rules used to calculate the mobility are simple sentences with some logical expressions. The list of all the rules used in the fuzzy logic model is listed in Table 4. Here, each input membership function is connected with AND statement. Such as the following:

(i) If terrain is *dry* and vehicle type is *wheel* and weight is *high* and speed is *low* then mobility is *low*.

(ii) If terrain is *dry* and vehicle type is *wheel* and weight is *low* and speed is *high* then mobility is *high*.

For each input combination, the mobility output is defined using these rules. The inference engine selects the rules that are applicable for the input values and produce the corresponding output fuzzy set. The fuzzy set is then converted to a crisp value by a defuzzifier. The surface diagram of the mobility output with respect to different set of inputs is given in Figure 6. The simulation of such a fuzzy model can be tested using Matlab simulation. The military vehicle mobility output obtained in such a simulation will also be in a normalized form.

After the simulation, the fuzzy rules can be converted to a Verilog code. This is simple and straightforward. The Verilog program can be written in a behavioral level with *if* statements. The linguistic parameters used in the fuzzy rules are converted to digital numbers while writing the Verilog code. For instance, the speed input variable linguistic parameters *low, medium,* and *high* can be converted to 00, 01, and 10, respectively.

After completing the Verilog code, it should be tested with some tools. The simulation and hardware testing for the code are possible. The simulation of the Verilog code can be done using Cadence NClaunch or some other software which can do the simulation. For hardware testing, the Verilog code can be implemented on an FPGA. If both the simulation and hardware results are satisfactory, the next step is to design the fuzzy chip using this Verilog code.

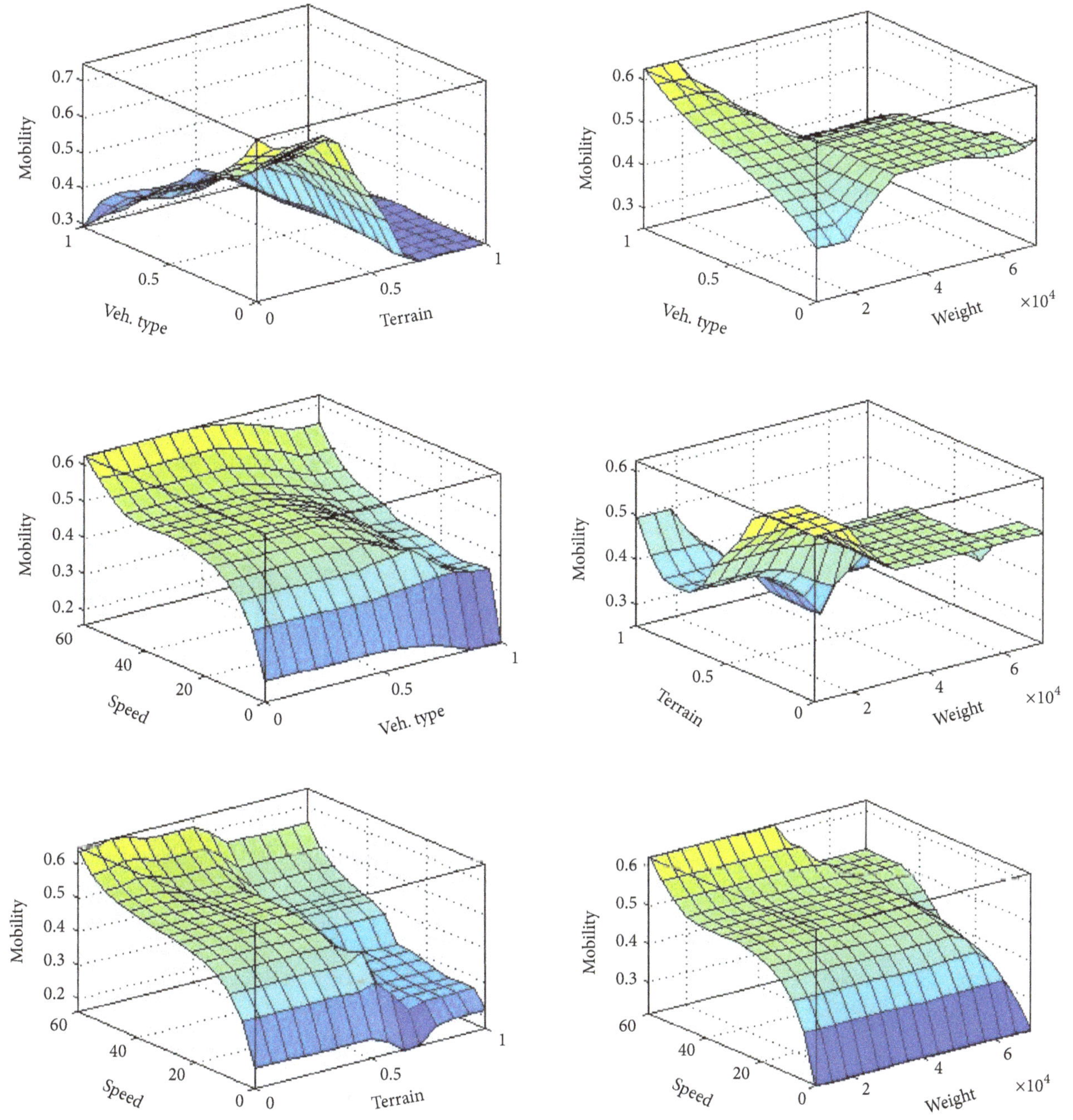

FIGURE 6: Surface plot obtained for different set of inputs (vehicle type-terrain, vehicle type-weight, vehicle type-speed, terrain-weight, terrain-speed, and weight-speed) versus mobility.

For building the chip from the Verilog code, the first step is to convert the behavioral level code to a netlist or gate level code. This can be done by Cadence synthesizer. After completing the gate level code, a layout for the circuit can be produced using Cadence Encounter. The Encounter uses some standard cell libraries to complete the layout. Once the layout is ready, the next step is to do the padding for the chip. A Cadence Virtuoso software is used for this purpose. After completing the padding the design can be sent to the manufacturer to fabricate the final chip. The process flow chart for the proposed scheme is shown in Figure 7.

Once the chip is ready to use, it can be installed in the military vehicle and the mobility output can be displayed to the driver in a range of 0 to 100%. The inputs for the chip can be from a sensor and the output can be displayed inside the vehicle which is shown in Figure 8 [30]. For a

Table 4: Fuzzy logic rule set.

Terrain	Veh. type	Weight	Speed	Mobility
Dry	Wheel	Low	Low	Low
Dry	Wheel	Low	Medium	MediumHigh
Dry	Wheel	Low	High	High
Dry	Wheel	Medium	Low	MediumLow
Dry	Wheel	Medium	Medium	MediumHigh
Dry	Wheel	Medium	High	High
Dry	Wheel	High	Low	Low
Dry	Wheel	High	Medium	MediumHigh
Dry	Wheel	High	High	High
Dry	Track	Low	Low	MediumLow
Dry	Track	Low	Medium	Medium
Dry	Track	Low	High	MediumHigh
Dry	Track	Medium	Low	Low
Dry	Track	Medium	Medium	MediumLow
Dry	Track	Medium	High	Medium
Dry	Track	High	Low	Low
Dry	Track	High	Medium	MediumLow
Dry	Track	High	High	MediumLow
Wet	Wheel	Low	Low	Low
Wet	Wheel	Low	Medium	Medium
Wet	Wheel	Low	High	MediumHigh
Wet	Wheel	Medium	Low	MediumLow
Wet	Wheel	Medium	Medium	MediumHigh
Wet	Wheel	Medium	High	MediumHigh
Wet	Wheel	High	Low	Low
Wet	Wheel	High	Medium	Medium
Wet	Wheel	High	High	MediumHigh
Wet	Track	Low	Low	MediumLow
Wet	Track	Low	Medium	MediumHigh
Wet	Track	Low	High	MediumHigh
Wet	Track	Medium	Low	Low
Wet	Track	Medium	Medium	MediumLow
Wet	Track	Medium	High	Medium
Wet	Track	High	Low	Low
Wet	Track	High	Medium	MediumLow
Wet	Track	High	High	Medium
Snow	Wheel	Low	Low	Low
Snow	Wheel	Low	Medium	MediumLow
Snow	Wheel	Low	High	Medium
Snow	Wheel	Medium	Low	Low
Snow	Wheel	Medium	Medium	MediumLow
Snow	Wheel	Medium	High	Medium
Snow	Wheel	High	Low	MediumLow
Snow	Wheel	High	Medium	Medium
Snow	Wheel	High	High	MediumHigh
Snow	Track	Low	Low	MediumLow
Snow	Track	Low	Medium	Medium
Snow	Track	Low	High	MediumHigh
Snow	Track	Medium	Low	Low
Snow	Track	Medium	Medium	MediumLow
Snow	Track	Medium	High	Medium

Table 4: Continued.

Terrain	Veh. type	Weight	Speed	Mobility
Snow	Track	High	Low	Low
Snow	Track	High	Medium	MediumLow
Snow	Track	High	High	MediumLow
Sand	Wheel	Low	Low	MediumLow
Sand	Wheel	Low	Medium	Medium
Sand	Wheel	Low	High	MediumHigh
Sand	Wheel	Medium	Low	MediumLow
Sand	Wheel	Medium	Medium	MediumLow
Sand	Wheel	Medium	High	Medium
Sand	Wheel	High	Low	Low
Sand	Wheel	High	Medium	MediumLow
Sand	Wheel	High	High	Medium
Sand	Track	Low	Low	MediumLow
Sand	Track	Low	Medium	Medium
Sand	Track	Low	High	MediumHigh
Sand	Track	Medium	Low	Low
Sand	Track	Medium	Medium	MediumLow
Sand	Track	Medium	High	Medium
Sand	Track	High	Low	Low
Sand	Track	High	Medium	MediumLow
Sand	Track	High	High	MediumLow

Table 5: Simulation results of fuzzy mobility model.

Terrain	Veh. type	Weight	Speed	Mobility
Snow	Track	30000	20	0.3617
Sand	Track	30000	40	0.4782
Dry	Wheel	50000	50	0.8089
Wet	Wheel	70000	20	0.4077
Dry	Wheel	20000	60	0.9139
Sand	Wheel	50000	40	0.3568
Dry	Track	25000	45	0.5469

given operational scenario, the parameters of the vehicle are constant. However, the speed, terrain, and weight may vary depending on a mission. The fuzzy electronic chip can estimate instantaneous mobility of the military vehicle and display it on the vehicle.

5. Testing of the Fuzzy Based Method

The simulation of the proposed fuzzy mobility model in Matlab is carried out with different set of input combinations. The results are given in Table 5. Furthermore, the simulation of the proposed method is done by setting the weight of the vehicle to 30,000 lb and varying the speed from 0 to 50 mph. The mobility of the vehicle is plotted under different terrains with track and wheel vehicle. The graph is shown in Figure 9.

The validation of the results can be done in different ways. One way of testing is by comparing the simulation model with user data obtained from similar real life experience. The proposed method can be verified by taking the feedback from vehicle users. Their feedbacks can be used to optimize the model by comparing the user data with the outputs of the fuzzy model for the same set of inputs.

In the second method, the human factor is introduced in predicting the mobility of a vehicle. The block diagram of the simulation model is given in Figure 10. This diagram shows the flow of mobility prediction from the visual information of a vehicle mobility in a computer simulation. The different attributes that decide the mobility are given as input to the computer model and the computer model gives simulation display, where the given type of vehicle moves through a particular terrain with specific weight and speed.

5.1. Overview of Simulation Based Testing. The cost for making hardware is massive. To improve the accuracy of the hardware, more emphasis should be given to modeling and simulation. For the simulation based testing, Autonomie software can be used [31]. This software has a large database of vehicles, and these data can be used to make the simulation more effective. Autonomy helps to simulate the model with plug-and-play functionalities. The models built in standard format can be added to the autonomy software for simulation. The library and tested models available with the software package make the simulation more accurate.

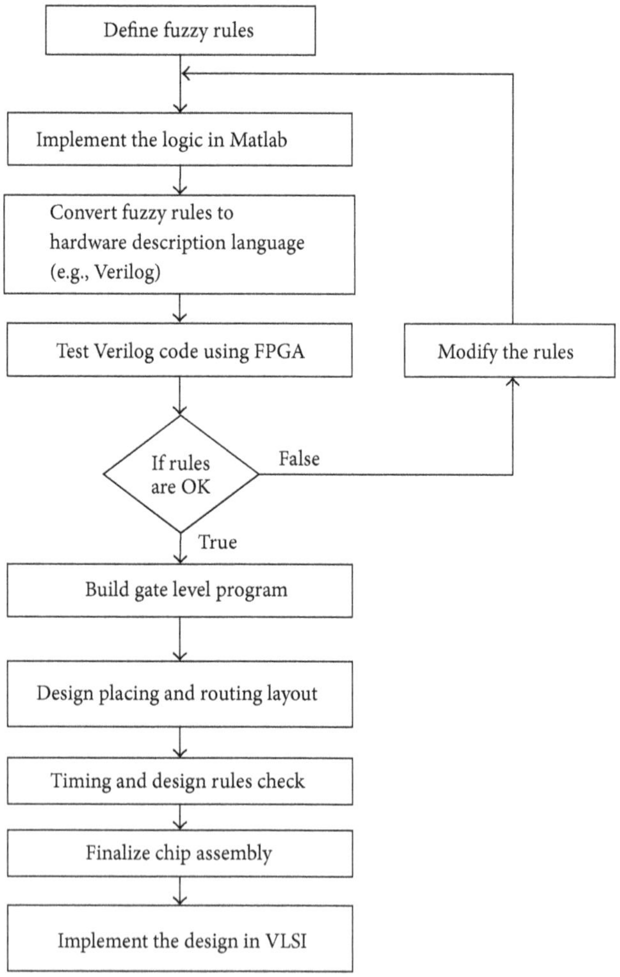

FIGURE 7: Flowchart showing the design of proposed fuzzy model chip.

The algorithm for testing the the fuzzy model can be summarized as follows:

(i) The Simulink software can be used to create the simulation of the vehicle by using the proposed fuzzy model. This simulation model can be converted to S function which can be used in the Autonomie software.

(ii) Some signal formatting is necessary while integrating the simulation S function with the Autonomie software. Some of the vehicle parameters are already available in the Autonomie software and the *s*-function can use these input variables.

(iii) Interface the modified *s*-function with the Autonomie software. The Autonomie software has the ability to identify the *s*-function parameters. Normally the input and output variables used with the *s*-function may not be documented in the standard format. The Autonomie converts these into a standard format and ensures compatibility of input output variables with the rest of the vehicle. Instead of *s*-function it is also possible to use the vehicle simulation programs such as TruckSim.

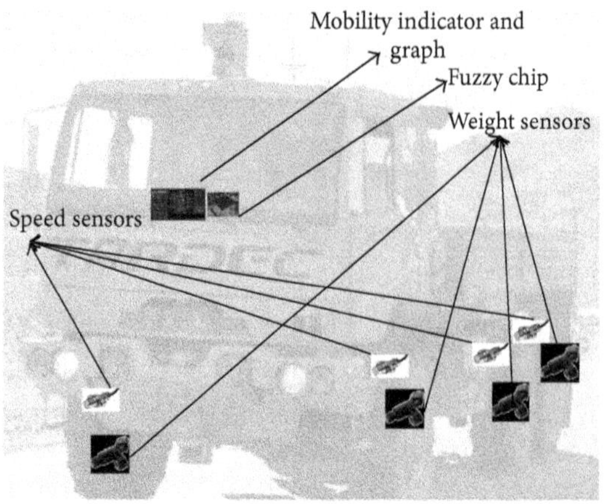

FIGURE 8: Diagram showing the position of sensors and displays in a military vehicle.

(iv) Run the simulation with a set of input data. For example, the performance can be tested with a tracked vehicle in a snow terrain with a medium weight and medium speed. The simulation is repeated with different set of inputs.

(v) For each simulation, let the vehicle users view the simulation and take their comments on vehicle mobility. The mobility can be described in linguistic parameters such as *Low, MediumLow, Medium, MediumHigh,* and *High*.

(vi) From the set of inputs getting from the vehicle users, compute a correlation between the mobility data provided by the vehicle user with the data calculated by the fuzzy chip.

(vii) Change the rules of the fuzzy model if required and repeat the experiment to achieve optimum results.

6. Conclusion

In this paper, the existing mobility models of military vehicles are reviewed. These methods use complex mathematical equations and, hence, are difficult to adapt. A fuzzy based mobility assessment model is developed here. Different attributes which affect the mobility such as terrain, vehicle type, speed, and weigh of the vehicle were taken as inputs. Mobility is defined in a range from 0 to 100%. The fuzzy mobility assessment model is simulated with different input combinations. The implementation of the proposed model in an integrated chip is discussed. The algorithm for testing the proposed fuzzy model is presented. The fuzzy based mobility model gives the designer the freedom to optimize the model without much effort by taking feedback from the user. The paper essentially describes an approach which can possibly be used for developing a chip to access the mobility of military

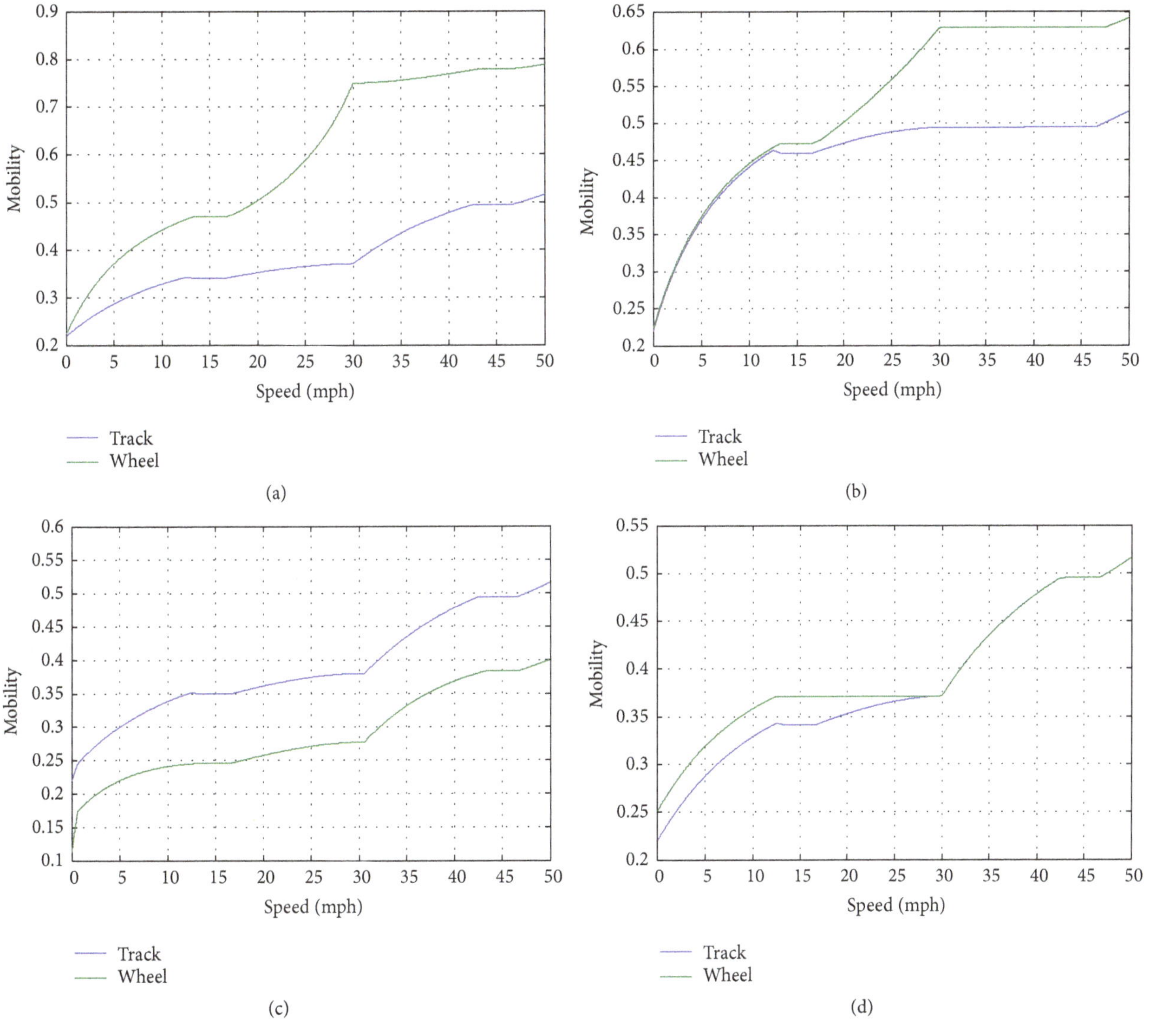

FIGURE 9: Speed versus mobility graph for track and wheel vehicles with weight = 30,000 lb in different terrains: (a) dry, (b) wet, (c) snow, and (d) sand.

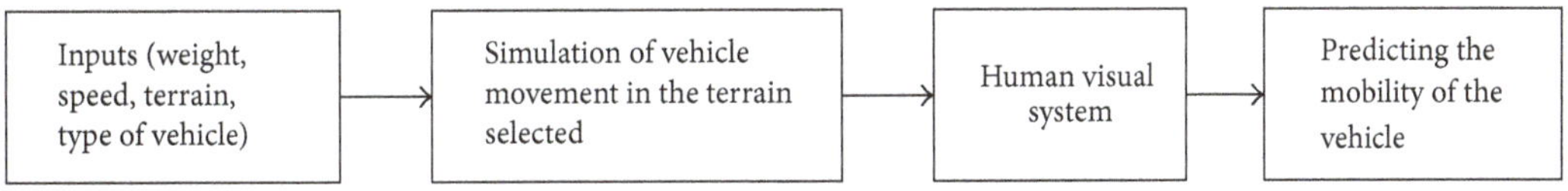

FIGURE 10: Block diagram showing the testing of the proposed method.

vehicle. More work is needed to modify the model so as to implement a real life chip which would be used ultimately in the military vehicle.

Disclosure

Unclassified: Distribution Statement A. Approved for public release; distribution is unlimited.

Conflicts of Interest

The authors declare that there are no conflicts of interest regarding the publication of this article.

Acknowledgments

This work is supported by TARDEC Grant.

References

[1] B. Kempinski and C. Murphy, "Technical challenges of the us armys ground combat vehicle program," Tech. Rep., 2012.

[2] M. S. Dattathreya and H. Singh, "Mission aware energy efficiency in stationary combat vehicles," *IEEE Transactions on Aerospace and Electronic Systems*, vol. 50, no. 2, pp. 1108–1117, 2014.

[3] M. S. Dattathreya and H. Singh, "Silent-watch and energy management strategy in combat vehicles," *IEEE Transactions on Aerospace and Electronic Systems*, vol. 50, no. 1, pp. 418–428, 2014.

[4] M. S. Dattathreya and H. Singh, "A novel approach for combat vehicle mobility definition and assessment," *SAE Technical Papers*, 2012.

[5] A. Lessem, G. Mason, and R. Ahlvin, "Stochastic vehicle mobility forecasts using the NATO reference mobility model," *Journal of Terramechanics*, vol. 33, no. 6, pp. 273–280, 1996.

[6] P. Nuñez, A. Reid, R. Jones, and S. Shoop, "A virtual evaluation suite for military ground vehicle dynamic performance and mobility," *SAE Technical Papers*, 2002.

[7] P. Hornback, "The wheel versus track dilemma," *Armor Magazine*, vol. 26, pp. 33-34, 1998.

[8] R. F. Unger, "Mobility analysis for the TRADOC wheeled versus tracked vehicle study," in *US Army Engineer Waterways Experiment Station*, 1988.

[9] "Trafficability of soils, vehicle classification," Tech. Rep., Ninth Supplement, US Army Engineer Waterways Experiment Station, Vicksburg, May 1951.

[10] J. S. Kennedy and E. S. Rush, "Trafficability of soils, development of revised mobility index formula for self-propelled wheeled vehicles in fine-grained soils," Tech. Rep., Eighteenth Supplement. US Army Engineer Waterways Experiment, Station, Vicksburg, M. S, March 1968.

[11] B. G. Schreiner, "Comments on the VCI system, unpublished memorandum for records," *US Army Engineer Waterways Experiment*, 1994.

[12] R. González, P. Jayakumar, and K. Iagnemma, "Stochastic mobility prediction of ground vehicles over large spatial regions: a geostatistical approach," *Autonomous Robots*, vol. 41, no. 2, pp. 311–331, 2017.

[13] M. T. Stevens, G. B. McKinley, and F. Vahedifard, "A comparison of ground vehicle mobility analysis based on soil moisture time series datasets from WindSat, LIS, and in situ sensors," *Journal of Terramechanics*, vol. 65, pp. 49–59, 2016.

[14] P. Thulasiraman, G. A. Clark, and T. M. Beach, "Mobility estimation using an extended Kalman filter for unmanned ground vehicle networks," in *Proceedings of the 2014 IEEE International Inter-Disciplinary Conference on Cognitive Methods in Situation Awareness and Decision Support, CogSIMA 2014*, pp. 223–229, San Antonio, TX, USA, March 2014.

[15] A. Recuero, R. Serban, B. Peterson, H. Sugiyama, P. Jayakumar, and D. Negrut, "A high-fidelity approach for vehicle mobility simulation: Nonlinear finite element tires operating on granular material," *Journal of Terramechanics*, vol. 72, pp. 39–54, 2017.

[16] D. Negrut, D. Melanz, H. Mazhar, D. Lamb, P. Jayakumar, and M. Letherwood, "Investigating Through Simulation the Mobility of Light Tracked Vehicles Operating on Discrete Granular Terrain," *SAE International Journal of Passenger Cars - Mechanical Systems*, vol. 6, no. 1, pp. 369–381, 2013.

[17] *Family of medium tactical vehicles (fmtv)*, 2015, http://www.dote.osd.mil/pub/reports/FY1999/army/99fmtv.html.

[18] "How the IED changed the US military," USA Today, 2015, http://www.usatoday.com/story/news/nation/2013/12/18/ied-10-years-blast-wounds-amputations/.

[19] J. D. Priddy and W. E. Willoughby, "Clarification of vehicle cone index with reference to mean maximum pressure," *Journal of Terramechanics*, vol. 43, no. 2, pp. 85–96, 2006.

[20] J. Wong and W. Huang, ""wheels vs. tracks"—a fundamental evaluation from the traction perspective," *Journal of Terramechanics*, vol. 43, no. 1, pp. 27–42, 2006.

[21] Wheeled versus track vehicle study, final report, Studies and Analysis Activity, Headquarters U.S. Army Training and Doctrine Command.

[22] P. W. Haley, M. Jurkat, and P. Brady, "Nato reference mobility model, edition 1 users guide , US Army Tank-Automotive Research and Development Command," *Techn. Rep*, vol. no. 12503, 1979.

[23] M. P. Meyer, I. R. Ehrlich, D. Sloss, N. Murphy, R. D. Wismer, and T. Czako, "International society for terrain-vehicle systems standards," *Journal of Terramechanics*, vol. 14, no. 3, pp. 153–182, 1977.

[24] J. C. Larminie, "Modifications to the mean maximum pressure system," *Journal of Terramechanics*, vol. 29, no. 2, pp. 239–255, 1992.

[25] D. Freitag, "A dimensional analysis of the performance of pneumatic tires on clay," *Journal of Terramechanics*, vol. 3, no. 3, pp. 51–68, 1966.

[26] E. Maclaurin, "Proposed revisions to mmp based on the results of tractive performance trials with single pneumatic tyres and a modular track system," *DERA/LS4/TR970122/1.0. Defence Evaluation and Research Agency*, 1997.

[27] D. Rowland, "Tracked vehicle ground pressure and its effect on soft ground performance," in *Proceedings of the Proc. 4th Int. Conf. of Int. Soc*, 1972.

[28] L. A. Zadeh, "Outline of a New Approach to the Analysis of Complex Systems and Decision Processes," *IEEE Transactions on Systems, Man and Cybernetics*, vol. 3, no. 1, pp. 28–44, 1973.

[29] H. Singh, M. M. Gupta, T. Meitzler et al., "Real-life applications of fuzzy logic," *Advances in Fuzzy Systems*, vol. 2013, Article ID 581879, 3 pages, 2013.

[30] "8th americas regional conference detroit," in *Proceedings of the USA - 12-14 september*, pp. 12–14, 2016, http://conference.istvs.org/detroit-military/57ulcdctmnjaxge0be6fkan9xrzsot.

[31] R. Vijayagopal, L. Michaels, A. P. Rousseau, S. Halbach, and N. Shidore, "Automated model based design process to evaluate advanced component technologies," *SAE Technical Papers*, 2010.

16

Common Fixed Points of Intuitionistic Fuzzy Maps for Meir-Keeler Type Contractions

Shazia Kanwal [1] **and Akbar Azam** [2]

[1] *Department of Mathematics, GC University, Faisalabad-38000, Pakistan*
[2] *Department of Mathematics, COMSATS University, Islamabad-44000, Pakistan*

Correspondence should be addressed to Shazia Kanwal; shaziakanwal690@yahoo.com

Academic Editor: Jose A. Sanz

The main purpose of this paper is to establish and prove some new common fixed point theorems for intuitionistic fuzzy maps in the context of (α, β)-cut sets of intuitionistic fuzzy sets on a complete metric space in association with the Hausdorff metric. Furthermore, the technique of Meir-Keeler (shortly, M-K) contraction is applied to obtain common fixed point of intuitionistic fuzzy compatible maps and fixed points of Kannan type intuitionistic fuzzy set-valued contractive mappings. Our results generalize M-K type fixed point theorem along with its various generalizations. Some nontrivial examples have been furnished in the support of the main results.

1. Introduction

The theory of fixed points has been revealed as a very powerful and important tool in the study of nonlinear phenomena. In particular, fixed point techniques have been applied in such diverse fields as biology, chemistry, economics, engineering, game theory, and physics. The Banach fixed point theorem [1] (also known as a contraction mapping principle) is an important tool in nonlinear analysis. It guarantees the existence and uniqueness of fixed points of self-mappings on complete metric spaces and provides a constructive method to find fixed points. Many extensions of this principle have been done up to now. In 1976, Jungck [2] studied coincidence and common fixed points of commuting mappings and improved the Banach contraction principle. In 1986, Jungck [3] introduced the notion of compatible maps for a pair of self-mappings and existence of common fixed points. In 1969, Meir and Keeler [4] obtained a valuable fixed point theorem for single valued mappings $\Phi : X \longrightarrow X$ that satisfies the following condition:

Given $\epsilon > 0$, there exists a $\delta > 0$ such that

$$\epsilon \leq d(x, y) < \epsilon + \delta \text{ implies } d(\Phi x, \Phi y) < \epsilon. \tag{1}$$

In 1981, Park and Bae [5] extended it to a pair of commuting single valued mappings. A variety of extensions, generalizations, and applications of this followed; e.g., see [6, 7]. In 1993, Beg and Shahzad [8] derived and proved random fixed point of two random multivalued operators satisfying the M-K [4] condition in Polish spaces. In 2001, Lim [9] wrote on characterization of M-K-contractive maps. In 2012, Abdeljawad et al.'s paper [10] contains a study of M-K type coupled fixed point on ordered partial metric spaces and Chen et al. [11] established and proved common fixed point theorems for the stronger M-K cone-type function in cone ball-metric spaces. In 2013, Karapinar et al. [12] studied the existence and uniqueness of a fixed point of the multidimensional operators in partially ordered metric space which satisfied M-K type contraction condition and improved the results mentioned above and the recent results on these topics in the literature. In the same year, Abdeljawad [13] established and proved M-K-contractive fixed point and common fixed point theorems. Patel et al. [14] formulated and proved a more generalized version of [13]. In 2014, Singh et al. [15] derived a new common fixed point theorem for Suzuki-M-K contractions. In 2015, Redjel et al. [16] proved fixed point theorems for (α, ψ)-Meir-Keeler-Khan mappings.

Abtahi [17, 18] established and proved fixed point theorems in 2016 and common fixed point theorems in 2017 for M-K type contractions in metric spaces. Popa and Patriciu [19] derived and proved a general theorem of M-K type for mappings satisfying an implicit relation in partial metric spaces in 2017.

Fuzzy sets were introduced by Zadeh [20] in 1965 to represent/manipulate data and information possessing nonstatistical uncertainties. In 1986, the concept of an intuitionistic fuzzy set (IFS) was put forward by Atanassov [21], which can be viewed as an extension of fuzzy set. Intuitionistic fuzzy sets not only define the degree of membership of an element, but also characterize the degree of nonmembership. IFS has much attention due to its significance to remove the vagueness or uncertainty in decision-making. IFS is a tool in modeling real life problems such as psychological investigation and career determination. Abbasizadeh and Davvaz [22] introduced intuitionistic fuzzy topological polygroups. In 2017, Azam et al. [23] proved coincidence and fixed point theorems of intuitionistic fuzzy mappings with applications. In the same year, Azam and Tabassum [24] established and proved fixed point theorems of intuitionistic fuzzy mappings in quasi-pseudometric spaces. Recently, Kumam et al. [25] and Shoaib et al. [26] derived and proved some fuzzy fixed point results for fuzzy mappings in complete b-metric spaces. Humaira et al. [27] established and proved fuzzy fixed point results for $\Phi-$contractive mapping and gave some applications. Ertrk and Karakaya [28] stated and proved n-tuplet coincidence point theorems in intuitionistic fuzzy normed spaces. Xia Li et al. [29] worked on the intuitionistic fuzzy metric spaces and the intuitionistic fuzzy normed spaces (see also [30–32]).

In this paper, the main focus is to establish the existence of fixed point and common fixed point theorems of M-K type contraction for intuitionistic fuzzy set-valued maps in complete metric spaces. Some nontrivial examples have been furnished in the support of the main results.

2. Preliminaries

We start this section by recalling some pertinent concepts.

Definition 1 (see [33]). Let (X, d) be a metric space. The set of all nonempty closed and bounded subsets of X is denoted by $CB(X)$. The function H defined on $CB(X) \times CB(X)$ by

$$H(A, B) = \max\left(\sup_{a\in A} d(a, B), \sup_{b\in B} d(A, b)\right) \quad (2)$$

for all $A, B \in CB(X)$ is a metric on $CB(X)$ called the Hausdorff metric of d,

where

$$d(x, A) = \inf_{y\in A} d(x, y). \quad (3)$$

Definition 2 (see [20]). Let X be an arbitrary nonempty set. A fuzzy set in X is a function with domain X and values in $[0, 1]$. If A is a fuzzy set and $x \in X$, then the function-value $A(x)$ is called the grade of membership of x in A. I^X stands for the collection of all fuzzy sets in X unless and until stated otherwise.

Definition 3 (see [21]). Let X be a nonempty set. An intuitionistic fuzzy set is defined as

$$A = \{\langle x, \mu_A(x), \nu_A(x)\rangle : x \in X\}, \quad (4)$$

where $\mu_A : X \longrightarrow [0, 1]$ and $\nu_A : X \longrightarrow [0, 1]$ denote the degree of membership and degree of nonmembership of each element x to the set A, respectively, such that

$$0 \le \mu_A(x) + \nu_A(x) \le 1, \quad \forall x, y \in X. \quad (5)$$

The collection of all intuitionistic fuzzy sets in X is denoted by $(IFS)^X$.

Definition 4 (see [34]). Let A be an intuitionistic fuzzy set and $x \in X$; then $\alpha-$level set of an intuitionistic fuzzy set A is denoted by $[A]_\alpha$ and is defined as

$$[A]_\alpha = \{x \in X : \mu_A(x) \ge \alpha, \ \nu_A(x) \le 1-\alpha\}, \quad \text{if } \alpha \in (0, 1]. \quad (6)$$

A generalized version of $\alpha-$level set of an intuitionistic fuzzy set A was investigated in [35, 36].

Definition 5 (see [35, 36]). Let $L = \{(\alpha, \beta) : \alpha+\beta \le 1, \ (\alpha, \beta) \in (0, 1] \times [0, 1)\}$ and let A be an IFS on X; then $(\alpha, \beta)-$cut set of A is defined as

$$A_{(\alpha,\beta)} = \{x \in X : \mu_A(x) \ge \alpha, \ \nu_A(x) \le \beta\}. \quad (7)$$

Definition 6 (see [23]). Let X be an arbitrary set and let Y be a metric space. A mapping $S : X \longrightarrow (IFS)^Y$ is called an intuitionistic fuzzy mapping.

Definition 7. Mappings $\Phi : X \longrightarrow (IFS)^X$ and $\psi : X \longrightarrow X$ are said to be (α, β) compatible if whenever there is a sequence $\{x_n\} \subseteq X$ satisfying $\lim_{n\longrightarrow\infty}\psi x_n \in \lim_{n\longrightarrow\infty}[\Phi x_n]_{(\alpha,\beta)}$ (provided $\lim_{n\longrightarrow\infty}\psi x_n$ and $\lim_{n\longrightarrow\infty}[\Phi x_n]_{(\alpha,\beta)}$ exist and $\psi[\Phi x_n]_{(\alpha,\beta)} \in CB(X)$), then $\lim_{n\longrightarrow\infty}H(\psi[\Phi x_n]_{(\alpha,\beta)}, [\Phi\psi x_n]_{(\alpha,\beta)}) = 0$.

Lemma 8 (see [37]). *Let $\{\Upsilon_n\}$ be a sequence in $CB(X)$ and $H(\Upsilon_n, \Upsilon) \longrightarrow 0$ for $\Upsilon \in CB(X)$. If $x_n \in \Upsilon_n$ and $\lim_{n\longrightarrow\infty}d(x_n, x) = 0$, then $x \in \Upsilon$.*

3. Main Results

Theorem 9. *Let X be a complete metric space and let $\Phi : X \longrightarrow (IFS)^X$, $\psi : X \longrightarrow X$ be (α, β) compatible mappings. Suppose for each $x \in X$ there exists $(\alpha, \beta) \in (0, 1] \times [0, 1)$*

such that $[\Phi x]_{(\alpha,\beta)} \in CB(X)$ *and* $\cup_{x\in X}[\Phi x]_{(\alpha,\beta)} \subseteq \psi X$ *and the following condition is satisfied:*

$$\text{for } \epsilon > 0 \text{ there exists a } \delta > 0 \text{ such that} \tag{8}$$

$$\epsilon \le d(\psi x, \psi y) < \epsilon + \delta \text{ implies } d(u,v) < \epsilon, \tag{9}$$

$$\begin{gathered} u \in [\Phi x]_{(\alpha,\beta)}, \\ v \in [\Phi y]_{(\alpha,\beta)}, \\ [\Phi x]_{(\alpha,\beta)} = [\Phi y]_{(\alpha,\beta)} \quad \text{when } \psi x = \psi y. \end{gathered} \tag{10}$$

If ψ *is continuous, then* Φ *and* ψ *have a common fixed point.*

Proof. Let $x_0 \in X$, and consider the following sequences x_n and y_n in X and $\Upsilon_n \in CB(X)$, $y_n = \psi x_n \in [\Phi x_{n-1}]_{(\alpha,\beta)}$, $n \ge 0$ (which is possible due to the hypothesis $\cup_{x\in X}[\Phi x]_{(\alpha,\beta)} \subseteq \psi X$). Then for each $\epsilon > 0$ there exists $\delta > 0$ such that $\epsilon \le (\psi x_m, \psi x_n) < \epsilon + \delta$ implies $d(\psi x_{m+1}, \psi x_{n+1}) < \epsilon$. It follows that $d(y_n, y_{n+1}) < d(y_{n-1}, y_n)$. Thus, the sequence $\{d(y_n, y_{n+1})\}$ is nonincreasing and converges to the greatest lower bound of its range, which we denote by l.

Now $l \ge 0$; in fact, $l = 0$. Otherwise, if $l > 0$, take N so that $n \ge N$ implies $l \le d(y_n, y_{n+1}) < l + \delta$. It implies that $d(y_{n+1}, y_{n+2}) < l$ which contradicts the fact that $l = \inf_n d(y_{n+1}, y_{n+2})$. Hence $d(\psi x_n, [\Phi x_n]_{(\alpha,\beta)}) \le d(\psi x_n, \psi x_{n+1}) \longrightarrow 0$. Now it is to prove that $\{y_n\}$ is a Cauchy sequence. Suppose that $d(y_n, y_{n+1}) = 0$ for some $n > 0$. Then $d(y_m, y_{m+1}) = 0$ for all $m > n$; otherwise $d(y_n, y_{n+1}) = 0 < d(y_{n+1}, y_{n+2})$, a contradiction. Hence, $\{y_n\}$ is a Cauchy sequence.

Now assume that $d(y_n, y_{n+1}) \neq 0$ for each n. Define $\zeta = 2\epsilon$ and choose (without loss of generality) δ, $0 < \delta < \epsilon$, such that (9) is satisfied. Since $d(y_n, y_{n+1}) \longrightarrow 0$, there exists an integer N such that $(y_i, y_{i+1}) < \delta/6$ for $i \ge N$. We now let $q > p > N$ and show that $d(y_p, y_q) \le \zeta$, to prove that $\{y_n\}$ is indeed Cauchy. Suppose that

$$d(y_p, y_q) \ge 2\epsilon = \zeta. \tag{11}$$

First, we show that there exists an integer $m > p$ such that

$$\epsilon + \frac{\delta}{3} < d(y_p, y_m) < \epsilon + \delta, \tag{12}$$

where p and m are of opposite parity. Let k be the smallest integer greater than p such that

$$d(y_p, y_k) > \epsilon + \frac{\delta}{2} \tag{13}$$

(which is possible due to (11) as $\delta < \epsilon$). Moreover,

$$d(y_p, y_k) < \epsilon + \frac{2\delta}{3}. \tag{14}$$

For otherwise,

$$\epsilon + \frac{2\delta}{3} \le d(y_p, y_{k-1}) + d(y_{k-1}, y_k). \tag{15}$$

Since $k - 1 \ge p \ge N$, therefore $d(y_{k-1}, y_k) < \delta/6$. It implies that

$$d(y_p, y_{k-1}) > \epsilon + \frac{\delta}{2}, \tag{16}$$

which contradicts the fact that k is the smallest such that (13) is satisfied. Thus,

$$\epsilon + \frac{\delta}{2} < d(y_p, y_k) < \epsilon + \frac{2\delta}{3}. \tag{17}$$

If p and k are of opposite parity, we can take $k = m$ in (17) to obtain (12). If p and k are of the same parity, p and $k + 1$ are of opposite parity. In this case,

$$\begin{aligned} d(y_p, y_{k+1}) &\le d(y_p, y_k) + d(y_k, y_{k+1}) \le \epsilon + \frac{2\delta}{3} + \frac{\delta}{6} \\ &= \epsilon + \frac{5\delta}{6}. \end{aligned} \tag{18}$$

Moreover,

$$\begin{gathered} d(y_p, y_k) \le d(y_p, y_{k+1}) + d(y_{k+1}, y_k), \\ d(y_p, y_k) - d(y_{k+1}, y_k) \le d(y_p, y_{k+1}), \\ \epsilon + \frac{\delta}{2} - \frac{\delta}{6} < d(y_p, y_{k+1}), \\ \epsilon + \frac{\delta}{3} < d(y_p, y_{k+1}). \end{gathered} \tag{19}$$

Thus,

$$\epsilon + \frac{\delta}{3} < d(y_p, y_{k+1}) < \epsilon + \frac{5\delta}{6}. \tag{20}$$

Putting $m = k + 1$, we obtain (12). Hence (12) holds. Now,

$$\begin{aligned} \epsilon + \frac{\delta}{3} &< d(y_p, y_m) \\ &\le d(y_p, y_{p+1}) + d(y_{p+1}, y_{m+1}) + d(y_{m+1}, y_m) \\ &< \frac{\delta}{6} + \epsilon + \frac{\delta}{6} = \epsilon + \frac{\delta}{3}, \end{aligned} \tag{21}$$

a contradiction.

Hence $\{y_n\} = \{\psi x_n\}$ is a Cauchy sequence. By completeness of the space, there exists an element $z \in X$ such that $d(y_n, z) \longrightarrow 0$; continuity of ψ implies that $d(\psi y_n, \psi z) \longrightarrow 0$. Hence, $H([\Phi y_n]_{(\alpha,\beta)}, [\Phi z]_{(\alpha,\beta)}) \le \sup\{d(u,v) : u \in [\Phi y_n]_{(\alpha,\beta)}, v \in [\Phi z]_{(\alpha,\beta)}\} < d(\psi y_n, \psi z) \longrightarrow 0$.

Since $\{\psi x_n\}$ is a Cauchy sequence in X and

$$\begin{aligned} H(\Upsilon_m, \Upsilon_n) &= H\left([\Phi x_m]_{(\alpha,\beta)}, [\Phi x_n]_{(\alpha,\beta)}\right) \\ &\le \sup\left\{d(u,v) : u \in [\Phi x_m]_{(\alpha,\beta)}, v \in [\Phi x_n]_{(\alpha,\beta)}\right\} \\ &< d(\psi x_m, \psi x_n), \end{aligned} \tag{22}$$

it follows that Υ_n is a Cauchy sequence in $CB(X)$. By completeness of $CB(X)$, there exists $\Upsilon \in CB(X)$ such that

$H(Y_n, Y) \longrightarrow 0$. Since $y_{n+1} \in Y_n$ and $d(y_{n+1}, z) \longrightarrow 0$, Lemma 8 implies that $z \in Y$, that is, $\lim_{n\longrightarrow\infty}\psi x_n \in \lim_{n\longrightarrow\infty}[\Phi x_n]_{(\alpha,\beta)}$. Compatibility of ψ and Φ further implies that

$$\lim_{n\longrightarrow\infty} H\left(\psi\left[\Phi x_n\right]_{(\alpha,\beta)}, \left[\Phi\psi x_n\right]_{(\alpha,\beta)}\right) = 0. \tag{23}$$

Since $d(\psi y_{n+1}, [\Phi y_n]_{\alpha,\beta}) \leq H(\psi[\Phi x_n]_{(\alpha,\beta)}, [\Phi\psi x_n]_{(\alpha,\beta)})$, therefore $\psi z \in [\Phi z]_{\alpha,\beta}$, that is, $\lim_{n\longrightarrow\infty}\psi y_n \in \lim_{n\longrightarrow\infty}[\Phi y_n]_{(\alpha,\beta)}$ and $\lim_{n\longrightarrow\infty}H(\psi[\Phi x_n]_{(\alpha,\beta)}, [\Phi\psi x_n]_{(\alpha,\beta)}) = H(\psi[\Phi z]_{(\alpha,\beta)}, [\Phi\psi z]_{(\alpha,\beta)}) = 0$.

Let $b = \psi z$; then, by (9) we have

$$\begin{aligned} d(b, \psi b) &\leq \sup\left\{d(u,v) : u \in [\Phi z]_{(\alpha,\beta)},\ v \in \psi[\Phi z]_{(\alpha,\beta)}\right\} \\ &\leq \sup\left\{d(u,v) : u \in [\Phi z]_{(\alpha,\beta)},\ v \in [\Phi\psi z]_{(\alpha,\beta)}\right\} \\ &< d(\psi z, \psi\psi z) = d(b, \psi b). \end{aligned} \tag{24}$$

Thus $b = \psi b$.

Now,

$$\begin{aligned} d\left(b, [\Phi b]_{(\alpha,\beta)}\right) &\leq d\left(\psi z, [\Phi\psi z]_{(\alpha,\beta)}\right) \\ &\leq \sup\left\{d(u,v) : u \in [\Phi z]_{(\alpha,\beta)},\ v \in [\Phi\psi z]_{(\alpha,\beta)}\right\} \\ &< d(\psi z, \psi\psi z) = d(b, \psi b) = 0. \end{aligned} \tag{25}$$

Hence, $b \in [\Phi b]_{(\alpha,\beta)}$.

Definition 10 (see [38]). Let (X, d) be a metric space and let $\Phi : X \longrightarrow (IFS)^X$ be an intuitionistic fuzzy map. A single valued map $\psi : X \longrightarrow X$ is said to be a selection of $\Phi : X \longrightarrow (IFS)^X$, if there exists $(\alpha, \beta) \in (0, 1] \times [0, 1)$ such that

$$\psi x \in [\Phi x]_{(\alpha,\beta)}, \quad x \in X. \tag{26}$$

Theorem 11. *Let Y be a compact subset of a complete metric space X and let $\Phi : Y \longrightarrow (IFX)^Y$ be a mapping which satisfies the following conditions:*

$$\begin{aligned} &\text{Given } \epsilon > 0, \text{ there exists a } \delta > 0 \\ &\text{such that for all } x, y \in Y, \end{aligned} \tag{27}$$

$$\begin{aligned} &\epsilon \leq \max\left\{d(x, [\Phi x])_{(\alpha,\beta)}, d\left(y, [\Phi y]_{(\alpha,\beta)}\right)\right\} \\ &\quad < \epsilon + \delta \\ &\textit{implies } H\left([\Phi x]_{(\alpha,\beta)}, [\Phi y]_{(\alpha,\beta)}\right) < \epsilon. \end{aligned} \tag{28}$$

Then, there exists a subset W of Y such that $[\Phi w]_{(\alpha,\beta)} = W$ for each $w \in W$. Moreover, for each $w \in W$ there exists a selection of Φ having w as a unique fixed point.

Proof. Let x_0 be an arbitrary fixed element of X. Two sequences $\{x_n\}$ and $\{r_n\}$ of elements in X and R, respectively, will be constructed. $[\Phi x_0]_{(\alpha,\beta)}$ is a closed subset of Y and therefore is compact. There exists a point $x_1 \in [\Phi x_0]_{(\alpha,\beta)}$ such that $d(x_0, x_1) = d(x_0, [\Phi x_0]_{(\alpha,\beta)}) = r_0$. Similarly, there exists $x_2 \in [\Phi x_1]_{(\alpha,\beta)}$ such that $d(x_1, x_2) = d(x_1, [\Phi x_1]_{(\alpha,\beta)}) = r_1$. By induction, we prove that sequences $\{x_n\}$ and $\{r_n\}$ are such that $x_n \in [\Phi x_{n-1}]_{(\alpha,\beta)}$, $d(x_n, x_{n+1}) = d(x_n, [\Phi x_n]_{(\alpha,\beta)}) = r_n$, $n \geq 0$. From inequality (28), we have

$$\begin{aligned} d\left(x_n, [\Phi x_n]_{(\alpha,\beta)}\right) &\leq H\left([\Phi x_{n-1}]_{(\alpha,\beta)}, [\Phi x_n]_{(\alpha,\beta)}\right) \\ &< \max\left\{d\left(x_{n-1}, [\Phi x_{n-1}]_{(\alpha,\beta)}\right), d\left(x_n, [\Phi x_n]_{(\alpha,\beta)}\right)\right\}. \end{aligned} \tag{29}$$

If $d(x_n, [\Phi x_n]_{(\alpha,\beta)}) > d(x_{n-1}, [\Phi x_{n-1}]_{(\alpha,\beta)})$, then (29) implies that $d(x_n, [\Phi x_n]_{(\alpha,\beta)}) < d(x_n, [\Phi x_n]_{(\alpha,\beta)})$, a contradiction. Hence,

$$d\left(x_n, [\Phi x_n]_{(\alpha,\beta)}\right) < d\left(x_{n-1}, [\Phi x_{n-1}]_{(\alpha,\beta)}\right). \tag{30}$$

Thus, $\{r_n\}$ is a monotone nonincreasing sequence of nonnegative real numbers. Therefore, $\{r_n\}$ converges to $\inf\{r_n : n \geq 0\}$. Suppose $\inf\{r_n : n \geq 0\} = r > 0$. Take N so that $n \geq N$ implies that

$$r \leq r_n < r + \delta. \tag{31}$$

It follows that

$$r_{n+1} \leq H\left([\Phi x_n]_{(\alpha,\beta)}, [\Phi x_{n+1}]_{(\alpha,\beta)}\right) < r, \tag{32}$$

which is a contradiction to the assumption that $\inf\{r_n : n \geq 0\} = r > 0$. Hence $r_n \longrightarrow 0$.

That is,

$$d\left(x_n, [\Phi x_n]_{(\alpha,\beta)}\right) \longrightarrow 0. \tag{33}$$

It follows that $H([\Phi x_n]_{(\alpha,\beta)}, [\Phi x_m]_{(\alpha,\beta)}) \longrightarrow 0$. By completeness of $(CB(Y), H)$ (see [39]), there exists a set $W \in CB(Y)$ such that $H([\Phi x_n]_{(\alpha,\beta)}, W) \longrightarrow 0$. Let $w \in W$; then $w \in [\Phi w]_{(\alpha,\beta)}$. If not, let $d(w, [\Phi w]_{(\alpha,\beta)}) = c > 0$; then

$$\begin{aligned} c = d\left(w, [\Phi w]_{(\alpha,\beta)}\right) &\leq H\left([\Phi w]_{(\alpha,\beta)}, W\right) \\ &\leq H\left([\Phi w]_{(\alpha,\beta)}, [\Phi x_n]_{(\alpha,\beta)}\right) + H\left([\Phi x_n]_{(\alpha,\beta)}, W\right) \\ &< \max\left\{d\left(w, [\Phi w]_{(\alpha,\beta)}\right), d\left(x_n, [\Phi x_n]_{(\alpha,\beta)}\right)\right\} \\ &\quad + H\left([\Phi x_n]_{(\alpha,\beta)}, W\right). \end{aligned} \tag{34}$$

In a limiting case when $n \longrightarrow \infty$, we have $c < c$, a contradiction. Hence,

$$w \in [\Phi w]_{(\alpha,\beta)}. \tag{35}$$

Now,

$$\begin{aligned} H\left([\Phi w]_{(\alpha,\beta)}, W\right) &= \lim_{n\longrightarrow\infty} H\left([\Phi w]_{(\alpha,\beta)}, [\Phi x_n]_{(\alpha,\beta)}\right) \\ &< \lim_{n\longrightarrow\infty} \max\left\{d(w, [\Phi w])_{(\alpha,\beta)}, d\left(x_n, [\Phi x_n]_{(\alpha,\beta)}\right)\right\} \\ &= 0. \end{aligned} \tag{36}$$

Hence, $[\Phi w]_{(\alpha,\beta)} = W$ for all $w \in W$.

Next, we will prove that there exists a selection of Φ which has a unique fixed point. For each $u \in Y$, $[\Phi u]_{(\alpha,\beta)}$ is compact. Therefore, for $w \in Y$ there exists $u_w \in [\Phi u]_{(\alpha,\beta)}$ such that

$$d(w, u_w) = d\left(w, [\Phi u]_{(\alpha,\beta)}\right). \quad (37)$$

Let $\psi : Y \longrightarrow Y$ defined as $\psi u = u_w$ be a selection of $\Phi : Y \longrightarrow (IFX)^Y$. Then, for each $u \in Y$ we have $\psi u = u_w \in [\Phi u]_{(\alpha,\beta)}$. Let $\psi w = v(= w_x)$; then $d(w,v) = d(w, [\Phi w]_{(\alpha,\beta)}) = 0$. This implies that

$$v = w = \psi w. \quad (38)$$

Now,

$$\begin{aligned} d(\psi u, \psi v) &\le d(\psi u, w) + d(w, \psi v) \\ &\le d(u_w, w) + d(w, v_w) \\ &\le d\left(w, [\Phi u]_{(\alpha,\beta)}\right) + d\left(w, [\Phi v]_{(\alpha,\beta)}\right) \\ &\le H\left([\Phi w]_{(\alpha,\beta)}, [\Phi u]_{(\alpha,\beta)}\right) \\ &\quad + H\left([\Phi w]_{(\alpha,\beta)}, [\Phi v]_{(\alpha,\beta)}\right) \\ &< d\left(u, [\Phi u]_{(\alpha,\beta)}\right) + d\left(v, [\Phi v]_{(\alpha,\beta)}\right) \\ &< d(u, \psi u) + d(v, \psi v). \end{aligned} \quad (39)$$

It follows that the fixed point of ψ is unique.

The following examples show that our results generalize a number of previous theorems.

Example 12. Let X be the set of all nonnegative integers with the Euclidean metric. Let $\psi : X \longrightarrow X$ be defined as $\psi x = 2x^2$ and let $\Phi : X \longrightarrow (IFS)^X$ be an intuitionistic fuzzy map defined as

$$\mu_{\Phi x}(t) = \begin{cases} \frac{3}{4}, & \text{if } t \in \Omega_x, \\ \frac{1}{4}, & \text{if } t \notin \Omega_x. \end{cases}$$
$$\nu_{\Phi x}(t) = \begin{cases} \frac{1}{5}, & \text{if } t \in \Omega_x, \\ \frac{4}{5}, & \text{if } t \notin \Omega_x, \end{cases} \quad (40)$$

where

$$\Omega_x = \{u \in \psi X : u \le x\}. \quad (41)$$

For $\alpha = 3/4$ and $\beta = 1/5$,

$$[\Phi x]_{(3/4,1/5)} = \{t \in \psi X : t \le x\}. \quad (42)$$

For $\epsilon > 0$, there exists $\delta(= \epsilon)$ such that all the hypotheses of Theorem 9 are valid to obtain common fixed point of ψ and Φ. Previously known results are not applicable to this example (even in the case when Φ is single valued, that is, $\Phi x = \max\{t \in \psi X : \ t \le x\}$) since $\psi \Phi x \ne \Phi \psi x$ at $x \ne 0$.

Example 13. Let $X = R$ with the Euclidean metric, $Y = [-20, 20]$, and $A =]10, 20[$.

For $x \in Y$, define

$$\Gamma_x = \left\{t : 2 - \frac{1}{x} \le t \le 4 - \frac{1}{x}\right\}. \quad (43)$$

Define intuitionistic fuzzy map $\Phi : Y \longrightarrow (IFS)^Y$ as follows: when $x \in A$,

$$\mu_{\Phi x}(t) = \begin{cases} \frac{1}{2}, & \text{if } t \in \Gamma_x, \\ \frac{1}{3}, & \text{if } t \notin \Gamma_x. \end{cases}$$
$$\nu_{\Phi x}(t) = \begin{cases} \frac{2}{3}, & \text{if } t \in \Gamma_x, \\ \frac{3}{4}, & \text{if } t \notin \Gamma_x. \end{cases} \quad (44)$$

When $x \notin A$,

$$\mu_{\Phi x}(t) = \begin{cases} \frac{1}{2}, & \text{if } t \in [2,4], \\ \frac{1}{3}, & \text{if } t \notin [2,4]. \end{cases}$$
$$\nu_{\Phi x}(t) = \begin{cases} \frac{2}{3}, & \text{if } t \in [2,4], \\ \frac{3}{4}, & \text{if } t \notin [2,4]. \end{cases} \quad (45)$$

For $\alpha = 1/2$ and $\beta = 2/3$,

$$[\Phi x]_{(1/2,2/3)} = \begin{cases} \Gamma_x, & \text{if } x \in A, \\ [2,4], & \text{if } x \notin A. \end{cases} \quad (46)$$

For $\epsilon > 0$, there exists $\delta(= 59\epsilon)$, such that Φ satisfies all the assumptions of Theorem 11. In this case, $W = [2,4] \in CB(Y)$ such that $[\Phi w]_{(1/2,2/3)} \in W$ for all $w \in W$ and corresponding to $w \in W$ the mapping $\psi : Y \longrightarrow Y$ defined as

$$\psi x = \begin{cases} w, & \text{if } w \in [\Phi u]_{(1/2,2/3)}, \\ 4 - \frac{1}{u}, & \text{if } w \notin [\Phi u]_{(1/2,2/3)}, \end{cases} \quad (47)$$

is a selection of Φ.

Conflicts of Interest

The authors declare that they have no conflicts of interest.

References

[1] S. Banach, "Sur les opérations dans les ensembles abstraits et leur application aux équations intégrales," *Fundamenta Mathematicae*, vol. 3, pp. 133–181, 1922.

[2] G. Jungck, "Commuting maps and fixed points," *The American Mathematical Monthly*, vol. 83, pp. 261–263, 1976.

[3] G. Jungck, "Compatible mappings and common fixed points," *International Journal of Mathematics and Mathematical Sciences*, vol. 9, pp. 771–779, 1986.

[4] A. Meir and E. Keeler, "A theorem on contraction mappings," *Journal of Mathematical Analysis and Applications*, vol. 28, no. 2, pp. 526–529, 1969.

[5] S. Park and J. S. Bae, "Extensions of a fixed point theorem of Meir-Keeler," *Arkiv for Matematika*, vol. 19, pp. 223–228, 1981.

[6] O. Hadzic, "Common fixed point theorem for family of mappings in complete metric spaces," *Mathematica Japonica*, vol. 29, pp. 127–134, 1984.

[7] B. E. Rhoades, S. Park, and K. B. Moon, "On generalizations of the Meir-Keeler type contraction maps," *Journal of Mathematical Analysis and Applications*, vol. 146, no. 2, pp. 482–494, 1990.

[8] I. Beg and N. Shahzad, "Random fixed points of random multivalued operators on Polish spaces," *Nonlinear Analysis: Theory, Methods & Applications*, vol. 20, no. 7, pp. 835–847, 1993.

[9] T.-C. Lim, "On characterizations of Meir-Keeler contractive maps," *Nonlinear Analysis: Theory, Methods & Applications*, vol. 46, no. 1, pp. 113–120, 2001.

[10] T. Abdeljawad, E. Karapinar, and H. Aydi, "A new Meir-Keeler type coupled fixed point on ordered partial metric spaces," *Mathematical Problems in Engineering*, vol. 2012, Article ID 327273, 20 pages, 2012.

[11] C. M. Chen, T. H. Chang, and K. S. Juang, "Common fixed point theorems for the stronger Meir-Keeler cone-type function in cone ball-metric spaces," *Applied Mathematics Letters*, vol. 25, no. 4, pp. 692–697, 2012.

[12] E. Karapınar, A. Roldán, J. Martínez-Moreno, and C. Roldán, "Meir-keeler type multidimensional fixed point theorems in partially ordered metric spaces," *Abstract and Applied Analysis*, vol. 2013, 9 pages, 2013.

[13] T. Abdeljawad, "Meir-Keeler α-contractive fixed and common fixed point theorems," *Fixed Point Theory And Applications*, vol. 19, 2013.

[14] D. K. Patel, T. Abdeljawad, and D. Gopal, "Common fixed points of generalized Meir-Keeler α-contractions," *Fixed Point Theory and Applications*, vol. 2013, article 260, 2013.

[15] S. L. Singh, R. Chugh, R. Kamal, and A. Kumar, "A new common fixed point theorem for Suzuki-Meir-Keeler contractions," *Filomat*, vol. 28, no. 2, pp. 257–262, 2014.

[16] N. Redjel, A. Dehici, E. Karapinar, and I. M. Erhan, "Fixed point theorems for (α, Ψ)-Meir-Keeler-Khan mappings," *Journal of Nonlinear Sciences and Applications*, vol. 8, pp. 955–964, 2015.

[17] M. Abtahi, "Fixed point theorems for Meir-Keeler type contractions in metric spaces," *Fixed Point Theory. An International Journal on Fixed Point Theory, Computation and Applications*, vol. 17, no. 2, pp. 225–236, 2016.

[18] M. Abtahi, "Common fixed point theorems of Meir-Keeler type in metric spaces," *Fixed Point Theory. An International Journal on Fixed Point Theory, Computation and Applications*, vol. 18, no. 1, pp. 17–26, 2017.

[19] V. Popa and A.-M. Patriciu, "A general fixed point theorem of Meir-Keeler type for mappings satisfying an implicit relation in partial metric spaces," *Functional Analysis, Approximation and Computation*, vol. 9, no. 1, pp. 53–60, 2017.

[20] L. A. Zadeh, "Fuzzy sets," *Information and Control*, vol. 8, no. 3, pp. 338–353, 1965.

[21] K. T. Atanassov, "Intuitionistic fuzzy sets," *Fuzzy Sets and Systems*, vol. 20, no. 1, pp. 87–96, 1986.

[22] N. Abbasizadeh and B. Davvaz, "Intuitionistic fuzzy topological polygroups," *International Journal of Analysis and Applications*, vol. 12, no. 2, pp. 163–179, 2016.

[23] A. Azam, R. Tabassum, and M. Rashid, "Coincidence and fixed point theorems of intuitionistic fuzzy mappings with applications," *Journal of Mathematical Analysis*, vol. 8, no. 4, pp. 56–77, 2017.

[24] A. Azam and R. Tabassum, "Fixed point theorems of intuitionistic fuzzy mappings in quasi-pseudo metric spaces," *Bulletin of Mathematical Analysis and Applications*, vol. 9, no. 4, pp. 42–57, 2017.

[25] W. Kumam, P. Sukprasert, P. Kumam, A. Shoaib, A. Shahzad, and Q. Mahmood, "Some fuzzy fixed point results for fuzzy mappings in complete b-metric spaces," *Cogent Mathematics and Statistics*, vol. 5, pp. 1–12, 2018.

[26] A. Shoaib, P. Kumam, A. Shahzad, S. Phiangsungnoen, and Q. Mahmood, "Fixed point results for fuzzy mappings in a b-metric space," *Fixed Point Theory and Applications*, vol. 1, pp. 1–12, 2018.

[27] Humaira, Muhammad Sarwar, and G. N. V. Kishore, "Fuzzy fixed point results for Φ-contractive mapping with applications," *Complexity*, vol. 2018, Article ID 5303815, 12 pages, 2018.

[28] M. Ertürk and V. Karakaya, "n-tuplet coincidence point theorems in intuitionistic fuzzy normed spaces," *Journal of Function Spaces*, vol. 2014, 14 pages, 2014.

[29] X. Li, M. Guo, and Y. Su, "On the intuitionistic fuzzy metric spaces and the intuitionistic fuzzy normed spaces," *Journal of Nonlinear Sciences and Applications. JNSA*, vol. 9, no. 9, pp. 5441–5448, 2016.

[30] R. Rani and S. Manro, "Fixed point theorem in intuitionistic fuzzy metric spaces using compatible mappings of type (A)," *Mathematical Sciences Letters*, vol. 7, no. 1, pp. 49–53, 2018.

[31] A. Garg, Z. K. Ansari, and P. Kumar, "Fixed point theorems in intuitionistics fuzzy metric spaces using implicit relations," *Applied Mathematics*, vol. 07, no. 06, pp. 569–577, 2016.

[32] V. Gupta, R. K. Sainib, and A. Kanwar, "Some common coupled fixed point results on modified intuitionistic fuzzy metric spaces," *Procedia Computer Science*, vol. 79, pp. 32–40, 2016.

[33] G. Beer, *Topologies on Closed And Closed Convex Sets*, Kluwer Academic Publishers, Dordrecht, Netherlands, 1993.

[34] K. T. Atanassov, *Intuitionistic Fuzzy Sets*, 11-37, Physica-Verlag, Heidelberg, Germany, 1999.

[35] K. T. Atanassov, "More on intuitionistic fuzzy sets," *Fuzzy Sets and Systems*, vol. 33, no. 1, pp. 37–45, 1989.

[36] P. K. Sharma, "Cut of intuitionistic fuzzy groups," *International Mathematics Forum*, vol. 6, no. 53, pp. 2605–2614, 2011.

[37] T. Hu, "Fixed point theorems for multivalued mappings," *Canadian Mathematical Bulletin*, vol. 23, no. 2, pp. 193–197, 1980.

[38] Calogero Vetro and Francesca Vetro, "Caristi type selections of multivalued mappings," *Journal of Function Spaces*, vol. 2015, Article ID 941856, 6 pages, 2015.

[39] J. P. Aubin, *Applied Abstract Analysis*, John Wiley and Sons, New York, USA, 1977.

Context Adaptation of Fuzzy Inference System-Based Construction Labor Productivity Models

Abraham Assefa Tsehayae[1] and Aminah Robinson Fayek[2]

[1]*School of Civil and Environmental Engineering, Addis Ababa Institute of Technology, Addis Ababa University, Room 206, AAiT Main Building, P.O. Box 385, Addis Ababa, Ethiopia*
[2]*Department of Civil and Environmental Engineering, Hole School of Construction Engineering, University of Alberta, 7-287 Donadeo Innovation Centre for Engineering, Edmonton, AB, Canada T6G 1H9*

Correspondence should be addressed to Aminah Robinson Fayek; aminah.robinson@ualberta.ca

Academic Editor: Stella Morris

Construction labor productivity (CLP) is one of the most studied areas in the construction research field, and several context-specific predictive models have been developed. However, CLP model development remains a challenge, as the complex impact of multiple subjective and objective influencing variables have to be examined in various project contexts while dealing with limited data availability. On the other hand, lack of a framework for adapting existing or original models from one context to other contexts limits the possibility of reusing existing models. Such challenges are addressed in this paper through the development of a context adaptation framework. The framework is used to transfer the knowledge represented in fuzzy inference (FIS) based CLP models from one context to another, by using linear and nonlinear evolutionary based transformation of the membership functions combined with sensitivity analysis of fuzzy operators and defuzzification methods. Using four context-specific CLP models developed for concreting activity under industrial, warehouse, high-rise, and institutional building project contexts, the framework was implemented, and the prediction capability of the adapted models was evaluated based on their prediction similarity with the original models. The results showed that linearly adapted CLP models for industrial and institutional contexts and nonlinearly adapted CLP models for warehouse and high-rise contexts provide a similar prediction capability with the original models. The proposed context adaptation framework and findings from this paper address the limitations in past context adaptation research by examining a practical context-sensitive application problem and further examining the role of fuzzy operators and defuzzification methods. The findings assist researchers and industry practitioners to take full advantage of existing FIS-based models in the study of new contexts, for which data availability might be limited.

1. Introduction and Background

Construction labor productivity (CLP) has a direct and significant influence on success of construction projects; therefore, CLP has been well studied [1]. As a result, numerous predictive CLP models have been tested and developed, even though, collecting and modeling productivity data are known to require significant financial investment [2]. Existing CLP models have used various data analysis methods, including regression analysis, neural networks, expert systems, and fuzzy inference systems [1]. CLP modeling deals with a complex problem involving a large number of subjective and objective variables and is faced with limited data availability, thus, making CLP modeling an exceptional target for fuzzy inference systems. Fuzzy inference systems (FISs) have been proven to be effective tools for solving many engineering problems in biomedicine, robotics, pattern recognition, image processing, and control application areas [3]. More importantly, FISs are suitable for context adaption, whereby they can be adapted to suit other project contexts and enable users to take full advantage of existing FIS in the analysis of new contexts. Such a context adaptation process cannot be utilized with other CLP modeling approaches such as regression equations or neural networks, as the existing models have to be redeveloped or retrained before use in new contexts. Mao [4] used FIS to model labor productivity of concrete wall formwork activity. Pan [5] studied the effect of rainfall on productivity and duration of highway activities using a FIS. Fayek and Oduba [6] also used FISs to model the labor productivity of industrial pipe rigging and welding activities. Tsehayae and Fayek [7], using a system model approach, studied CLP of concreting activity under industrial, warehouse, high-rise, and institutional building project contexts; the authors used a fuzzy hybrid approach, which incorporated FISs, developed using a data-driven fuzzy clustering technique, combined with a genetic algorithm-based model tuning process. However, model input variables (i.e., CLP influencing variables) used in the past FIS-based CLP models and their associated impact on CLP varied from project to project, thus, making all of the developed CLP models context-specific.

Context plays an essential role in FIS-based CLP research, as it defines in which scenarios the findings of the FIS-based CLP models are applicable, which implies that, under changed contexts, existing models cannot be used without some adaptation. Moreover, context adaptation of FISs has been recognized as an effective method for generating interpretable and accurate FIS models through tuning of the parameters contained in the data sets and based on the collected context-specific information [8]. However, in the CLP modeling field, an approach for adapting such models from one context to another context is missing. Model adaptation is of particular importance for Mamdani-type FIS-based models, as the fuzzy sets used in such models are highly context-specific. In FIS, the if-then rules are composed of fuzzy conditions (represented by the MFs of the input variables) and fuzzy conclusions (represented by the MFs of the output variables). Fuzzy sets are used to describe the respective linguistic variables, which characterize a CLP influencing variable like *crew size* using linguistic variables such as **small** *crew size*, **medium** *crew size*, and **large** *crew size*. An example of a CLP focused fuzzy rule is shown in the following, where the words in italics are the influencing variables or features and the linguistic variables are shown in bold. "If the *crew size* is **small** and the *cooperation among craftsperson* is **very good** and the *level of interruption and disruption* is **low** and the *direct work proportion* is **high** then *construction labor productivity* is **high**."

A fuzzy set A, representing a linguistic variable, is characterized using its membership function (MF), which represents numerically the degree to which an element x belongs to the fuzzy set and fits the linguistic variable over a continuous range $A : \mathbf{X} \rightarrow \mu = [0,1]$. In the FIS-based CLP models, the input (key factors, practices, and work sampling proportions) and output (CLP) variables are partitioned using linguistic variables over their respective universe of discourse. For example, the linguistic variables could be represented using Gaussian membership functions. The function for a Gaussian MF $A(x,\sigma,\mu)$ is shown in (1), where σ represents the standard deviation, denoting the spread of A, and μ represents the modal value, denoting the typical element of A:

$$A(x,\sigma,\mu) = e^{-(x-\mu)^2/2\sigma^2}. \quad (1)$$

Membership functions (MFs) have many important descriptors. MFs employed in modeling endeavours are required to be normal; that is, at least one element of $\mathbf{X}$ attains full membership ($\mu = 1$) and represents a typical value of the fuzzy set. Support of fuzzy set A represents all elements of $\mathbf{X}$ that exhibit some association with the fuzzy set by having nonzero membership degrees and core of fuzzy set A represents all elements of the universe $\mathbf{X}$ that are typical to A. Fuzzy inference systems consist of two main parts: a rule-base (RB) composed of the linguistic if-then rules and a database (DB) which associates membership functions to the linguistic variables used in the RB, for example, **small** for crew size or **high** for CLP. However, it is worth noting that, in the DB parts of FISs, the concept of "**small** *crew size*" cannot be uniquely and universally defined, while concepts like "*crew size* under 5" can be. It is natural that construction professionals could define such fuzzy sets differently depending on the context of use, and the context itself must be properly defined using context attributes such as the contractor organization's experience, project location, and type of projects (e.g., commercial or industrial). Thus, the exact definition of a fuzzy set like **small** *crew size* depends on a context attribute like type of project [9]. For example, industrial projects tend to have larger crew sizes as compared to commercial ones; thus, a **small** *crew size* in an industrial project could be equivalent to a **medium** *crew size* in commercial projects. On the other hand, the rule-base is assumed to be context-free and is usually valid in any context, as the context of use affects only the meaning associated with each linguistic term used in the rules rather than the logic of the rules themselves [8, 10]. Therefore, in the context adaptation of FISs, the focus is on the linguistic variables and their respective membership functions.

Accordingly, the main objectives of this research are to (1) develop a framework for context adaption of FIS-based CLP models, (2) test the framework using real-world problem, and (3) improve on limitations with existing context adaptation approaches which relied on normalized MFs. This objective is critical in CLP research domain as context adaptation will enable the reuse of existing FIS models in new contexts, thereby saving model developers the considerable effort required to collect data and to develop new CLP models; it will also improve the application of existing models by construction organizations as existing models can be adapted to suit the organizations' specific needs or contexts.

This paper begins with a literature review of the general context adaptation approaches. Then, a framework for adapting CLP models is formulated. Next, the paper illustrates the application of the framework by adapting a series of context-specific CLP models to suit new contexts based on the field data collected for the same model features (i.e., model input and output variables) of the respective original CLP models. Then, a comparative assessment in terms of prediction accuracy and similarity between the adapted and original context-specific models is carried out, and the effectiveness of the context adaptation framework is discussed. Finally, conclusions and areas for future research are presented.

2. Literature Review of Context Adaptation Approaches

Context plays an active part in construction research analysis as it is invariably dynamic and imperative for the development of meaningful findings [11]. Thus, context adaptation has an important application in deriving new models from existing ones. In the general computing field, most of the context adaptation research has been carried out on Mamdani-type FISs, which have been used in a wide range of areas due to their ability to handle linguistic concepts and perform accurate modeling of input-output relations [3]. In the context adaptation of FISs, the focus has been mainly on the linguistic variables and their respective membership functions as the rule-base is considered to be a context-free model [12, 13]. According to Botta et al. [8], the following principles have been followed in adapting FISs: (1) context adaptation will not modify the rule-base as the rule-base is considered to be a context-free and universal knowledge; (2) context adaptation will not change the number of linguistic variables and, consequently, the number of corresponding fuzzy sets defined in the rule-base; and (3) context adaptation will not affect the sematic ordering of linguistic variables.

Context adaptation of FISs has been carried using either transformation functions or adaptive operators. Most context adaptation studies focused on the use of transformation functions [12–15]. According to Botta [3], a transformation function serves to adapt a database (DB) or fuzzy partition, which is made up of a group of MFs, each representing a linguistic term. The transformation function maps the universe of discourse of the base or original context MF to the universe of discourse of the adapted context, thereby modifying the distribution (i.e., support and core) and the shape of the MFs. Transformation functions have been commonly applied to adapt the base partition defined over a normalized universe of discourse $[0, 1]$ as shown in Figure 1. However, the use of a normalized universe of discourse for the base or original MFs makes the database more general or context independent as the MFs in the DB are defined over a normalized $[0, 1]$ range [12, 13]. In such a case, the universe of discourse is partitioned using common MFs like triangular, trapezoidal, or Gaussian [15] and is then adapted to another context partition defined over a universe of discourse of $[a, b]$ as shown in Figure 1. Commonly, linear and nonlinear transformation functions are used. Linear transformation functions are applied on the overall partition of the fuzzy sets and will either linearly expand or contract all the fuzzy sets (see Figure 1). However, nonlinear transformation functions are applied either to the overall partition or selected points of the fuzzy sets (refer to Figure 1), thus, changing all fuzzy sets or the breakpoints (i.e., points of intersection) of selected fuzzy sets [8].

The parameters of nonlinear transformation functions are derived using genetic algorithm or neural network based optimization approach over data collected for the new or target context using either field experiments or experts [12, 13]. Several nonlinear transformation functions have been applied. Magdalena [14] proposed a sign function for nonlinear transformation of fuzzy sets and provided an application example using cart-pole balancing system control problem. Gudwin et al. [12] used a linear combination of sigmoidal functions and demonstrated the development of the functions using assumed data. Pedrycz and Gomide [16] suggested the use of piecewise linear function. According to Gudwin et al. [12], transformation functions should meet certain requirements so as to preserve ordering and normality of the base linguistic variables. In particular, they are required to fulfil continuity, nondecreasing monotonicity, and boundary conditions.

Adaptive operators are specifically designed operators that adjust the universe of discourse of the fuzzy sets and modify the core, support, and shape of fuzzy sets. According to Botta [3], the context adaptation process using adaptive operators is based on a flexible nonlinear transformation function and four orthogonal fuzzy modifiers: core-position modifier, core-width modifier, support-width modifier, and membership function shape modifier. The fuzzy modifiers are formulated using genetic algorithm based optimization process. Botta [3] tested the approach using four datasets, including a context-aware benchmarking dataset that arbitrarily assigns MFs to a universe of discourse, structure of wage dataset which includes years of experience and wage, a synthetic dataset generated using a parametric function, and fuel consumption dataset. However, adaptive operators modified the fuzzy partition in way that the order of the fuzzy sets was affected, resulting in reduced interpretability of the adapted fuzzy partition [8].

The interpretability of the adapted fuzzy partitions can be verified using coverage, normality, and distinguishability properties [8]. The adapted fuzzy partition has to cover the new or adapted context's universe of discourse $U = [a, b]$. The adapted MFs of the fuzzy sets are also required to be normal so that at least one element of the universe of discourse will have full membership. The overlap of the fuzzy sets in the adapted fuzzy partitions has to be kept to a level that each couple of fuzzy sets are distinguishable enough [3]. The overlap between fuzzy sets (A_1, A_2) can be measured using a possibility measure Π (see (2)). In order to ensure interpretability of the adapted fuzzy partition, the overlap (Π) among the fuzzy sets should not exceed 0.8 [17].

$$\Pi(A_1, A_2) = \sup_{x \in U} \min \left\{ \mu_{A_1}(x), \mu_{A_2}(x) \right\}. \quad (2)$$

In the general computing field, several FIS context adaptation studies have been carried out [8, 12, 14, 18]; however, the following limitations are observed. First, transformation function based studies were applied on normalized base or original MFs defined over $[0, 1]$ range, resulting in the adaptation of theoretical or context independent MFs to contexts-specific MFs. Second, most studies lacked practical application and rather focused on demonstrating the context adaptation method using benchmark datasets. Third, in FISs, not only the MFs but also the fuzzy operators and defuzzification methods are context-dependent [19]. However, past

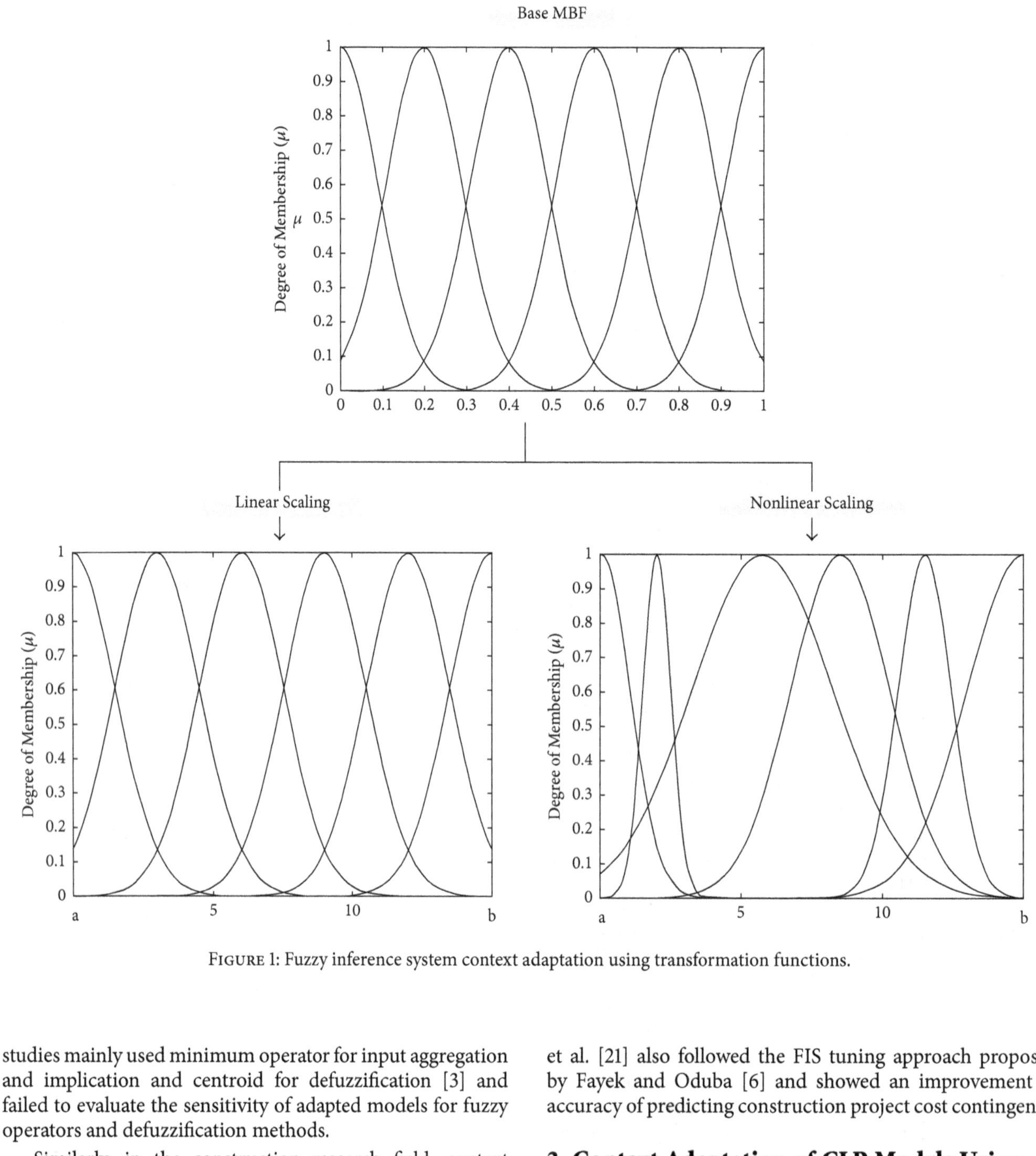

FIGURE 1: Fuzzy inference system context adaptation using transformation functions.

studies mainly used minimum operator for input aggregation and implication and centroid for defuzzification [3] and failed to evaluate the sensitivity of adapted models for fuzzy operators and defuzzification methods.

Similarly, in the construction research field, context adaptation is also scantly explored. Past studies have not studied context adaptation of FIS models as an approach for reusing existing models in new contexts but rather focused on the fine tuning of the FIS model parameters with the sole aim of improving the model's accuracy. According to Awad and Fayek [20], the fine tuning of a FIS model improved the accuracy of predicting contractor's default by tuning the MFs and weight of rules of the FIS using neural network and genetic algorithm techniques. Fayek and Oduba [6] improved the accuracy of FIS-based CLP models for industrial pipe rigging and welding activities, by shifting the right, left, and both legs of the triangular and trapezoidal output MFs. Idrus et al. [21] also followed the FIS tuning approach proposed by Fayek and Oduba [6] and showed an improvement in accuracy of predicting construction project cost contingency.

3. Context Adaptation of CLP Models Using Transformation Functions

This paper provides a framework that addresses CLP modeling challenges through the use of a context adaptation framework. The context adaptation framework is based on linear and nonlinear transformation functions rather than adaptive operators, as transformation functions provide a transparent context adaptation framework for the MFs of FIS-based CLP models while adaptive operators will modify the fuzzy partition, resulting in reduced interpretability of the adapted fuzzy partition. The framework, unlike past studies which focused on context independent base MFs, will focus

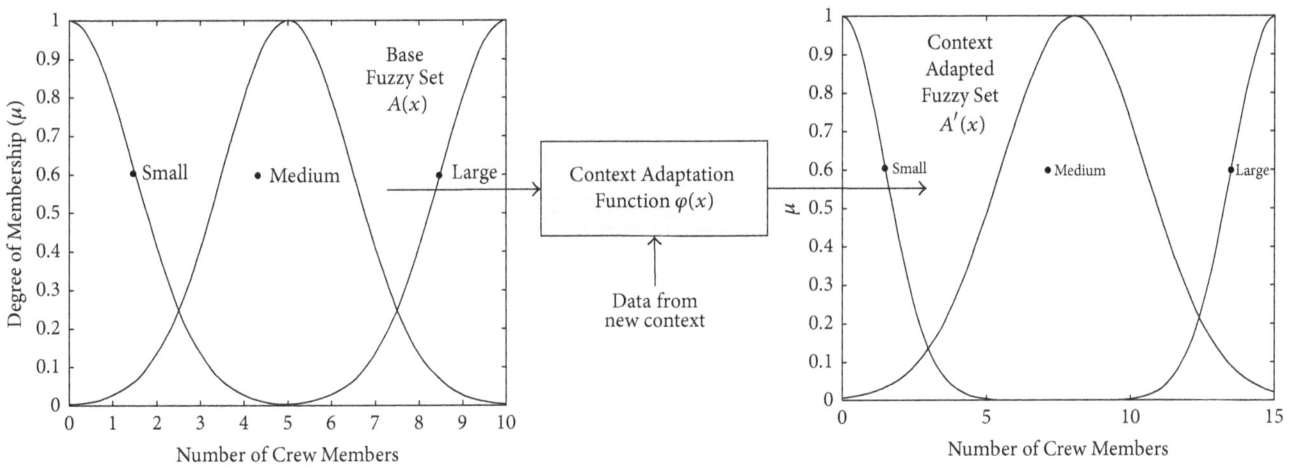

FIGURE 2: Context adaptation procedure.

on adapting context-specific MFs that are defined over universe of discourse $B = [l, u]$ to another context defined over a universe of discourse $U = [a, b]$. In addition to adapting the MFs, the framework also tests the effect of fuzzy operators and defuzzification methods on the adapted models. The following procedure is developed for context adaptation of FIS-based CLP models. The underlying process in the context adaptation procedure involves the determination of a context adaptation or transformation function, using data collected from the new context, for adapting the base fuzzy set A, represented by the base or base MF, to an adapted fuzzy set A', represented by the adapted MF as shown in Figure 2.

The procedure for context adaptation is summarized in the following steps:

(1) Identify the base fuzzy sets for a given context and for each model feature (model input and output variables) of the base context-specific CLP model. The universe of discourse of each feature $B = [l, u]$ and the parameters of the MFs such as the standard deviation and modal values of Gaussian MF's of the base fuzzy set $A(x, \sigma, \mu)$ are documented.

(2) Collect data from the new context for each feature of the base model, using field data collection or experts. The documented data set $(d_1, d_2, \ldots, d_N)$, where N represents the total number of data instances of a given feature, will then be used to determine the upper and lower limits of the adapted fuzzy set $A'(x)$, the preliminary MFs of the adapted fuzzy set $A'_0(x)$, and then the appropriate context adaptation function $\varphi_h(x)$.

(3) Determine the boundary or upper and lower limits of the adapted fuzzy set $U = [a, b]$. For example, using the absolute limit context determination approach [12], the lower bound a is taken as the minimum data value; thus $a = \min(d_1, d_2, \ldots, d_N)$ and the upper bound b is taken as the maximum data value; thus $b = \max(d_1, d_2, \ldots, d_N)$.

(4) Determine the preliminary MFs of the adapted fuzzy set $A'_0(x)$ using the collect data from the new context using either expert- or data-driven MF development approaches. Expert-driven approaches such as direct (horizontal or vertical), exemplification, and pairwise comparison or data-driven approaches such as inductive reasoning and fuzzy clustering can be used. The preliminary MFs will be used to determine the nonlinear transformation function.

(5) Determine the parameters of the context adaptation functions $\varphi_1(x)$ for linear and $\varphi_2(x)$ for nonlinear transformation of MFs.

(6) Develop the MF of the adapted fuzzy set $A'(x)$ using either the linear or nonlinear adaptation functions as expressed in

$$A'_i(x) = A(\varphi_h(x)), \quad x \subset [a, b],$$
$$\varphi_h(x) \in [l, u], \quad h \in \{1, 2\}. \tag{3}$$

(7) Evaluate the overlap among the adapted MFs using possibility measure (see (2)) in order to ensure the distinguishability of adapted fuzzy sets.

(8) Adapt the base CLP model by replacing the base MFs $A(x)$ with the adapted MFs $A'(x)$ for each model feature (input and output variables) and evaluate the prediction ability of the adapted CLP model. FIS model accuracies can be determined using the root mean square error (RMSE) [22]. The RMSE of the adapted CLP model is calculated using (4), where t_i is the new context's target CLP value for the ith data instance, z_i is the corresponding predicted CLP value, and N is the total number of data instances.

$$\mathrm{RMSE}_i = \frac{1}{N}\sum_{i=1}^{N}\sqrt{(t_i - z_i)^2}. \tag{4}$$

(9) Evaluate the sensitivity of the adapted CLP model for fuzzy operators and defuzzification methods and summarize the improvement in prediction ability.

(10) Determine the similarity between the adapted and base model of the new context using model agreement measures. The use of the modified Willmott agreement index is recommended to determine the similarity between models. The index is dimensionless, bounded, less sensitive to extreme values and outliers, and suitable for cross-comparison between models [23]. The Willmott agreement index WI_i is shown in (5), where PA_i represents the predicted values obtained from the adapted model, PB_i is the predicted values obtained from the base or original model of the new context, $\overline{PB}$ is the mean value of the predicted values from the base model of the new context, and N is the total number of data instances. The value WI_i varies from 0 to 1, and a value of 1 indicates a perfect agreement between the adapted and base models.

$$WI_i = 1 - \frac{\sum_{i=1}^{N} |PA_i - PB_i|}{\sum_{i=1}^{N} \left(|PA_i - \overline{PB}| + |PB_i - PB|\right)}. \quad (5)$$

(11) Compare and contrast the agreement indices and prediction ability of the resulting adapted models for both linear and nonlinear transformation of MFs and combinations of fuzzy operators and defuzzification methods and identify the most appropriate context adaptation approach.

4. Context Adaptation of Context-Specific Concreting Activity CLP Models

The preceding context adaptation procedure was tested using the field data collected for concrete pouring activity. The data were gathered from six building projects in the greater Edmonton area of Alberta, Canada, from June 2012 to October 2014. The projects included (1) a mixed-use office and staff facility building, (2) an industrial warehouse building, (3) a commercial warehouse building, (4) a mixed residential and community center building, (5) a mixed commercial-residential building, and (6) an institutional building. Concrete pouring activity was studied in three data collection cycles, where each cycle extended over a month-long period and encompassed different weather seasons. For each data collection case, WS observations were made for the crew under study, and influencing variables (factors and practices), total man-hours, and installed quantities were documented. CLP was defined as the ratio of units of output—in terms of installed quantity (m^3)—to units of input—in terms of total labor work hours (MH)—where higher CLP values are desirable.

First, an operational definition of context for CLP modeling was developed, which was based on the context attributes that constrain the four elements of a CLP model (user, model developer, model, and prevailing environment of the model) without intervening in the model development process explicitly. Accordingly, the context attributes for each of the studied six projects were generated using the 5W1H questions approach: Who, What, Where, When, Why, and How. Among the generated context attributes, the following context attributes distinguished the six projects that were studied and, thus, were used as the key context attributes in comparing and identifying the similarity of the projects: "Who" attributes, related to the project owner's primary driver (schedule, cost, quality, or safety), contractor team's experience, and contractor organization's experience; "What" attributes related to project (i.e., building) type, site layout, project safety practice, and project productivity measurement and tracking practice; and "Where" attributes related to project location.

Accordingly, projects having identical values to key context attributes were grouped together, and the project nature (building type) context variable was used to name the four unique contexts. Context 1, representing concreting in industrial buildings, includes the datasets of the first two projects (the mixed-use office and staff facility building and the industrial warehouse building). Context 2, representing concreting in Warehouse buildings, includes the dataset of the third project (a commercial warehouse building). Context 3, representing concreting in high-rise buildings, includes the datasets of the fourth and fifth projects (the mixed residential and community center building and the mixed commercial-residential building). Context 4, representing concreting in institutional buildings, includes the data set of the sixth project (the institutional building). A more detailed discussion of the context attributes can be found in the study presented in the thesis work of the first author [24].

Then, four context-specific CLP models, summarized in Table 1, were developed and optimized for predicting labor productivity of concreting (concrete pouring) activity using Mamdani-type FIS models, which were used as base or original CLP models. The models were developed using Fuzzy C-Means clustering and Gaussian membership functions (MFs) were used. The following FIS model parameters were optimized: (1) the fuzzification coefficient m in FCM clustering, (2) membership function parameters, (3) number of rules, and (4) fuzzy operators and defuzzification methods, resulting in improved accuracy and interpretability of the four context-specific CLP models. A more detailed discussion of the context-specific model development process can be found in the study presented by Tsehayae and Fayek [7].

The properties of the optimized FIS-based CLP models, shown in Table 1, indicate the context-specific nature of models as they had distinctly different key influencing input variables (made up of factors, practices, and work sampling proportions), number of membership functions or number of rules, fuzzification coefficients, fuzzy operators, and defuzzification methods. The context adaptation framework presented in this paper was tested by adapting the four context-specific CLP base models from one context to another, as shown in Figure 3. Accordingly, the industrial context CLP model is adapted to suit the warehouse, high-rise, and institutional contexts using linear and nonlinear transformation functions, and a similar process is repeated for the other three contexts. The adaptation process, for each context, resulted in six adapted models: three linearly adapted models and three nonlinearly adapted models (see Figure 3). The adapted models are compared with the base

Table 1: Base context-specific CLP models: features, structure, and model parameters.

Features, FIS structure, and model parameters	Context 1 Concreting, industrial buildings	Context 2 Concreting, warehouse buildings	Context 3 Concreting, high-rise buildings	Context 4 Concreting, institutional buildings
Number of input features	16	7	8	11
Number of data instances	23	16	28	25
Model features	$[x_2, x_5, x_{20}, x_{21}, x_9, x_{11}, x_{13}, x_{14}, x_{15}, x_{16}, x_{17}, x_{18}, x_{19}, x_{22}, x_{23}, x_{24}, y]$	$[x_{20}, x_{21}, x_8, x_{10}, x_{14}, x_{22}, x_{27}, y]$	$[x_{20}, x_{21}, x_{11}, x_{22}, x_{23}, x_{24}, x_{25}, x_{26}, y]$	$[x_1, x_3, x_4, x_6, x_7, x_{21}, x_{12}, x_{22}, x_{24}, x_{25}, x_{26}, y]$
Fuzzification coefficient	1.5	2.5	2.0	2.0
Number of rules	6	7	6	7
Input aggregation operator	PROD	MIN	PROD	PROD
Implication method	PROD	MIN	PROD	PROD
Rule aggregation operator	MAX	SUM	PROBOR	PROBOR
Defuzzification method	MOM	BISECTOR	CENTROID	BISECTOR
Accuracy (RMSE)	1.162	0.467	0.992	0.671

Note. Model features represent: x_1-crew size, x_2-craftsperson on-job training, x_3-crew composition, x_4-cooperation among craftspersons, x_5-team spirit of crew, x_6-level of interruption and disruption, x_7-fairness of work assignment, x_8-location of work scope (distance), x_9-location of work scope (elevation), x_{10}-congestion of work area, x_{11}-fairness in performance review of crew by foreman, x_{12}-site congestion, x_{13}-treatment of foremen by superintendent and project manager, x_{14}-uniformity of work rules by superintendent, x_{15}-out-of-sequence inspection, x_{16}-safety training, x_{17}-project safety administration and reporting, x_{18}-oil price fluctuation, x_{19}-natural gas price, x_{20}-concrete placement technique, x_{21}-structural element type, x_{22}-direct work proportion, x_{23}-preparatory work proportion, x_{24}-tools and equipment proportion, x_{25}-material handling proportion, x_{26}-travel proportion, and x_{27}-personal proportion, y-CLP.

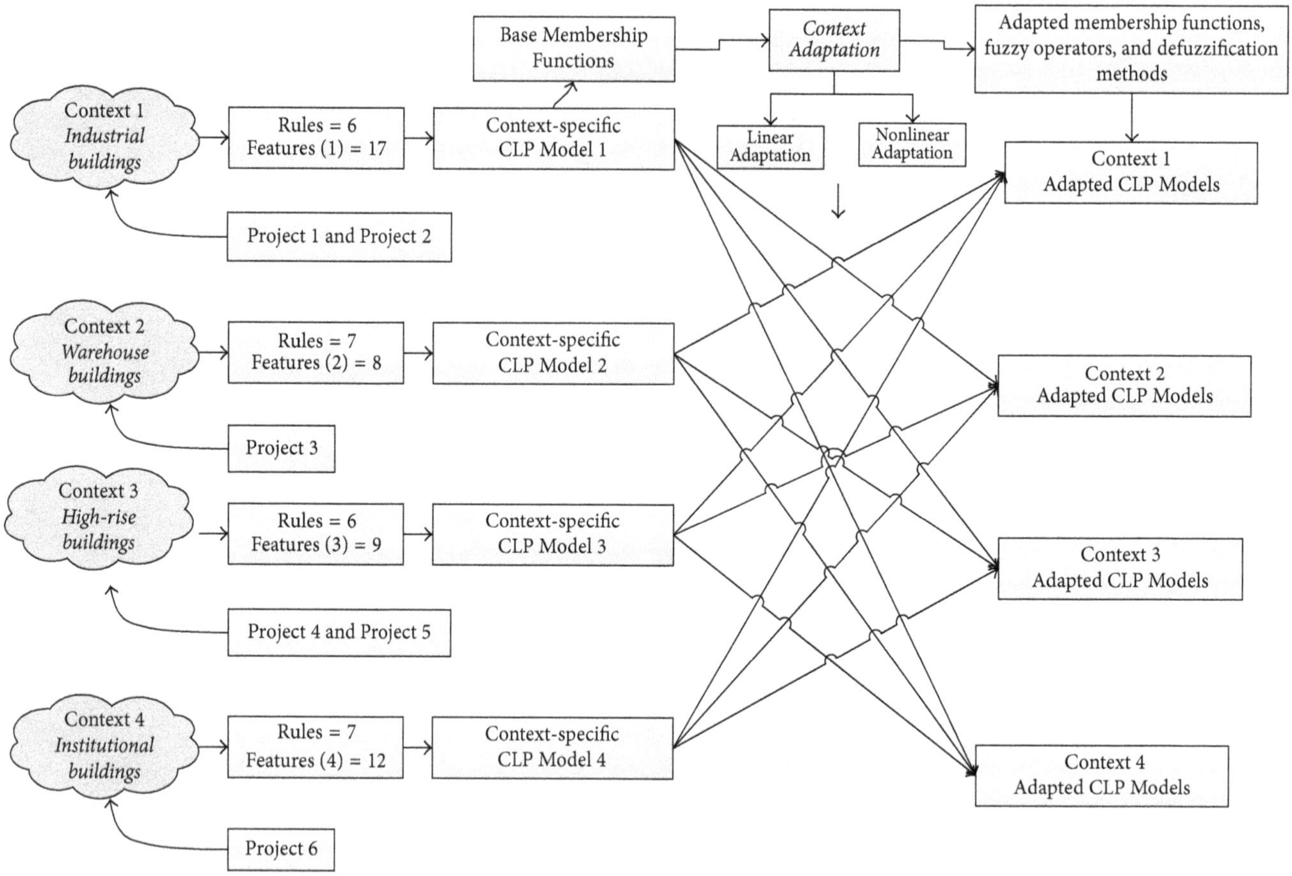

FIGURE 3: Context-specific CLP models adaptation.

model developed for the given context (refer to Table 1) using model accuracy in terms of RMSE and modified Willmott agreement indices.

In the following sections, a numerical illustration of the context adaptation framework is discussed based on the adaptation of the model output (CLP) feature from institutional to industrial context. The institutional (context 4) CLP model has 12 model features (11 input and one model, i.e., CLP variable) and seven rules, implying an equal number MFs representing the fuzzy sets $A_j = \{A_1, A_2, \ldots, A_7\}$. The base MFs for model output (CLP) had a universe of discourse $B = [1.80, 11.25]$ and the mean CLP was 4.25 m^3/MH, with standard deviation of 2.21.

For given context-specific CLP model, all model features' MFs are adapted using linear and nonlinear transformation functions to three other contexts, and initial model accuracies, $RMSE_{CA\text{-}L}$ (where CA-L represents the context adapted CLP model using linear adaptation) and $RMSE_{CA\text{-}NL}$ (where CA-NL represents the context adapted CLP model using nonlinear adaptation), respectively, are established. Then, the sensitivity of the adapted models for fuzzy operators and defuzzification methods is investigated for both linear and nonlinear adapted models and final model accuracies, $RMSE_{CA\text{-}LS}$ (where CA-L represents the context adapted CLP model using linear adaptation after sensitivity analysis) and $RMSE_{CA\text{-}NLS}$ (where CA-L represents the context adapted CLP model using nonlinear adaptation after sensitivity analysis), respectively, are established. Next, the agreement indices (WI_i) between adapted and base models were computed for identifying the appropriate linear or nonlinear adaptation approach. Finally, the model accuracies of the adapted models are compared and the best performing context adaptation approach is identified. For the adapted MFs the following linguistic labels are used: A_1 (very low), A_2 (low), A_3 (medium low), A_4 (medium), A_5 (medium high), A_6 (high), and A_7 (very high). The parameters of the seven Gaussian MFs $A(x, \sigma, \mu)$ of the base or base model output variable are as follows: $A_1(x) = G(x, 0.973, 1.875)$, $A_2(x) = G(x, 0.573, 3.514)$, $A_3(x) = G(x, 0.786, 4.569)$, $A_4(x) = G(x, 0.864, 5.990)$, $A_5(x) = G(x, 0.877, 7.673)$, $A_6(x) = G(x, 0.707, 9.101)$, and $A_7(x) = G(x, 1.101, 11.105)$.

The CLP data from industrial context (context 3) was then retrieved from collected data and had values of universe of discourse $U = [0.03, 8.66]$ and the mean CLP was 2.69 m^3/MH, with standard deviation of 1.83. Next, the parameters of the context adaptation functions $\varphi_1(x)$ for linear adaptation and $\varphi_2(x)$ for nonlinear adaptation are developed, as discussed in the following subsections.

4.1. Linear Adaptation. In linear adaptation, the base MFs defined over a universe of discourse of $B = [l, u]$ are adapted to the context-adapted universe of discourse by means of a

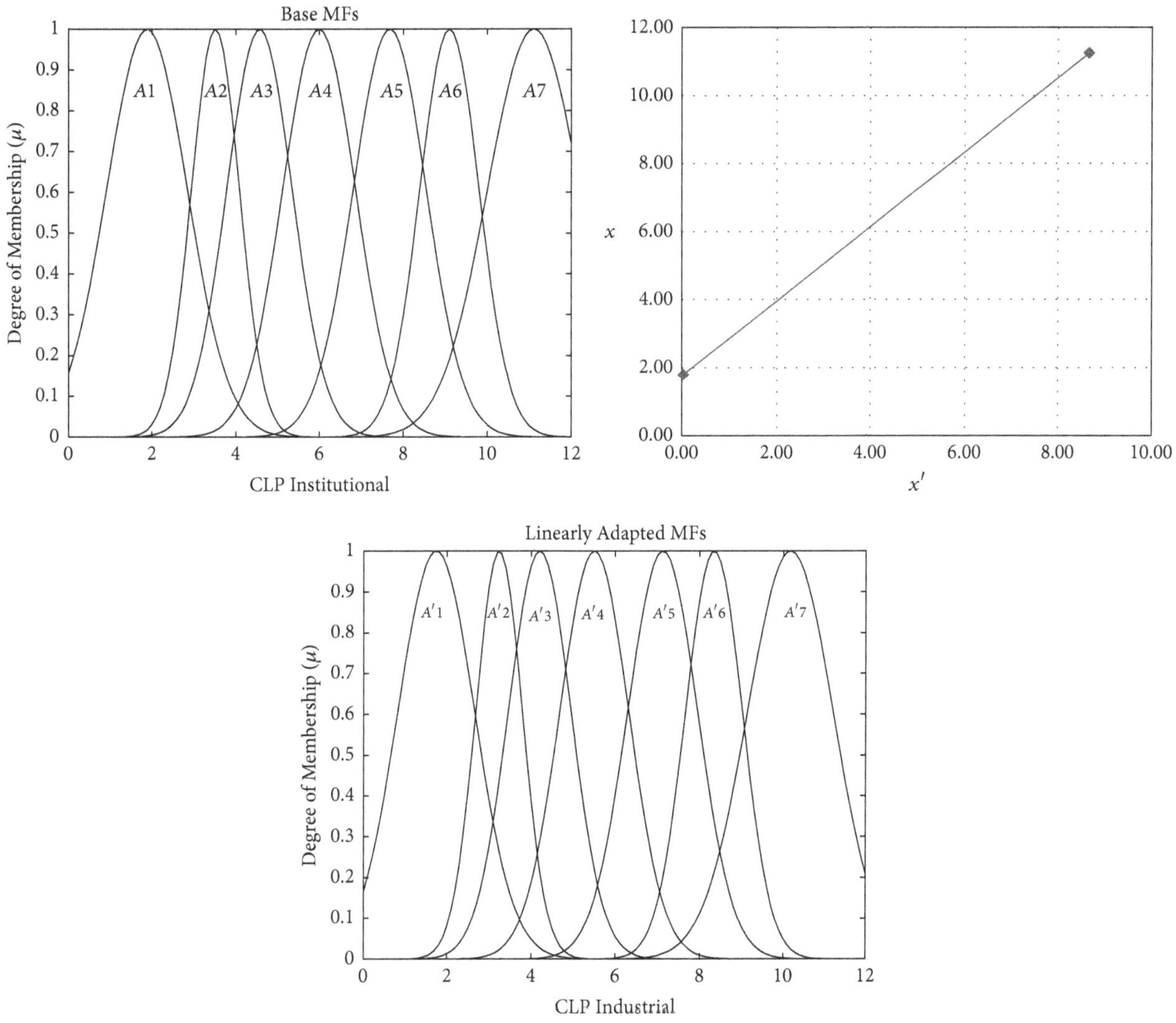

FIGURE 4: Linear context adaptation of CLP feature from Institutional to industrial context.

linear transformation function shown in (6), where $U = [a, b]$ is used to represent the bounds of the adapted MF:

$$\varphi_1 (x, a, b) = \frac{(b - a)}{(u - l)} x + a. \quad (6)$$

Accordingly, the respective institutional CLP context linguistic variables $A(x)$ are adapted to industrial CLP context linguistics variables $A'(x)$ using the linear transformation function shown in the following:

$$\varphi_1 (x) = x' = \frac{(8.66 - 0.03)}{(11.25 - 1.80)} x + 0.03 = 0.913x + 0.03. \quad (7)$$

Thus, the seven MFs $A'(x)$ for CLP feature in the adapted context (industrial context) are determined by replacing the parameters of the membership function $[\sigma, \mu]$ of the base fuzzy sets $A_j = \{A_1, A_2, \ldots, A_c\}$ with adapted values based on $\varphi_1(x)$. The parameters of the seven adapted Gaussian MFs $A'(x, \sigma', \mu')$ of the CLP variable are $A'_{11}(x) = G(x, 0.918, 1.742)$, $A'_{12}(x) = G(x, 0.553, 3.239)$, $A'_{13}(x) = G(x, 0.748, 4.203)$, $A'_{14}(x) = G(x, 0.819, 5.500)$, $A'_{15}(x) = G(x, 0.831, 7.123)$, $A'_{16}(x) = G(x, 0.676, 8.341)$, and $A'_{17}(x) = G(x, 1.035, 10.171)$.

The results of the linear adaptation of the MFs are shown in Figure 4. The degree of overlap among the adapted MFs is not exceeding the limiting value of 0.8 [17]. A similar linear adaptation procedure was applied to the other 11 input variables of the Institutional CLP model.

Then, the MFs of the Institutional context model were replaced with the adapted ones, resulting in the linearly adapted Institutional context model for use in industrial context. The linearly adapted CLP model was used to predict CLP values of the industrial context, which had 23 data instances (refer to Table 1). The adapted model had an initial $\text{RMSE}_{\text{CA-L}}$ value of 1.832.

4.2. Nonlinear Adaptation. Nonlinear adaption involves the use of a nonlinear transformation function that changes the universe of discourse of the base or base MFs and also modifies the shape and distribution of the MFs in the space of the adapted universe of discourse [3]. Accordingly, the base or base MFs defined over a universe of discourse of $B = [l, u]$

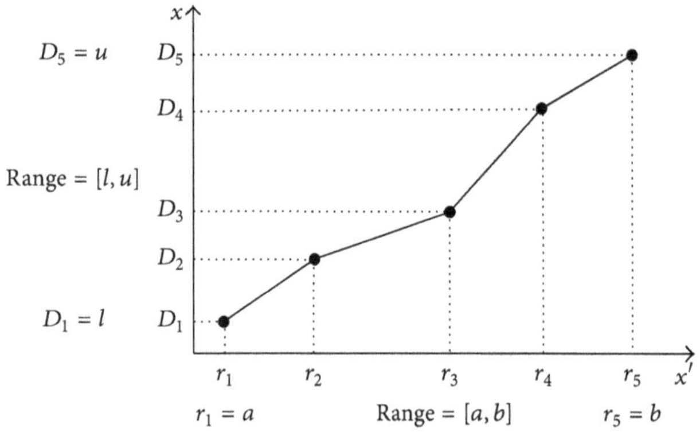

FIGURE 5: Nonlinear context adaptation function: φ_2.

are adapted in the context-adapted universe of discourse by means of a nonlinear transformation function φ_2, where $U = [a, b]$ represents the identified bounds of the adapted MFs (refer to Figure 5). The determination of the parameters of the nonlinear transformation function requires an optimization process [12].

In this research, a piecewise linear transformation function is used in order to develop an interpretable, logical, and fully invertible, where the inverse is also a function, nonlinear context adaptation process [16]. For piecewise linear transformation function φ_2, shown in Figure 5, the set of the adjustable parameters $\mathbf{p}$ is made up of a collection of the split points $r_1, r_2, \ldots, r_5$ and associated differences $D_1, D_2, \ldots, D_5$, represented as $\mathbf{p} = [r_1, r_2, \ldots, r_5, D_1, D_2, \ldots, D_5]$. The piecewise functions, which will be determined using an optimization process discussed below, will result in nonlinear mapping as some regions of x will be contracted and some of them will be expanded, resulting in modification of the shape and distribution of the MFs in the space of the adapted universe of discourse [16].

In order to improve the effectiveness of the optimization process required for determining the parameters of the piecewise linear transformation function with limited data from new contexts, the number of split points is kept to five points, as setting a higher number will have required increased amount of data and on the other hand setting a lower number will reduce the nonlinearity of the function. According to Gudwin et al. [12], context transformation functions are expected to fulfil the following requirements: continuity, monotonicity, and boundary conditions. The use of specifically nondecreasing monotonic piecewise functions ensures that the meaning and order of the linguistic terms is not changed [10]. Additionally, setting the boundary conditions $\varphi_2(l) = a$ and $\varphi_2(u) = b$ allows for the coverage of the new context data.

4.2.1. Computation of Nonlinear Transformation Function. Once the format or type of the nonlinear transfer function φ_2 is selected, the determination of the parameters of φ_2 in terms of $\mathbf{p} = [r_1, r_2, \ldots, r_5, D_1, D_2, \ldots, D_5]$ was carried out via optimization computations. The optimization process begins with the collection of MFs (linguistic terms) $A_j = \{A_1, A_2, \ldots, A_c\}$ as the base MFs of a given model feature (model input or output variable) and the dataset $(d_1, d_2, \ldots, d_N)$ collected for the same feature, but from the new context. Then, using Fuzzy C-Means (FCM) clustering, the preliminary MFs representing the adapted fuzzy sets $A'_0 = \{A'_{10}, A'_{20}, \ldots, A'_{c0}\}$ were developed using the collected dataset $(d_1, d_2, \ldots, d_N)$. The numbers of prototypes or cluster centres for FCM clustering are set equal to the number of base or base MFs, and the commonly used fuzzification coefficient of 2.0 is used [16]. The resulting degrees of memberships of each data instance in the preliminary adapted fuzzy sets are arranged in the form of $N(c+1)$ – tuples, as shown below, where the kth tuple consists of d_k that denotes a point, under consideration, in adapted universe of discourse x' where $\mu_{k1}, \mu_{k2}, \ldots, \mu_{kc}$ are the numeric values of the corresponding membership degrees of d_k in preliminary adapted fuzzy sets $A'_{10}, A'_{20}, \ldots, A'_{c0}$, respectively:

$$\begin{gathered}(d_1, (\mu_{11}, \mu_{12}, \ldots, \mu_{1c})) \\ (d_2, (\mu_{21}, \mu_{22}, \ldots, \mu_{2c})) \\ \vdots \\ (d_N, (\mu_{N1}, \mu_{N2}, \ldots, \mu_{Nc})).\end{gathered} \tag{8}$$

Then, the difference between the degree of membership of d_k from the preliminary adapted MF and the degree of membership of d_k computed using the transformation process was computed for each MF and for the respective data instance. This difference between the preliminary adapted fuzzy sets and the adapted MFs developed using the nonlinear transformation functions formed the objective function of the optimization process. The objective function Q was calculated using sum of squared differences as shown in the following:

$$\begin{aligned}Q(p) = {} & \sum_{i=1}^{c} (A_i(\varphi(d_1, \mathbf{p}) - \mu_{1i}))^2 \\ & + \sum_{i=1}^{c} (A_i(\varphi(d_2, \mathbf{p}) - \mu_{2i}))^2 + \ldots \\ & + \sum_{i=1}^{c} (A_i(\varphi(d_N, \mathbf{p}) - \mu_{Ni}))^2\end{aligned} \tag{9}$$

Thus, the determination of the nonlinear transformation function involved the minimization of the objective function with respect to parameters of $\mathbf{p}$. The solution of this constrained nonlinear minimization optimization problem can be effectively developed using genetic algorithm, as the objective function is nonlinear and nonconvex (has multiple feasible regions), as the use of the traditional gradient based optimization techniques, such as generalized reduced gradient approach, will only lead to local optimum solutions [12]. Accordingly, an optimization process using genetic algorithm was used, and the parameters of the nonlinear transformation function $\varphi_2 = \mathbf{p} = [r_1, r_2, \ldots, r_5, D_1, D_2, \ldots, D_5]$ were used for real coding of the chromosome in the genetic optimization process. The objective of the genetic search was to minimize the objective function Q, and the fitness value of each solution was determined by calculating Q. Then, the genetic operations of reproduction, crossover, and mutation were performed. Each operation generated new sets of chromosomes, representing a new solution that meets the optimization constraints. The solution chromosomes are checked according to the following nonlinear context adaptation constraints. (1) The parameters of the nonlinear transformation function $\mathbf{p}$ must be greater than zero. (2) Boundary conditions for coverage of the new context data are defined over $U = [a, b]$ and using the base or base MF range of $B = [l, u]$: $r_1 = l$, $r_5 = u$, $D_1 = a$, and $D_5 = b$. (3) The jth split value r_j must not be greater than that of the $j+1$th split value r_{j+1}. (4) The jth difference value D_j must not be greater than that of the $j+1$th difference value D_{j+1}. In the genetic optimization process, initial population of 100 solutions was randomly generated, and mutation rate of 0.075 and stopping criteria based on convergence value of 0.0001 was used.

4.2.2. Computational Results for Nonlinear Adaptation of CLP Models. The numerical illustration of the nonlinear context adaptation of the output feature (CLP) from institutional to industrial context is discussed here. As discussed above, the seven base fuzzy sets $A_j = \{A_1, A_2, \dots, A_7\}$ for CLP variable and associated MFs from institutional context were first documented. Next the CLP dataset $(d_1, d_2, \dots, d_N)$ collected from the industrial context is retrieved and had values of universe of discourse of $U = [0.03, 8.66]$. Using Fuzzy C-Means (FCM) clustering, the preliminary MFs of the adapted fuzzy sets $A'_0 = \{A'_{01}, A'_{02}, \dots, A'_{07}\}$ were developed, where the number of prototypes for FCM clustering was set at seven, equal to the original number of MFs. Then, using genetic algorithm based optimization, the parameters of the nonlinear piecewise transformation function $\mathbf{p}$ were developed, and the resulting parameters were $\varphi_2 = [r_1 = 0.03, r_2 = 2.05, r_3 = 5.91, r_4 = 6.36, r_5 = 8.66, D_1 = 1.80, D_2 = 5.52, D_3 = 7.76, D_4 = 9.63, D_5 = 11.25]$.

Thus, the seven MFs for CLP variable in the adapted context (industrial context) $A'_{2j}(x) = \{A'_{21}, A'_{22}, \dots, A'_{2c}\}$ were determined by replacing the parameters of the membership function $[\sigma, \mu]$ of the base fuzzy sets $A_j = \{A_1, A_2, \dots, A_c\}$ with adapted values based on $\varphi_2(x)$. The results of the nonlinear adaptation of the MFs are shown in Figure 6.

The parameters of the seven nonlinearly adapted Gaussian MFs $A'_{2j}(x, \sigma', \mu')$ of the CLP variable were $A'_{21}(x) = G(x', 0.528, 1.018)$, $A'_{22}(x) = G(x, 0.311, 1.908)$, $A'_{23}(x) = G(x, 0.427, 2.481)$, $A'_{24}(x) = G(x, 0.469, 2.860)$, $A'_{25}(x) = G(x, 0.476, 5.760)$, $A'_{26}(x) = G(x, 0.384, 6.233)$, and $A'_{27}(x) = G(x, 0.598, 8.454)$. The results shown in Figure 6 indicate that the shape of some of the MFs has contracted, the distribution of the MFs over the universe of discourse have been modified, and in some cases the degree of overlap among the adapted sets was higher than the recommended maximum value of 0.8 [17]. Such changes of MFs will naturally reduce the interpretability of the adapted models; however, this is a common problem witnessed in nonlinear adaptation of fuzzy systems [15].

A similar nonlinear adaptation procedure was applied to the remaining 11 input variables of the institutional CLP model. Then, the MFs of the institutional context model were replaced with the adapted ones, resulting in the nonlinearly adapted institutional context model for use in industrial context. The nonlinearly adapted CLP model was used to predict CLP values of the industrial context, which had 23 data instances (refer to Table 1). The adapted model had an initial $\text{RMSE}_{\text{CA-NL}}$ value of 2.742.

4.3. Sensitivity Analysis of Adapted Models for Fuzzy Operators and Defuzzification Methods. The sensitivity of the linearly and nonlinearly adapted CLP models was then further evaluated by changing the fuzzy operators and defuzzification methods. The following options of fuzzy operators and defuzzification methods were tested: for input aggregation [MIN (minimum) and PROD (product)], for implication [MIN (minimum) and PROD (product)], for rule aggregation [MAX (maximum), SUM (sum of each rule's output set), and PROBOR (probabilistic OR)], and for defuzzification [CENTROID, BISECTOR, MOM (middle of maximum), LOM (largest of maximum), and SOM (smallest of maximum)]. The sensitivity options were varied one at a time, and a total of 30 unique combinations were tested. The options and results for linearly adapted institutional context model, which was linearly adapted to suit industrial context, are shown in Table 2.

For each linear and nonlinear adapted CLP model, the final adapted model's (i.e., after sensitivity analysis) accuracy measures of $\text{RMSE}_{\text{CA-LS}}$ and $\text{RMSE}_{\text{CA-NLS}}$ of 1.832 and 2.738, respectively, were determined based on the best combination of the listed options of fuzzy operators and defuzzification methods. The results yielding the lowest $\text{RMSE}_{\text{CA-LS}}$ and $\text{RMSE}_{\text{CA-NLS}}$ values, shown in bold in Table 2, provided the best adapted fuzzy operators and defuzzification methods for adapted CLP models. Finally, the agreement index WI_i between the adapted and the base model, described in Table 3, was computed using the modified Willmott index (see (5)). Accordingly, the appropriate context adaptation approach was the linearly adapted model, as it had an agreement index of 0.340, while the nonlinearly adapted model had an agreement index of only 0.142.

5. Results and Discussion

The linear and nonlinear context adaptation framework was tested and validated using the four context-specific CLP models, as shown in Figure 3, by comparing the prediction results of the context adapted models with the actual field data of the new context. The results of the context adaptation process, summarized for each context, are presented in the following sections.

5.1. Industrial Context CLP Models. For the industrial context, six adapted models were developed from warehouse, high-rise, and institutional contexts based on linear and nonlinear adaptation process. The base CLP model for industrial context had an RMSE value of 1.162. In Table 3

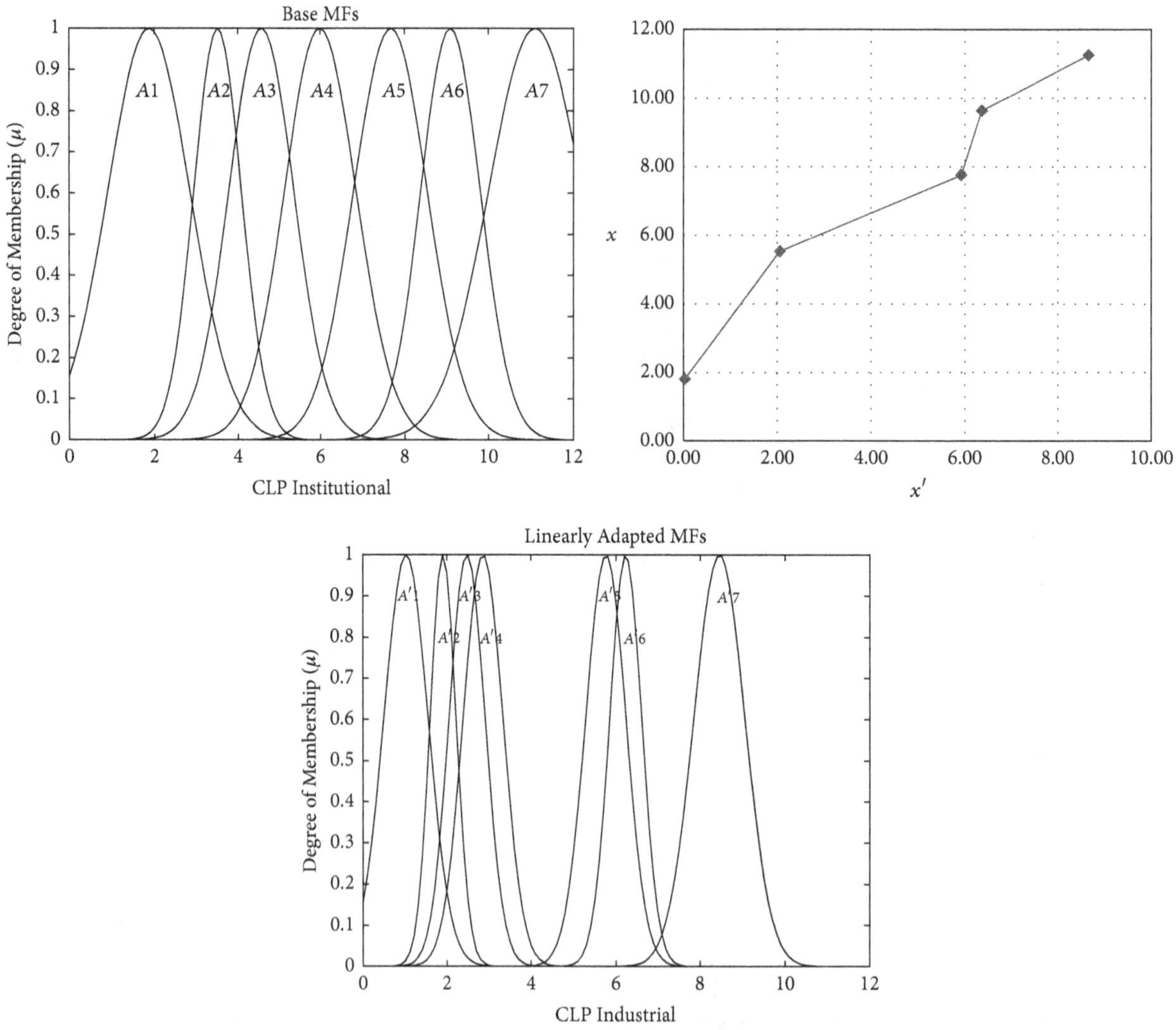

FIGURE 6: Nonlinear context adaptation of CLP feature from institutional to industrial context.

the results of the adaptation process are shown, where the initial RMSE values represent the accuracy of the adapted models in predicting the CLP values of the Industrial context, and the final RMSE values represent the accuracy of the adapted models after sensitivity analysis. The best performing fuzzy operators and defuzzification methods are also shown in Table 3.

Based on the agreement indices of the six adapted models, as shown in Table 3, the model linearly adapted from institutional context has the highest agreement with the base model of the industrial context with an agreement index value of 0.340 and RMSE value of 1.832. The model nonlinearly adapted from institutional context has the least agreement with the base model with an agreement index value of 0.142 and RMSE value of 2.738. Additionally, the linearly adapted models from all three contexts, as compared to the nonlinearly adapted models, had better agreement indices with the base model; the models also had a higher accuracy based on RMSE values. However, it should be noted that, in terms of model accuracy, none of the adapted models performed better than the base model.

5.2. Warehouse Context CLP Models. For the warehouse context, six adapted models were developed from industrial, high-rise, and institutional contexts and based on linear and nonlinear adaptation process. The base CLP model for warehouse context had an RMSE value of 0.467. Based on the agreement indices of the six adapted models, the model nonlinearly adapted from industrial context has the highest agreement to the base model of the warehouse context with an agreement index value of 0.459 and RMSE value of 0.939. The model linearly adapted from institutional context has the least agreement with the base model with an agreement index value of 0.144 and RMSE value of 2.267. Additionally, the nonlinearly adapted models from industrial and high-rise contexts, as compared to respective the linear adapted models, had better agreement with the base model; the models also had a higher accuracy based on RMSE values.

Table 2: Context adapted models' sensitivity analysis: linearly and nonlinearly adapted institutional model.

Case	Fuzzy operators and defuzzification methods					
	Input aggregation	Implication method	Rule aggregation	Defuzzification method	Linearly adapted model accuracy ($RMSE_{CA-LS}$)	Nonlinearly adapted model accuracy ($RMSE_{CA-NLS}$)
1	MIN	MIN	MAX	CENTROID	1.840	2.742
2	MIN	MIN	MAX	BISECTOR	1.840	2.742
3	MIN	MIN	MAX	MOM	1.955	2.766
4	MIN	MIN	MAX	LOM	2.030	2.762
5	MIN	MIN	MAX	SOM	1.955	2.766
6	MIN	MIN	SUM	CENTROID	1.840	2.742
7	MIN	MIN	SUM	BISECTOR	1.840	2.742
8	MIN	MIN	SUM	MOM	1.955	2.766
9	MIN	MIN	SUM	LOM	2.030	2.762
10	MIN	MIN	SUM	SOM	1.955	2.766
11	MIN	MIN	PROBOR	CENTROID	1.840	2.742
12	MIN	MIN	PROBOR	BISECTOR	1.840	2.742
13	MIN	MIN	PROBOR	MOM	1.955	2.766
14	MIN	MIN	PROBOR	LOM	2.030	2.762
15	MIN	MIN	PROBOR	SOM	1.955	2.766
16	PROD	PROD	MAX	CENTROID	**1.832**	**2.738**
17	PROD	PROD	MAX	BISECTOR	1.840	2.740
18	PROD	PROD	MAX	MOM	1.955	2.765
19	PROD	PROD	MAX	LOM	2.030	2.762
20	PROD	PROD	MAX	SOM	1.955	2.765
21	PROD	PROD	SUM	CENTROID	**1.832**	**2.738**
22	PROD	PROD	SUM	BISECTOR	1.840	2.740
23	PROD	PROD	SUM	MOM	1.840	2.740
24	PROD	PROD	SUM	LOM	2.030	2.762
25	PROD	PROD	SUM	SOM	1.955	2.765
26	PROD	PROD	PROBOR	CENTROID	**1.832**	**2.738**
27	PROD	PROD	PROBOR	BISECTOR	1.840	2.742
28	PROD	PROD	PROBOR	MOM	1.955	2.766
29	PROD	PROD	PROBOR	LOM	2.030	2.762
30	PROD	PROD	PROBOR	SOM	1.955	2.766

However, in terms of model accuracy, none of the adapted models performed better than the base model.

5.3. High-Rise Context CLP Models. For the high-rise context, six adapted models were developed from industrial, warehouse, and institutional contexts and based on linear and nonlinear adaptation process. The base or base CLP model for high-rise context had an RMSE value of 0.992. Based on the agreement indices of the six adapted models, the model nonlinearly adapted from warehouse context has the highest agreement to the base model of the high-rise context with an agreement index value of 0.427 and RMSE value of 3.851. The model linearly adapted from institutional context has the least agreement with the base model with an agreement index value of 0.181 and RMSE value of 4.627. Additionally, the nonlinearly adapted models from all three contexts, as compared to the linearly adapted models, had better or equal agreement with the base model. However, both linearly and nonlinearly adapted models had similar accuracy based on RMSE values. Similar to the contexts discussed above, none of the adapted models performed better than the base model.

5.4. Institutional Context CLP Models. For the institutional context, six adapted models were developed from industrial, warehouse, and high-rise contexts and based on linear and nonlinear adaptation process. The base CLP model for institutional context had an RMSE value of 0.671. Based on the agreement indices of the six adapted models, the model linearly adapted from industrial context has the highest agreement to the base model of the institutional context with an agreement index value of 0.398 and RMSE value of 2.552. The model nonlinearly adapted from industrial context has

TABLE 3: Context adaptation results for industrial context (C1).

Adapted models	Adapted from warehouse: C2		Adapted from high-rise: C3		Adapted from institutional: C4	
	Linear	Nonlinear	Linear	Nonlinear	Linear	Nonlinear
RMSE (initial)	1.892	2.079	1.880	1.880	1.832	2.742
RMSE (final)	1.831	2.028	1.873	1.873	1.832	2.738
Sensitivity improvement (%)	3.26	2.47	0.32	0.32	0.00	0.14
Agreement index	**0.203**	0.188	**0.282**	**0.282**	**0.340**	0.142
Input aggregation operator	PROD	PROD	PROD	PROD	PROD	PROD
Implication method	PROD	PROD	PROD	PROD	PROD	PROD
Rule aggregation operator	SUM	MAX	PROBOR	PROBOR	PROBOR	PROBOR
Defuzzification method	BISECTOR	MOM	MOM	MOM	CENTROID	CENTROID

the least agreement with the base model with an agreement index value of 0.199 and RMSE value of 2.552. However, a clear difference between linear and nonlinear adaptation options based on agreement or accuracy measures were not found in the institutional context case. Also, in terms of model accuracy, none of the adapted models performed better than the base model.

In summary, the review of the linear and nonlinear adaptation approaches indicated that linear adapted models were in better agreement with those of the base models for industrial and institutional contexts, and nonlinear adapted models were in better agreement with those of the base models for warehouse and high-rise contexts. However, the use of nonlinear adaptation approach has resulted in reduced interpretability of the adapted models, due to the change in the shape and distribution and higher overlap of adapted MFs. Notably, the sensitivity analysis on fuzzy operators and defuzzification methods did not show significant improvement in adapted model's accuracy. This finding supports the option used in past studies, such as Botta [3], which only used minimum operator for input aggregation and implication and centroid for defuzzification.

In terms of prediction accuracy, in all four contexts, none of the adapted models performed better than the base models. This is expected as the base models have been developed and further optimized to improve models' accuracy [7]. The comparison of the adapted models' accuracy with those of the base context-specific models, before optimization process was carried out to fine tune the base model's parameters as described in Tsehayae and Fayek [7], showed promising results for the industrial and warehouse contexts. The base industrial context model before optimization had an RMSE value of 1.582, and the most accurate linearly adapted model from warehouse context has an RMSE value of 1.831. Similarly, the base warehouse context model before optimization had an RMSE value of 0.586 and the most accurate linearly adapted model from high-rise context has an RMSE value of 0.719.

Considering the effort and cost required for collecting data on all CLP influencing variables and to develop and optimize new models, the use of context adaptation framework presented in this paper, which enables the reuse of existing CLP models through the transfer of knowledge from one context to another will provide a simpler and efficient alternative for developing CLP models.

6. Conclusions and Future Research

Construction labor productivity is one of the most studied areas in the construction research field, and several-FIS-based predictive models have been developed; however, a framework for adapting such FIS-based models from one context to another is missing. This paper developed a context adaptation framework for transferring the knowledge represented in FIS-based CLP models from one context to another. The framework employed linear and genetic optimization-based nonlinear MF transformation functions accompanied by sensitivity analysis for fuzzy operators and defuzzification methods of the adapted FIS models. Using four context-specific CLP models for concreting activity under industrial, warehouse, high-rise, and institutional building contexts, the developed context adaptation framework was implemented. Comparisons of the adapted models, using agreement indices and RMSE accuracy measures, with the base model of the four contexts uncovered many findings. The results of the investigation indicated that linearly adapted CLP models were similar to the base CLP models in industrial and institutional contexts, while the nonlinearly adapted CLP models were similar to the base CLP models in warehouse and high-rise contexts. However, in terms of model accuracy, none of the adapted models performed better than the base model of a given context. Additionally, the sensitivity analysis on fuzzy operators and defuzzification methods did not show significant improvement in adapted model accuracy. Furthermore, the best adapted model for each context was validated and contextual similarities in terms of CLP prediction agreement indices between base and adapted models were examined.

The contributions of this paper can be grouped into three areas. First, the paper showed the value of context adaptation approaches using a practical context-sensitive application problem (CLP modeling). While past studies manly focused on demonstrating the value of context adaptation approaches using benchmark datasets and theoretical context independent MFs defined over a normalized universe of discourse, this study tested the context adaptation approaches using real world CLP datasets collected from six construction projects and context-dependent MFs defined over the actual universes of discourses of FIS model features. This study also investigated the effect of fuzzy operators and defuzzification methods in the context adaptation of FIS and showed their limited effect on the accuracy of the adapted models. Second, the paper has advanced the state of the art in fuzzy modeling in CLP studies by introducing and testing an approach for adapting MF using linear and genetic optimization-driven nonlinear transformation functions, which provided a transparent approach for adapting MFs from base to adapted contexts. Finally, the paper developed a context adaptation framework, which enables researchers and industry practitioners to take full advantage of existing FIS-based CLP models in the study of new contexts for which data availability is limited. The framework also improves the application of existing CLP models by construction companies as existing models can be adapted to suit the companies' specific need or context. Also, the comparison of the adapted and original models demonstrated that, for the sake of CLP model development, similarities among unique construction contexts do exist. Such similarities will assist project managers to easily transfer past experiences and knowledge bases and lead to the better management of future projects.

Currently, additional data are being collected from ongoing construction projects to further test this study and its findings and to examine other nonlinear transformation functions so as to improve the prediction accuracy of adapted CLP models. Future research is focusing on the development of a framework for abstraction of various context-specific CLP models to develop a universal or granular CLP model capable of predicting CLP in any context. This approach will explore the practicality and performance of a variety of granular approaches, including case-based reasoning, fuzzy clustering, and fuzzy regression, to identify and aggregate the most important relationships represented in the context-specific models and produce a universal model that is better able to predict CLP.

Conflicts of Interest

The authors declare that there are no conflicts of interest regarding the publication of this paper.

Acknowledgments

The authors would like to express their sincere appreciation to all of the companies that participated in this study for their cooperation, time, and the valuable information they provided. The authors particularly recognize the graduate research assistants who diligently helped with data collection. This research was conducted under the NSERC Industrial Research Chair in Strategic Construction Modeling and Delivery (NSERC IRCPJ 428226–15), held by Dr. Aminah Robinson Fayek. The financial support of the industrial partners to this chair and of the Natural Sciences and Engineering Research Council of Canada is gratefully acknowledged.

References

[1] W. Yi and A. Chan, "Critical review of labor productivity research in construction journals," *Journal of Management in Engineering*, vol. 30, no. 2, pp. 214–225, 2014.

[2] K. M. El-Gohary, R. F. Aziz, and H. A. Abdel-Khalek, "Engineering approach using ANN to improve and predict construction labor productivity under different influences," *Journal of Construction Engineering and Management*, vol. 143, no. 8, Article ID 04017045, 2017.

[3] A. Botta, *Automatic context adaptation of fuzzy systems [Ph.D. thesis]*, IMT Institute for Advanced Studies, Lucca, Italy, 2008.

[4] H. Mao, *Estimating labour productivity using fuzzy set theory [M.Sc. thesis]*, University of Alberta, Edmonton, Alberta, Canada, 1999.

[5] N.-F. Pan, "Assessment of productivity and duration of highway construction activities subject to impact of rain," *Expert Systems with Applications*, vol. 28, no. 2, pp. 313–326, 2005.

[6] A. R. Fayek and A. Oduba, "Predicting industrial construction labor productivity using fuzzy expert systems," *Journal of Construction Engineering and Management*, vol. 131, no. 8, pp. 938–941, 2005.

[7] A. A. Tsehayae and A. R. Fayek, "Developing and Optimizing Context-Specific Fuzzy Inference System-Based Construction Labor Productivity Models," *Journal of Construction Engineering and Management*, vol. 142, no. 7, Article ID 4016017, 2016.

[8] A. Botta, B. Lazzerini, F. Marcelloni, and D. C. Stefanescu, "Context adaptation of fuzzy systems through a multi-objective evolutionary approach based on a novel interpretability index," *Soft Computing*, vol. 13, no. 5, pp. 437–449, 2009.

[9] O. Cordón, F. Herrera, L. Magdalena, and P. Villar, "A genetic learning process for the scaling factors, granularity and contexts of the fuzzy rule-based system data base," *Information Sciences*, vol. 136, no. 1-4, pp. 85–107, 2001.

[10] W. Pedrycz, R. R. Gudwin, and F. A. C. Gomide, "Nonlinear context adaptation in the calibration of fuzzy sets," *Fuzzy Sets and Systems*, vol. 88, no. 1, pp. 91–97, 1997.

[11] M. Engwall, "No project is an island: Linking projects to history and context," *Research Policy*, vol. 32, no. 5, pp. 789–808, 2003.

[12] R. R. Gudwin, F. A. Gomide, and W. Pedrycz, "Context adaptation in fuzzy processing and genetic algorithms," *International Journal of Intelligent Systems*, vol. 13, no. 10-11, pp. 929–948, 1998.

[13] R. R. Gudwin and F. Gomide, "Context adaptation in fuzzy processing," in *Proceedings of Brazil-Japan Joint Symposium on Fuzzy Systems*, Campinas, Brazil, 1994.

[14] L. Magdalena, "Adapting the gain of an FLC with genetic algorithms," *International Journal of Approximate Reasoning*, vol. 17, no. 4, pp. 327–349, 1997.

[15] D. T. Ho, *Context dependent fuzzy modelling and its application [Ph.D. thesis]*, University of Nottingham, Nottingham, UK, 2013.

[16] W. Pedrycz and F. Gomide, *Fuzzy Systems Engineering: toward Human-Centric Computing*, John Wiley and Sons, Hoboken, New Jersey, US, 2007.

[17] P. Pulkkinen and H. Koivisto, "A dynamically constrained multiobjective genetic fuzzy system for regression problems," *IEEE Transactions on Fuzzy Systems*, vol. 18, no. 1, pp. 161–177, 2010.

[18] L. Magdalena, "On the role of context in hierarchical fuzzy controllers," *International Journal of Intelligent Systems*, vol. 17, no. 5, pp. 471–493, 2002.

[19] G. Klir and B. Yuan, *Fuzzy Sets and Fuzzy Logic*, Prentice Hall, Upper Saddle River, New Jersey, US, 1995.

[20] A. Awad and A. R. Fayek, "Adaptive learning of contractor default prediction model for surety bonding," *Journal of Construction Engineering and Management*, vol. 139, no. 6, pp. 694–704, 2013.

[21] A. Idrus, M. F. Fadhil Nuruddin, and M. A. Rohman, "Development of project cost contingency estimation model using risk analysis and fuzzy expert system," *Expert Systems with Applications*, vol. 38, no. 3, pp. 1501–1508, 2011.

[22] S. S. S. Ahmad, S. Ahmad, and W. Pedrycz, "Data and feature reduction in fuzzy modeling through particle swarm optimization," *Applied Computational Intelligence and Soft Computing*, vol. 10, pp. 1–21, 2012.

[23] C. Willmott, S. Robeson, and K. Matsuura, "A refined index of model performance," *International Journal of Climatology*, vol. 32, no. 13, pp. 2088–2094, 2012.

[24] A. A. Tsehayae, *Developing and optimizing context-specific and universal construction labour productivity models [Ph.D. thesis]*, University of Alberta, Edmonton, Canada, 2015.

18

FCM-Type Fuzzy Coclustering for Three-Mode Cooccurrence Data: 3FCCM and 3Fuzzy CoDoK

Katsuhiro Honda,[1] Yurina Suzuki,[1] Seiki Ubukata,[1] and Akira Notsu[2]

[1] *Graduate School of Engineering, Osaka Prefecture University, Sakai, Osaka 599-8531, Japan*
[2] *Graduate School of Humanities and Sustainable System Sciences, Osaka Prefecture University, Sakai, Osaka 599-8531, Japan*

Correspondence should be addressed to Katsuhiro Honda; honda@cs.osakafu-u.ac.jp

Academic Editor: Ferdinando Di Martino

Cocluster structure analysis is a basic technique for revealing intrinsic structural information from cooccurrence data among objects and items, in which coclusters are composed of mutually familiar pairs of objects and items. In many real applications, it is also the case that we have not only cooccurrence information among objects and items but also intrinsic relation among items and other ingredients. For example, in food preference analysis, users' preferences on foods should be found considering not only user-food cooccurrences but also the implicit relation among users and cooking ingredients. In this paper, two FCM-type fuzzy coclustering models, that is, FCCM and Fuzzy CoDoK, are extended for revealing intrinsic cocluster structures from three-mode cooccurrence data, where the aggregation degree of three elements in each cocluster is maximized through iterative updating of three types of fuzzy memberships for objects, items, and ingredients. The characteristic features of the proposed methods are demonstrated through a numerical experiment.

1. Introduction

In many web data analyses, we often have cooccurrence information among objects and items instead of multidimensional observations on objects. For example, web document summarization and web market purchase summarization are reduced to document-keyword cooccurrence analysis and customer-product basket analysis, respectively. FCM-type fuzzy coclustering is an extension of fuzzy c-Means (FCM) [1], where the degree of belongingness to clusters is represented by fuzzy memberships under the fuzzy partition concept [2]. Fuzzy clustering for categorical multivariate data (FCCM) [3] replaced the FCM clustering criterion with the aggregation degree of objects and items in coclusters by adopting entropy-based fuzzification [4, 5]. In fuzzy coclustering of documents and keywords (fuzzy CoDoK) [6], the FCCM criterion was maximized with quadratic regularization-based fuzzification [7], so that it can be applied to large data sets.

Besides their usefulness in many applications, it is also the case that the conventional fuzzy coclustering models cannot work well under severe influences of other intrinsic features. For example, in food preference analysis, users' preferences on foods cannot be revealed considering only user-food cooccurrences but should be found considering implicit relation among users and cooking ingredients, which compose the foods. Then, when we have not only cooccurrence information among objects and items but also intrinsic relation among items and other ingredients; we can expect to find more useful cocluster structures in three-mode cooccurrence information data.

In this paper, two FCM-type fuzzy coclustering models are extended for analyzing three-mode cooccurrence information data, in which FCM-like alternative optimization schemes are performed considering cooccurrence relation among objects, items, and other ingredients. First, the FCCM algorithm is extended to the three-mode FCCM (3FCCM) algorithm by utilizing three types of fuzzy memberships for objects, items, and ingredients, where the aggregation degree of three features in each cocluster is maximized through iterative updating of memberships supported by the entropy-based fuzzification. Second, the 3FCCM algorithm is further extended to the three-mode Fuzzy CoDoK (3Fuzzy CoDoK) by introducing the quadratic regularization-based fuzzification. The characteristic features of the proposed methods are demonstrated through a numerical experiment.

The remainder of this paper is organized as follows: Section 2 gives a brief review on the conventional FCM-type fuzzy coclustering models and Section 3 proposes the novel extensions of the FCM-type coclustering models for three-mode cooccurrence information data. The experimental result is shown in Section 4 and a summary conclusion is presented in Section 5.

2. FCM-Type Fuzzy Coclustering

Fuzzy c-Means (FCM) [1, 5] is a fuzzy extension of the conventional crisp k-Means [8] by introducing fuzzy partition concept [2]. When we have multidimensional observations on n objects $\mathbf{x}_i$, $i = 1, \ldots, n$, they are partitioned into C fuzzy clusters by estimating fuzzy memberships u_{ci} for each object, where u_{ci} represents the degree of belongingness of object i to cluster c and is generally calculated under the probabilistic constraint of $\sum_{c=1}^{C} u_{ci} = 1$. In FCM, each cluster is represented by prototypical centroids and objects are partitioned so that the membership-weighted within-cluster errors from prototypes are minimized in the multidimensional data space. On the other hand, in the coclustering context, we have only relational information among elements but do not use any cluster prototypes in multidimensional space. In this paper, two variants of FCM-type fuzzy coclustering are considered.

2.1. FCCM. Assume that we have $n \times m$ cooccurrence information $R = \{r_{ij}\}$ among n objects and m items; for example, in document-keyword analysis, r_{ij} can be the frequency of keyword (item) j in document (object) i. The goal is to extract coclusters composed of mutually familiar pairs of objects and items by simultaneously estimating fuzzy memberships of objects u_{ci} and items w_{cj} such that mutually familiar objects and items with large r_{ij} tend to have large memberships in the same cluster considering the aggregation degree of each cocluster. The sum of aggregation degrees to be maximized is defined as [3]

$$L = \sum_{c=1}^{C} \sum_{i=1}^{n} \sum_{j=1}^{m} u_{ci} w_{cj} r_{ij}. \tag{1}$$

This objective function is based on the similar concept to such relational matrix decomposition methods as corresponding analysis (CA) [9] and nonnegative matrix factorization (NMF) [10], where relational matrices $R = \{r_{ij}\}$ are decomposed into two component matrices having orthogonal columns. Beside both objects and items are equally forced to be exclusive in the matrix decomposition methods, FCM-type coclustering models adopt different kinds of partition constraints [11]. Here, object memberships u_{ci} have a similar role to those of FCM under the same condition, such that $\sum_{c=1}^{C} u_{ci} = 1$. If item memberships w_{cj} also obey a similar condition of $\sum_{c=1}^{C} w_{cj} = 1$, the aggregation criterion has a trivial maximum of $u_{ci} = w_{cj} = 1$, $\forall i, j$ in a particular cluster c. Then, in order to avoid trivial solutions, w_{cj} are forced to be exclusive in each cluster, such that $\sum_{j=1}^{m} w_{cj} = 1$, and, so, w_{cj} represent the relative typicalities of items in each cluster. As a result, object partitioning is mainly targeted in FCM-type coclustering while CA and NMF equally force exclusive nature to partitions of both objects and items.

Because of the linear nature with respect to u_{ci} and w_{cj}, (1) is maximized with crisp memberships of $u_{ci} \in \{0, 1\}$ and $w_{cj} \in \{0, 1\}$ in a similar manner to k-Means. In order to find fuzzy partition, some fuzzification mechanism must be introduced like FCM.

In [3], the linear aggregation criterion of (1) was nonlinearized with respect to u_{ci} and w_{cj} by entropy-based penalties [4, 5] for fuzzification of two-types of memberships and the objective function for Fuzzy Clustering for Categorical Multivariate data (FCCM) was proposed as

$$L_{\text{fccm}} = \sum_{c=1}^{C} \sum_{i=1}^{n} \sum_{j=1}^{m} u_{ci} w_{cj} r_{ij} - \lambda_u \sum_{c=1}^{C} \sum_{i=1}^{n} u_{ci} \log u_{ci} - \lambda_w \sum_{c=1}^{C} \sum_{j=1}^{m} w_{cj} \log w_{cj}, \tag{2}$$

where λ_u and λ_w are the fuzzification weights for object memberships and item memberships, respectively. Larger λ_u and λ_w bring fuzzier partitions of objects and items.

Based on the alternative optimization principle, u_{ci} and w_{cj} are iteratively updated until convergent using the following updating rules:

$$u_{ci} = \frac{\exp\left(\lambda_u^{-1} \sum_{j=1}^{m} w_{cj} r_{ij}\right)}{\sum_{l=1}^{C} \exp\left(\lambda_u^{-1} \sum_{j=1}^{m} w_{lj} r_{ij}\right)},$$
$$w_{cj} = \frac{\exp\left(\lambda_w^{-1} \sum_{i=1}^{n} u_{ci} r_{ij}\right)}{\sum_{l=1}^{m} \exp\left(\lambda_w^{-1} \sum_{i=1}^{n} u_{ci} r_{il}\right)}. \tag{3}$$

Although the two updating rules are always fair under the constraints, they can be numerically unstable due to overflows because $\exp(\cdot)$ function can take extremely large values with very large n or m.

2.2. Fuzzy CoDoK. As an alternative approach, Kummamuru et al. [6] extended FCCM by introducing the quadric term-based fuzzification mechanism [7] instead of the entropy-based fuzzification, so that it can handle larger data sets. The objective function of fuzzy coclustering of documents and keywords (Fuzzy CoDoK) was proposed as

$$L_{\text{fcdk}} = \sum_{c=1}^{C} \sum_{i=1}^{n} \sum_{j=1}^{m} u_{ci} w_{cj} r_{ij} - \lambda_u \sum_{c=1}^{C} \sum_{i=1}^{n} u_{ci}^2 - \lambda_w \sum_{c=1}^{C} \sum_{j=1}^{m} w_{cj}^2, \tag{4}$$

where λ_u and λ_w play similar roles to FCCM.

Based on the Lagrangian multiplier method, the updating rules are obtained as

$$u_{ci} = \frac{1}{C} + \frac{1}{2\lambda_u}\left\{\sum_{j=1}^{m} w_{cj} r_{ij} - \frac{1}{C}\left(\sum_{l=1}^{C}\sum_{j=1}^{m} w_{lj} r_{ij}\right)\right\},$$
$$w_{cj} = \frac{1}{m} + \frac{1}{2\lambda_w}\left\{\sum_{i=1}^{n} u_{ci} r_{ij} - \frac{1}{m}\left(\sum_{l=1}^{m}\sum_{i=1}^{n} u_{ci} r_{il}\right)\right\}. \quad (5)$$

The updating rules are more numerically stable than those of FCCM because their calculation ranges are in linear orders with respect to n and m. However, u_{ci} and w_{cj} can be negative and are not fair under the constraints. Then, in practice, the negative memberships are set to zero, and the remaining positive memberships are renormalized so that their sum is one.

Besides the usefulness of these fuzzy coclustering models in handling two-modes cooccurrence information, their cocluster structures may be influenced by other third elements. Specifically, if each item is related to some other ingredients, the partition quality is expected to be improved by considering the intrinsic relation among three-mode elements. In the following section, the FCM-type coclustering algorithms are extended for analyzing such three-mode cooccurrence information data.

3. Extension of FCM-Type Coclustering for Three-Mode Cooccurrence Data Analysis

Assume that we have $n \times m$ cooccurrence information $R = \{r_{ij}\}$ among n objects and m items, and the items are characterized with other ingredients, where cooccurrence information among m items and p other ingredients are summarized in $m \times p$ matrix $S = \{s_{jk}\}$ with s_{jk} representing the cooccurrence degree of item j and ingredient k. For example, in food preference analysis, R can be an evaluation matrix by n users on m foods and S may be appearance/absence of p cooking ingredients in m foods. The goal of three-mode cocluster analysis is to reveal the cocluster structures among the objects, items, and ingredients considering R and S and intrinsic relation among objects and ingredients.

In order to extend the conventional FCCM and Fuzzy CoDoK algorithms to three-mode cocluster analysis, additional memberships z_{ck} are introduced for representing the membership degree of ingredients k to cocluster c. Besides the familiar pairs of objects and items simultaneously occur in the same cluster; typical ingredients of the items should also belong to the same cluster. Then, the aggregation degree to be maximized in the three-mode coclustering can be as

$$L = \sum_{c=1}^{C}\sum_{i=1}^{n}\sum_{j=1}^{m}\sum_{k=1}^{p} u_{ci} w_{cj} z_{ck} r_{ij} s_{jk}, \quad (6)$$

where each cluster should be composed of the familiar group of objects, items, and ingredients such that they are assigned to the same cluster when object i cooccurs with item j composed of ingredient k by implying an intrinsic connection between object i and ingredient k.

In the following parts of this section, the conventional FCCM and Fuzzy CoDoK algorithms are extended to their three-mode versions utilizing the above aggregation criterion.

3.1. Three-Mode Extension of FCCM. First, the FCCM algorithm is extended by using the modified aggregation criterion of (6) supported by the entropy-based fuzzification scheme. The objective function for three-mode FCCM (3FCCM) is constructed by modifying the FCCM objective function of (2) as

$$L_{3\text{fccm}} = \sum_{c=1}^{C}\sum_{i=1}^{n}\sum_{j=1}^{m}\sum_{k=1}^{p} u_{ci} w_{cj} z_{ck} r_{ij} s_{jk} - \lambda_u \sum_{c=1}^{C}\sum_{i=1}^{n} u_{ci}\log u_{ci} - \lambda_w \sum_{c=1}^{C}\sum_{j=1}^{m} w_{cj}\log w_{cj} - \lambda_z \sum_{c=1}^{C}\sum_{k=1}^{p} z_{ck}\log z_{ck}, \quad (7)$$

where λ_z is the additional penalty weight for fuzzification of ingredient memberships z_{ck}. The larger the value of λ_z is, the fuzzier the ingredient memberships are.

Here, it should be noted that we can adopt two different types of constraints to ingredient memberships z_{ck}, such that object-type probabilistic constraint $\sum_{c=1}^{C} z_{ck} = 1$, $\forall k$ or item-type typicality constraint $\sum_{k=1}^{p} z_{ck} = 1$, $\forall c$. In such cases as food preference analysis, some common ingredients may be widely used in many foods while other rare ingredients can be negligible in all clusters. Then, from the view point of typical ingredient selection for characterizing cocluster features, item-type typicality constraint is adopted in this paper, such that $\sum_{k=1}^{p} z_{ck} = 1$, $\forall c$.

The clustering algorithm is an iterative process of updating u_{ci}, w_{cj}, and z_{ck} under the alternative optimization principle. Considering the necessary conditions for the optimality $\partial L_{3\text{fccm}}/\partial u_{ci} = 0$, $\partial L_{3\text{fccm}}/\partial w_{cj} = 0$, and $\partial L_{3\text{fccm}}/\partial z_{ck} = 0$ under the sum-to-one constraints, the updating rules for three memberships are given as

$$u_{ci} = \frac{\exp\left((1/\lambda_u)\sum_{j=1}^{m}\sum_{k=1}^{p} w_{cj} z_{ck} r_{ij} s_{jk}\right)}{\sum_{\ell=1}^{C}\exp\left((1/\lambda_u)\sum_{j=1}^{m}\sum_{k=1}^{p} w_{\ell j} z_{\ell k} r_{ij} s_{jk}\right)}, \quad (8)$$

$$w_{cj} = \frac{\exp\left((1/\lambda_w)\sum_{i=1}^{n}\sum_{k=1}^{p} u_{ci} z_{ck} r_{ij} s_{jk}\right)}{\sum_{\ell=1}^{m}\exp\left((1/\lambda_w)\sum_{i=1}^{n}\sum_{k=1}^{p} u_{ci} z_{ck} r_{i\ell} s_{\ell k}\right)}, \quad (9)$$

$$z_{ck} = \frac{\exp\left((1/\lambda_z)\sum_{i=1}^{n}\sum_{j=1}^{m} u_{ci} w_{cj} r_{ij} s_{jk}\right)}{\sum_{\ell=1}^{p}\exp\left((1/\lambda_z)\sum_{i=1}^{n}\sum_{j=1}^{m} u_{ci} w_{cj} r_{ij} s_{j\ell}\right)}. \quad (10)$$

3.2. Three-Mode Extension of Fuzzy CoDoK. Next, Fuzzy CoDoK is extended to the three-mode coclustering model named three-mode Fuzzy CoDoK (3Fuzzy CoDoK). The objective function of (4) is modified as

$$L_{3\text{fcdk}} = \sum_{c=1}^{C}\sum_{i=1}^{n}\sum_{j=1}^{m}\sum_{k=1}^{p} u_{ci}w_{cj}z_{ck}r_{ij}s_{jk} - \lambda_u \sum_{c=1}^{C}\sum_{i=1}^{n} u_{ci}^2 - \lambda_w \sum_{c=1}^{C}\sum_{j=1}^{m} w_{cj}^2 - \lambda_z \sum_{c=1}^{C}\sum_{k=1}^{p} z_{ck}^2, \quad (11)$$

where λ_z play a similar role to that in 3FCCM and the three types of fuzzy memberships also follow the same constraints with 3FCCM.

The updating rules are given in the similar manner to the previous section as follows:

$$u_{ci} = \frac{1}{C} + \frac{1}{2\lambda_u}\left\{\sum_{j=1}^{m}\sum_{k=1}^{p} w_{cj}z_{ck}r_{ij}s_{jk} - \frac{1}{C}\left(\sum_{l=1}^{C}\sum_{j=1}^{m}\sum_{k=1}^{p} w_{lj}z_{lk}r_{ij}s_{jk}\right)\right\}, \quad (12)$$

$$w_{cj} = \frac{1}{m} + \frac{1}{2\lambda_w}\left\{\sum_{i=1}^{n}\sum_{k=1}^{p} u_{ci}z_{ck}r_{ij}s_{jk} - \frac{1}{m}\left(\sum_{l=1}^{m}\sum_{i=1}^{n}\sum_{k=1}^{p} u_{ci}z_{ck}r_{il}s_{lk}\right)\right\}, \quad (13)$$

$$z_{ck} = \frac{1}{p} + \frac{1}{2\lambda_z}\left\{\sum_{i=1}^{n}\sum_{j=1}^{m} u_{ci}w_{cj}r_{ij}s_{jk} - \frac{1}{p}\left(\sum_{l=1}^{p}\sum_{i=1}^{n}\sum_{j=1}^{m} u_{ci}w_{cj}r_{ij}s_{jl}\right)\right\}. \quad (14)$$

In a similar manner to Fuzzy CoDoK, the above updating rules are computationally more stable than 3FCCM because of the lack of exp(·) function. However, u_{ci}, w_{cj}, and z_{ck} can be negative. Then, in practice, the negative memberships should be set to zero, and the remaining positive memberships can be renormalized so that their sum is one.

3.3. A Sample Algorithm for FCM-Type Fuzzy Coclustering for Three-Mode Cooccurrence Data. Following the above derivation, a sample algorithm is represented as follows:

[FCM-Type Fuzzy Coclustering for Three-Mode Cooccurrence Data: 3FCCM and 3Fuzzy CoDoK]

(1) Given $n \times m$ cooccurrence matrix R and $m \times p$ cooccurrence matrix S, let C be the number of clusters. Choose the fuzzification weights λ_u, λ_w, and λ_z.

(2) *[Initialization]* Randomly initialize u_{ci}, w_{cj}, and z_{ck}, such that $\sum_{c=1}^{C} u_{ci} = 1$, $\sum_{j=1}^{m} w_{cj} = 1$, and $\sum_{k=1}^{p} z_{ck} = 1$.

(3) *[Iterative process]* Iterate the following process until convergence of all u_{ci}.

(a) Update u_{ci} with (8) for 3FCCM or (12) for 3Fuzzy CoDoK.

(b) Update w_{cj} with (9) for 3FCCM or (13) for 3Fuzzy CoDoK.

(c) Update z_{ck} with (10) for 3FCCM or (14) for 3Fuzzy CoDoK.

4. Experimental Results

4.1. Experimental Design. In order to demonstrate the characteristics of the proposed algorithms, a numerical experiment was performed with an artificially generated three-mode data set, in which 40 objects ($n = 40$) have relational connection with 50 items ($m = 50$) and the items are related to 30 ingredients ($p = 30$). The artificial three-mode cooccurrence matrices were generated under the assumption that objects and ingredients have intrinsic (unknown) connections, as shown in the 40×30 matrix $X = \{x_{ik}\}$ of Figure 1(a), where black and white cells represent full-connection ($x_{ik} = 1$) and no-connection ($x_{ik} = 0$), respectively. (Note that all the following gray-scale figures depict visual images of matrices, where black and white cells represent maximum and minimum values.)

50×30 cooccurrence matrix $S = \{s_{jk}\}$ among items and ingredients was constructed, as shown in Figure 1(b), where ingredients k randomly occurred ($s_{jk} = 1$) in each item with 10% probability, whereas others remained as $s_{jk} = 0$. Then, 40×50 cooccurrence matrix $R = \{r_{ij}\}$ among objects and items was generated, such that $r_{ij} = 1$ if item j cooccurs with several ingredients, which are connected with object i in X, and $r_{ij} = 0$ otherwise. Figure 1(c) shows the cooccurrence matrix R. For example, in food preference analysis, 50 foods are made from 30 cooking ingredients (matrix S) and each of 40 user chooses some foods (matrix R) considering their intrinsic preferences on cooking ingredients (matrix X).

The goal of this experiment is to extract the intrinsic cocluster structure among objects, items, and ingredients from cooccurrence matrices R and S without utilizing the intrinsic (unknown) connection X among objects and ingredients; that is, X is withheld in the following experiments.

4.2. Cocluster Extraction by 3FCCM and 3Fuzzy CoDoK. First, the proposed 3FCCM and 3Fuzzy CoDoK algorithms were applied to R and S with $C = 3$ and their results are compared. Fuzziness penalties were $\lambda_u = 0.1$, $\lambda_w = 0.2$ and $\lambda_z = 0.3$ for 3FCCM, and $\lambda_u = 0.05$, $\lambda_w = 10.0$, and $\lambda_z = 15.0$ for 3Fuzzy CoDoK. The derived three types of memberships, which were the most frequent solutions in 100 trials with different random initializations, are depicted in the gray-scale figures of Figures 2 and 3, where each row represents the membership degree of objects, items, or ingredients for a cluster.

Figures 2(a) and 2(b) indicate that the 40 objects were successfully partitioned into three clusters by the 3FCCM algorithm, in which some meaningful ingredients, that is, cluster-wise typical ingredients in Figure 1(a), have large memberships for characterizing each cocluster, even though the intrinsic information X was withheld in the experiment.

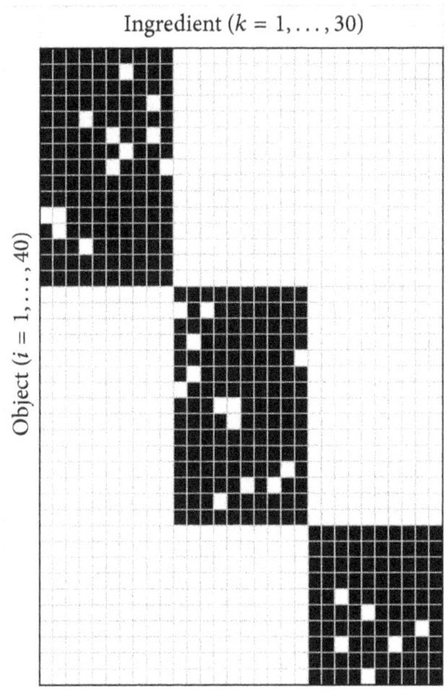

(a) Intrinsic connection $X = \{x_{ik}\}$ among objects and items (to be withheld in experiments)

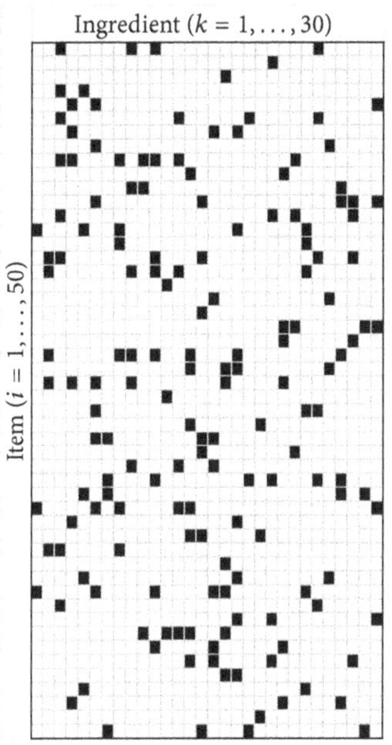

(b) Cooccurrence matrix $S = \{s_{jk}\}$ among items and ingredients

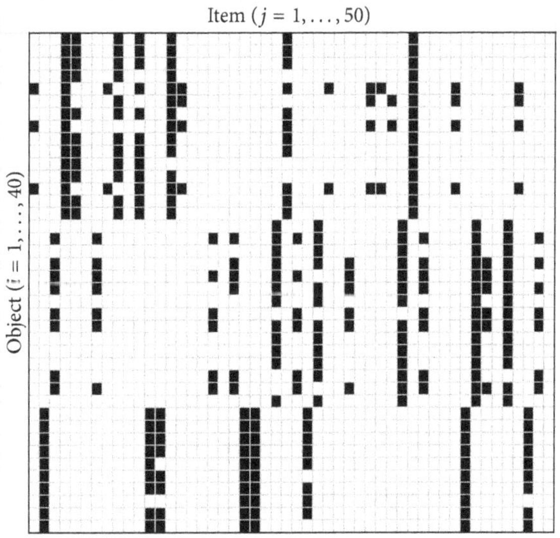

(c) Cooccurrence matrix $R = \{r_{ij}\}$ among objects and items

Figure 1: Artificial three-mode cooccurrence information data.

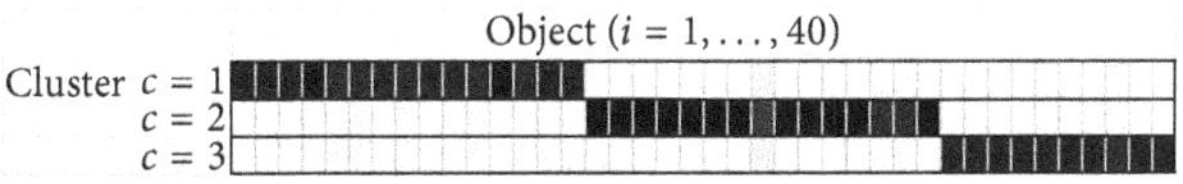

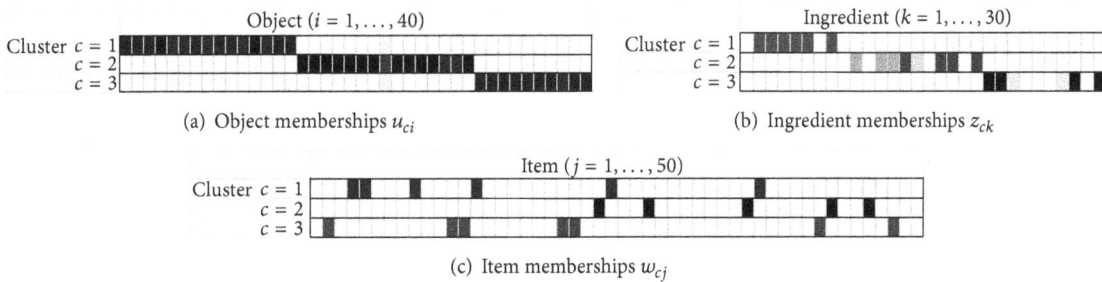

(a) Object memberships u_{ci}

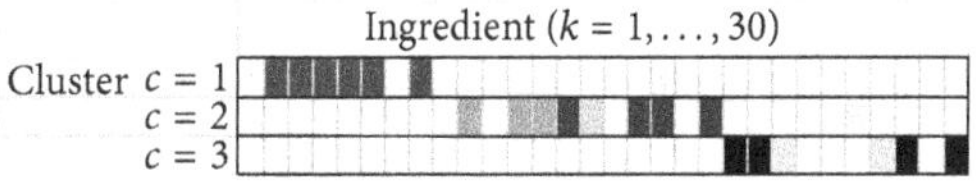

(b) Ingredient memberships z_{ck}

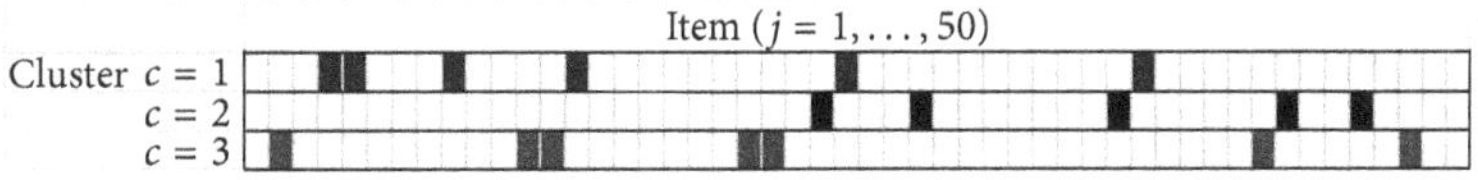

(c) Item memberships w_{cj}

Figure 2: Derived memberships by proposed 3FCCM.

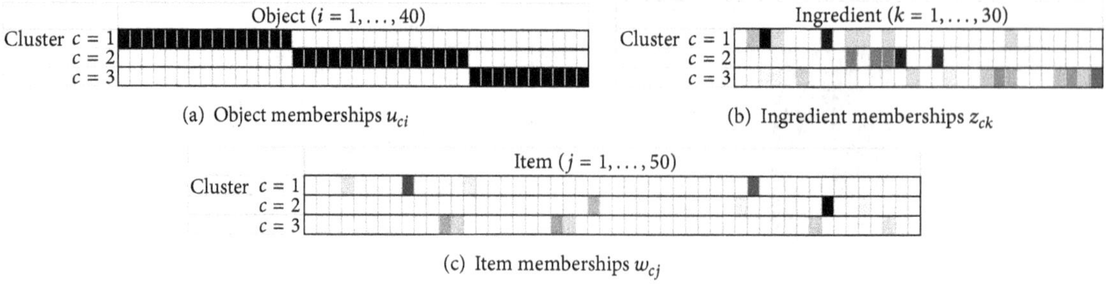

FIGURE 3: Derived memberships by proposed 3Fuzzy CoDoK.

Additionally, some typical items of each cluster were also indicated by large w_{cj}, as shown in Figure 2(c); for example, items 4, 5, 9, 14, 25, and 37 are typical in cluster 1.

By the way, Figure 3(a) indicates that the 3Fuzzy CoDoK algorithm also extracted almost same object clusters with Figure 2(a), but the ingredient memberships shown in Figure 3(b) have slightly different features from Figure 2(b). Only a few ingredients have very large memberships while many other ones have completely zero memberships because of negativity of (14). Additionally, some meaningless ingredients had nonzero memberships in contrast to the result of 3FCCM. The similar feature can be also seen in Figure 3(c).

These results imply that the 3FCCM algorithm is more suitable for clearly capturing the intrinsic connections although 3Fuzzy CoDoK has an advantage in computational stability.

4.3. Comparison with Conventional Two-Mode Fuzzy Coclustering. Second, the above clustering results are compared with the conventional FCCM and Fuzzy CoDoK, which are designed only for two-mode cooccurrence information. Although the intrinsic connection X is withheld in this experiment, a similar intrinsic information can be reconstructed by multiplying two cooccurrence matrices R and S, such that $R \times S$ gives an $n \times p$ relational matrix on objects and ingredients. Figure 4 shows the estimated 40×30 intrinsic connection matrix $\widetilde{X} = R \times S$.

The conventional FCCM and Fuzzy CoDoK were applied to $\widetilde{X}$. Fuzziness penalties were $\lambda_u = 0.05$ and $\lambda_w = 50.0$ for FCCM and $\lambda_u = 0.1$ and $\lambda_w = 100.0$ for Fuzzy CoDoK. Here, item memberships w_{ck} are identified with ingredient memberships z_{ck} in the algorithms. Figures 5 and 6 show the derived memberships, which most frequently appeared in 100 trials with different random initializations. The figures imply that 40 objects were partitioned into similar three clusters to those of 3FCCM and 3Fuzzy CoDoK. However, ingredient memberships z_{ck} were slightly contaminated and it is hard to intuitively select meaningful ingredients comparing with the result of 3FCCM. It may be because all items are embedded into $\widetilde{X}$ with equal responsibilities and the estimated $\widetilde{X} = R \times S$ was contaminated by noise as shown in Figure 4 rather than X of Figure 1(a). In contrast, the typical ingredients can be extracted in 3FCCM by selecting only meaningful items in each cluster.

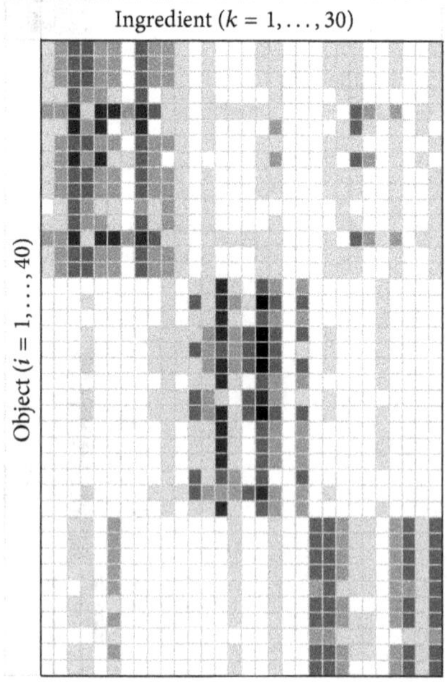

FIGURE 4: Estimated intrinsic connection matrix $\widetilde{X} = R \times S$.

TABLE 1: Comparison of frequencies of plausible solutions in 100 trials.

Algorithm	Two-mode		Three-mode	
	FCCM	Fuzzy CoDoK	3FCCM	3Fuzzy CoDoK
Frequency	78	47	90	100

Next, the robustness of the algorithms against random initialization is studied by comparing the frequencies of the plausible solutions. Table 1 compares the frequencies of the above results in 100 trials with different random initializations and indicates that the proposed three-mode coclustering models are more robust to random initialization than the conventional two-modes ones by utilizing three-mode cooccurrence information. That is, the optimal selections of both items and ingredients contribute to reduction of influences of randomness.

Therefore, the proposed algorithms are useful in analyzing three-mode cooccurrence information, which simultaneously consider the typicality of three elements.

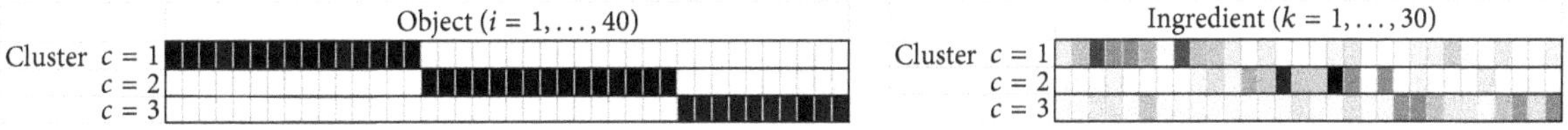

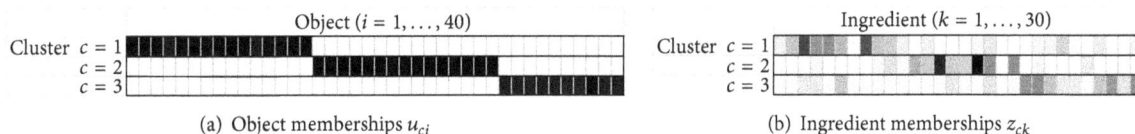

(a) Object memberships u_{ci}

(b) Ingredient memberships z_{ck}

FIGURE 5: Derived memberships by conventional FCCM.

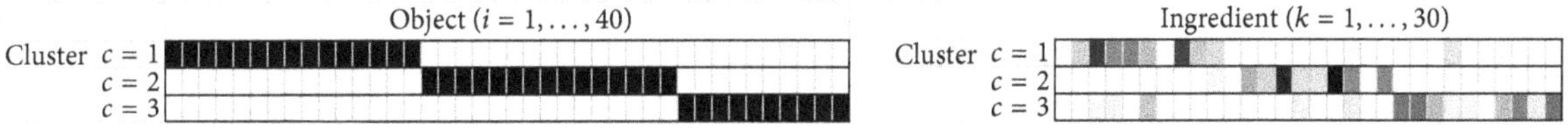

(a) Object memberships u_{ci}

(b) Ingredient memberships z_{ck}

FIGURE 6: Derived memberships by conventional Fuzzy CoDoK.

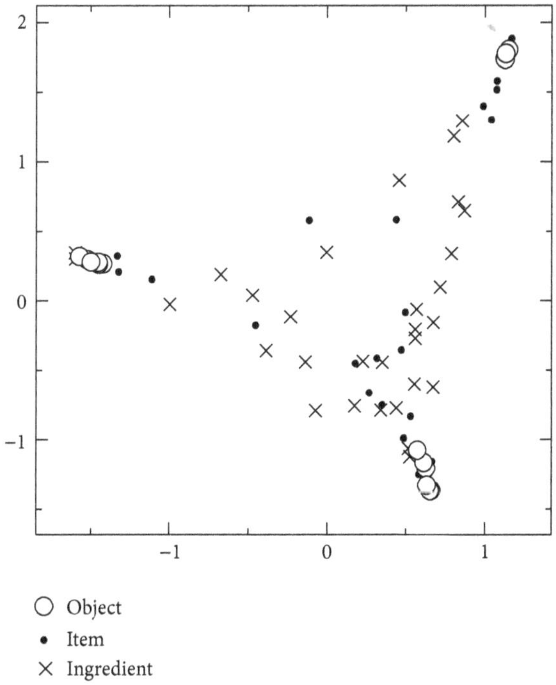

FIGURE 7: 2D Plots given by multicorresponding analysis.

4.4. Comparison with Multiple Corresponding Analysis. Finally, the partition characteristic of the proposed coclustering models is compared with the relational matrix decomposition method. Multiple correspondence analysis (MCA) [9] is a technique for revealing the structural information of categorical data, where mutual relations among objects and multiple categories are summarized into low-dimensional plots. In this experiment, an enlarged cross-tabulation was constructed by combining two cooccurrence matrices R and Z into $m \times (n + s)$ matrix $[R^{\top}, Z]$ so that the three elements are summarized on a plots figure. Figure 7 shows the 2D plots figure given by MCA. Although MCA does not necessarily aim at object-targeting partition, n objects were clearly separated into three subgroups in a similar manner to the proposed coclustering models because the objects had almost crisp boundaries. However, many other items and ingredients were distributed in the middle area and their contribution to the clusters was not emphasized as in the case of two-mode fuzzy coclustering of the previous subsection.

The proposed algorithms have advantages in handling three-mode elements by emphasizing their contributions to each coclusters. Additionally, while the implicit fuzziness degree of MCA is fixed (unchangeable), the proposed coclustering model can improve the interpretability of cluster partition by tuning the fuzziness degrees.

5. Conclusion

In this paper, novel coclustering models were proposed for analyzing three-mode cooccurrence information with the goal being to improve the partition quality of the conventional two-modes analysis. The proposed 3FCCM and 3Fuzzy CoDoK algorithms extended the conventional FCCM and Fuzzy CoDoK algorithms by introducing an additional membership for ingredients into the aggregation degree of three elements: objects, items, and ingredients. A numerical experiment with an artificial data set demonstrated that 3FCCM is more useful in capturing the intrinsic connection among objects and ingredients while 3Fuzzy CoDoK is suitable for handling large data sets with its computational stability.

Besides the simplicity of FCM-type coclustering, FCCM and fuzzy CoDoK sometimes have the difficulty in tuning of fuzziness degrees. In the conventional two-modes coclustering, an MMMs-induced model [12] showed a better utility than FCCM and fuzzy CoDoK. A potential future work is to improve the proposed FCM-type three-mode coclustering by introducing a statistical concept for easy tuning of fuzziness degrees. Another direction of future work is to develop a validity measure [13] for selecting the optimal cluster partitions.

Conflicts of Interest

The authors declare that there are no conflicts of interest regarding the publication of this paper.

Acknowledgments

This work was supported in part by Tateisi Science and Technology Foundation, Japan, under Research Grant 2017.

References

[1] J. C. Bezdek, *Pattern Recognition with Fuzzy Objective Function Algorithms*, Plenum Press, NY, USA, 1981.

[2] E. H. Ruspini, "A new approach to clustering," *Information and Control*, vol. 15, no. 1, pp. 22–32, 1969.

[3] C.-H. Oh, K. Honda, and H. Ichihashi, "Fuzzy clustering for categorical multivariate data," in *Proceedings of the Joint 9th IFSA World Congress and 20th NAFIPS International Conference*, vol. 4, pp. 2154–2159, July 2001.

[4] S. Miyamoto and M. Mukaidono, "Fuzzy c-means as a regularization and maximum entropy approach," in *Proceedings of the 7th International Fuzzy Systems Association World Congress*, vol. 2, pp. 86–92, 1997.

[5] S. Miyamoto, H. Ichihashi, and K. Honda, *Algorithms for Fuzzy Clustering: Methods in C-Means Clustering with Applications*, vol. 229 of *Studies in Fuzziness and Soft Computing*, Springer, Berlin, Germany, 2008.

[6] K. Kummamuru, A. Dhawale, and R. Krishnapuram, "Fuzzy co-clustering of documents and keywords," in *Proceedings of the IEEE International Conference on Fuzzy Systems*, vol. 2, pp. 772–777, May 2003.

[7] S. Miyamoto and K. Umayahara, "Fuzzy clustering by quadratic regularization," in *Proceedings of the 1998 IEEE International Conference on Fuzzy Systems IEEE World Congress on Computational Intelligence*, vol. 2, pp. 1394–1399, Anchorage, AK, USA.

[8] J. MacQueen, "Some methods for classification and analysis of multivariate observations," in *Proceedings of the 5th Berkeley Symposium on Mathematical Statistics and Probability*, pp. 281–297, University of California Press, Berkeley, Calif, USA, 1967.

[9] M. Tenenhaus and F. W. Young, "An analysis and synthesis of multiple correspondence analysis, optimal scaling, dual scaling, homogeneity analysis and other methods for quantifying categorical multivariate data," *Psychometrika*, vol. 50, no. 1, pp. 91–119, 1985.

[10] D. D. Lee and H. S. Seung, "Learning the parts of objects by non-negative matrix factorization," *Nature*, vol. 401, no. 6755, pp. 788–791, 1999.

[11] K. Honda, "Fuzzy co-clustering and application to collaborative filtering," in *Integrated Uncertainty in Knowledge Modelling and Decision Making*, V. N. Huynh, M. Inuiguchi, B. Le, and T. Denoeux, Eds., vol. 9978 of *Lecture Notes in Computer Science*, pp. 16–23, Springer International Publishing, Cham, Switzerland, 2016.

[12] K. Honda, S. Oshio, and A. Notsu, "Fuzzy co-clustering induced by multinomial mixture models," *Journal of Advanced Computational Intelligence and Intelligent Informatics*, vol. 19, no. 6, pp. 717–726, 2015.

[13] W. Wang and Y. Zhang, "On fuzzy cluster validity indices," *Fuzzy Sets and Systems*, vol. 158, no. 19, pp. 2095–2117, 2007.

19

Quaternionic Serret-Frenet Frames for Fuzzy Split Quaternion Numbers

Cansel Yormaz, Simge Simsek, and Serife Naz Elmas

Department of Mathematics, Pamukkale University, Denizli 20070, Turkey

Correspondence should be addressed to Cansel Yormaz; c_aycan@pau.edu.tr

Academic Editor: Rustom M. Mamlook

We build the concept of fuzzy split quaternion numbers of a natural extension of fuzzy real numbers in this study. Then, we give some differential geometric properties of this fuzzy quaternion. Moreover, we construct the Frenet frame for fuzzy split quaternions. We investigate Serret-Frenet derivation formulas by using fuzzy quaternion numbers.

1. Introduction

The Serret-Frenet formulas describe the kinematic properties of a particle moving along a continuous and differentiable curve in Euclidean space E^3 or Minkowski space E_1^3. These formulas are used in many areas such as mathematics, physics (especially in relative theory), medicine, and computer graphics.

Quaternions were discovered by Sir William R. Hamilton in 1843. The most widely used and most important feature of quaternions is that each unit quaternion represents a transformation. This representation has a special and important role on turns in 3-dimensional vector spaces. This situation is detailed in the study [1]. Nowadays, quaternions are used in many areas such as physics, computer graphics, and animation. For example, visualizing and translating with computer graphics are much easier with quaternions. It is known by especially mathematicians and physicists that any unit (split) quaternion corresponds to a rotation in Euclidean and Minkowski spaces.

The notion of a fuzzy subset was introduced by Zadeh [2] and later applied in various mathematical branches. According to the standard condition, a fuzzy number is a convex and a normalized fuzzy subset of real numbers. Basic operations on fuzzy quaternion numbers can be seen in study [3]. There are many applications of quaternions. In physics, we have highlighted applications in quantum mechanics [4] and theory of relativity [5]. In addition, there are applications in aviation projects and flight simulators [6]. On the other hand, the study [7] is a basic study for quaternionic fibonacci forms. All of references that we reviewed guided us to studying the geometry of quaternions.

In this paper, we have described the basic operations of fuzzy split quaternions. With this number of structures we aimed to achieve the frenet frame equation. Previously, frenet frame has been created by split quaternions in [8].

In these studies, we obtained Frenet frame by the fuzzy split quaternion.

2. Serret-Frenet Frame

The Serret-Frenet frame is defined as follows [8].

Let $\overrightarrow{\alpha}(t)$ be any second-order differentiable space curve with nonvanishing second derivative. We can choose this local coordinate system to be the Serret-Frenet frame consisting of the tangent vector $\overrightarrow{T}(t)$, the binormal vector $\overrightarrow{B}(t)$, and the normal vector $\overrightarrow{N}(t)$ vectors at any point on the curve given by

$$\overrightarrow{T}(t) = \frac{\overrightarrow{\alpha}(t)}{\left\| \overrightarrow{\alpha'}(t) \right\|}$$

$$\vec{B}(t) = \frac{\vec{\alpha'}(t) \times \vec{\alpha''}(t)}{\left\|\vec{\alpha'}(t) \times \vec{\alpha''}(t)\right\|}$$

$$\vec{N}(t) = \vec{B}(t) \times \vec{T}(t) \tag{1}$$

The Serret-Frenet frame for the curve $\vec{\alpha}(t)$ is given as the following differential equation. Writing this frame with matrices is easily for the mathematical calculations.

$$\begin{bmatrix} \vec{T'}(t) \\ \vec{B'}(t) \\ \vec{N'}(t) \end{bmatrix} = v(t) \begin{bmatrix} 0 & \kappa(t) & 0 \\ -\kappa(t) & 0 & \tau(t) \\ 0 & -\tau(t) & 0 \end{bmatrix} \begin{bmatrix} \vec{T}(t) \\ \vec{B}(t) \\ \vec{N}(t) \end{bmatrix} \tag{2}$$

The speed value of the curve $\vec{\alpha}(t)$ is denoted by $v(t) = \|\vec{\alpha'}(t)\|$. The scalar curvature of $\vec{\alpha}(t)$ is symbolized as $\kappa(t)$ and the torsion value of the curve $\vec{\alpha}(t)$ is symbolized as $\tau(t)$. The torsion of the curve $\vec{\alpha}(t)$ measures how sharply it is twisting out of the plane of curvature. The curvature of $\vec{\alpha}(t)$ is the magnitude of the acceleration of a particle moving along this curve. The torsion of curvature is related by the Serret-Frenet formulas and their generalization. These can be expressed with following formulas:

$$\kappa(t) = \frac{\left\|\vec{\alpha'}(t) \times \vec{\alpha''}(t)\right\|}{\left\|\vec{\alpha'}(t)\right\|^3}$$

$$\tau(t) = \frac{\vec{\alpha'}(t) \times \vec{\alpha''}(t) \times \vec{\alpha'''}(t)}{\left\|\vec{\alpha'}(t) \times \vec{\alpha''}(t)\right\|^2} \tag{3}$$

3. Split Quaternion Frames

In this section, firstly we will give the split quaternions definition and their characteristics properties.

Definition 1. The set $H' = \{q = q_0 1 + q_1 i + q_2 j + q_3 k,\ q_0, q_1, q_2, q_3 \in R\}$ is a vector space over R having basis $\{1, i, j, k\}$ with the following properties:

$$\begin{aligned} i^2 &= -1, \\ j^2 &= k^2 = 1 \\ ij &= -ji = k \\ kj &= -jk = -i \\ ki &= -ik = j \end{aligned} \tag{4}$$

Every element of the set H' is called a split quaternion. [9].

Definition 2. Let two split quaternions be $q = q_0 1 + q_1 i + q_2 j + q_3 k$ and $p = p_0 1 + p_1 i + p_2 j + p_3 k$. These two split quaternions multiplication is calculated as

$$\begin{aligned} q.p = {} & (q_0 p_0 - q_1 p_1 + q_2 p_2 + q_3 p_3) \\ & + (q_0 p_1 + q_1 p_0 - q_2 p_3 + q_3 p_2)\, i \\ & + (q_0 p_2 + q_2 p_0 + q_3 p_1 - q_1 p_3)\, j \\ & + (q_0 p_3 + q_3 p_0 + q_1 p_2 - q_2 p_1)\, k \end{aligned} \tag{5}$$

Definition 3. The conjugate of the split quaternion $q = q_0 1 + q_1 i + q_2 j + q_3 k$ is defined as

$$\bar{q} = q_0 1 - q_1 i - q_2 j - q_3 k \tag{6}$$

Definition 4. A unit-length split quaternion's norm is

$$N_q = q\bar{q} = \bar{q}q = (q_0)^2 + (q_1)^2 - (q_2)^2 - (q_3)^2 = 1 \tag{7}$$

Definition 5. Because of $H' \simeq E_2^4$, we can define the timelike, spacelike, and lightlike quaternions for $q = (q_0, q_1, q_2, q_3)$ as follows:

(i) Spacelike quaternion for $I_q < 0$

(ii) Timelike quaternion for $I_q > 0$

(iii) Lightlike quaternion for $I_q = 0$

Here, $I_q = q\bar{q} = \bar{q}q$. [1].

We can add to Definition 5 following descriptions. Timelike, spacelike, and lightlike vectors are important for the Minkowski space E_1^3. The Minkowski space E_1^3 is the accepted common space for the physical reality. We know that the general properties of the quaternions are similar to Minkowski space E_2^4. The Minkowski space E_2^4 is a vector space with real dimension $'4'$ and index $'2'$. Elements of Minkowski space E_2^4 are called events or four vectors. On Minkowski space E_2^4, there is an inner product of signature two "plus" and two "minus". Also, we prefer to define the vector structure of Minkowski space with quaternions.

Every possible rotation R (a 3×3 special split orthogonal matrix) can be constructed from either one of the two related split quaternions $q = q_0 1 + q_1 i + q_2 j + q_3 k$ or $-q = -q_0 1 - q_1 i - q_2 j - q_3 k$ using the transformation law [8]:

$$\begin{aligned} q w \bar{q} &= R.w \\ [q w \bar{q}]_i &= \sum_{j=1}^{3} R_{ij}.w_j \end{aligned} \tag{8}$$

where $w = v_1 i + v_2 j + v_3 k$ k is a pure split quaternion. We compute R_{ij} directly from (5)

$$R = \begin{bmatrix} (q_0)^2 + (q_1)^2 - (q_2)^2 - (q_3)^2 & 2q_1q_2 - 2q_0q_3 & 2q_0q_2 + 2q_1q_3 \\ 2q_0q_3 + 2q_1q_2 & -(q_0)^2 + (q_1)^2 + (q_2)^2 - (q_3)^2 & 2q_2q_3 + 2q_0q_1 \\ 2q_1q_3 - 2q_0q_2 & -2q_0q_1 + 2q_2q_3 & -(q_0)^2 + (q_1)^2 - (q_2)^2 + (q_3)^2 \end{bmatrix} \quad (9)$$

All columns of this matrix expressed in this form are orthogonal but not orthonormal. This matrix form is a special orthogonal group $SO(1,2)$. On the other hand, the matrix R can be obtained by the unit split quaternions q and $-q$. There are two unit timelike quaternions for every rotation in Minkowski 3-space. These timelike quaternions are q and $-q$. For this reason, a timelike quaternion R_q can be supposed as a 3×3 dimensional orthogonal rotation matrix.

The equations obtained as a result of this coincidence are quaternion valued linear equations. If we derive the column equation of (9), respectively, then we obtain the following results:

$$d\vec{T} = 2\begin{bmatrix} q_0 & q_1 & q_2 & q_3 \\ q_3 & q_2 & q_1 & q_0 \\ -q_2 & q_3 & -q_0 & q_1 \end{bmatrix}\begin{bmatrix} dq_0 \\ dq_1 \\ dq_2 \\ dq_3 \end{bmatrix} = 2\,[A]\,[q']$$

$$d\vec{N} = 2\begin{bmatrix} -q_3 & q_2 & q_1 & -q_0 \\ -q_0 & q_1 & q_2 & -q_3 \\ -q_1 & -q_0 & q_3 & q_2 \end{bmatrix}\begin{bmatrix} dq_0 \\ dq_1 \\ dq_2 \\ dq_3 \end{bmatrix} = 2\,[B]\,[q'] \quad (10)$$

$$d\vec{B} = 2\begin{bmatrix} q_2 & q_3 & q_0 & q_1 \\ q_1 & q_0 & q_3 & q_2 \\ -q_0 & q_1 & -q_2 & q_3 \end{bmatrix}\begin{bmatrix} dq_0 \\ dq_1 \\ dq_2 \\ dq_3 \end{bmatrix} = 2\,[C]\,[q']$$

4. Serret-Frenet Frames of Split Quaternions

In this section, we give the Serret-Frenet Frame equations for split quaternions. If we calculate the differential equations corresponding to Serret-Frenet Frames with split quaternions, we can obtain the following differential equations. These equations are the formulas Serret-Frenet frames with split quaternions.

$$2\,[A]\,[q'] = \vec{T'} = \nu\kappa\vec{N'} \quad (11)$$

$$2\,[B]\,[q'] = \vec{N'} = -\nu\kappa\vec{T'} + \nu\tau\vec{T'} \quad (12)$$

$$2\,[C]\,[q'] = \vec{B'} = -\nu\tau\vec{N'} \quad (13)$$

where

$$[q'] = \begin{bmatrix} da_0 \\ da_1 \\ da_2 \\ da_3 \end{bmatrix} = \begin{bmatrix} b_0 & b_1 & b_2 & b_3 \\ c_0 & c_1 & c_2 & c_3 \\ d_0 & d_1 & d_2 & d_3 \\ e_0 & e_1 & e_2 & e_3 \end{bmatrix}\begin{bmatrix} a_0 \\ a_1 \\ a_2 \\ a_3 \end{bmatrix} \quad (14)$$

Therefore, with using (11), (12), and (13) we obtain the H' split quaternion Frenet frame equations as [8]

$$[q'] = \begin{bmatrix} da_0 \\ da_1 \\ da_2 \\ da_3 \end{bmatrix} = \frac{\nu}{2}\begin{bmatrix} 0 & -\tau & 0 & -\kappa \\ \tau & 0 & \kappa & 0 \\ 0 & \kappa & 0 & \tau \\ -\kappa & 0 & -\tau & 0 \end{bmatrix}\begin{bmatrix} a_0 \\ a_1 \\ a_2 \\ a_3 \end{bmatrix} \quad (15)$$

5. Serret-Frenet Frames of Fuzzy Split Quaternions

In this section, we study obtaining the Frenet frame equations with split quaternions in the fuzzy space. For this, firstly we define a fuzzy real set and fuzzy real numbers.

Definition 6. The real number's set is denoted by R and let H be a set of quaternion numbers. A fuzzy real set is a function $\overline{A} : R \to [0,1]$.

A fuzzy real set $\overline{A}$ is a fuzzy real numbers set $\Leftrightarrow$.

(i) $\overline{A}$ is normal, i.e., there exists $x \in R$ whose $\overline{A} = 1$.

(ii) For all $\alpha \in (0,1]$, the set $\overline{A}[\alpha] = \{x \in R : \overline{A}(x) \geq \alpha\}$ is a limited set.

The set of all fuzzy real numbers is denoted by R_F. We can see that $R \subset R_F$, since every $\alpha \in R$ can be written as $\alpha : R \to [0,1]$, where $\alpha(x) = 1$ if $x = \alpha$ and $\alpha(x) = 0$ if $x \neq \alpha$. [3]

Now, we define fuzzy numbers with quaternionic forms.

Definition 7. A fuzzy quaternion number is defined by a function $h : \mathrm{H} \to [0,1]$, where $h(a_0 1 + a_1 i + a_2 j + a_3 k) = \min\{\overline{A}_0(a_0), \overline{A}_1(a_1), \overline{A}_2(a_2), \overline{A}_3(a_3)\}$, for $\overline{A}_0, \overline{A}_1, \overline{A}_2, \overline{A}_3 \in R_F$ [3].

Similarly, a fuzzy split quaternion number is given by $h' : \mathrm{H}' \to [0,1]$ such that $h'(a_0 1 + a_1 i + a_2 j + a_3 k) = \min\{\overline{A}_0(a_0), \overline{A}_1(a_1), \overline{A}_2(a_2), \overline{A}_3(a_3)\}$, for $\overline{A}_0, \overline{A}_1, \overline{A}_2, \overline{A}_3 \in R_F$.

The fuzzy quaternion number's set is denoted by H_F and the set of all fuzzy split quaternion numbers is denoted by H'_F and identified as R_F^4, where every element h' is associated with $(\overline{A}, \overline{B}, \overline{C}, \overline{D})$.

We can define the fuzzy split quaternion numbers as follows:

$h' = (\overline{A}_0, \overline{A}_1, \overline{A}_2, \overline{A}_3) \in H'_F$, where $Re(h') = \overline{A}_0$ is called the real part and $Im1(h') = \overline{A}_1, Im2(h') = \overline{A}_2,\ Im3(h') = \overline{A}_3$ are called imaginary parts.

Let $h = a_0 1 + a_1 i + a_2 j + a_3 k \in H'$ and the function $h' : H' \to [0,1]$ is given by

$$h'(b_0 1 + b_1 i + b_2 j + b_3 k) = \begin{cases} 1, & \text{if } a_0 = b_0 \text{ and } a_1 = b_1 \text{ and } a_2 = b_2 \text{ and } a_3 = b_3 \\ 0, & \text{if } a_0 \neq b_0 \text{ or } a_1 \neq b_1 \text{ or } a_2 \neq b_2 \text{ or } a_3 \neq b_3 \end{cases} \quad (16)$$

Definition 8. In the fuzzy split quaternion numbers H'_F, we can define the addition and multiplication operations as follows [3].

Let $s', h' \in H'_F$, where $s' = (\overline{B}_0, \overline{B}_1, \overline{B}_2, \overline{B}_3)$ and $h' = (\overline{A}_0, \overline{A}_1, \overline{A}_2, \overline{A}_3)$; then,

$$\begin{aligned} s' + h' &= \left(\overline{B}_0 + \overline{A}_0, \overline{B}_1 + \overline{A}_1, \overline{B}_2 + \overline{A}_2, \overline{B}_3 + \overline{A}_3\right) \\ s'.h' &= \left(\overline{B}_0\overline{A}_0 - \overline{B}_1\overline{A}_1 + \overline{B}_2\overline{A}_2 + \overline{B}_3\overline{A}_3,\ \overline{B}_0\overline{A}_1 + \overline{B}_1\overline{A}_0 \right. \\ &\quad - \overline{B}_2\overline{A}_3 + \overline{B}_3\overline{A}_2,\ \overline{B}_0\overline{A}_2 + \overline{B}_1\overline{A}_3 + \overline{B}_2\overline{A}_0 \\ &\quad \left. - \overline{B}_3\overline{A}_1,\ \overline{B}_0\overline{A}_3 + \overline{B}_1\overline{A}_2 + \overline{B}_2\overline{A}_1 - \overline{B}_3\overline{A}_0\right) \end{aligned} \quad (17)$$

Definition 9. Let R be the field of real numbers and (R, τ) be a fuzzy topological vector space over the field R.

$f : R \to R, a \in R$; the function f is said to be fuzzy differentiable at the point a if there is a function ϕ that is fuzzy continuous at the point a and have

$$f(x) - f(a) = \phi(x)(x - a) \quad (18)$$

for all $x \in R$. $\phi(a)$ is said to be fuzzy derivative of f at and denote

$$f'(a) = \phi(a) \quad (19)$$

[10].

Definition 10. Let $h' = (\overline{A}_0, \overline{A}_1, \overline{A}_2, \overline{A}_3)$; the conjugate of h' is defined as

$$\overline{h'} = \left(\overline{A}_0, -\overline{A}_1, -\overline{A}_2, -\overline{A}_3\right) \quad (20)$$

The norm of h' is defined as

$$N_{h'} = \overline{h'}h' = h'\overline{h'} = \left(\overline{A}_0\right)^2 + \left(\overline{A}_1\right)^2 - \left(\overline{A}_2\right)^2 - \left(\overline{A}_3\right)^2 \quad (21)$$

Because of $H' \subset H'_F$, the following equation can be written:

$$\left[h'.w'.\overline{h'}\right]_i = \sum_{j=1}^{3} R_{ij} w_j \quad (22)$$

where $w' = (V_1, V_2, V_3)$.

Here, R_{ij} is the component of the matrix R and the matrix is calculated from (17) as follows:

$$R = \begin{bmatrix} \left(\overline{A}_0\right)^2 + \left(\overline{A}_1\right)^2 + \left(\overline{A}_2\right)^2 + \left(\overline{A}_3\right)^2 & 2\overline{A}_1\overline{A}_2 - 2\overline{A}_0\overline{A}_3 & 2\overline{A}_0\overline{A}_2 + 2\overline{A}_1\overline{A}_3 \\ 2\overline{A}_0\overline{A}_3 + 2\overline{A}_1\overline{A}_2 & -\left(\overline{A}_0\right)^2 + \left(\overline{A}_1\right)^2 + \left(\overline{A}_2\right)^2 - \left(\overline{A}_3\right)^2 & 2\overline{A}_2\overline{A}_3 + 2\overline{A}_0\overline{A}_1 \\ 2\overline{A}_1\overline{A}_3 - 2\overline{A}_0\overline{A}_2 & -2\overline{A}_0\overline{A}_1 + 2\overline{A}_2\overline{A}_3 & -\left(\overline{A}_0\right)^2 + \left(\overline{A}_1\right)^2 - \left(\overline{A}_2\right)^2 + \left(\overline{A}_3\right)^2 \end{bmatrix} \quad (23)$$

In this matrix (23), we calculate the derivative of the columns, respectively, to the elements $\overline{A}_0, \overline{A}_1, \overline{A}_2$, and $\overline{A}_3$. We will get the Fuzzy tangent vector $\overrightarrow{T'}$ to the derivation from the first column to the elements $\overline{A}_0, \overline{A}_1, \overline{A}_2$, and $\overline{A}_3$:

$$\overrightarrow{T'} = d\overrightarrow{T} = 2\begin{bmatrix} \overline{A}_0 & \overline{A}_1 & \overline{A}_2 & \overline{A}_3 \\ \overline{A}_3 & \overline{A}_2 & \overline{A}_1 & \overline{A}_0 \\ -\overline{A}_2 & \overline{A}_3 & -\overline{A}_0 & \overline{A}_1 \end{bmatrix} \begin{bmatrix} d\overline{A}_0 \\ d\overline{A}_1 \\ d\overline{A}_2 \\ d\overline{A}_3 \end{bmatrix} = 2[X]\left[d\left(h'\right)\right] \quad (24)$$

We will get the fuzzy normal vector $\overrightarrow{N'}$ to the derivation from the second column to the elements $\overline{A}_0, \overline{A}_1, \overline{A}_2$, and $\overline{A}_3$:

$$\overrightarrow{N'} = d\overrightarrow{N} = 2\begin{bmatrix} -\overline{A}_3 & \overline{A}_2 & \overline{A}_1 & -\overline{A}_0 \\ -\overline{A}_0 & \overline{A}_1 & \overline{A}_2 & -\overline{A}_3 \\ -\overline{A}_1 & -\overline{A}_0 & \overline{A}_3 & \overline{A}_2 \end{bmatrix} \begin{bmatrix} d\overline{A}_0 \\ d\overline{A}_1 \\ d\overline{A}_2 \\ d\overline{A}_3 \end{bmatrix} = 2[Y]\left[d\left(h'\right)\right] \quad (25)$$

We will get the fuzzy binormal vector $\overrightarrow{B'}$ to the derivation from the third column to the elements $\overline{A}_0, \overline{A}_1, \overline{A}_2$, and $\overline{A}_3$:

$$\overrightarrow{B'} = d\overrightarrow{B} = 2\begin{bmatrix} \overline{A}_2 & \overline{A}_3 & \overline{A}_0 & \overline{A}_1 \\ \overline{A}_1 & \overline{A}_0 & \overline{A}_3 & \overline{A}_2 \\ -\overline{A}_0 & \overline{A}_1 & -\overline{A}_2 & \overline{A}_3 \end{bmatrix} \begin{bmatrix} d\overline{A}_0 \\ d\overline{A}_1 \\ d\overline{A}_2 \\ d\overline{A}_3 \end{bmatrix}$$

$$= 2[Z]\left[d\left(h'\right)\right] \tag{26}$$

If we write, respectively, these founded matrices in (11), (12), and (13), we can obtain the following equalities for Serret-Frenet frame equations:

$$2[X]\left[d\left(h'\right)\right] = \overrightarrow{T'} = \nu\kappa\overrightarrow{N'} \tag{27}$$

$$2[Y]\left[d\left(h'\right)\right] = \overrightarrow{N'} = -\nu\kappa\overrightarrow{T'} + \nu\tau\overrightarrow{T'} \tag{28}$$

$$2[Z]\left[d\left(h'\right)\right] = \overrightarrow{B'} = -\nu\tau\overrightarrow{N'} \tag{29}$$

The differential of fuzzy split quaternion h' is expressed with matrix form as follows:

$$\left[d\left(h'\right)\right] = \begin{bmatrix} d\overline{A}_0 \\ d\overline{A}_1 \\ d\overline{A}_2 \\ d\overline{A}_3 \end{bmatrix} = \begin{bmatrix} \overline{B}_0 & \overline{B}_1 & \overline{B}_2 & \overline{B}_3 \\ \overline{C}_0 & \overline{C}_1 & \overline{C}_2 & \overline{C}_3 \\ \overline{D}_0 & \overline{D}_1 & \overline{D}_2 & \overline{D}_3 \\ \overline{E}_0 & \overline{E}_1 & \overline{E}_2 & \overline{E}_3 \end{bmatrix} \begin{bmatrix} \overline{A}_0 \\ \overline{A}_1 \\ \overline{A}_2 \\ \overline{A}_3 \end{bmatrix} \tag{30}$$

Here, $(\overline{A}_0, \overline{A}_1, \overline{A}_2, \overline{A}_3)$ is the real and imaginary elements of the fuzzy split quaternionic vector. Now, we must need to calculate the elements $\overline{B}_i, \overline{C}_i, \overline{D}_i, \overline{E}_i$, $(0 \le i \le 3)$ of the coefficient matrix. We need solutions of (27), (28), and (29) to obtain the elements $\overline{B}_i, \overline{C}_i, \overline{D}_i, \overline{E}_i$, $(0 \le i \le 3)$. For this reason, we put the differential of fuzzy split quaternion h', fuzzy tangent vector $\overrightarrow{T'}$, fuzzy normal vector $\overrightarrow{N'}$, and fuzzy binormal vector $\overrightarrow{B'}$ in (27), (28), and (29) in its places. When we make the needed calculations, we can obtain the following results:

$$\begin{aligned} &\overline{B}_0\overline{A}_0\overline{A}_3 + \overline{B}_1\overline{A}_1\overline{A}_3 + \overline{B}_2\overline{A}_2\overline{A}_3 + \overline{B}_3\left(\overline{A}_3\right)^2 + \overline{C}_0\overline{A}_0\overline{A}_2 \\ &\quad + \overline{C}_1\overline{A}_1\overline{A}_2 + \overline{C}_2\left(\overline{A}_2\right)^2 + \overline{C}_3\overline{A}_2\overline{A}_3 + \overline{D}_0\overline{A}_0\overline{A}_1 \\ &\quad + \overline{D}_1\left(\overline{A}_1\right)^2 + \overline{D}_2\overline{A}_1\overline{A}_2 + \overline{D}_3\overline{A}_1\overline{A}_3 + \overline{E}_0\left(\overline{A}_0\right)^2 \\ &\quad + \overline{E}_1\overline{A}_0\overline{A}_1 + \overline{E}_2\overline{A}_0\overline{A}_2 + \overline{E}_3\overline{A}_0\overline{A}_3 \\ &\quad = \frac{\nu}{2}\kappa\left(\left(\overline{A}_0\right)^2 + \left(\overline{A}_1\right)^2 + \left(\overline{A}_2\right)^2 - \left(\overline{A}_3\right)^2\right) \end{aligned} \tag{31}$$

$$\begin{aligned} &-\overline{B}_0\overline{A}_0\overline{A}_3 - \overline{B}_1\overline{A}_1\overline{A}_3 - \overline{B}_2\overline{A}_2\overline{A}_3 - \overline{B}_3\left(\overline{A}_3\right)^2 \\ &\quad + \overline{C}_0\overline{A}_0\overline{A}_2 + \overline{C}_1\overline{A}_1\overline{A}_2 + \overline{C}_2\left(\overline{A}_2\right)^2 + \overline{C}_3\overline{A}_2\overline{A}_3 \\ &\quad + \overline{D}_0\overline{A}_0\overline{A}_1 + \overline{D}_1\left(\overline{A}_1\right)^2 + \overline{D}_2\overline{A}_1\overline{A}_2 + \overline{D}_3\overline{A}_1\overline{A}_3 \\ &\quad - \overline{E}_0\left(\overline{A}_0\right)^2 - \overline{E}_1\overline{A}_0\overline{A}_1 - \overline{E}_2\overline{A}_0\overline{A}_2 - \overline{E}_3\overline{A}_0\overline{A}_3 \\ &\quad = -\frac{\nu}{2}\kappa\left(\left(\overline{A}_0\right)^2 + \left(\overline{A}_1\right)^2 + \left(\overline{A}_2\right)^2 + \left(\overline{A}_3\right)^2\right) \\ &\quad + \frac{\nu}{2}\tau\left(2\overline{A}_0\overline{A}_2 + 2\overline{A}_1\overline{A}_3\right) \end{aligned} \tag{32}$$

$$\begin{aligned} &-\overline{B}_0\overline{A}_0\overline{A}_1 - \overline{B}_1\left(\overline{A}_1\right)^2 - \overline{B}_2\overline{A}_1\overline{A}_2 - \overline{B}_3\overline{A}_1\overline{A}_3 \\ &\quad - \overline{C}_0\left(\overline{A}_2\right)^2 - \overline{C}_1\overline{A}_0\overline{A}_1 - \overline{C}_2\overline{A}_0\overline{A}_2 - \overline{C}_3\overline{A}_1\overline{A}_3 \\ &\quad + \overline{D}_0\overline{A}_0\overline{A}_3 + \overline{D}_1\overline{A}_1\overline{A}_3 + \overline{D}_2\overline{A}_1\overline{A}_3 + \overline{D}_3\left(\overline{A}_3\right)^2 \\ &\quad + \overline{E}_0\overline{A}_0\overline{A}_2 + \overline{E}_1\overline{A}_1\overline{A}_2 + \overline{E}_2\left(\overline{A}_2\right)^2 + \overline{E}_3\overline{A}_2\overline{A}_3 \\ &\quad = -\frac{\nu}{2}\kappa\left(2\overline{A}_1\overline{A}_3 - 2\overline{A}_0\overline{A}_2\right) \\ &\quad + \frac{\nu}{2}\tau\left(-\left(\overline{A}_0\right)^2 + \left(\overline{A}_1\right)^2 - \left(\overline{A}_2\right)^2 + \left(\overline{A}_3\right)^2\right) \end{aligned} \tag{33}$$

$$\begin{aligned} &\overline{B}_0\overline{A}_0\overline{A}_1 + \overline{B}_1\left(\overline{A}_1\right)^2 + \overline{B}_2\overline{A}_1\overline{A}_2 + \overline{B}_3\overline{A}_1\overline{A}_3 + \overline{C}_0\left(\overline{A}_2\right)^2 \\ &\quad + \overline{C}_1\overline{A}_0\overline{A}_1 + \overline{C}_2\overline{A}_0\overline{A}_2 + \overline{C}_3\overline{A}_1\overline{A}_3 + \overline{D}_0\overline{A}_0\overline{A}_3 \\ &\quad + \overline{D}_1\overline{A}_1\overline{A}_3 + \overline{D}_2\overline{A}_1\overline{A}_3 + \overline{D}_3\left(\overline{A}_3\right)^2 + \overline{E}_0\overline{A}_0\overline{A}_2 \\ &\quad + \overline{E}_1\overline{A}_1\overline{A}_2 + \overline{E}_2\left(\overline{A}_2\right)^2 + \overline{E}_3\overline{A}_2\overline{A}_3 \\ &\quad = -\frac{\nu}{2}\tau\left(-\left(\overline{A}_0\right)^2 + \left(\overline{A}_1\right)^2 + \left(\overline{A}_2\right)^2 - \left(\overline{A}_3\right)^2\right) \end{aligned} \tag{34}$$

Finally, we get results for the elements $\overline{B}_i, \overline{C}_i, \overline{D}_i, \overline{E}_i$, $(0 \le i \le 3)$ as follows:

$$\begin{aligned} \overline{B}_0 &= 0, \\ \overline{B}_1 &= -\frac{\nu\tau}{2}, \\ \overline{B}_2 &= 0, \\ \overline{B}_3 &= -\frac{\nu\kappa}{2} \\ \overline{C}_0 &= \frac{\nu\tau}{2}, \\ \overline{C}_1 &= 0, \\ \overline{C}_2 &= \frac{\nu\kappa}{2}, \\ \overline{C}_3 &= 0 \\ \overline{D}_0 &= 0, \\ \overline{D}_1 &= \frac{\nu\tau}{2}, \\ \overline{D}_2 &= 0, \\ \overline{D}_3 &= \frac{\nu\kappa}{2} \end{aligned}$$

$$\overline{E}_0 = -\frac{\nu\tau}{2},$$
$$\overline{E}_1 = 0,$$
$$\overline{E}_2 = -\frac{\nu\kappa}{2},$$
$$\overline{E}_3 = 0 \tag{35}$$

Therefore, by using these values (35) we obtain the fuzzy split quaternionic Serret-Frenet frame equation as

$$\left[d\left(h'\right)\right] = \begin{bmatrix} d\overline{A}_0 \\ d\overline{A}_1 \\ d\overline{A}_2 \\ d\overline{A}_3 \end{bmatrix} = \frac{\nu}{2}\begin{bmatrix} 0 & \tau & 0 & -\kappa \\ \tau & 0 & \kappa & 0 \\ 0 & \kappa & 0 & \tau \\ -\kappa & 0 & -\tau & 0 \end{bmatrix}\begin{bmatrix} \overline{A}_0 \\ \overline{A}_1 \\ \overline{A}_2 \\ \overline{A}_3 \end{bmatrix} \tag{36}$$

6. Conclusion and Discussion

In this study, we redefined the algebraic operations for split quaternions on fuzzy split quaternions. The set of split quaternions is a subset of fuzzy split quaternions ($H' \subset H'_F$). This condition is important because the given definitions for fuzzy split quaternions are provided with it. As a result of this, given definitions are similar to definitions for split quaternions. We have seen that these definitions are similar to the split quaternion structures. We have obtained in this study fuzzy tangent vector $\overrightarrow{T'}$, fuzzy normal vector $\overrightarrow{N'}$, and fuzzy binormal vector $\overrightarrow{B'}$. These vector forms are a new description and calculation. Also, we have redefined these Serret-Frenet frames for fuzzy split quaternions on familiar Serret-Frenet frames. For fuzzy quaternionic forms the torsion and curvature functions are defined as

$$\tau : I \subset R \longrightarrow [0,1]$$
$$\kappa : I \subset R \longrightarrow [0,1] \tag{37}$$

For this reason, Serret-Frenet frame elements in (36) for fuzzy split quaternions get values in the range $[-1, 1]$. In Definition 7, we can see that if we take equal fuzzy split quaternion to the split quaternion, the function $h' \in H'$ can take the value $'1'$ and if we take not equal fuzzy split quaternion to the split quaternion, the function h' can take the value $'0'$. Hence, for calculating (27), (28), and (29), the necessary rule is

$$h'\left(b_0 1 + b_1 i + b_2 j + b_3 k\right) = 1 \tag{38}$$

Conflicts of Interest

The authors declare that they have no conflicts of interest.

Acknowledgments

The basic properties and required features of this study are provided in the 15th International Geometry Symposium Amasya University, Amasya, Turkey, July 3-6.

References

[1] M. Özdemir and A. A. Ergin, "Rotations with unit timelike quaternions in Minkowski 3-space," *Journal of Geometry and Physics*, vol. 56, no. 2, pp. 322–336, 2006.

[2] L. A. Zadeh, "Fuzzy sets," *Information and Control*, vol. 8, no. 3, pp. 338–353, 1965.

[3] R. Moura, F. Bergamaschi, R. Santiago, and B. Bedregal, "Rotation of triangular fuzzy numbers via quaternion," in *Proceedings of the 2014 IEEE International Conference on Fuzzy Systems (FUZZ-IEEE)*, pp. 2538–2543, Beijing, China, July 2014.

[4] H. Flint, "XLIII. Applications of Quaternions To The Theory Of Relativity," *The London, Edinburgh, and Dublin Philosophical Magazine and Journal of Science*, vol. 39, no. 232, pp. 439–449, 2009.

[5] J. B. Kuipers, *Quaternions and rotation sequences: A primer with applications to orbits, aerospace and virtual reality author: Jb kui*, 2002.

[6] J. M. Cooke, *Flight simulation dynamic modeling using quaternions, Ph.D. dissertation*, Naval Postgraduate Schoo, Monterey, Calif, USA, 1992.

[7] S. Halici, "On Fibonacci quaternions," *Advances in Applied Clifford Algebras (AACA)*, vol. 22, no. 2, pp. 321–327, 2012.

[8] E. Ata, Y. Kemer, and A. Atasoy, "Generalized Quaternions Serret-Frenet and Bishop Frames , Dumlup nar Universty," *Science Institute*, 2012.

[9] A. J. Hanson, *Quaternion Frenet Frames Making Optimal Tubes and Ribbons from Curves*, Computer Science Department, Indiana University Bloomington, In 47405.

[10] S. Fuhua, *The Basic Properties of Fuzzy Derivative*, Daqing Petroleum Institute, Anda, China.

Gaussian Qualitative Trigonometric Functions in a Fuzzy Circle

M. Clement Joe Anand and Janani Bharatraj

Department of Mathematics, Hindustan Institute of Technology and Science (Deemed to be University), Chennai 603103, India

Correspondence should be addressed to M. Clement Joe Anand; arjoemi@yahoo.com

Academic Editor: Erich Peter Klement

We build a bridge between qualitative representation and quantitative representation using fuzzy qualitative trigonometry. A unit circle obtained from fuzzy qualitative representation replaces the quantitative unit circle. Namely, we have developed the concept of a qualitative unit circle from the view of fuzzy theory using Gaussian membership functions, which play a key role in shaping the fuzzy circle and help in obtaining sharper boundaries. We have also developed the trigonometric identities based on qualitative representation by defining trigonometric functions qualitatively and applied the concept to fuzzy particle swarm optimization using α-cuts.

1. Introduction

The term "Trigonometry" was first coined as the title of a book *(Trigonometria)*, which translates to "triangle's measurement." Although trigonometry is now taught with an emphasis on right triangles, its origin goes back to an era where it was used to determine the positions of celestial bodies and the distances between them and to understand the concept of chords in circles. Euclidean geometry, that is, planar geometry, deals with two-dimensional figures. The study of planar geometry provides constructions for planar figures, their properties, and the relationships between points, lines, and figures. The planar figure formed when a point traces a path at a fixed distance with respect to another fixed point called a circle, where the circle divides a plane into interior and exterior regions. Various theorems and properties on circles have been developed since time immemorial.

Gradually, researchers have developed theories on intersecting circles, which led to divergent properties between circular triangles and spherical triangles. The introduction of fuzzy sets and systems by Zadeh [1] changed the face of research in trigonometry. Fuzzy trigonometry was introduced by Buckley and Eslami [2], wherein continuous fuzzy numbers and sets were defined using the principle of extension. This method formed the basis of fuzzy trigonometry but failed to satisfy many criteria and identities. Furthermore, Ress [3] developed an approach for mapping standard trigonometric functions into the fuzzy realm. Using these modified fuzzy trigonometric functions, the proofs of a few inverse trigonometric identities, in addition to the standard identities, were given. A breakthrough in the study of fuzzy trigonometry was achieved by Liu et al. [4], with an aim of connecting symbolic cognitive functions to qualitative functions. The basic identities were satisfied, but a few properties could not be achieved. Ghosh and Chakraborty [5] proposed two methodologies for describing a fuzzy circle. The first methodology defines a circle as a set of points which are equidistant from a fixed point. The second methodology describes a circle using three fuzzy points. The definitions using both methods are as follows.

(1) Considering fuzzy numbers plotted along various line segments which pass through a specific point $\widetilde{P}(a,b)$ which is fuzzy and located at a distance of $\widetilde{r}$ from the fixed fuzzy point, then the fuzzy circle is defined as

$$\widetilde{C}_1 = \bigvee_{\theta} \{\widetilde{B}_\theta\}, \tag{1}$$

where the distance between $\widetilde{B}_\theta$ and a random point on the support is always fixed.

(2) Considering three fuzzy points, the fuzzy circle is drawn by passing the circle through these three points, and the supremum of the membership values defines the fuzzy circle which was proposed.

Technology in today's world needs advanced level procedures or heuristics which involve few or zero assumptions about the problem being optimized. The need for metaheuristics has many advantages over other algorithms, with one advantage being its ability to search very large candidate solutions spaces. Particle swarm optimization (PSO) is one such metaheuristic method which allows optimization based on iterations, which in turn helps in improving the candidate solution. The solution is improved by creating a population of candidate solutions (i.e., particles) and making them move around in the search-space, taking into consideration their positions and velocities. The movement of the particle is influenced by its local best known position, simultaneously being guided towards the best known position in the search area. These are updated time-to-time for all other particles to form a swarm. PSO was first introduced by Kennedy et al. with an intention of simulating the social behaviour in a flock of birds or a school of fish. Shi and Eberhart [6] developed an optimization technique for fuzzy systems by dynamically adjusting the inertia weight, improving PSO's performance. Pang et al. [7] utilized fuzzy discrete PSO to solve the Travelling salesman problem., Abdelbar et al. [8] compared the behaviour of the Gaussian and Cauchy membership functions in fuzzy PSO and concluded that the Cauchy membership functions are best suited for fuzzy PSO. It was noted that traditional PSO converges prematurely when applied as a global optimization technique. To prevent this downfall, Anantathanavit and Munlin [9] proposed radius (R-PSO) as an extension of standard PSO. R-PSO regroups the particles into a circle of the given radius and determines the best agent particle of the group. The experiment results proved that R-PSO performed better than traditional PSO when solving multimodal complex problems.

To produce wholesome proofs for identities based on fuzzy trigonometric functions, we introduce the Gaussian qualitative trigonometric functions (GQTFs) on a fuzzy circle. The unit fuzzy circle is defined using Gaussian membership function (GMF) partitions. Using the GMF circumference, fuzzy centre, and fuzzy point on the circle, trigonometric functions and their identities are successfully developed. In this paper, we use the advantages of R-PSO and GMFs to define a fuzzy R-PSO based on a fuzzy qualitative circle. Section 2 gives the preliminaries and introduces the concept of GQTFs on the fuzzy circle. Section 3 provides insight on the formation of the fuzzy centre and fuzzy point on the circle and provides a broad perspective of the redefined trigonometric functions, their ratio identities, their Pythagorean identities, their Taylor's series expansions, their differentials, and laws for obtaining the solutions of triangles. Section 4 describes the properties of intersecting fuzzy circles, and Section 5 describes the PSO with fuzzy matrices. Section 6 describes fuzzy R-PSO with GMF, which makes an efficient model for finding the best agent. Finally, Section 7 concludes the paper.

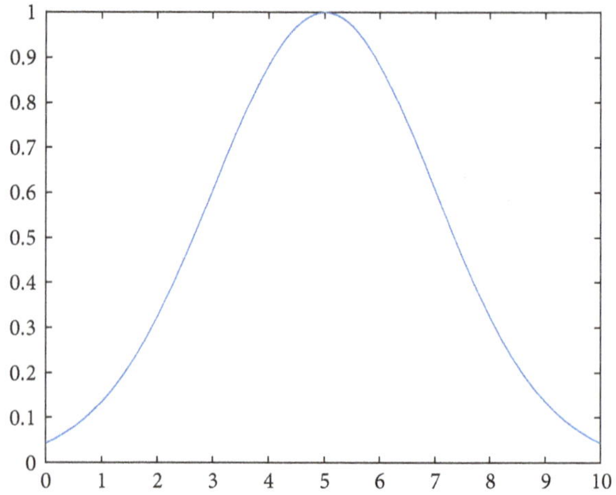

Figure 1: GMF with $c = 5$ and $\sigma = 2$.

2. Prerequisites

The prerequisites required for the developed concept have been taken from various research articles which are cited in the references. Slightly modified versions of the concepts have been given in this section.

2.1. Fuzzy Subset. Considering a set E which is either finite or infinite, for every element x in E, the set consisting of all ordered pairs of the form $\{(x, \mu_{\widetilde{A}}(x))\}$ is called a fuzzy subset, denoted by $\widetilde{A}$, in E. Here, $\mu_{\widetilde{A}}(x)$ is called the membership function. The membership function gives a mapping of the elements in set E to the membership set M.

2.2. Zadeh's Extension Principle. Considering a fuzzy subset of a universal set E, if $f : E \to Z$ is a function, then the extension principle generates a function $\widehat{f}$ whose membership function is defined over the supremum of all $f^{-1}(z) \neq \emptyset$, that is,

$$\mu_{\widehat{f}}(z) = \sup_{f^{-1}(z)} \mu_A(x), \quad x \in E. \tag{2}$$

Here, f^{-1} is the preimage of z in E.

2.3. Gaussian Membership Function (GMF). The GMF is given by

$$G_{mf}(x : c, \sigma) = \exp\left[-\frac{(c-x)^2}{2\sigma^2}\right], \tag{3}$$

where c and σ are the centre and width of the fuzzy set, respectively. An example of this function is shown in Figure 1. This membership function is also defined in terms of the interval $[m, j]$ as follows:

$$G_{mf}(x : m, j) = \exp\left[-\frac{(j-x)^2}{2m^2}\right]. \tag{4}$$

Table 1: Arithmetic operations.

Sl.no.	Operation	Notation	Result
1.	Sum	$[m, c] + [j, a]$	$[m + j, c + a]$
2	Difference	$[m, c] - [j, a]$	$[m - a, c - j]$
3	Product	$[m, c] \times [j, a]$	$[m \times j, c \times a]$
4	Quotient	$[m, c] \div [j, a]$	$[m \div a, c \div j]$

The GMF with fuzzification factor m is given by

$$G_{mf}(x : c, \sigma, m) = \exp\left[-\frac{(c - x)^m}{2\sigma^m}\right]. \quad (5)$$

2.4. Arithmetic Operations on GMFs. The Arithmetic Operations on GMFs can be summarised as follows.

Letting $G_{mf}(x : m, c)$ and $G_{mf}(x : j, a)$ be two GMFs, their arithmetic operations are shown in Table 1.

3. Proposed Method

Fuzzy Gaussian qualitative coordinates are proposed to facilitate the geometrical interpretation of GQTFs. To achieve this, we first define a fuzzy circle obtained using GMFs.

3.1. Gaussian Qualitative Coordinates. The x- and y-axes are defined for a unit circle. For the sake of simplicity and ease of use, the abscissa and ordinate are first considered as real lines. Figure 2 describes the circumference of the fuzzy unit circle which is obtained by partitioning the circumference into GMFs. The cloud formation, as described by Ghosh and Chakraborty [10], is now refined to obtain sharp boundaries for the circle. The crisp centre of the circle has a neighbourhood which is formed by extending and converging the core area of every GMF towards the centre of the circle. Thus the centre of the derived fuzzy circle will have infinitely many sides depending on the number of partitions. We can henceforth generalise the centre to be an infinitesimal circle with an infinitesimal radius, that is, l.

3.2. Fuzzy Centre and Position of a Point. The values of a and b obtained on the circumference of the fuzzy circle are now extended towards the core of the circle. Without loss of generality, a point on the proposed fuzzy circle can be assumed to be the centre of some GMF. The position of a point can be mapped as shown in Figure 3.

3.3. Trigonometric Functions. The graphs of sine, cosine, and tangent functions under the proposed concept have been obtained and verified to be in coherence with the crisp trigonometric functions. Various identities and laws for triangles are presented as follows.

3.3.1. Ratio Identities

(i) $\sin(G_{mf}[-\delta\ \ \delta]) = G_{mf}[a_y\ \ b_y]/G_{mf}[\sqrt{(a_x - a_0)^2 + a_y^2}\ \ \sqrt{(b_x - b_0)^2 + b_y^2}]$

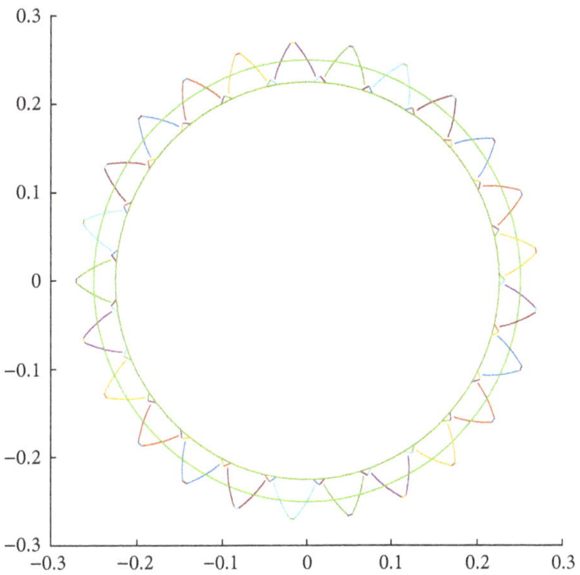

Figure 2: Fuzzy circle with Gaussian membership function.

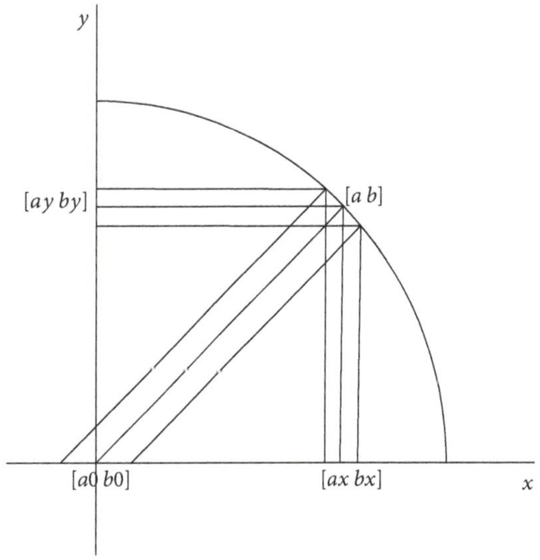

Figure 3: Coordinates of a point on the fuzzy circle.

(ii) $\cos(G_{mf}[-\delta\ \ \delta]) = G_{mf}[a_0 - b_x\ \ b_0 - a_x]/G_{mf}[\sqrt{(a_x - a_0)^2 + a_y^2}\ \ \sqrt{(b_x - b_0)^2 + b_y^2}]$,

(iii) $\tan(G_{mf}[-\delta\ \ \delta]) = G_{mf}[a_y\ \ b_y]/G_{mf}[a_0 - b_x\ \ b_0 - a_x]$.

3.3.2. Pythagorean Identities

(i) $\sin^2(G_{mf}[-\delta\ \ \delta]) + \cos^2(G_{mf}[-\delta\ \ \delta]) = G_{mf}[1 - \delta\ \ 1 + \delta]$,

(ii) $\sec^2(G_{mf}[-\delta\ \ \delta]) - \tan^2(G_{mf}[-\delta\ \ \delta]) = G_{mf}[1 - \delta\ \ 1 + \delta]$,

(iii) $\csc^2(G_{mf}[-\delta\ \ \delta]) - \cot^2(G_{mf}[-\delta\ \ \delta]) = G_{mf}[1 - \delta\ \ 1 + \delta]$.

3.3.3. Laws for Obtaining Solutions to Triangles

(i) The law of sines gives the relationship between the sines of the angles of a triangle and the lengths of the sides of the triangle:

$$\frac{\sin\left(G_{mf}\left[\theta_y-\delta_f\ \ \theta_y+\delta_f\right]\right)}{G_{mf}\left[a_y\ \ b_y\right]} = \frac{\sin\left(G_{mf}\left[\theta_x-\delta_f\ \ \theta_x+\delta_f\right]\right)}{G_{mf}\left[a_x\ \ b_x\right]} = \frac{\sin\left(G_{mf}\left[\theta_z-\delta_f\ \ \theta_z+\delta_f\right]\right)}{G_{mf}\left[a_z\ \ b_z\right]}. \quad (6)$$

(ii) The law of cosines relates the sides of a triangle with the cosine of an angle in the triangle:

(1) $G_{mf}[a_x-b_y\ \ a_y-b_x]^2 = G_{mf}[c_x-b_y\ \ c_y-b_x]^2 + G_{mf}[a_x-c_y\ \ a_y-c_x]^2 - 2G_{mf}[c_x-b_y\ \ c_y-b_x]\times G_{mf}[a_x-c_y\ \ a_y-c_x]\times\cos(G_{mf}[C-\delta\ \ C+\delta])$

(2) $G_{mf}[c_x-b_y\ \ c_y-b_x]^2 = G_{mf}[a_x-c_y\ \ a_y-c_x]^2 + G_{mf}[a_x-b_y\ \ a_y-b_x]^2 - 2\times G_{mf}[a_x-c_y\ \ a_y-c_x]\times G_{mf}[a_x-b_y\ \ a_y-b_x]\times\cos(G_{mf}[A-\delta\ \ A+\delta])$,

(3) $G_{mf}[a_x-c_y\ \ a_y-c_x]^2 = G_{mf}[b_x-c_y\ \ b_y-c_x]^2 + G_{mf}[a_x-b_y\ \ a_y-b_x]^2 - 2\times G_{mf}[b_x-c_y\ \ b_y-c_x]\times G_{mf}[a_x-b_y\ \ a_y-b_x]\times\cos(G_{mf}[B-\delta\ \ B+\delta])$.

(iii) The law of tangents gives the relationship between the sides and tangents of the angles of a triangle:

$$\frac{G_{mf}\left[b_x+c_x-\left(a_y+c_y\right)\ \ b_y+c_y-\left(a_x+c_x\right)\right]}{G_{mf}\left[b_x+a_x-2c_y\ \ b_y+a_y-2c_x\right]} = \frac{\tan\left((1/2)\,G_{mf}\left[A-B-2\delta\ \ A-B+2\delta\right]\right)}{\tan\left((1/2)\,G_{mf}\left[A+B-2\delta\ \ A+B+2\delta\right]\right)}. \quad (7)$$

3.3.4. Mollweide's Identity. This identity is a tool for checking the solutions of triangles. It uses all sides and angles of the triangle.

$$\frac{G_{mf}\left[a_x+b_x-2c_y\ \ a_y+b_y-2c_x\right]}{G_{mf}\left[a_x-b_y\ \ a_y-b_x\right]} = \frac{\cos\left((1/2)\,G_{mf}\left[A-B-2\delta\ \ A-B+2\delta\right]\right)}{\sin\left((1/2)\,G_{mf}\left[C-\delta\ \ C+\delta\right]\right)}. \quad (8)$$

3.4. Taylor's Series Expansions of Trigonometric Functions. The Taylor's series expansions of the trigonometric functions are as follows:

(i) $\cos(G_{mf}[x_{mf}-\delta_{mf}\ \ x_{mf}+\delta_{mf}]) = [1-(G_{mf}[x_{mf}-\delta_{mf}\ \ x_{mf}+\delta_{mf}])^2/2! + (G_{mf}[x_{mf}-\delta_{mf}\ \ x_{mf}+\delta_{mf}])^4/4! \mp\cdots]$

Thus, $\cos(G_{mf}[x_{mf}-\delta_{mf}\ \ x_{mf}+\delta_{mf}]) = G_{mf}[\cos(x_{mf}-\delta_{mf})-\Delta_{mf}\ \ \cos(x_{mf}-\delta_{mf})+\Delta_{mf}]$, where $\Delta_{mf} = 2\sum_{n=2,4,6,\ldots}^{\infty}(((x_{mf}+\delta_{mf})^n+(x_{mf}-\delta_{mf})^n)/n!)$

(ii) $[\sin(G_{mf}[x_{mf}-\delta_{mf}\ \ x_{mf}+\delta_{mf}]) = G_{mf}[x_{mf}-\delta_{mf}\ \ x_{mf}+\delta_{mf}]-(G_{mf}[x_{mf}-\delta_{mf}\ \ x_{mf}+\delta_{mf}])^3/3! + (G_{mf}[x_{mf}-\delta_{mf}\ \ x_{mf}+\delta_{mf}])^5/5!\mp\cdots]$

Thus, $[\sin(G_{mf}[x_{mf}-\delta_{mf}\ \ x_{mf}+\delta_{mf}]) = G_{mf}[\sin(x_{mf}-\delta_{mf})-\Delta_1\ \ \sin(x_{mf}-\delta_{mf})+\Delta_1]]$, where $\Delta_1 = 2\sum_{n=1,3,5,\ldots}^{\infty}(((x_{mf}+\delta_{mf})^n+(x_{mf}-\delta_{mf})^n)/n!)$

(iii) $\tan(G_{mf}[x_{mf}-\delta_{mf}\ \ x_{mf}+\delta_{mf}]) = G_{mf}[x_{mf}-\delta_{mf}\ \ x_{mf}+\delta_{mf}] + (1/3)G_{mf}[(x_{mf}-\delta_{mf})^3\ \ (x_{mf}+\delta_{mf})^3] + (2/5)G_{mf}[(x_{mf}-\delta_{mf})^5\ \ (x_{mf}+\delta_{mf})^5]+\cdots$

Thus, $[\tan(G_{mf}[x_{mf}-\delta_{mf}\ \ x_{mf}+\delta_{mf}]) = G_{mf}[\tan(x_{mf}-\delta_{mf})\ \ \tan(x_{mf}+\delta_{mf})]]$.

3.5. Differentiation of the Gaussian Trigonometric Functions. The trigonometric functions thus obtained are also fuzzy differentiable. Supposing that $F(x) = f(G_{mf}[c\ \ \sigma])$ is a Gaussian fuzzy trigonometric function, then the derivatives of the function are as follows:

First Derivative: $F'(x) = f'(G_{mf}[c\ \ \sigma])\times\exp(-x^2/2\sigma^2)\times(-x/\sigma^2)$.

The second and higher order derivatives are obtained by using the chain rule for the above functions.

4. Intersection of Fuzzy Circles

We now introduce the concept of intersecting fuzzy circles and study their structure and properties while comparing them with regular intersecting circles.

4.1. Circles Touching at Exactly One Point

Case 1. Two circles have an equal number of partitions, implying that the radii of both circles are equal. Figure 4 shows the fuzzified point of intersection.

In Figure 4, we observe that the centres of both Gaussian curves intersect at a crisp point. The fuzzified areas of the intersection are congruent triangles. The area of each triangle is given by $F_{ta} = 1/2\times(2r)\times\sigma = r\times\sigma$. Thus, the total fuzzy area is $2r\sigma$.

Case 2. The number of partitions on the intersecting circles is unequal, as shown in Figure 5.

We observe that the crisp point of intersection is obtained where the centres of both Gaussian curves meet. The point of intersection is bounded by triangles whose areas are given by

$$\text{Area of }\Delta_1 = \frac{1}{2}\times\sigma_1\times\left(r_1+\sigma_1\right) = \frac{\sigma_1\left(r_1+\sigma_1\right)}{2}$$

$$\text{Area of }\Delta_2 = \frac{1}{2}\times\sigma_1\times\left(r_1-\sigma_1\right) = \frac{\sigma_1\left(r_1-\sigma_1\right)}{2}$$

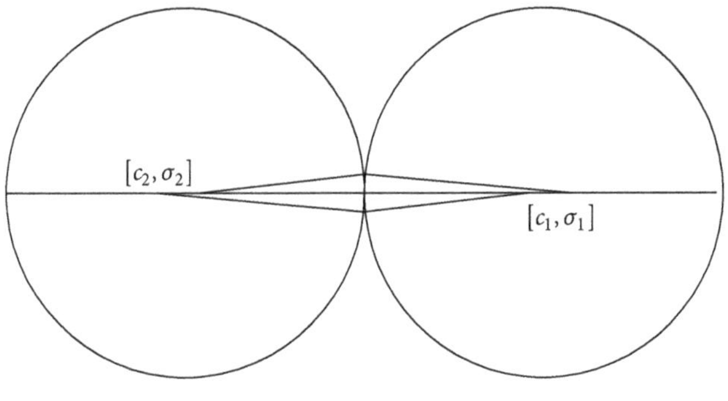

FIGURE 4: Circles of equal radii touching at a single point.

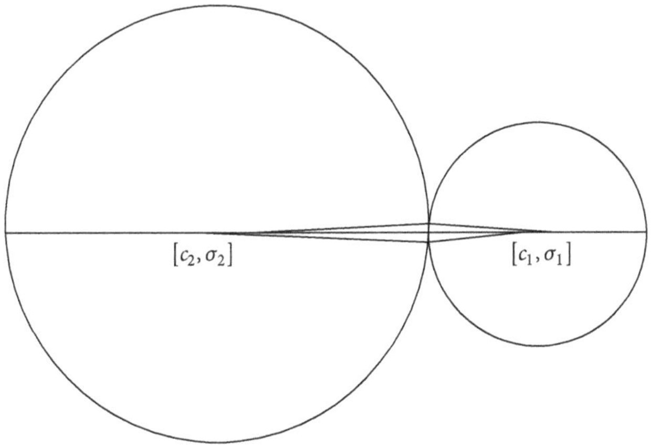

FIGURE 5: Circles of unequal radii touching at a single point.

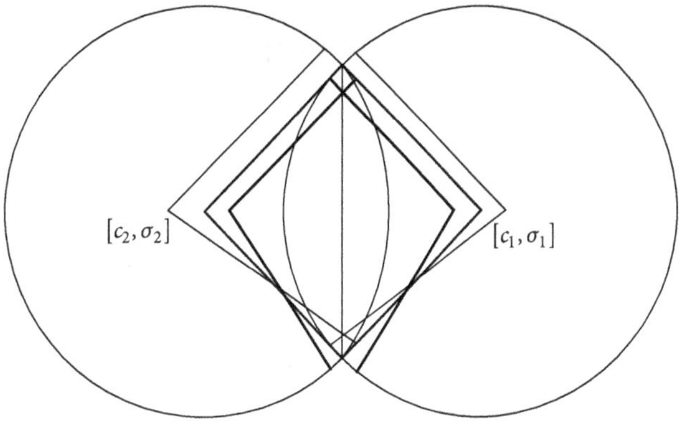

FIGURE 6: Intersecting circles of equal radii.

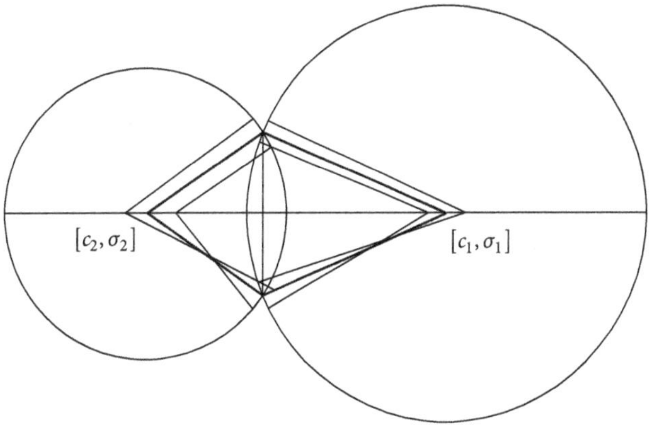

FIGURE 7: Intersecting circles of unequal radii.

$$\text{Area of } \Delta_3 = \frac{1}{2} \times \sigma_2 \times (r_2 + \sigma_2) = \frac{\sigma_2 (r_2 + \sigma_2)}{2}$$
$$\text{Area of } \Delta_4 = \frac{1}{2} \times \sigma_2 \times (r_2 - \sigma_2) = \frac{\sigma_2 (r_2 - \sigma_2)}{2}. \tag{9}$$

Hence, the total area is $= r_1\sigma_1 + r_2\sigma_2$. The area obtained is in coherence with the traditional way of determining the area of intersection.

The intersection of circles becomes interesting when we consider circles intersecting at two points. In such a case, the area of intersection can be either symmetric or dominate a single circle with a small portion of the second circle as a common area.

4.2. Fuzzy Circles Intersecting at Two Points

Case 1. Two circles have an equal number of partitions. Figure 6 shows the intersecting fuzzy circles. From the following diagram, we observe that the crisp points of intersection and the common chord are obtained during the fuzzification process. We obtain congruent triangles in the fuzzy area of intersection. The centres of the Gaussian curves meet to produce the crisp point of intersection. The two points, when joined, produce the crisp common chord for the intersecting circles.

The area of the overlapping region is calculated as follows:

(i) The area of each sector is $(r_- \times \theta)/2$.

(ii) The length of the arc of the sector is given by $r \times \theta$.

(iii) Thus, the area of the fuzzified sector is $(r_+^2 \times \theta)/2$.

(iv) The area of the smaller sector is $(r_-^2 \times \theta)/2$. Here, $r_+ = r + \sigma$ and $r_- = r - \sigma$.

(v) Supposing the line r_+ cuts the smaller radius in the ratio $m : n$, then $m = r_+/2$. Thus, the area is $(r_+^2 \times \theta_1)/4$, and the total area is $r_+^2/2 \times (2\theta + \theta_1)$.

Case 2. Two circles have an unequal number of partitions, which implies that the radii of both circles are different (Figure 7).

The area of the bigger fuzzy sector is $(r^2 \times \theta)/2 + (r^2 \times \theta_1)/4 = (r^2/4)(2\theta + \theta_1)$.

Similarly, the area of the smaller fuzzy sector is $(r_1^2/4)(2\delta + \delta_1)$.

Thus, in both cases, we obtain a fuzzy diagram bounded by ten points. The area of the overlapping section is obtained using the following formulas:

$$\begin{aligned} \text{Area}_{\text{fuzzy}} &= \frac{1}{\text{Dis}} \sqrt{C \times J \times A \times M}, \\ C &= (-\text{Dis} + \text{rad}_1 + \text{rad}_2), \\ J &= (\text{Dis} - \text{rad}_1 + \text{rad}_2), \\ A &= (\text{Dis} + \text{rad}_1 - \text{rad}_2), \\ M &= (\text{Dis} + \text{rad}_1 + \text{rad}_2), \end{aligned} \tag{10}$$

where $\text{Dis} = [O_1 - O_2 - (\sigma_1 + \sigma_2) \;\; O_1 - O_2 + (\sigma_1 + \sigma_2)]$ is the distance between the centres and $\text{rad}_1 = [r_1 - \sigma_1 \;\; r_1 + \sigma_1]$ and $\text{rad}_2 = [r_2 - \sigma_2 \;\; r_2 + \sigma_2]$ are the Gaussian radii of the fuzzy circles.

4.3. Intersection of Three Fuzzy Circles. Three circles can intersect in many ways, but we consider two special cases wherein we obtain circular triangles and irregular convex quadrilaterals. However, we must check whether an overlapping area exists between the considered circles. To verify, we check if either of the following conditions are satisfied.

(i) A point of intersection of two circles should be a point inside the third circle;

(ii) One of the circles should be engulfed completely inside another circle.

Case 1. Three fuzzy circles with equal numbers of partitions are considered, implying that the radii of all three circles are equal. The overlapping area formed by the three segments and an inner triangle whose sides are the arcs of the three circles is called a circular triangle (Figure 8).

The area of the triangle is calculated using Heron's formula:

$$\text{Area}_{abc} = \sqrt{S(S-A)(S-B)(S-C)}, \tag{11}$$

where $S = (1/2)(A + B + C)$, $A = G_{mf}[a_x \;\; a_y]$, $B = G_{mf}[b_x \;\; b_y]$, and $C = G_{mf}[c_x \;\; c_y]$. The area of the combined overlap is given by

$$\begin{aligned} \text{Area}_{\text{seg}} &= \sum_{n=1}^{3} R_n^2 \sin^{-1}\left(\frac{A}{2R_n}\right) - \sum_{n=1}^{3} \frac{A}{4}\sqrt{4 \times R_n^2 - A^2} \\ &\quad + \sqrt{S(S-A)(S-B)(S-C)}. \end{aligned} \tag{12}$$

Case 2. Three fuzzy circles with unequal numbers of partitions will have different radii. In this case, we must recalculate the overlapping segment area using the area of the bigger circle (Figure 9).

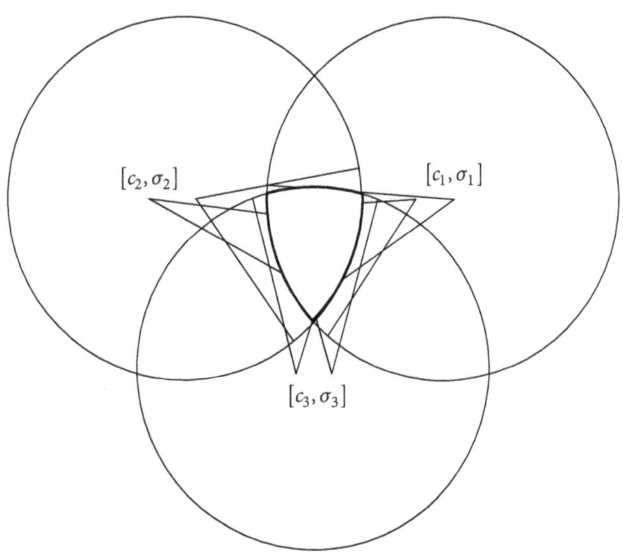

FIGURE 8: Circles of equal radii intersecting and forming a circular triangle.

Step 1. We find the midpoint of the chord in the segment, which is given by

$$\begin{aligned} x_{\text{mid}} &= \frac{x_{i12} + x_{i13}}{2}, \\ y_{\text{mid}} &= \frac{y_{i12} + y_{i13}}{2}. \end{aligned} \tag{13}$$

Step 2. We then find the equation of the line joining the midpoint and the centre of the circle, which is given by

$$y = \frac{B_1 - y_{\text{mid}}}{A_1 - x_{\text{mid}}} \times x - \frac{x_{\text{mid}}(B_1 - y_{\text{mid}})}{A_1 - x_{\text{mid}}} + y_{\text{mid}}. \tag{14}$$

Step 3. Next, we determine the points of intersection of the above line and circle, which are given by $X_{ic1,ic2} = (a_1 + b_1 \times m - m \times n \pm \sqrt{\delta})/(1 + m^2)$, $Y_{ic1,ic2} = (n + a_1 \times m + b_1 \times m^2 \pm m \times \sqrt{\delta})/(1 + m^2)$, where $m = (B_1 - y_{\text{mid}})/(A_1 - y_{\text{mid}})$, $n = (-x_{\text{mid}} \times (B_1 - y_{\text{mid}}))/(A_1 - x_{\text{mid}}) + y_{\text{mid}}$.

Step 4. Finally, we check whether the other two radii of the circles are greater than the distance between the centre and the midpoint.

Thus, the area of the bigger segment completely encompassed by the bigger circle is $(R_1^2/2)(\theta_1 - \sin\theta_1)$.

Results

(i) The regions of intersection of three equal fuzzy circles have been considered. Hence the common value of radii has been denoted by R. Here R denotes the common radii.

(ii) Given three intersecting fuzzy circles, the lines joining the points of intersection outside the circular triangle satisfy $ACE = BDF$ which defines Haruki's theorem.

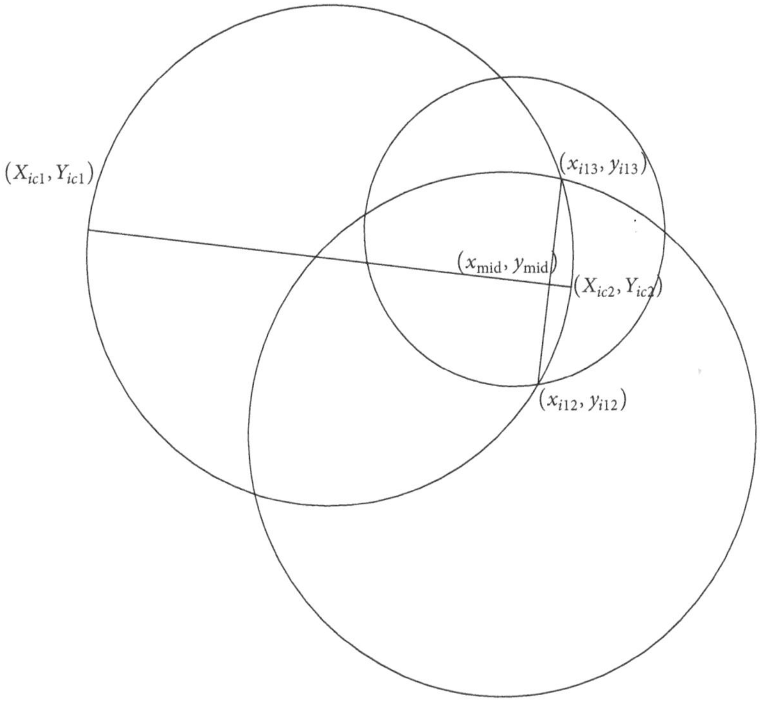

FIGURE 9: Three intersecting circles of unequal radii.

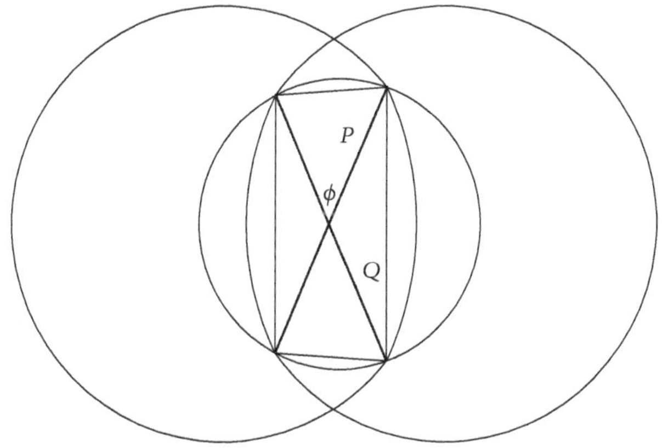

FIGURE 10: Formation of irregular convex quadrilateral.

(iii) The crisp circle which touches the inner circles of the circular triangle and its neighbouring triangles is either tangent to the innermost circle and internally tangent to the inner circles of the neighbouring triangles or vice versa. Such a circle is called the "Hart's circle." We can construct 8 such Hart circles for a given circular triangle.

Case 3. The most complex case is that of an overlapping area which is obtained when the circular segments form the boundary shown in Figure 10. In the process, we obtain an irregular convex quadrilateral.

The area of the irregular convex quadrilateral is given by $\text{Area}_{\text{quad}} = (1/2)P \times Q \times \sin\phi$, where ϕ is the angle between the diagonals P and Q.

Hence the combined area of the segments is given by

$$\text{Area} = \sum_{n=1} 4R_n^2 \sin^{-1}\left(\frac{A_n}{2 \times R_n}\right) - \sum_{n=1}^{4} \frac{A_n}{4}\sqrt{4 \times R_n^2 - A_n^2}. \quad (15)$$

Here A_n is the length of the chord joining the points of intersection of the circles. There are four chords which form the edges of the convex quadrilateral. And, R_n is the radii of the circles.

5. Fuzzy Particle Swarm Optimization (PSO) with Fuzzy Matrices and GMFs

5.1. Fuzzy Matrices. A matrix of the form $P = [p_{ij\mu}]_{m\times n}$ where $p_{ij\mu}$ is the membership degree of the element $p_{ij} \in \widetilde{P}$ is called a fuzzy matrix.

In PSO, a finite number of particles define a swarm. The particles move in the n-dimensional space with certain velocity. The particle's velocity and position are updated using the following formula:

$$V_i^{k+1} = w \otimes V_i^k \oplus (c_1 \otimes r(\cdot)) \otimes \left(P_i^k \ominus X_i^k\right) \oplus \sum_{h\in\eta(i,k)} c_2 \otimes r(\cdot) \otimes \psi(p_h)\left(P_g^k \ominus X_i^k\right), \quad (16)$$

where $\eta(i,k)$ denotes the k best particles in the neighbourhood of the particle and $\psi(p_h)$ denotes the matrix containing the corresponding membership degrees of the particles defined using the GMF.

5.2. Particle Position Initialization. The positions of the particles are initialized and represented in the form of a fuzzy matrix. The position matrix is given by

$$P_{rij} = \begin{bmatrix} p_{r11} & p_{r12} & \cdots & p_{r1n} \\ \vdots & \ddots & & \vdots \\ p_{rn1} & p_{rn2} & \cdots & p_{rnn} \end{bmatrix}. \quad (17)$$

The positions of the particles for the fuzzy matrix are generated randomly and satisfy the following conditions:

$$\sum_{j=1}^{n} p_{rij} = 1, \quad P_{rij} \in [0,1], \ i = 1,2,\ldots,n. \quad (18)$$

5.3. Particle Velocity Initialization. The velocity matrix is given by

$$V_{rij} = \begin{bmatrix} v_{r11} & v_{r12} & \cdots & v_{r1n} \\ \vdots & \ddots & & \vdots \\ v_{rn1} & v_{rn2} & \cdots & v_{rnn} \end{bmatrix}, \quad (19)$$

where the velocities of the particles are generated according to the randomly generated position of the particles with conditions

$$\sum_{j=1}^{n} v_{rij} = 0, \quad i = 1,2,\ldots,n. \quad (20)$$

5.4. Normalization of the Particle Position Matrix. The particle position matrix generated using random positions may sometimes not adhere to the constraint $P_{rij} \in [0,1]$. Hence, we need to normalize the particle position matrix. The first step in normalization involves replacing all negative values in the matrix with zeros. Then, the matrix is transformed into

$$P_{rij} = \begin{bmatrix} \frac{p_{r11}}{\sum_{i=1}^{n} p_{r1i}} & \frac{p_{r12}}{\sum_{i=1}^{n} p_{r1i}} & \cdots & \frac{p_{r1n}}{\sum_{i=1}^{n} p_{r1i}} \\ \vdots & \ddots & & \vdots \\ \frac{p_{rn1}}{\sum_{i=1}^{n} p_{rni}} & \frac{p_{rn2}}{\sum_{i=1}^{n} p_{rni}} & \cdots & \frac{p_{rnn}}{\sum_{i=1}^{n} p_{rni}} \end{bmatrix}. \quad (21)$$

5.5. Choosing the Best Agents with α-levels. The agent particles are now selected based on the value of α, where α is a real number in the interval $[0,1]$. Once the agent particles are selected, they are placed on the circumference of the circle of radius α. Furthermore, the global best position of the particle is calculated using the distance formula

$$d\left(p_{rij}, p_{rwj}\right) = \frac{1}{8(\sigma)^2}\sum_{i=1}^{n}\left[\left(\mu_{p_{rij}}(x_i) - \mu_{p_{rwj}}(x_i)\right)^2 + (c_1 - c_2)^2\right]; \quad (22)$$

$$d \le 2\alpha.$$

This process is illustrated in Figure 11.

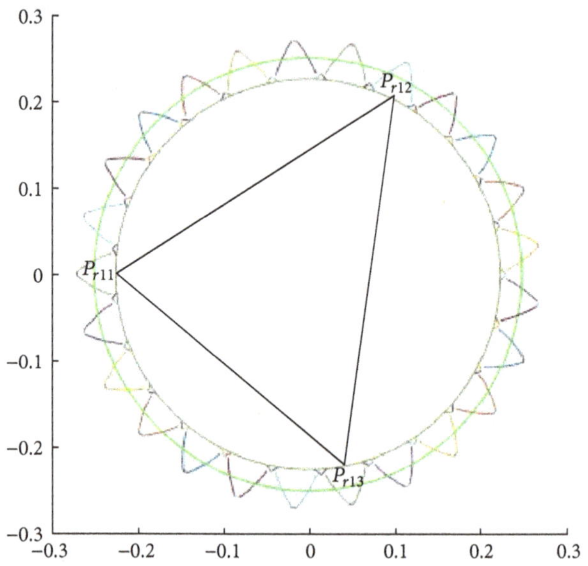

FIGURE 11: Neighbourhood for obtaining the global best agent.

6. Algorithm for Fuzzy R-PSO with GMFs

The proposed algorithm provides the local best position of particles based on the threshold, as given in Figure 12.

(1) Start.

(2) Initialize: maximum number of particles in the target neighbourhood: $\max_p$; maximum number of iterations: $\max_n$.

(3) Assign random position and velocity matrices for the particles.

(4) For $i = 0$ to $\max_p$

   (a) Calculate positions and velocities of the particles as given in the equation.

   (b) Normalize the particle position matrix using the equation.

(5) Fix the α-level for the normalized particle position matrix.

(6) If a particle's position in the normalized matrix is $\le\alpha$, then update the first neighbourhood circle.

(7) Calculate the distance between the particles and locate the global best position.

(8) Stop.

7. Discussion and Conclusion

The proposed methodology of constructing a fuzzy qualitative circle with GMF provides a smoother graphs of fuzzy trigonometric functions when compared to previously described methods. The graphs of the curves as obtained by Kaufmann and Gupta [11] flatten either the min- or max-curve depending upon the point of inflection, which led to a restriction on the extent of the spread in terms of angular

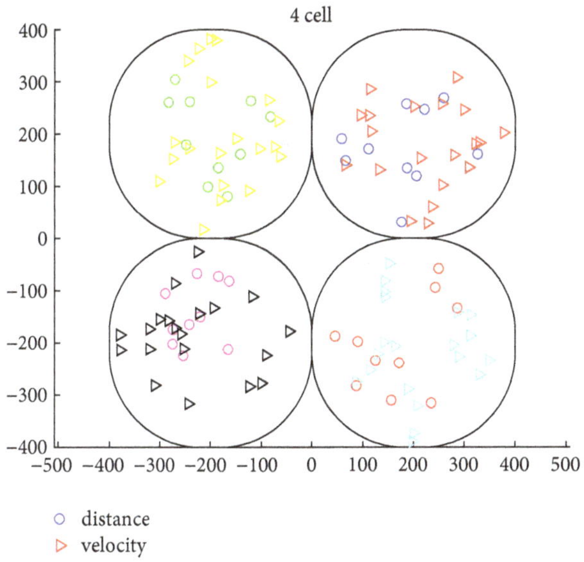

FIGURE 12: Neighbourhood obtained using α-cuts.

measurements. This angular spread has now been defined in terms of σ which is the width of the GMF. When $\sigma = 0$, the Gaussian Fuzzy Circle reduces to the conventional crisp circle. The centre of the fuzzy circle is also fuzzy and the radius is a linear combination of fuzzy numbers. Also, we have demonstrated geometrical interpretation of the GQTFs and the representation of a fuzzy circle with $n = 25$ GMFs. The position of a point on the fuzzy circle has been explained. Trigonometrical identities, Pythagorean identities, and laws for obtaining solutions to triangles (i.e., law of sines, law of cosines, and law of tangents) have been derived. The GMFs on a fuzzy circle and Taylor's series expansions have been verified. The relationship between qualitative and quantitative trigonometric states is very difficult to describe; however, Gaussian fuzzy qualitative trigonometry helps in describing their relationship. In this paper, we have successfully described the intersection of fuzzy circles when the circle is divided into either equal or unequal numbers of partitions, and a study of circular triangles has been given. The results on circular triangles have also been verified. This research article gives a theoretical interpretation of applying R-PSO inside a fuzzy circle. The R-PSO with fuzzy circle of radius α (developed using α-cuts) reduced the premature convergence of the standard PSO algorithm. By fixing various levels of α, global best position of the particles can be obtained. Furthermore, the development of fuzzy circles with partitions provides a sharper boundary when compared to the existing ideas of fuzzy circles and fuzzy qualitative circles.

7.1. Future Direction. The authors intend to implement this idea in designing a navigation stick using the concept of radar. We propose to utilize this concept in designing navigation systems for people with special needs. Further, this research, being an extension to fuzzy qualitative trigonometry, will be implemented in extensive research towards robot kinematics and human movement analysis.

Conflicts of Interest

The authors declare that they have no conflicts of interest.

Acknowledgments

The authors would like to thank Editage (https://www.editage.com) for English language editing and Mr. Mohamed E. for his timely help in developing the R-PSO for fuzzy circle algorithm using Matlab.

References

[1] L. A. Zadeh, "Fuzzy sets," *Information and Computation*, vol. 8, pp. 338–353, 1965.

[2] J. J. Buckley and E. Eslami, *An Introduction to Fuzzy Logic and Fuzzy Sets*, Physica-Verlag Heidelberg, NewYork, NY, USA, 2002.

[3] D. A. Ress, *Development of Fuzzy Trigonometric Functions to support design and manufacturing [Ph.D. Thesis]*, North Carolina University, North Carolina, NC, USA, 1999.

[4] H. Liu, G. M. Coghill, and D. P. Barnes, "Fuzzy qualitative trigonometry," *International Journal of Approximate Reasoning*, vol. 51, pp. 371–388, 2009.

[5] D. Ghosh and D. Chakraborty, "Analytical fuzzy plane geometry I," *Fuzzy Sets and Systems*, vol. 209, pp. 66–83, 2012.

[6] Y. Shi and R. Eberhart, "Fuzzy Adaptive Particle Swarm Optimization," in *Proceedings of the Congress on Evolutionary Computation*, pp. 101–106, 2001.

[7] W. Pang, K. Wang, C. Zhou, and L. Dong, "Fuzzy discrete particle swarm optimization for solving traveling salesman problem," in *Proceedings of the 4th International Conference on Computer and Information Technology (CIT '04)*, pp. 796–800, 2004.

[8] A. M. Abdelbar, S. Abdelshahid, and D. C. Wunsch, "Gaussian versus Cauchy membership functions in fuzzy PSO," in *Proceedings of the 2007 International Joint Conference on Neural Networks, IJCNN '07*, August 2007.

[9] M. Anantathanavit and M.-A. Munlin, "Radius particle swarm optimization," in *Proceedings of the 17th International Computer Science and Engineering Conference, ICSEC '13*, pp. 126–130, September 2013.

[10] D. Ghosh and D. Chakraborty, "Analytical fuzzy plane geometry III," *Fuzzy Sets and Systems*, vol. 283, pp. 83–107, 2016.

[11] A. Kaufmann and M. M. Gupta, *Introduction to Fuzzy Arithmetic: Theory and Applications*, Van Nostrand Reinhold, New York, NY, USA, 1985.

21

Ultra-Short-Term Prediction of Wind Power Based on Fuzzy Clustering and RBF Neural Network

Huang Hui[1,2] **Jia Rong,**[1] **and Wang Songkai**[1]

[1]*Institute of Water Resources and Hydro-Electric Engineering, Xi'an University of Technology, Xi'an 710048, China*
[2]*School of Electric Power, North China University of Water Resources and Electric Power, Zhengzhou 450011, China*

Correspondence should be addressed to Huang Hui; cghuanghui2000@163.com

Academic Editor: Hanbo Zheng

High-precision wind power forecast can reduce the volatility and intermittency of wind power output, which is conducive to the stable operation of the power system and improves the system's effective capacity for large-scale wind power consumption. In the wind farm, the wind turbines are located in different space locations, and its output characteristics are also affected by wind direction, wake effect, and operation conditions. Based on this, two-step ultra-short-term forecast model was proposed. Firstly, fuzzy C-means clustering (FCM) theory was used to cluster the units according to the out characteristics of wind turbines. Secondly, a prediction model of RBF neural network is established for the classification clusters, respectively, and the ultra-short-term power forecast is performed for each unit. Finally, the above results are compared with the RBF single prediction model established by unclassified g wind turbines. A case study of a wind farm in northern China is carried out. The results show that the proposed method can effectively improve the prediction accuracy of wind power and prove the effectiveness of the method.

1. Introduction

In order to solve the challenges of current energy development, such as resources, environmental pollution, and climate change, the future electric power system adopts low-carbon, green, and clean development as the development direction, which will increase the installed capacity of renewable energy represented by wind energy. Due to the randomness and volatility of wind energy, grid-connected wind power generation results in fluctuations in grid voltage and frequency, which directly affect the stability of the power quality of the grid and the operation of the power system and also bring many uncertainties to the power grid dispatching [1]. Accurate wind power prediction is one of the effective ways to reduce the above factors.

At present, approaches for short-term wind power forecasting is mainly divided into two categories: statistical models based purely on historical data and physical numerical weather prediction (NWP) models. They are better for short-term forecasting 6–72 hours ahead. Though the former model is relatively concise and the calculation speed is fast, the prediction accuracy decreases sharply with the increase of prediction time. Models based on numerical weather prediction can obtain wind power prediction values for the next 1 to 3 days. The prediction accuracy is relatively stable, but the calculation volume is huge and often requires supercomputers to continuously operate for hours [2]. Statistical models with NWP data as additional exogenous inputs, considering spatial relationships is one of the main research methods for improving wind power prediction accuracy in the future. Considering the research of spatial correlation, most of the literature is based on the analysis of wind velocity and spatial correlation to establish wind speed prediction model [2–5]. In [4], based on the statistical data of wind speed in the history year, in order to study the temporal evolution and spatial extent of the statistics of the data, using a data-coupled clustering method (SODCC algorithm) to calculate the cluster size probabilities and the node to cluster size probabilities, a spatiotemporal model of wind speed is established. The method provides guidance for forecasting analysis of wind farm output. Assume that there exist low-dimensional structures governing the interactions among a

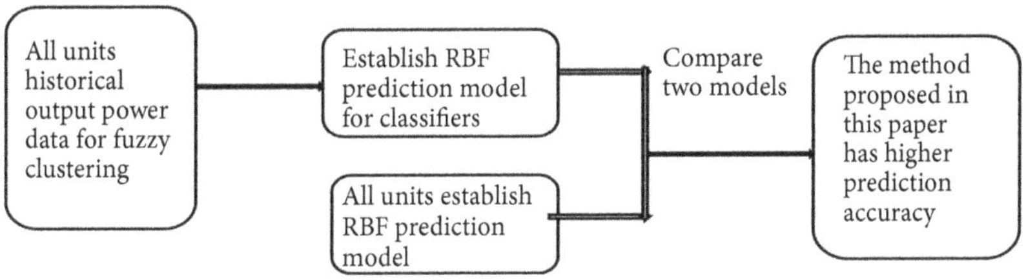

FIGURE 1: Fuzzy clustering and RBF network prediction model flow diagram.

set of historical data from meteorological stations, and we utilize Wavelet Transform (WT) for decomposition of the wind speed data into more stationary components. Based on Compressive Sensing (CS) and structured-sparse recovery algorithms, a spatiotemporal model of each subsequence is established to predict wind speed. Another study is about the spatial distribution of regional wind farms, analyzing the spatial-temporal correlation of the output of wind farms and establishing wind power forecasting models based on measured historical data. In [6], based on the geographic spatial distribution information of multiple wind farms and the historical time series of statistics, the power probability prediction of wind farms with parameters and nonparametric regression is carried out by using the correlation of wind power output in different locations. Using EFO decomposition to extract the characteristics of the regional wind farm, the representative unit of each wind farm is selected to predict the output of the wind farm. Finally, the statistical upscaling method is used to predict the total power value of the regional wind farm. At present, most of the research based on spatial correlation is mainly considering the changes of wind speed, caused by, for example, the geographical location of different wind farms, the terrain data of wind farms, wind direction, roughness, temperature, atmospheric pressure, and so on. The literature that directly analyzes the influence of spatial correlation on the output characteristics of a wind farm with different wind turbines is rarely seen.

Ultra-short-term (typically minutes to hours) forecasting is time series based models, which rely on historical wind speed or power measurements and take the predicted variable itself as explanatory variables. They can capture the hidden stochastic characteristics of wind speed or wind power [7]. Reference [8] applied a hybrid model to develop multipoint prediction and single-point prediction for ultra-short-term wind power prediction. Reference [9] proposed a novel hybrid wind power time series prediction model to improve accuracy of ultra-short-term wind power forecasting. There are also some literatures studying the impact of wind speed or wind direction on output power [10, 11]. For the same NWP data, in fact, the output power of wind turbines in different geographical locations is related to the above factors, but also to the geographical location and its own structural characteristics. In this paper, the spatial distribution information of the unit in the wind farm is considered. First, according to the historical data of the wind turbine output as the sample, the fuzzy mean clustering method is used to classify the units in the wind farm. Secondly, the RBF neural network prediction model is set up for the classified units, and the prediction results are added up to obtain the total wind power forecast power.

2. Fuzzy Clustering and RBF Network Forecasting Model

2.1. Flow Chart of Two-Step Forecasting Model. The proposed method considers the wind turbines at different space positions have different contributions to output power of wind farm. Fuzzy clustering is performed based on the measured historical power data of the wind turbines, and the classification of the unit is realized using the advantages of the nonlinear fitting of the RBF; a sample of historical data of 33 wind turbines in a wind farm is trained and tested. The specific flow chart is characterized in Figure 1.

2.2. Fuzzy C-Means Clustering (FCM). Fuzzy clustering is regarded as one of the commonly used approaches for data analysis. Fuzzy C-means clustering is adopted to classify the historical power data of wind turbines to discover the output characteristics of different turbines. Take a sample set of the n typhoon unit in the wind farm, the j-th sample has a set of eigenvectors, where m is the characterizing time series output characteristics of the j-th unit. All samples are classified into category c by fuzzy clustering algorithm [12, 13]. The sample set X can be expressed as follows:

$$\begin{gathered} X = X_1 \cup X_2 \cup \ldots X_c \\ X_i \cup X_j = \varnothing, \quad i \neq j \end{gathered} \tag{1}$$

where $X_i (i = 1, 2, \cdots, c)$ is the set of classification for the i-th crew and represents the i-th vector or cluster prototype vector, $X_i = (x_{i1}, x_{i2}, \cdots, x_{im})$.

The relationship between each sample and all clusters is represented by membership matrix. U_{ij} represents the degree of membership of the jth sample for the ith cluster center. In the clustering process, the distance weighted squared sum of each sample to all cluster centers is taken as the objective function, defined as follows:

$$J(U, X) = \sum_{j=1}^{n} \sum_{i=1}^{c} (u_{ij})^m (d_{ij})^2 \tag{2}$$

where m is the fuzzy coefficient, take 2 in this article; d_{ij} is the distance between the wind power history data and the cluster prototype in the i-th classification. $J(U, X)$ is the sum of squared errors of the sample data of each classifier and the prototype of the cluster.

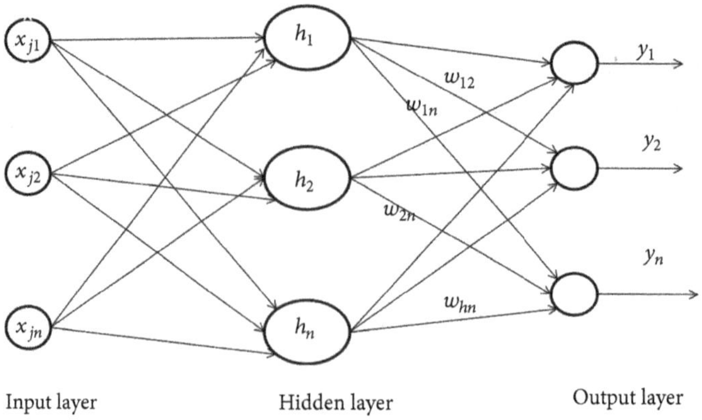

FIGURE 2: RBF neural network structure.

The specific steps of the FCM algorithm [14] are as follows:

Step 1. Update membership matrix $U^{(b)}$, and the matrix indicates the membership values of each cluster sample data belonging to the corresponding cluster prototype:

$\forall i, j$, if $\exists {d_{ij}}^{(b)} > 0$,

$$u^{(b)}{}_{ij} = \left\{ \sum_{k=1}^{c} \left(\frac{{d_{ij}}^{(b)}}{{d_{kj}}^{(b)}} \right)^{2/(m-1)} \right\}^{-1} \tag{3}$$

If $\exists i, r$, make ${d_{ir}}^{(b)} = 0$,

$$\begin{aligned} {u_{ir}}^{(b)} &= 1, \\ j &\neq r, \\ {u_{ij}}^{(b)} &= 0 \end{aligned} \tag{4}$$

Step 2. Update cluster prototype matrix $X^{(b+1)}$:

$$X_i^{(b+1)} = \frac{\sum_{j=1}^{n} \left({u_{ij}}^{(b+1)} \right)^m \cdot x_j}{\sum_{j=1}^{n} \left({u_{ij}}^{(b+1)} \right)^m}, \quad i = 1, 2 \cdots, c \tag{5}$$

Step 3. Repeated iteratively, if $\| X^{(b)} - X^{(b+1)} \| < \varepsilon$, the algorithm stops, and the membership degree matrix U and the cluster prototype matrix X are output. Otherwise, turn to the first step.

In order to evaluate the clustering results of wind turbines and determine the optimal number of clusters, two evaluation indexes, partition coefficient K_{PC} and classified entropy K_{CE}, were introduced [15].

$$\begin{aligned} K_{PC} &= \frac{1}{N} \sum_{i=1}^{c} \sum_{j=1}^{N} \left(u_{ij} \right)^2 \\ K_{CE} &= -\frac{1}{N} \sum_{i=1}^{c} \sum_{j=1}^{N} \left(u_{ij} \right) \log \left(u_{ij} \right) \end{aligned} \tag{6}$$

K_{PC} is used to evaluate the degree of separation between clusters of different units. The larger the value, the better. K_{CE} is used to evaluate the degree of fuzzy clustering among wind turbines. The smaller the value, the better.

2.3. RBF Neural Network Prediction Model. The RBF neural network is a highly efficient multilayer feed forward neural network. Using the multidimensional spatial interpolation technique, it can approximate any nonlinear function. Compared with other feed forward neural networks, the neural network has good optimal approximation performance and global optimal characteristics. The RBF neural network is composed of three layers of input layer, hidden layer, and output layer, as shown in Figure 1. $x_j = \{x_{j1}, x_{j2}, \ldots, x_{jm}\}$ is the jth input sample, $j = 1, 2 \cdots, n$, n is the total number of units. w is the connection weight between the output layer and the hidden layer; h is the number of hidden layer neurons [16, 17]. This is also demonstrated in Figure 2.

The determination of the RBF network structure requires three key parameters: the center of the basis function, the variance, and the connection weight from the hidden layer to the output layer. The parameters are solved as follows:

Step 1. The center of the basis function is obtained by the K-means clustering method. Firstly, the network is initialized, k training samples are randomly selected as the initial cluster center $c_i (i = 1, 2, \cdots, k)$, the Euclidean distance between x_j and the initial cluster center c_i is calculated, and clustering is performed according to the nearest neighbor rule. Secondly, the cluster center is readjusted and calculated. The average value of the samples in the clustering set thus obtains a new clustering center. If the new clustering center no longer changes, the calculation is stopped; otherwise, it returns to the previous step to continue to determine the center of the basis function.

Step 2. The function of RBF network is Gauss basis function, and the solution of its variance can be solved as follows.

$$\sigma_i = \frac{c_{\max}}{\sqrt{2k}} \tag{7}$$

where $i = 1, 2, \cdots, k$; $c_{\max}$ is the maximum distance from the selected basis function center.

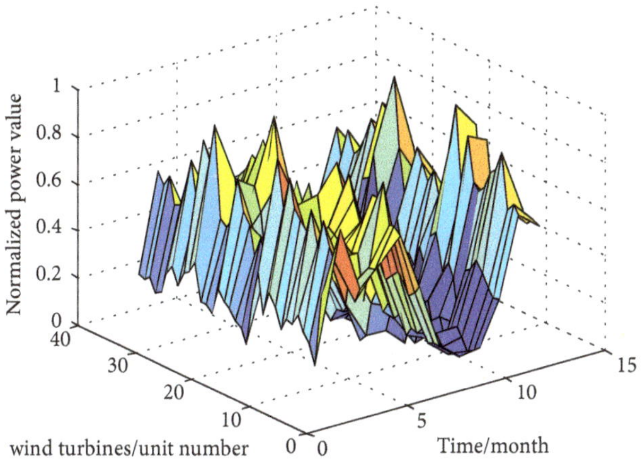

FIGURE 3: Wind turbine output power curve.

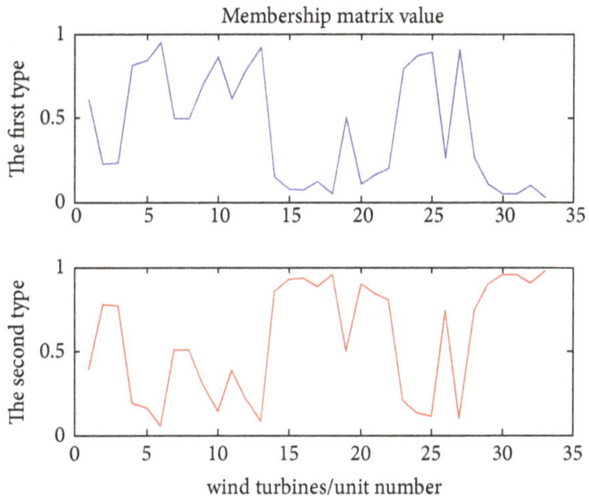

FIGURE 4: Membership matrix value curves of two types of units.

Step 3. The connection weights from the hidden layer to the output layer can be calculated directly using the least square method. The formula is described below.

$$w = \exp\left(\frac{k}{c^2_{\max}}\left\|x_j - c_i\right\|^2\right) \quad (8)$$

where $i = 1, 2, \cdots, k$; $j = 1, 2 \cdots, n$, n is the total number of samples.

The input layer to hidden layer mapping of the RBF network is nonlinear, and the hidden layer to output layer is a linear mapping. The parameter centers c_i and weights w are adjusted by the input and output errors, and then the internal layer coefficients of the network are adjusted accordingly, through repeated iteration calculations. When the output to network error of the network reaches the preset accuracy requirement, the network terminates the calculation and outputs the predicted value.

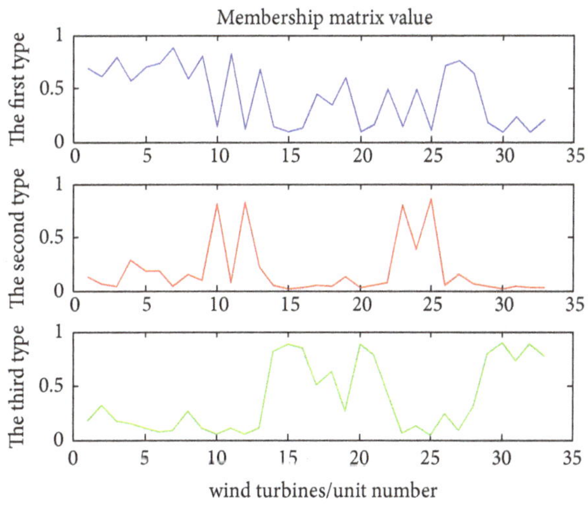

FIGURE 5: Membership matrix value curves of three types of units.

3. Case Study

The 12-month historical power data of 33 wind turbines measured at a northern wind farm was selected, and the single-unit capacity was 1.5 MW. The power curve is shown in Figure 3. It can be seen that the generating power of the 33 wind turbines horizontally related to the time sequence and has a certain correlation with the spatial distribution in the longitudinal direction.

Taking 12 months of historical power data as inputs of fuzzy clustering, clustering and grouping wind turbines are carried out. Figure 4 describes the membership matrix curve of the units divided into two clusters, and Figure 5 describes the membership matrix curve of the units divided into 3 clusters.

Select the number of different clusters c, and the membership matrix values are shown in Figures 3 and 4. Two index values are calculated by formula (6) and as shown in Table 1.

According to the membership matrix and Table 1, when the number of clusters is 2, the cluster evaluation index K_{PC} is large and K_{CE} is small. Therefore, it is better to divide the wind turbines into 2 clusters. The first group includes 14 units, and the second group includes 19 units.

TABLE 1: Evaluation indicators of clustering results.

c	K_{PC}	K_{CE}
2	0.73	0.42
3	0.611	0.68

The 10-minute historical data of the wind farm in March 2017, 733 sampling points, are adopted to set the RBF neural network modeling and prediction. The objective function error is set to 0.001, and sc is 3, where the MN is 20, and the DF is 1. The prediction of the RBF neural network is carried out for the cluster group and the entire wind farm unit, respectively. The prediction curve is shown separately in the next following figures. Figure 6 is the RBF prediction curve for the first cluster, and Figure 7 is the RBF prediction curve for the second cluster. Figure 8 depicts the RBF prediction curve for all the wind turbines in the wind farm.

In the forecast of wind power generation, the commonly evaluation indexes are the root mean squared error (RMSE)

TABLE 2: Wind power forecast error comparison analysis.

Wind turbines	RMSE	MAE
The first group wind turbines	0.055	0.0096
The second group wind turbines	0.075	0.014
The combination model with two groups	0.085	0.013
A single model with all wind turbines	0.119	0.016

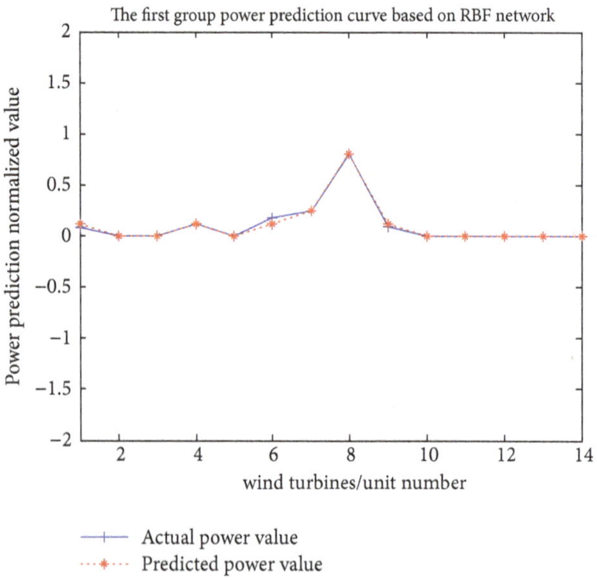

FIGURE 6: The first group power forecast curve.

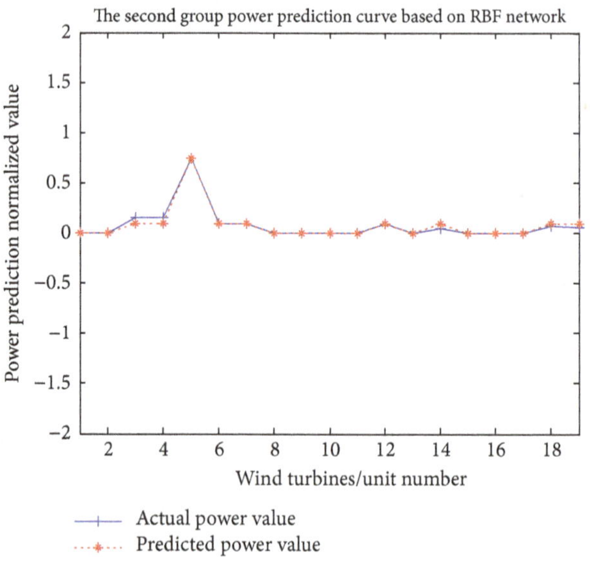

FIGURE 7: The second group power forecast curve.

and the absolute error (MAE). The specific definitions are as follows [18–20].

$$RMSE = \sqrt{\frac{1}{N}\sum_{t=1}^{n} e^2{}_t} \quad (9)$$

$$MAE = \frac{1}{n}\sum_{t=1}^{n} |e_t| \quad (10)$$

where $e_t = y_t - \hat{y}_t$, y_t and $\hat{y}_t$ are actual value and predicted value, respectively.

The RBF neural network prediction model is set up for different units, and the prediction error analysis is shown in Table 2. The predicted values of the two groups wind turbines are added by the equal weights to obtain the output of the combined model of the wind farm. From Table 2, the error based on combination model with two groups is lower than the single model.

Compared to RBF neural network prediction model, the ARIMA forecast model error curves are illustrated in Figure 9.

According to above comparison and analysis, the prediction error based on the ARIMA model is more than ultra-short time prediction model in this paper. The accuracy of wind power forecasting can be effectively improved by the two-step ultra-short-time prediction approach.

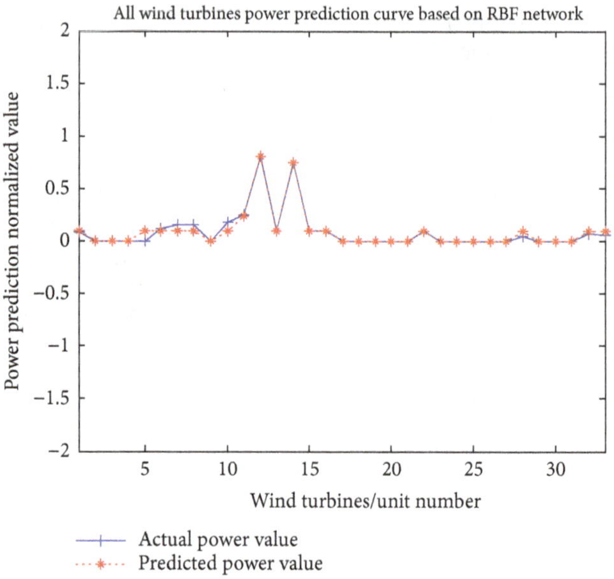

FIGURE 8: All wind turbines power forecast curve.

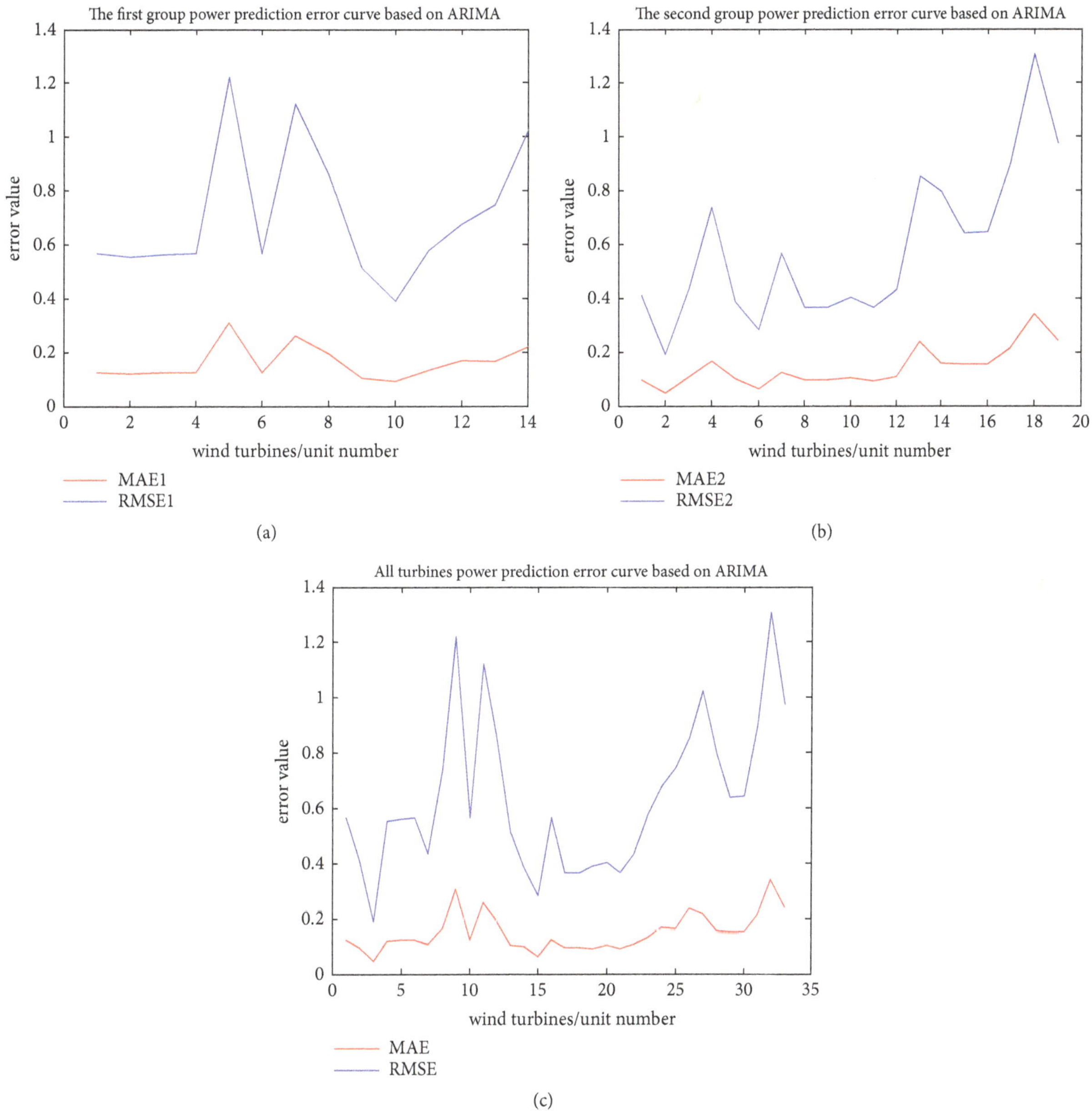

FIGURE 9: Wind power forecast error curves based on ARIMA model.

4. Conclusions

In this study, the power of the generators in the wind farm is derived from wind energy. The power output of the wind farm is affected by, for example, wind speed, wind direction, the tail flow effect of unit, and so on. Each unit's output has a certain influence on each other. According to the output of the wind turbine and taking into account the uncertain relationship between these factors, fuzzy clustering and RBF neural network are combined to establish the two-step prediction model. Different contributions of the wind turbines at different space positions to the power of wind farm, and the correlation of wind power time series are also considered. Compared to the ARIMA forecast model and single RBF model, the case verified that the two-step forecasting method proposed in this paper can effectively improve the precision in the ultra-short-term power prediction and has obtained certain practical value in engineering.

Conflicts of Interest

The authors declare that they have no conflicts of interest.

Acknowledgments

This work is financially supported in part by The Key Research Project of Henan Higher Education Institution (70556).

References

[1] Z. Liang, J. Liang, C. Wang, X. Dong, and X. Miao, "Short-term wind power combined forecasting based on error forecast correction," *Energy Conversion and Management*, vol. 119, pp. 215–226, 2016.

[2] N. Chen, Z. Qian, X. Meng, and K. Meng, "Multi-step ahead wind speed forecasting model based on spatial correlation and support vector machine," *Transactions of China Electrotechnical Society*, vol. 28, no. 5, pp. 15–21, 2013.

[3] M. I. Chidean, A. J. Caamaño, J. Ramiro-Bargueño, C. Casanova-Mateo, and S. Salcedo-Sanz, "Spatio-temporal analysis of wind resource in the Iberian Peninsula with data-coupled clustering," *Renewable & Sustainable Energy Reviews*, vol. 81, pp. 2684–2694, 2018.

[4] A. Tascikaraoglu, B. M. Sanandaji, K. Poolla, and P. Varaiya, "Exploiting sparsity of interconnections in spatio-temporal wind speed forecasting using Wavelet Transform," *Applied Energy*, vol. 165, pp. 735–747, 2016.

[5] Z. Yang, Y. Liu, Z. Zhang, X. Zhu, and J. Zhang, "Ultra-Short-Term Wind Speed Prediction With Spatial Correlation Using Recent Historical Observations and PLSR," *Power System Technology*, vol. 41, no. 6, pp. 1815–1822, 2017.

[6] J. Tastu, P. Pinson, P.-J. Trombe, and H. Madsen, "Probabilistic forecasts of wind power generation accounting for geographically dispersed information," *IEEE Transactions on Smart Grid*, vol. 5, no. 1, pp. 480–489, 2014.

[7] Y. Zhao, L. Ye, Z. Li, X. Song, Y. Lang, and J. Su, "A novel bidirectional mechanism based on time series model for wind power forecasting," *Applied Energy*, vol. 177, pp. 793–803, 2016.

[8] E. Mohammed, S. Wang, and J. Yu, "Ultra-short-term wind power prediction using a hybrid model," *IOP Conference Series: Earth and Environmental Science*, vol. 63, 2017.

[9] P. Lu, L. Ye, B. Sun, C. Zhang, Y. Zhao, and T. Zhu, "A new hybrid prediction method of ultra-short-term wind power forecasting based on EEMD-PE and LSSVM optimized by the GSA," *Energies*, vol. 11, no. 4, p. 697, 2018.

[10] L. Ye and Y. Zhao, "A review on wind power prediction based on spatial correlation approach," *Automation of Electric Power Systems*, vol. 38, no. 14, pp. 126–135, 2014.

[11] G. W. Chang, H. J. Lu, L. Y. Hsu, and Y. Y. Chen, "A hybrid model for forecasting wind speed and wind power generation," in *Proceedings of the 2016 IEEE Power and Energy Society General Meeting (PESGM '16)*, pp. 1–5, IEEE, 2016.

[12] G. W. Chang, H. J. Lu, Y. R. Chang, and Y. D. Lee, "An improved neural network-based approach for short-term wind speed and power forecast," *Journal of Renewable Energy*, vol. 105, pp. 301–311, 2017.

[13] L. Zhang, W. Zhong, C. Zhong, W. Lu, X. Liu, and W. Pedrycz, "Fuzzy C-means clustering based on dual expression between cluster prototypes and reconstructed data," *International Journal of Approximate Reasoning*, vol. 90, pp. 389–410, 2017.

[14] X. Kong, Q. Hu, X. Dong, Y. Zeng, and Z. Wu, "Load Data Identification and Correction Method with Improved Fuzzy C-means Clustering Algorithm," *Automation of Electric Power Systems*, vol. 41, no. 9, pp. 90–95, 2017.

[15] X. Peng, *Load Forecasting Method Research Based on Fuzzy Clustering and RBF Neural Network*, Guangxi University, 2012.

[16] Z. Cai, L. Deng, D. Li, X. Yao, D. Cox, and H. Wang, "A FCM cluster: cloud networking model for intelligent transportation in the city of Macau," *Cluster Computing*, vol. 3, pp. 1–10, 2017.

[17] X. Bai, L. Qingyong et al., "A spatial load forecasting method based on RBF neural network and cellular load characteristics analysis," *Power System Technology*, vol. 42, no. 1, pp. 301–307, 2018.

[18] G. Sideratos and N. D. Hatziargyriou, "Probabilistic wind power forecasting using radial basis function neural networks," *IEEE Transactions on Power Systems*, vol. 27, no. 4, pp. 1788–1796, 2012.

[19] Y. Jiang, X. Chen, K. Yu, and Y. Liao, "Short-term wind power forecasting using hybrid method based on enhanced boosting algorithm," *Journal of Modern Power Systems and Clean Energy*, vol. 5, no. 1, pp. 1–8, 2017.

[20] J. Zhang, M. Cui, B.-M. Hodge, A. Florita, and J. Freedman, "Ramp forecasting performance from improved short-term wind power forecasting over multiple spatial and temporal scales," *Energy*, vol. 122, pp. 528–541, 2017.

22

An Extension of the Fuzzy Possibilistic Clustering Algorithm Using Type-2 Fuzzy Logic Techniques

Elid Rubio, Oscar Castillo, Fevrier Valdez, Patricia Melin, Claudia I. Gonzalez, and Gabriela Martinez

Tijuana Institute of Technology, Tijuana, BC, Mexico

Correspondence should be addressed to Oscar Castillo; ocastillo@tectijuana.mx

Academic Editor: Ning Xiong

In this work an extension of the Fuzzy Possibilistic C-Means (FPCM) algorithm using Type-2 Fuzzy Logic Techniques is presented, and this is done in order to improve the efficiency of FPCM algorithm. With the purpose of observing the performance of the proposal against the Interval Type-2 Fuzzy C-Means algorithm, several experiments were made using both algorithms with well-known datasets, such as Wine, WDBC, Iris Flower, Ionosphere, Abalone, and Cover type. In addition some experiments were performed using another set of test images to observe the behavior of both of the above-mentioned algorithms in image preprocessing. Some comparisons are performed between the proposed algorithm and the Interval Type-2 Fuzzy C-Means (IT2FCM) algorithm to observe if the proposed approach has better performance than this algorithm.

1. Introduction

Different areas of research have widely used clustering algorithms for different purposes, such as image segmentation [1, 2], data mining [3], pattern recognition [4], classification [5], and modeling [6]. Clustering algorithms arise due to need to find data groups that share similar features in a given dataset; at this time there are several fuzzy clustering algorithms, such as FCM [4], PCM [7], FPCM [8], and PFCM [8]. The acceptance of these algorithms is due to the fact that they permit a datum to belong to different data clusters into a given dataset.

However, the algorithms mentioned above do not have the capability to handle the uncertainty that lies within a dataset during the clustering process; because of this, some of these algorithms (FCM and PCM) have been improved using Type-2 Fuzzy Logic Techniques [9, 10], and the improvement of these algorithms has been called Interval Type-2 Fuzzy C-Means (IT2FCM) [11, 12] and Interval Type-2 Possibilistic C-Means (IT2PCM) [12, 13], respectively. These algorithms have been used for different purposes, such as modeling [14–17], creation of membership functions [18, 19], image processing [20, 21], and classification [22]. In recent years research has also been performed in the extension of other clustering algorithms using Type-2 Fuzzy Logic Techniques, such as the ones proposed in [13, 23–27].

In this work we are presenting the extension of the FPCM using Type-2 Fuzzy Logic Techniques to provide this method with the capability of handling a higher degree of uncertainty in a dataset to solve real world problems where data clustering is involved. Other clustering algorithms have been extended using Type-2 Fuzzy Logic Techniques, but the FPCM algorithm has not been previously extended using these techniques.

This paper is organized as follows. Section 2 describes the extension of the FPCM algorithm presented in this paper, Section 3 shows the concept of cluster validation index to measure the performance of the clustering algorithm, Section 4 shows the results obtained by the IT2FPCM algorithm and its comparison with the IT2FCM algorithm, and Section 5 contains the conclusions and future work.

2. Interval Type-2 Fuzzy Possibilistic C-Means Algorithm

This is an extension of the FPCM algorithm proposed by N. R. Pal et al. in 1997, using Type-2 Fuzzy Logic Techniques, and in the same way that FPCM algorithm produces membership and possibilities using the weight exponents m and η for the fuzziness and possibility, respectively, this may now be represented by a range rather than a precise value; that is, $m = [m_1, m_2]$, where m_1 and m_2 represent the lower and upper limit of weighting exponent for fuzziness and $\eta = [\eta_1, \eta_2]$, where η_1 and η_2 represent the lower and upper limit of weighting exponent for possibility.

Because the m value is represented by an interval, the fuzzy partition matrix $\mu_i(x_k)$ must be calculated for the interval $[m_1, m_2]$; for this reason $\mu_i(x_k)$ would be given by the belonging interval $[\underline{\mu}_i(x_k), \overline{\mu}_i(x_k)]$, where $\underline{\mu}_i(x_k)$ and $\overline{\mu}_i(x_k)$ represent the lower and upper limit of the belonging interval of datum x_j to a clustering v_i, and updating the lower and upper limits of the range of the fuzzy membership matrix can be expressed as

$$\underline{\mu}_i(x_k) = \min\left\{\left(\sum_{j=1}^{c}\left(\frac{d_{ik}}{d_{jk}}\right)^{2/(m_1-1)}\right)^{-1}, \left(\sum_{j=1}^{c}\left(\frac{d_{ik}}{d_{jk}}\right)^{2/(m_2-1)}\right)^{-1}\right\},$$
$$\overline{\mu}_i(x_k) = \max\left\{\left(\sum_{j=1}^{c}\left(\frac{d_{ik}}{d_{jk}}\right)^{2/(m_1-1)}\right)^{-1}, \left(\sum_{j=1}^{c}\left(\frac{d_{ik}}{d_{jk}}\right)^{2/(m_2-1)}\right)^{-1}\right\}. \quad (1)$$

Because the η value is represented by an interval, the possibilistic partition matrix $\tau_i(x_k)$ must be calculated for the interval $[\eta_1, \eta_2]$, and for this reason $\tau_i(x_k)$ would be given by the belonging interval $[\underline{\tau}_i(x_k), \overline{\tau}_i(x_k)]$, where $\underline{\tau}_i(x_k)$ and $\overline{\tau}_i(x_k)$ represent the lower and upper limit of the belonging interval of datum x_j to a clustering v_i, and the update of the lower and upper limits of the range of the fuzzy membership matrix can be expressed as

$$\underline{\tau}_i(x_k) = \min\left\{\left(\sum_{j=1}^{c}\left(\frac{d_{ik}}{d_{jk}}\right)^{2/(\eta_1-1)}\right)^{-1}, \left(\sum_{j=1}^{c}\left(\frac{d_{ik}}{d_{jk}}\right)^{2/(\eta_2-1)}\right)^{-1}\right\},$$
$$\overline{\tau}_i(x_k) = \max\left\{\left(\sum_{j=1}^{c}\left(\frac{d_{ik}}{d_{jk}}\right)^{2/(\eta_1-1)}\right)^{-1}, \left(\sum_{j=1}^{c}\left(\frac{d_{ik}}{d_{jk}}\right)^{2/(\eta_2-1)}\right)^{-1}\right\}. \quad (2)$$

Updating the positions of the centroids of clusters should take into account the degree of belonging interval of the fuzzy and possibilistic matrices, resulting in a range of coordinates of the positions of the centroids of the clusters. The procedure for updating cluster prototypes in IT2FPCM requires calculating the centroids for the lower and upper of the limit of the interval using the fuzzy and possibilistic membership matrices, and these centroids are given by the following equations:

$$\underline{v}_i = \frac{\sum_{j=1}^{n}\left(\underline{\mu}_i(x_j) + \underline{\tau}_i(x_j)\right)^{m_1} x_j}{\sum_{j=1}^{n}\left(\underline{\mu}_i(x_j) + \underline{\tau}_i(x_j)\right)^{m_1}},$$
$$\overline{v}_i = \frac{\sum_{j=1}^{n}\left(\overline{\mu}_i(x_j) + \overline{\tau}_i(x_k)\right)^{m_1} x_j}{\sum_{j=1}^{n}\left(\overline{\mu}_i(x_j) + \overline{\tau}_i(x_k)\right)^{m_1}}. \quad (3)$$

The centroid calculation for the lower and upper limits of the interval results in an interval of coordinates of positions of the clusters centroids. Type-reduction and defuzzification use the type-2 fuzzy operations. The centroids matrix and the fuzzy partition matrix are obtained by the type-reduction operation as shown in the following equations:

$$v_j = \frac{\underline{v}_j + \overline{v}_j}{2},$$
$$\mu_i(x_j) = \frac{\underline{\mu}_i(x_j) + \overline{\mu}_i(x_j)}{2}. \quad (4)$$

This extension on the FPCM algorithm is intended to show that this algorithm is capable of handling uncertainty and is less susceptible to noise. Figure 3 shows the graphical representation of the steps FPCM algorithm in a block diagram where we can appreciate the operation of the Fuzzy Possibilistic C-Means algorithm step by step.

3. Cluster Validation

Cluster validation is one of the main topics in data clustering; this problem consists in finding and objective criterion to determine how good a partition generated by the clustering algorithm is. Nowadays there exist several index validation methods mentioned in [28–32], but these indices are proposed for validation of clusters found by Type-1 Fuzzy clustering algorithms. In order to evaluate the lower and upper bound of the interval of clusters found by the IT2FPCM and IT2FCM algorithms with some of the these indices of validation, we need to modify the following indices of validation to evaluate the partitions found by the Interval Type-2 Fuzzy clustering proposed in this work:

(i) Partition entropy index,

(ii) Xie-Beni Index,

(iii) MPE-DMFP index.

The partition entropy was proposed by Bezdek [2, 5, 6] as a validation index for the Fuzzy C-Means algorithm and was defined by the following equation:

$$\mathrm{PE} = -\frac{1}{n}\sum_{i=1}^{c}\sum_{j=1}^{n} u_{ij}\log_2 u_{ij}. \quad (5)$$

In a general we can define an optimal number of clusters c^* with the solution $\min 2 \leq c \leq n-1$ for PE to produce a better performance by grouping the dataset X. To make this index able to evaluate the lower and upper bounds we need to compute the following equations to the upper and lower bounds, respectively:

$$\begin{aligned} \text{PE}^{\text{lower}} &= -\frac{1}{n}\sum_{i=1}^{c}\sum_{j=1}^{n} \underline{u}_{ij}\log_2 \underline{u}_{ij}, \\ \text{PE}^{\text{upper}} &= -\frac{1}{n}\sum_{i=1}^{c}\sum_{j=1}^{n} \overline{u}_{ij}\log_2 \overline{u}_{ij}. \end{aligned} \tag{6}$$

Xie and Beni in 1991 proposed a validation index based on compactness and separation [2, 5, 6], which is defined by the following equation:

$$\text{XB} = \frac{\sum_{i=1}^{c}\sum_{j=1}^{n} u_{ij}^{m}\left\|x_j - v_i\right\|^2}{n \cdot \min_{\substack{ik \\ i\neq k}} \left\|v_i - v_k\right\|}. \tag{7}$$

In general, an optimal number of clusters c^* is found by solving $\min 2 \leq c \leq n-1$ for XB to produce a better clustering performance for the dataset X. To make this index able to evaluate the lower and upper bounds we compute the following equations to the upper and lower bounds, respectively:

$$\begin{aligned} \text{XB}^{\text{lower}} &= \frac{\sum_{i=1}^{c}\sum_{j=1}^{n} \underline{u}_{ij}^{m}\left\|x_j - \underline{v}_i\right\|^2}{n \cdot \min_{\substack{ik \\ i\neq k}} \left\|\underline{v}_i - \underline{v}_k\right\|}, \\ \text{XB}^{\text{upper}} &= \frac{\sum_{i=1}^{c}\sum_{j=1}^{n} \overline{u}_{ij}^{m}\left\|x_j - \overline{v}_i\right\|^2}{n \cdot \min_{\substack{ik \\ i\neq k}} \left\|\overline{v}_i - \overline{v}_k\right\|}. \end{aligned} \tag{8}$$

Elid Rubio et al. proposed the MPD-DFP index, which is composed of two metrics, the modified partition entropy index and the sum of the distances between the means of the fuzzy partitions. This validation index is represented by the following equation:

$$\text{MPE-DMPF} = I_{\text{MPE}} + D_M, \tag{9}$$

where the modified partition entropy I_{MPE} that represents the variation of the data in clusters of the dataset is represented by the following equations:

$$I_{\text{MPE}} = -\frac{1}{n}\sum_{i=1}^{c}\sum_{k=1}^{n} u_{ij}^2 \log_2 u_{ij}. \tag{10}$$

And the sum of the distances between the means of the fuzzy partition D_{M_k} that represents the separation between clusters in the dataset

$$D_{M_k} = \sum_{\substack{i,j=1 \\ i\neq j}}^{k} \left\|M_i - M_j\right\|^2, \quad k = 1,\dots,c, \tag{11}$$

where M_k is the mean of the fuzzy partitions generated by the Fuzzy C-Means algorithm. In general, we can define an optimal number of clusters c^* for the solution $\min 2 \leq c \leq n-1$ $I_{\text{MPE-DMFP}}$ to produce a better performance by grouping the dataset X. To make this index able to evaluate the lower and upper bounds of the interval cluster we compute the following equations to the upper and lower bounds, respectively:

$$\begin{aligned} \text{MPE-DMPF}^{\text{lower}} &= I_{\text{MPE}}^{\text{lower}} + D_M^{\text{lower}}, \\ \text{MPE-DMPF}^{\text{upper}} &= I_{\text{MPE}}^{\text{upper}} + D_M^{\text{upper}}, \end{aligned} \tag{12}$$

where $I_{\text{MPE}}^{\text{lower}}$ and $I_{\text{MPE}}^{\text{upper}}$ represent the variation of the data in clusters of the dataset for the upper and lower bounds of the interval of clusters, respectively, and are represented by the following equations:

$$\begin{aligned} I_{\text{MPE}}^{\text{lower}} &= -\frac{1}{n}\sum_{i=1}^{c}\sum_{k=1}^{n} \underline{u}_{ij}^2 \log_2 \underline{u}_{ij}, \\ I_{\text{MPE}}^{\text{upper}} &= -\frac{1}{n}\sum_{i=1}^{c}\sum_{k=1}^{n} \overline{u}_{ij}^2 \log_2 \overline{u}_{ij} \end{aligned} \tag{13}$$

and where D_M^{lower} and D_M^{upper} represent the separation between clusters in the dataset for the upper and lower bounds of the interval of clusters, respectively, and are represented by the following equations:

$$\begin{aligned} D_{M_k}^{\text{lower}} &= \sum_{\substack{i,j=1 \\ i\neq j}}^{k} \left\|M_i^{\text{lower}} - M_j^{\text{lower}}\right\|^2, \quad k = 1,\dots,c, \\ D_{M_k}^{\text{upper}} &= \sum_{\substack{i,j=1 \\ i\neq j}}^{k} \left\|M_i^{\text{upper}} - M_j^{\text{upper}}\right\|^2, \quad k = 1,\dots,c. \end{aligned} \tag{14}$$

4. Results of the Implementation of the IT2FPCM Algorithm

The IT2FPCM algorithm was tested with several benchmark datasets and images, in order to observe if the IT2FPCM algorithm is better than the IT2FCM algorithm. We perform 30 experiments using the Wine, WDBC, Iris Flower, Ionosphere, Abalone, and Cover type datasets. In order to observe the performance of the IT2FPCM algorithm against the IT2FCM algorithm we perform the data clustering of the datasets mentioned above with both algorithms mentioned above to compare the results obtained by these algorithms, and to measure the performance of these algorithms we use the validation indices mentioned in the previous section.

In Tables 1, 2, and 3, we show the results obtained for the WDBC dataset with 30 dimensions and 2 clusters with 569 samples; this dataset was tested with 2 to 10 clusters with the IT2FPCM and IT2FCM algorithms using different validation indices to evaluate the performance of both algorithms. The results that are shown are the mean of 30 experiments for each number of clusters tested in both algorithms. We can observe in Tables 1 and 2 that both algorithms find the correct

TABLE 1: Results of the IT2MPE-DMFP validation index for the WDBC dataset clustering using IT2FPCM and IT2FCM algorithm using $m = [1.5, 2]$ and $\eta = [1.5, 2.5]$ as the parameters.

Dataset	Clusters	Index of validation IT2MPE-DMFP					
		IT2FPCM			IT2FCM		
		Defuzz	Lower	Upper	Defuzz	Lower	Upper
WDBC	2	**0.59497**	**0.60327**	**0.58667**	**0.59374**	**0.60213**	**0.58536**
	3	1.38830	1.37432	1.40228	1.36790	1.35511	1.38069
	4	1.57493	1.49703	1.65282	1.57353	1.49646	1.65060
	5	2.02818	1.90438	2.15198	2.02736	1.90424	2.15048
	6	2.68035	2.48420	2.87650	2.66823	2.47214	2.86432
	7	2.85760	2.62283	3.09237	2.84878	2.61268	3.08488
	8	3.81363	3.46430	4.16295	3.80316	3.45317	4.15314
	9	4.63064	4.18376	5.07752	4.61201	4.16408	5.05994
	10	4.80744	4.30470	5.31017	4.75902	4.25342	5.26462

TABLE 2: Results of the IT2PE validation index for WDBC dataset clustering using IT2FPCM and IT2FCM algorithms with $m = [1.5, 2]$ and $\eta = [1.5, 2.5]$ as parameters.

Dataset	Clusters	Index validation IT2PE					
		IT2FPCM			IT2FCM		
		Defuzz	Lower	Upper	Defuzz	Lower	Upper
WDBC	2	**0.12504**	**0.10470**	**0.14539**	**0.12531**	**0.10476**	**0.14585**
	3	0.19601	0.15366	0.23835	0.20312	0.16011	0.24613
	4	0.29863	0.24209	0.35517	0.29892	0.24215	0.35569
	5	0.34488	0.26872	0.42105	0.34477	0.26853	0.42100
	6	0.39213	0.30051	0.48375	0.39355	0.30176	0.48533
	7	0.42646	0.32156	0.53137	0.42798	0.32287	0.53310
	8	0.43077	0.32277	0.53877	0.43215	0.32387	0.54044
	9	0.45241	0.33526	0.56956	0.45486	0.33697	0.57276
	10	0.48363	0.35382	0.61344	0.48387	0.35353	0.61421

TABLE 3: Results of the IT2XB validation index for WDBC dataset clustering using IT2FPCM and IT2FCM algorithm with $m = [1.5, 2]$ and $\eta = [1.5, 2.5]$ as parameters.

Dataset	Clusters	Index validation IT2XB					
		IT2FPCM			IT2FCM		
		Defuzz	Lower	Upper	Defuzz	Lower	Upper
WDBC	2	**0.06048**	0.05100	**0.06995**	**0.06137**	**0.05115**	**0.07159**
	3	0.06220	**0.05061**	0.07378	0.06586	0.05283	0.07889
	4	0.17967	0.13465	0.22470	0.18209	0.13502	0.22915
	5	0.18247	0.13453	0.23042	0.18533	0.13494	0.23572
	6	0.17079	0.12100	0.22059	0.17506	0.12244	0.22768
	7	0.22981	0.15569	0.30393	0.23242	0.15547	0.30937
	8	0.19547	0.13585	0.25509	0.19844	0.13601	0.26088
	9	0.17247	0.12025	0.22468	0.17563	0.12098	0.23028
	10	0.18893	0.12504	0.25282	0.19963	0.13171	0.26754

number of clusters for the lower and upper bound of the interval and its defuzzification using the IT2PE and IT2MPE-DMFP validation indices. In Table 3 we can observe that with the IT2XB validation index the IT2FPCM did not find the correct number of clusters for the lower bound of the interval, but for the upper bound and defuzzification of the lower and upper bound of the interval it found the correct number of clusters.

In order to observe if there exists significant difference between the IT2FPCM and IT2FCM algorithms we perform a statistical test with the results obtained with the 3 validation indices for the results obtained by the clustering algorithms

TABLE 4: Hypothesis testing for the IT2PE, IT2XB, and IT2MPE-DMFP indices of validation for the WDBC dataset clustering.

Dataset	Validation index	Algorithm	N	μ	σ^2	z-value	z-critical value	P value
WDBC	IT2PE	IT2FCM	30	0.125307301	$9.45E-30$	$-3.35094E+11$	1.645	0
		IT2FPCM		0.125044778	$8.96E-30$			
	IT2XB	IT2FCM	30	0.061368738	$1.75E-29$	$-8.46459E+11$	1.645	0
		IT2FPCM		0.060476715	$1.58E-29$			
	IT2MPE-DMFP	IT2FCM	30	0.593744494	$1.06E-27$	$1.45455E+11$	1.645	1
		IT2FPCM		0.594972091	$1.08E-27$			

TABLE 5: Results of the IT2MPE-DMFP validation index to Wine dataset clustering using IT2FPCM and IT2FCM algorithm using $m = [1.5, 2]$ and $\eta = [1.5, 2.5]$ as parameters.

Dataset	Clusters	Index of validation IT2MPE-DMFP					
		IT2FPCM			IT2FCM		
		Defuzz	Lower	Upper	Defuzz	Lower	Upper
Wine	2	0.40744	0.41801	0.39686	0.40619	0.41684	0.39554
	3	0.40225	0.39940	0.40509	0.40445	0.40146	0.40744
	4	0.89656	0.87046	0.92267	0.89553	0.86964	0.92141
	5	1.08796	1.03901	1.13691	1.08502	1.03721	1.13282
	6	1.67907	1.51748	1.84066	1.90199	1.75824	2.04574
	7	1.84502	1.67461	2.01543	2.86659	2.63321	3.09998
	8	2.81205	2.54975	3.07434	1.68624	1.47369	1.89880
	9	3.47078	3.14652	3.79504	3.46131	3.13748	3.78513
	10	3.30712	2.92591	3.68834	3.30032	2.92011	3.68053

mentioned above. The z-test was used with the following hypothesis for each validation index:

$$H_0;\ \text{IT2FPCM} \geq \text{IT2FCM}$$
$$H_1;\ \text{IT2FPCM} < \text{IT2FCM}. \quad (15)$$

The hypothesis testing is performed for the best number of clusters found by the mentioned clustering algorithms. Table 4 shows the results from the hypothesis testing realized for the defuzzification of Type-2 clusters of the WDBC dataset.

According to the assumptions made in (15), which arise in order to demonstrate that IT2FPCM algorithm is better than IT2FCM algorithm, in Table 4 we can observe the results of the z-test performed to data clustering of the WDBC dataset using the indices of validation mentioned in Section 4. In this case we can observe that the z-values of the hypothesis testing for the IT2PE, IT2XB, and IT2MPE-DMFP indices of validation are $-3.35094E + 11$, $-8.46459E + 11$, and $1.45455E + 11$, respectively. We can observe that the z-test shows that the IT2PE and IT2XB indices of validation are lower than the z-critical value that is equal to -1.645 with a significance level α of 0.05, whose z-values confirm the acceptance of the alternative hypothesis posed in (15) for these indices of validation. In this way we demonstrate that the IT2FPCM algorithm is better than IT2FPCM algorithm for the data clustering of the WDBC dataset using the IT2PE and IT2XB indices. The z-value for hypothesis testing with the IT2MPE-DMFP index is greater than the z-critical value that is equal to -1.645 with a significance level α of 0.05, whose z-value rejects the alternative hypothesis and accepts the null hypothesis posed in (15), demonstrating that there is no significant difference between the IT2FCM and IT2FPCM algorithms used in the z-test of the defuzzification.

In Tables 5, 6, and 7 we show the results obtained for the Wine dataset with 13 dimensions and 3 classes with 178 samples, and this dataset was tested with 2 to 10 clusters with the IT2FPCM and IT2FCM algorithms using different validation index to evaluate the performance of both algorithms. The results that are presented are the means of 30 experiments for each number of clusters used to test both algorithms; we can observe in Tables 8 and 9 that both algorithms did not find the correct number of clusters for the lower and upper bound of the interval and its defuzzification using the IT2PE and IT2XB validation index. In Table 7 we can observe that with the IT2MPE-DMFP validation index the IT2FPCM did not find the correct number of clusters for the upper bound of the interval, but to the lower bound and defuzzification of the lower and upper bound of the interval it did find the correct number of clusters.

According to the assumptions made in (15), which arise in order to demonstrate that the IT2FPCM algorithm is better than IT2FCM algorithm, in Table 8 we can observe the results of the z-test performed for data clustering of the Wine dataset using the indices of validation mentioned in Section 4. In this case we can observe that the z-values of the hypothesis testing for the IT2PE, IT2XB, and IT2MPE-DMFP indices of validation are $-1.01952E+11$, $-7.83335E+11$, and -22812613207, respectively. We can notice that these values are lower than the z-critical value that is equal to -1.645 with a significant level α of 0.05, and these z-values confirm the acceptance of the alternative hypothesis posed in (15) for

Table 6: Results of the IT2PE validation index to Wine dataset clustering using IT2FPCM and IT2FCM algorithm with $m = [1.5, 2]$ and $\eta = [1.5, 2.5]$ as parameters.

Dataset	Clusters	Index of validation IT2PE					
		IT2FPCM			IT2FCM		
		Defuzz	Lower	Upper	Defuzz	Lower	Upper
Wine	2	0.14977	0.12509	0.17446	0.15006	0.12522	0.17491
	3	0.26280	0.21554	0.31007	0.26287	0.21547	0.31028
	4	0.28775	0.22242	0.35308	0.28769	0.22226	0.35311
	5	0.33984	0.25551	0.42417	0.34005	0.25549	0.42461
	6	0.36657	0.27738	0.45575	0.34110	0.25173	0.43048
	7	0.33464	0.24323	0.42605	0.35228	0.25489	0.44966
	8	0.34198	0.24496	0.43900	0.37465	0.26445	0.48484
	9	0.36397	0.25674	0.47120	0.36531	0.25761	0.47300
	10	0.39186	0.27080	0.51292	0.39213	0.27098	0.51328

Table 7: Results of the IT2XB validation index to Wine dataset clustering using IT2FPCM and IT2FCM algorithm with $m = [1.5, 2]$ and $\eta = [1.5, 2.5]$ as parameters.

Dataset	Clusters	Index of validation IT2XB					
		IT2FPCM			IT2FCM		
		Defuzz	Lower	Upper	Defuzz	Lower	Upper
Wine	2	0.06009	0.05289	0.06730	0.06097	0.05301	0.06893
	3	0.13492	0.10564	0.16421	0.13602	0.10573	0.16631
	4	0.09865	0.07860	0.11871	0.09966	0.07883	0.12049
	5	0.10894	0.08126	0.13661	0.10966	0.08169	0.13762
	6	0.10862	0.08197	0.13526	0.08022	0.06075	0.09969
	7	0.09032	0.06714	0.11350	0.08072	0.05840	0.10304
	8	0.08295	0.06177	0.10414	0.12357	0.09168	0.15546
	9	0.11040	0.08227	0.13852	0.11182	0.08301	0.14062
	10	0.09347	0.06982	0.11712	0.09389	0.06992	0.11786

Table 8: Hypothesis test for IT2PE, IT2XB, and IT2MPE-DMFP indices of validation for the Wine dataset clustering.

Dataset	Validation Index	Algorithm	N	μ	σ^2	z-value	z-critical value	P value
Wine	IT2PE	IT2FCM	30	0.150064427	$1.14E-28$	$-1.01952E+11$	1.645	0
		IT2FPCM		0.149773266	$1.31E-28$			
	IT2XB	IT2FCM	30	0.060970001	$1.84E-29$	$-7.83335E+11$	1.645	0
		IT2FPCM		0.060093276	$1.92E-29$			
	IT2MPE-DMFP	IT2FCM	30	0.404445222	$1.37E-25$	−22812613207	1.645	0
		IT2FPCM		0.402246542	$1.42E-25$			

Table 9: Results of the IT2MPE-DMFP validation index to Iris Flower dataset clustering using IT2FPCM and IT2FCM algorithm with $m = [1.5, 2]$ and $\eta = [1.5, 2.5]$ as parameters.

Dataset	Clusters	Index validation IT2MPE-DMFP					
		IT2FPCM			IT2FCM		
		Defuzz	Lower	Upper	Defuzz	Lower	Upper
Iris	2	0.29224	0.30348	0.28101	0.29206	0.30341	0.28071
	3	0.27581	0.27769	0.27394	0.27186	0.27371	0.27001
	4	0.65772	0.70196	0.61348	0.78895	0.75105	0.82686
	5	0.65979	0.59070	0.72888	0.64700	0.57492	0.71908
	6	0.92039	0.79450	1.04628	1.66071	1.62370	1.69772
	7	1.51645	1.31241	1.72049	0.85814	0.89191	0.82436
	8	2.45753	2.32286	2.59220	1.41122	1.36998	1.45246
	9	2.25261	2.08606	2.41917	1.90568	1.78587	2.02548
	10	2.62804	2.40631	2.84977	2.60432	2.38812	2.82051

TABLE 10: Results of the IT2PE validation index to Iris Flower dataset clustering using IT2FPCM and IT2FCM algorithm with $m = [1.5, 2]$ and $\eta = [1.5, 2.5]$ as parameters.

Dataset	Clusters	IT2FPCM			IT2FCM		
		Defuzz	Lower	Upper	Defuzz	Lower	Upper
	2	0.12706	0.10034	0.15378	0.12776	0.10080	0.15472
	3	0.27001	0.21982	0.32020	0.27104	0.22074	0.32134
	4	0.38333	0.30413	0.46252	0.38442	0.30513	0.46370
	5	0.45027	0.34297	0.55758	0.45232	0.34465	0.55999
Iris	6	0.54534	0.42762	0.66305	0.54757	0.42954	0.66560
	7	0.63092	0.49076	0.77108	0.63321	0.49262	0.77379
	8	0.67371	0.51134	0.83607	0.68029	0.51647	0.84411
	9	0.74321	0.56287	0.92354	0.73528	0.53285	0.93771
	10	0.80024	0.60567	0.99480	0.80718	0.61146	1.00291

TABLE 11: Results of the IT2XB validation index to Iris Flower dataset clustering using IT2FPCM and IT2FCM algorithm with $m = [1.5, 2]$ and $\eta = [1.5, 2.5]$ as parameters.

Dataset	Clusters	IT2FPCM			IT2FCM		
		Defuzz	Lower	Upper	Defuzz	Lower	Upper
	2	0.05604	0.04877	0.06331	0.05656	0.04870	0.06442
	3	0.13410	0.11048	0.15772	0.13778	0.11084	0.16472
	4	0.18766	0.14543	0.22990	0.19211	0.14687	0.23736
	5	0.22753	0.16904	0.28602	0.23302	0.17094	0.29509
Iris	6	0.29483	0.21162	0.37804	0.30525	0.21204	0.39846
	7	0.31905	0.23313	0.40497	0.33896	0.23724	0.44067
	8	0.23901	0.17431	0.30371	0.24926	0.17482	0.32369
	9	0.39515	0.27792	0.51239	0.55028	0.38453	0.71604
	10	0.34780	0.23825	0.45735	0.35965	0.24133	0.47796

TABLE 12: Hypothesis testing for the IT2PE, IT2XB, and IT2MPE-DMFP indices of validation for the Iris Flower dataset clustering.

Dataset	Validation index	Algorithm	N	μ	σ^2	z-value	z-critical value	P value
	IT2PE	IT2FCM	30	0.127758962	$2.81E-24$	−1447853758	1.645	0
		IT2FPCM		0.127055881	$4.27E-24$			
Iris Flower	IT2XB	IT2FCM	30	0.056558597	$9.64E-28$	−38476101768	1.645	0
		IT2FPCM		0.056038216	$4.52E-27$			
	IT2MPE-DMFP	IT2FCM	30	0.271858146	$5.91E-21$	184969393.7	1.645	1
		IT2FPCM		0.275812862	$7.80E-21$			

all the tested indices of validation, demonstrating that the IT2FPCM algorithm is better than the IT2FPCM algorithm for the data clustering of the Wine dataset. In Tables 9, 10, and 11 we show the results obtained for a Iris Flower dataset with 4 dimensions and 3 classes with 150 samples, and this dataset was tested with 2 to 10 clusters with the IT2FPCM and IT2FCM algorithms using different validation indices to evaluate the performance of both algorithms and the results shown are the mean of 30 experiments for each number of clusters used tested in both algorithms. In Table 9 we can observe that both algorithms with the IT2MPE-DMFP index validation did find the correct number of clusters for the lower and upper bounds of the limit and its defuzzification. On the other hand, in Tables 10 and 11 the IT2PE and IT2XB algorithms did not find the correct number the clusters.

In Table 12 we can observe the results of the z-test performed to the data clustering of the Iris Flower dataset using the indices of validation mentioned in Section 4. In this case we can observe that the z-values of the hypothesis testing for the IT2PE, IT2XB, and IT2MPE-DMFP indices of validation are −1447853758, −38476101768, and 184969393.7, respectively. We can observe that z-test shows that the IT2PE and IT2XB indices of validation are lower than the z-critical value that is equal to −1.645 with a significance level α of 0.05, whose z-values confirm the acceptance of the alternative hypothesis posed in (15) for these indices of validation, demonstrating that the IT2FPCM algorithm is better than IT2FPCM algorithm for the data clustering of the Iris Flower dataset using the IT2PE and IT2XB indices. However, the z-value for the test with the IT2MPE-DMFP index is greater than the z-critical value that is equal to −1.645

TABLE 13: IT2MPE-DMFP validation index results to Ionosphere dataset clustering using IT2FPCM and IT2FCM algorithm with $m = [1.5, 2.5]$ and $\eta = [1.5, 2.5]$ as parameters.

Dataset	Clusters	Index of validation IT2MPEDFP					
		IT2FPCM			IT2FCM		
		Defuzz	Lower	Upper	Defuzz	Lower	Upper
Ionosphere	2	**0.37490**	**0.38189**	**0.36792**	**0.37872**	**0.38264**	**0.37479**
	3	0.67283	0.57794	0.76772	0.67487	0.57848	0.77126
	4	0.83519	0.69115	0.97923	0.83839	0.69166	0.98511
	5	1.01579	0.80954	1.22204	1.01694	0.80933	1.22456
	6	1.17398	0.89175	1.45622	1.17490	0.89191	1.45789
	7	1.31217	0.95649	1.66785	1.31246	0.95366	1.67126
	8	1.47017	1.03109	1.90924	1.46777	1.02669	1.90885
	9	1.64906	1.09276	2.20535	1.64527	1.08737	2.20317
	10	1.79078	1.15390	2.42766	1.78653	1.14450	2.42856

TABLE 14: Results of the IT2PE validation index to Ionosphere dataset clustering using IT2FPCM and IT2FCM algorithm with $m = [1.5, 2.5]$ and $\eta = [1.5, 2.5]$ as parameters.

Dataset	Clusters	Index of validation IT2PE					
		IT2FPCM			IT2FCM		
		Defuzz	Lower	Upper	Defuzz	Lower	Upper
Ionosphere	2	**0.46246**	**0.46354**	**0.46138**	**0.46629**	**0.46478**	**0.46780**
	3	0.80424	0.75328	0.85520	0.80622	0.75262	0.85982
	4	1.01893	0.92719	1.11067	1.02047	0.92553	1.11541
	5	1.20877	1.07092	1.34663	1.20802	1.06779	1.34826
	6	1.37426	1.20016	1.54836	1.37065	1.19522	1.54607
	7	1.52065	1.31404	1.72727	1.51381	1.30717	1.72044
	8	1.65658	1.41618	1.89699	1.64516	1.40690	1.88342
	9	1.77473	1.50879	2.04066	1.75803	1.49612	2.01995
	10	1.87674	1.58679	2.16670	1.85595	1.57139	2.14051

TABLE 15: Results of the IT2XB validation index to Ionosphere dataset clustering using IT2FPCM and IT2FCM algorithm with $m = [1.5, 2.5]$ and $\eta = [1.5, 2.5]$ as parameters.

Dataset	Clusters	Index of validation IT2XB					
		IT2FPCM			IT2FCM		
		Defuzz	Lower	Upper	Defuzz	Lower	Upper
Ionosphere	2	**0.62505**	**0.51872**	**0.73137**	**0.62236**	**0.51799**	**0.72674**
	3	2.36442	3.07170	1.65714	2.34434	3.05359	1.63509
	4	1.25146	1.18415	1.31877	1.24377	1.18394	1.30359
	5	4.05*E* + 04	5.90*E* + 04	2.20*E* + 04	1.93*E* + 06	2.82*E* + 06	1.03*E* + 06
	6	5.58*E* + 03	7.99*E* + 03	3.17*E* + 03	6.35*E* + 06	9.12*E* + 06	3.58*E* + 06
	7	2.46*E* + 11	3.52*E* + 11	1.39*E* + 11	5.22*E* + 11	7.51*E* + 11	2.92*E* + 11
	8	1.98*E* + 15	2.74*E* + 15	1.23*E* + 15	3.38*E* + 16	4.71*E* + 16	2.05*E* + 16
	9	1.69*E* + 14	2.29*E* + 14	1.08*E* + 14	1.70*E* + 14	2.32*E* + 14	1.08*E* + 14
	10	6.22*E* + 20	8.71*E* + 20	3.73*E* + 20	1.31*E* + 21	1.85*E* + 21	7.62*E* + 20

with a significance level α of 0.05, whose z-value rejects the alternative hypothesis and accepts the null hypothesis posed in (15), demonstrating that there is no significant difference between IT2FCM and IT2FPCM algorithms used for the z-test of the defuzzification.

In Tables 13, 14, and 15 we show the results obtained for the Ionosphere dataset with 34 dimensions and 2 classes with 351 samples. This dataset was tested with 2 to 10 clusters with the IT2FPCM and IT2FCM algorithms using different validation indices to evaluate the performance of both algorithms. The results presented are the means of 30 experiments for each number of clusters used in both algorithms. In Tables 13, 14, and 15 we can observe that both algorithms find the correct number of clusters for the Ionosphere dataset with all the

TABLE 16: Hypothesis test for IT2PE, IT2XB, and IT2MPE-DMFP indices of validation for the Ionosphere dataset clustering.

Dataset	Validation index	Algorithm	N	μ	σ^2	z-value	z-critical value	P value
Ionosphere	IT2PE	IT2FCM	30	0.466290024	$0.00E + 00$	$-7.55279E + 13$	1.645	0
		IT2FPCM		0.462462683	$7.70E - 32$			
	IT2XB	IT2FCM	30	0.622362066	$1.89E - 31$	$2.66339E + 13$	1.645	1
		IT2FPCM		0.625045136	$1.16E - 31$			
	IT2MPE-DMFP	IT2FCM	30	0.378715301	$7.70E - 32$	$-3.89081E + 13$	1.645	0
		IT2FPCM		0.374904357	$2.11E - 31$			

TABLE 17: IT2MPE-DMFP validation index results to Abalone dataset clustering using IT2FPCM and IT2FCM algorithm with $m = [1.5, 2.5]$ and $\eta = [1.5, 2.5]$ as parameters.

Dataset	Clusters	Index validation IT2MPEDFP					
		IT2FPCM			IT2FCM		
		Defuzz	Lower	Upper	Defuzz	Lower	Upper
Abalone	2	**0.49799**	**0.51768**	**0.47829**	**0.49841**	**0.51791**	**0.47891**
	3	0.81900	0.75233	0.88568	0.81958	0.75242	0.88675
	4	1.16915	0.99286	1.34545	1.16989	0.99317	1.34660
	5	1.49665	1.18659	1.80671	1.49743	1.18658	1.80828
	6	2.02515	1.55638	2.49393	2.00117	1.52847	2.47388
	7	2.34757	1.73028	2.96486	2.35907	1.73590	2.98224
	8	2.75194	1.97464	3.52923	2.71039	1.96388	3.45690
	9	2.89066	2.08773	3.69358	2.91788	2.10701	3.72875
	10	3.36064	2.41209	4.30919	3.46894	2.48156	4.45633

TABLE 18: Results of the IT2PE validation index to Abalone dataset clustering using IT2FPCM and IT2FCM algorithm with $m = [1.5, 2.5]$ and $\eta = [1.5, 2.5]$ as parameters.

Dataset	Clusters	Index of validation IT2PE					
		IT2FPCM			IT2FCM		
		Defuzz	Lower	Upper	Defuzz	Lower	Upper
Abalone	2	**0.28046**	**0.24561**	**0.31530**	**0.28088**	**0.24588**	**0.31587**
	3	0.41503	0.33056	0.49950	0.41555	0.33088	0.50022
	4	0.51390	0.38126	0.64654	0.51451	0.38167	0.64735
	5	0.58848	0.40296	0.77400	0.58910	0.40341	0.77479
	6	0.65438	0.42798	0.88077	0.65645	0.42848	0.88441
	7	0.72528	0.46266	0.98791	0.72530	0.46222	0.98837
	8	0.77107	0.47984	1.06230	0.77356	0.48179	1.06533
	9	0.79848	0.48202	1.11494	0.80140	0.48452	1.11828
	10	0.81290	0.47804	1.14776	0.81251	0.47777	1.14725

validation indices used to measure the performance of the both algorithms.

In Table 16 we can observe the results of the z-tests performed for the clustering of the Ionosphere dataset using the indices of validation mentioned in Section 4. In this case we can observe that the z-values of the hypothesis testing with the IT2PE, IT2XB, and IT2MPE-DMFP indices of validation are $-7.55279E + 13$, $2.66339E + 13$, and $-3.89081E + 13$, respectively. We can observe that z-test shows that the IT2PE and IT2MPE-DMFP indices of validation are lower than the z-critical value that is equal to -1.645 with a significance level α of 0.05, and these z-values confirm the acceptance of the alternative hypothesis posed in (15) for these indices of validation, demonstrating that the IT2FPCM algorithm is better than the IT2FPCM algorithm for the data clustering of the Ionosphere dataset using the IT2PE and IT2MPE-DMFP indices. The z-value for the test with the IT2XB index shows that is greater than the z-critical value that is equal to -1.645 with a significance level α of 0.05, and this z-value rejects the alternative hypothesis and accepts the null hypothesis posed in (15), demonstrating that there is no significant difference between the IT2FCM and IT2FPCM algorithms used in the z-test of the defuzzification for the IT2XB index.

In Tables 17, 18, and 19 we show the results obtained for the Abalone dataset with 8 dimensions and 3 classes according to the sex of the Abalone and with 4177 samples. This dataset was tested with 2 to 10 clusters with the IT2FPCM and IT2FCM algorithms using different validation index to

TABLE 19: Results of the IT2XB validation index to Abalone dataset clustering using IT2FPCM and IT2FCM algorithm with $m = [1.5, 2.5]$ and $\eta = [1.5, 2.5]$ as parameters.

Dataset	Clusters	Index of validation IT2XB					
		IT2FPCM			IT2FCM		
		Defuzz	Lower	Upper	Defuzz	Lower	Upper
Abalone	2	**0.12162**	0.08980	**0.15344**	**0.12160**	0.08980	**0.15340**
	3	0.15659	0.08965	0.22353	0.15651	0.08964	0.22338
	4	0.14604	**0.06983**	0.22224	0.14592	**0.06982**	0.22202
	5	0.19353	0.08216	0.30489	0.19328	0.08212	0.30445
	6	0.18472	0.07469	0.29474	0.18444	0.07514	0.29374
	7	0.18783	0.07297	0.30269	0.18551	0.07230	0.29873
	8	0.23117	0.08481	0.37753	0.23847	0.08663	0.39030
	9	0.34175	0.11190	0.57161	0.34974	0.11459	0.58489
	10	0.32883	0.10652	0.55115	0.32309	0.10437	0.54182

TABLE 20: Hypothesis test for IT2PE, IT2XB, and IT2MPE-DMFP indices of validation for the Abalone dataset clustering.

Dataset	Validation index	Algorithm	N	μ	σ^2	z-value	z-critical value	P value
Abalone	IT2PE	IT2FCM	30	0.280876194	$2.77E-32$	$-1.30990E+13$	1.645	0
		IT2FPCM		0.280456378	$3.08E-33$			
	IT2XB	IT2FCM	30	0.121597213	$3.08E-33$	$1.78710E+12$	1.645	1
		IT2FPCM		0.121619853	$1.73E-33$			
	IT2MPE-DMFP	IT2FCM	30	0.498409184	$9.18E-32$	$-5.21128E+12$	1.645	0
		IT2FPCM		0.497985339	$1.07E-31$			

evaluate the performance of both algorithms. The results presented in the tables are the means of 30 experiments for each number of clusters used in both algorithms. In Tables 17, 18, and 19 we can observe that both algorithms fail to find the correct number of clusters for the Abalone dataset with all validation indices used to measure the performance of the both algorithms.

In Table 20 we can observe the results of the z-test performed to data clustering for the Abalone dataset using the indices of validation mentioned in Section 4, where we can observe that the z-values of the hypothesis test to the indices of validation IT2PE, IT2XB, and IT2MPE-DMFP are $-1.30990E+13$, $1.78710E+12$, and $-5.21128E+12$, respectively. In this case we can observe that z-test shows that IT2PE and IT2MPE-DMFP indices of validation are lower than the z-critical value that is equal to -1.645 with a significance level α of 0.05, whose z-values confirm the acceptance of the alternative hypothesis posed in (15) for these indices of validation, demonstrating that the IT2FPCM algorithm is better than the IT2FPCM algorithm for the data clustering of the Ionosphere dataset using the IT2PE and IT2MPE-DMFP indices; the z-value for the hypothesis test of the IT2XB index is greater than the z-critical value that is equal to -1.645 with a significant level (α) of 0.05, whose z-value rejects the alternative hypothesis and accepts the null hypothesis posed in (15), demonstrating that there is no significant difference between IT2FCM and IT2FPCM algorithm used to z-test of the defuzzification for the IT2XB index.

In Tables 21, 22, and 23 we show the results obtained for a Cover type dataset with 54 dimensions and 7 classes with 581012 samples, and this dataset was tested with 2 to 9 clusters with the IT2FPCM and IT2FCM algorithm using different validation index to evaluate the performance of both algorithms. The results shown are the means of 30 experiments for each number of clusters used tested in both algorithms. In Tables 21, 22, and 23 we can observe that both algorithms fail in finding the correct number of clusters for the Cover type dataset with all validation index used to measure the performance of the both algorithms.

In Table 24 we can observe the results of the z-test performed for data clustering of the Cover type dataset using the indices of validation mentioned in Section 4, where we can observe that the z-values of the hypothesis testing for the IT2PE, IT2XB, and IT2MPE-DMFP indices of validation are $-8.15509E+10$, $6.67332E+02$, and $-3.12019E+10$, respectively. In this case we can observe that the z-test shows that the IT2PE and IT2MPE-DMFP indices of validation are lower than the z-critical value that is equal to -1.645 with a significance level α of 0.05, whose z-values confirm the acceptance of the alternative hypothesis posed in (15) for these indices of validation tested, demonstrating that the IT2FPCM algorithm is better than the IT2FPCM algorithm for the data clustering of the Ionosphere dataset using the IT2PE and IT2MPE-DMFP indices. The z-value for the hypothesis test for the IT2XB index is greater than the z-critical value that is equal to -1.645 with a significance level α of 0.05, whose z-value rejects the alternative hypothesis and

TABLE 21: IT2MPE-DMFP validation index results to Cover type dataset clustering using IT2FPCM and IT2FCM algorithm with $m = [1.5, 2.5]$ and $\eta = [1.5, 2.5]$ as parameters.

Dataset	Clusters	Index of validation IT2MPEDFP					
		IT2FPCM			IT2FCM		
		Defuzz	Lower	Upper	Defuzz	Lower	Upper
Cover type	2	**0.44077**	**0.45732**	**0.42422**	**0.44077**	**0.45732**	**0.42422**
	3	0.87629	0.82323	0.92935	0.87629	0.82323	0.92936
	4	1.04019	0.91650	1.16388	1.04019	0.91650	1.16389
	5	1.16102	0.94169	1.38036	1.16102	0.94168	1.38037
	6	1.32155	1.01723	1.62587	1.32183	1.01743	1.62623
	7	1.66573	1.24415	2.08731	1.67645	1.25221	2.10069
	8	1.79403	1.30555	2.28251	1.78721	1.29820	2.27623
	9	1.92141	1.34475	2.49807	1.93580	1.35667	2.51492

TABLE 22: Results of the IT2PE validation index to Cover type dataset clustering using IT2FPCM and IT2FCM algorithm with $m = [1.5, 2.5]$ and $\eta = [1.5, 2.5]$ as parameters.

Dataset	Clusters	Index of validation IT2PE					
		IT2FPCM			IT2FCM		
		Defuzz	Lower	Upper	Defuzz	Lower	Upper
Cover type	2	**0.35587**	**0.33093**	**0.38080**	**0.35587**	**0.33093**	**0.38081**
	3	0.56607	0.47652	0.65563	0.56608	0.47652	0.65563
	4	0.73183	0.58434	0.87933	0.73183	0.58434	0.87933
	5	0.85155	0.64428	1.05882	0.85155	0.64428	1.05883
	6	0.97154	0.70936	1.23372	0.97152	0.70934	1.23370
	7	1.05830	0.75102	1.36559	1.05713	0.74991	1.36436
	8	1.14555	0.79324	1.49786	1.14611	0.79369	1.49853
	9	1.22450	0.83102	1.61797	1.22440	0.83113	1.61766

TABLE 23: Results of the IT2XB validation index to Cover type dataset clustering using IT2FPCM and IT2FCM algorithm with $m = [1.5, 2.5]$ and $\eta = [1.5, 2.5]$ as parameters.

Dataset	Clusters	Index of validation IT2XB					
		IT2FPCM			IT2FCM		
		Defuzz	Lower	Upper	Defuzz	Lower	Upper
Cover Type	2	0.19894	0.14137	**0.25651**	0.19894	0.14137	**0.25651**
	3	**0.19557**	0.11641	0.27474	**0.19557**	0.11641	0.27474
	4	0.33846	0.19325	0.48366	0.33845	0.19325	0.48366
	5	0.24190	0.11492	0.36887	0.24189	0.11492	0.36886
	6	0.24645	**0.10625**	0.38665	0.24628	**0.10618**	0.38639
	7	0.29318	0.12373	0.46263	0.29165	0.12304	0.46026
	8	0.34262	0.14107	0.54418	0.33522	0.13744	0.53301
	9	0.28467	0.10933	0.46002	0.28741	0.11148	0.46335

TABLE 24: Hypothesis test for IT2PE, IT2XB, and IT2MPE-DMFP indices of validation for the Abalone dataset clustering.

Dataset	Validation index	Algorithm	N	μ	σ^2	z-value	z-critical value	P value
Abalone	IT2PE	IT2FCM	30	0.355868292	$2.77E-32$	$-8.15509E+10$	1.645	0
		IT2FPCM		0.355865678	$3.08E-33$			
	IT2XB	IT2FCM	30	0.195572668	$5.33E-17$	$6.67332E+02$	1.645	1
		IT2FPCM		0.195573745	$2.49E-17$			
	IT2MPE-DMFP	IT2FCM	30	0.440773577	$1.89E-31$	$-3.12019E+10$	1.645	0
		IT2FPCM		0.440771083	$3.08E-33$			

TABLE 25: Results of the IT2MPE-DMFP validation index to data clustering of image shown in Figure 2(a) using IT2FPCM and IT2FCM algorithm with $m = [1.5, 2.5]$ and $\eta = [1.5, 2.5]$ as parameters.

Image	Clusters	Index of validation IT2MPE-DMFP					
		IT2FPCM			IT2FCM		
		Defuzz	Lower	Upper	Defuzz	Lower	Upper
Figure 2(a)	2	**0.31330**	**0.32940**	**0.29720**	**0.31331**	**0.32941**	**0.29722**
	3	0.64736	0.61172	0.68301	0.64738	0.61172	0.68304
	4	0.87847	0.78421	0.97273	0.87849	0.78421	0.97278
	5	1.25053	1.08130	1.41975	1.25056	1.08131	1.41981
	6	1.52447	1.27572	1.77322	1.52451	1.27573	1.77329
	7	1.80474	1.48249	2.12700	1.80483	1.48250	2.12716
	8	2.01639	1.61788	2.41490	2.01654	1.61781	2.41527
	9	2.29257	1.81493	2.77020	2.29500	1.81652	2.77349
	10	2.60279	2.03320	3.17238	2.60219	2.03318	3.17121

TABLE 26: Results of the IT2PE validation index to the clustering of image shown in Figure 2(a) using IT2FPCM and IT2FCM algorithms with $m = [1.5, 2.5]$ and $\eta = [1.5, 2.5]$ as parameters.

Image	Clusters	Index of validation IT2PE					
		IT2FPCM			IT2FCM		
		Defuzz	Lower	Upper	Defuzz	Lower	Upper
Figure 2(a)	2	0.27752	0.24869	0.30635	0.27753	0.24869	0.30637
	3	0.38492	0.30127	0.46858	0.38494	0.30128	0.46861
	4	0.47242	0.34265	0.60219	0.47244	0.34267	0.60222
	5	0.52743	0.36020	0.69467	0.52746	0.36022	0.69470
	6	0.56679	0.36941	0.76416	0.56682	0.36943	0.76421
	7	0.60220	0.37799	0.82641	0.60223	0.37800	0.82646
	8	0.63271	0.38504	0.88037	0.63277	0.38508	0.88045
	9	0.66060	0.39199	0.92922	0.66067	0.39201	0.92933
	10	0.68623	0.39829	0.97417	0.68643	0.39838	0.97449

TABLE 27: Results of the IT2XB validation index for data clustering of image shown in Figure 2(a) using IT2FPCM and IT2FCM algorithms with $m = [1.5, 2.5]$ and $\eta = [1.5, 2.5]$ as parameters.

Image	Clusters	Index validation IT2XB					
		IT2FPCM			IT2FCM		
		Defuzz	Lower	Upper	Defuzz	Lower	Upper
Figure 2(a)	2	0.11253	0.07812	0.14694	0.11253	0.07812	0.14694
	3	0.11105	0.06516	0.15695	0.11105	0.06516	0.15694
	4	0.11980	0.06082	0.17877	0.11979	0.06082	0.17877
	5	0.11078	0.05214	0.16943	0.11078	0.05214	0.16943
	6	0.11347	0.04983	0.17711	0.11347	0.04983	0.17711
	7	0.11478	0.04732	0.18224	0.11478	0.04732	0.18223
	8	0.12419	0.04862	0.19975	0.12418	0.04862	0.19974
	9	0.12270	0.04573	0.19967	0.12257	0.04567	0.19946
	10	0.12837	0.04641	0.21032	0.12848	0.04643	0.21053

accepts the null hypothesis posed in (15), demonstrating that there is no significant difference between the IT2FCM and IT2FPCM algorithms using the z-test for the defuzzification for the IT2XB index.

Also we test both algorithms using images and perform 30 experiments validating the results with each one of the validation indices, in order to observe the behavior of the algorithm performing image segmentation. In this case for the image segmentation with the IT2FPCM and IT2FCM algorithms we perform the steps shown in Figure 1. Using these steps we are capable of making a segmentation of the image using the mentioned above algorithms. The images used for these experiments are shown in Figure 2.

In Tables 25, 26, and 27 we can observe the averages of the IT2MPE-DFPM, IT2PE, and IT2XB indices of validation, respectively, for 2 to 10 clusters, computed with the results

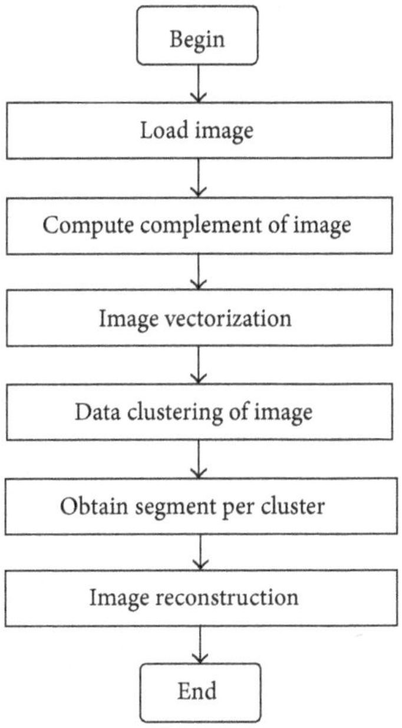

Figure 1: Block diagram for segmentation images using clustering algorithms.

Figure 2: Images for segmentations using the IT2FPCM and IT2FCM algorithms.

(defuzzification, lower and upper bounds of the interval) obtained by the IT2FPCM and IT2FCM algorithms of Figure 2(a). These tables show the number of clusters for the defuzzification, upper and lower, that each validation index found as the best one. Table 28 shows the hypothesis test for the 30 experiments performed in order to know if there exists significant difference between the algorithms using as null and alternative hypothesis the assumptions made in (15).

According to the assumptions made in (15), in Table 28, we can observe for IT2PE and IT2MPE-DMFP indices with the z-value of -2342856728 and -232549065.42, respectively, the hypothesis test that these indices are lower than the z-critical value that is equal to -1.645 with a significance level α of 0.05, whose z-value confirms the acceptance of the alternative hypothesis posed in (15), demonstrating that the IT2FPCM algorithm is better than IT2FCM algorithm

TABLE 28: Hypothesis test for IT2PE, IT2XB, and IT2MPE-DMFP indices of validation for Figure 2(a) clustering.

Dataset	Validation index	Algorithm	N	μ	σ^2	z-value	z-critical value	P value
Figure 2(a)	IT2PE	IT2FCM	30	0.27753175	$4.39E-28$	−2342856728	1.645	0
		IT2FPCM		0.27751942	$3.64E-28$			
	IT2XB	IT2FCM	30	0.11078111	$1.76E-16$	361.5239129	1.645	1
		IT2FPCM		0.11078219	$9.16E-17$			
	IT2MPE-DMFP	IT2FCM	30	0.31331411	$4.92E-26$	−232549065.42	1.645	0
		IT2FPCM		0.31330165	$3.40E-26$			

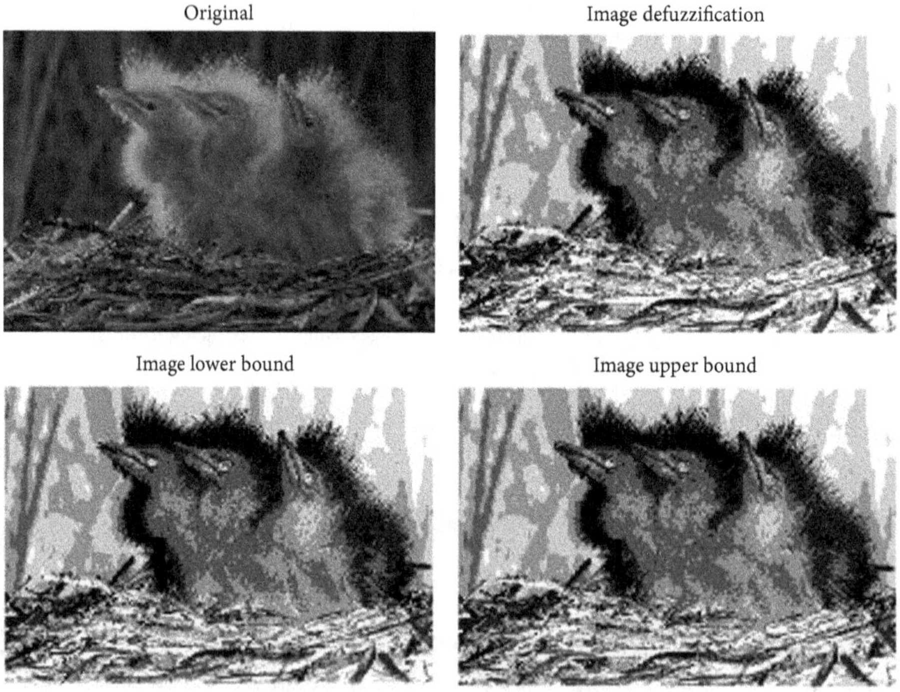

FIGURE 3: Resulting image clustering performed to Figure 2(a) by the IT2FPCM algorithm.

to z-test of the defuzzification according to IT2MPE-DMFP and IT2PE indices of validation index for the cluster found by the algorithms in the image shown in Figure 2(a). Also we observe that z-value for the IT2XB is 361.5239129 which is greater than the z-critical value that is equal to −1.645 and with this information the null hypothesis is accepted demonstrating that the IT2FPCM algorithm is not better than IT2FCM. Figure 3 shows the resulting image clustering performed by the IT2FPCM algorithm for 6 clusters for Figure 2(a) and this is because the gray levels containing the image.

In Tables 29, 30, and 31 we can observe the averages of the IT2MPE-DFPM, IT2PE, and IT2XB indices of validation respectively, for 2 to 10 clusters, computed with the results (defuzzification, lower and upper bounds of the interval) obtained by the IT2FPCM and IT2FCM algorithms for Figure 2(b). These tables show the number of clusters for the defuzzification, upper and lower, that each validation index found like better.

Table 32 shows the hypothesis test for the 30 experiments performed for each index validation mentioned in Section 4, in order to know if there exists significant difference between the algorithms by using as the null and alternative hypothesis the assumptions made in (15).

In Table 32 we can observe that z-values for the hypothesis test of the IT2PE and IT2MPE-DMPF are −10078102681 and −2621392303.80, respectively, which are lower than the z-critical value that is equal to −1.645 with a significance level α of 0.05, whose z-value confirms the acceptance of the alternative hypothesis posed in (15), demonstrating that the IT2FPCM algorithm is better than the IT2FCM algorithm for the z-test of the defuzzification according to IT2PE and IT2MPE-DMFP indices of validation indices for the clusters found by the algorithms in the image shown in Figure 2(b). Also we can observe that the z-value for the IT2XB validation index is 139756984.7, which is greater than the z-critical value that is equal to −1.645 and with this information the null hypothesis is accepted demonstrating that the IT2FPCM algorithm is not better than IT2FCM according to IT2XB validation index. Figure 4 shows the resulting image clustering performed by the IT2FPCM algorithm for 5 clusters to Figure 2(b) and this is because of the gray levels containing the image.

TABLE 29: Results of the IT2MPE-DMFP validation index to data clustering of image shown in Figure 2(b) using IT2FPCM and IT2FCM algorithm with $m = [1.5, 2.5]$ and $\eta = [1.5, 2.5]$ as parameters.

Image	Clusters	Index of validation IT2MPEDFP					
		IT2FPCM			IT2FCM		
		Defuzz	Lower	Upper	Defuzz	Lower	Upper
Figure 2(b)	2	0.20780	0.22583	0.18978	0.20781	0.22584	0.18979
	3	0.35850	0.37345	0.34355	0.35852	0.37346	0.34358
	4	0.54660	0.54745	0.54574	0.54662	0.54748	0.54577
	5	0.84354	0.81276	0.87432	0.84357	0.81278	0.87436
	6	0.76870	0.72207	0.81534	0.76874	0.72208	0.81540
	7	0.91927	0.84436	0.99419	0.91977	0.84473	0.99480
	8	1.21142	1.07707	1.34578	1.20963	1.07551	1.34375
	9	1.16203	1.02427	1.29979	1.16137	1.02366	1.29909
	10	1.51477	1.30270	1.72684	1.51163	1.29979	1.72348

TABLE 30: Results of the IT2PE validation index to the clustering of image shown in Figure 2(b) using IT2FPCM and IT2FCM algorithm with $m = [1.5, 2.5]$ and $\eta = [1.5, 2.5]$ as parameters.

Image	Clusters	Index of validation IT2PE					
		IT2FPCM			IT2FCM		
		Defuzz	Lower	Upper	Defuzz	Lower	Upper
Figure 2(b)	2	0.22538	0.18895	0.26181	0.22539	0.18895	0.26183
	3	0.32425	0.24288	0.40561	0.32427	0.24290	0.40564
	4	0.39445	0.27438	0.51452	0.39447	0.27440	0.51455
	5	0.44970	0.29589	0.60351	0.44973	0.29591	0.60355
	6	0.51026	0.32817	0.69236	0.51030	0.32819	0.69240
	7	0.55406	0.34735	0.76078	0.55410	0.34737	0.76083
	8	0.58858	0.36076	0.81639	0.58866	0.36084	0.81649
	9	0.61605	0.36984	0.86225	0.61608	0.36987	0.86229
	10	0.63779	0.37485	0.90074	0.63801	0.37496	0.90107

TABLE 31: Results of the IT2XB validation index to the clustering of image shown in Figure 2(b) using IT2FPCM and IT2FCM algorithm with $m = [1.5, 2.5]$ and $\eta = [1.5, 2.5]$ as parameters.

Image	Clusters	Index of validation IT2XB					
		IT2FPCM			IT2FCM		
		Defuzz	Lower	Upper	Defuzz	Lower	Upper
Figure 2(b)	2	**0.05906**	0.04269	**0.07544**	**0.05906**	0.04269	**0.07544**
	3	0.06108	0.03734	0.08483	0.06108	0.03734	0.08483
	4	0.06307	0.03463	0.09150	0.06307	0.03463	0.09150
	5	0.06633	**0.03419**	0.09846	0.06633	**0.03419**	0.09846
	6	0.10278	0.04910	0.15646	0.10278	0.04910	0.15646
	7	0.11868	0.05438	0.18298	0.11862	0.05434	0.18289
	8	0.11918	0.05262	0.18574	0.11938	0.05273	0.18604
	9	0.14904	0.06355	0.23454	0.14913	0.06358	0.23468
	10	0.13597	0.05632	0.21562	0.13620	0.05641	0.21599

In Tables 33, 34, and 35 we can observe the averages of the IT2MPE-DFPM, IT2PE, and IT2XB indices of validation, respectively, for 2 to 10 clusters, computed with the results (defuzzification, lower and upper bounds of the interval) obtained by the IT2FPCM and IT2FCM algorithms to Figure 2(c). These tables show the number of clusters for the defuzzification, the upper and lower values that each validation index found like better.

In Table 36, we can observe that z-values to the hypothesis test of the IT2PE and IT2MPE-DMPF are −7686717373 and −1117084835.71, respectively, which are less than the z-critical value that is equal to −1.645 with a significant level (α) of

TABLE 32: Hypothesis test for IT2PE, IT2XB, and IT2MPE-DMFP indices of validation for Figure 2(b) clustering.

Dataset	Validation index	Algorithm	N	μ	σ^2	z-value	z-critical value	P value
Figure 2(b)	IT2PE	IT2FCM	30	0.22539012	$4.13E-29$	−10078102681	1.645	0
		IT2FPCM		0.2253774	$6.51E-30$			
	IT2XB	IT2FCM	30	0.05906138	$7.70E-30$	139756984.7	1.645	1
		IT2FPCM		0.05906146	$2.49E-30$			
	IT2MPE-DMFP	IT2FCM	30	0.20781375	$3.03E-28$	−2621392303.80	1.645	0
		IT2FPCM		0.20780095	$4.13E-28$			

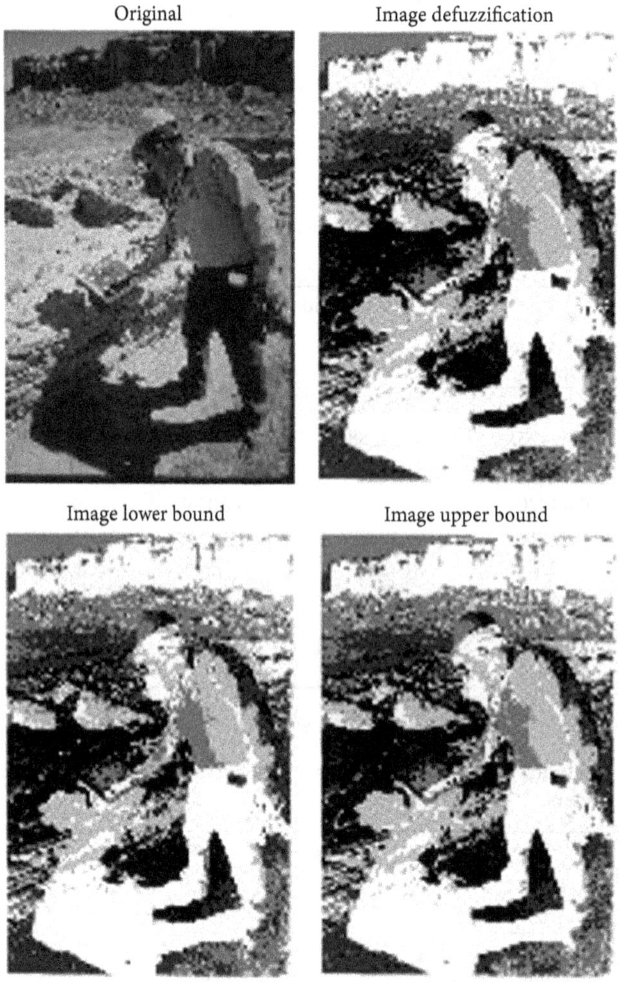

FIGURE 4: Resulting image clustering performed by the IT2FPCM algorithm for 5 clusters to Figure 2(b) because of the gray levels containing the image.

0.05, whose z-value confirms the acceptance of the alternative hypothesis posed in (15), demonstrating that IT2FPCM algorithm is better than IT2FCM algorithm to z-test of the defuzzification according to IT2PE and IT2MPE-DMFP indices of validation index for the cluster found by the algorithms in image shown in Figure 2(c). Also we can observe that z-value to the IT2XB validation index is 63042445.95, which is greater than the z-critical value that is equal to −1.645; with this information the null hypothesis is accepted demonstrating that the IT2FPCM algorithm is not better than the IT2FCM according to the IT2XB validation index.

Figure 5 shows the resulting image clustering performed by the IT2FPCM algorithm for 5 clusters to Figure 2(c) because of the gray levels containing the image.

In Tables 37, 38, and 39 we can observe the averages of the IT2MPE-DFPM, IT2PE, and IT2XB indices of validation, respectively, for 2 to 10 clusters, computed with the results (defuzzification, lower and upper bounds of the interval) obtained by the IT2FPCM and IT2FCM algorithms to Figure 2(d). These tables show the number of clusters for the defuzzification, the upper and lower values that each validation index found like better.

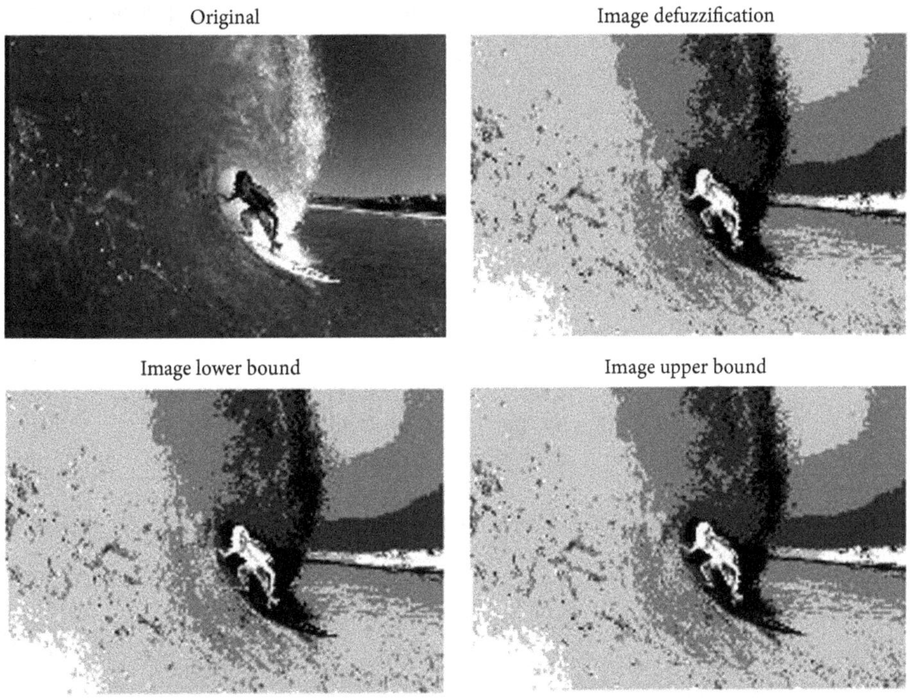

FIGURE 5: Resulting image clustering performed by the IT2FPCM algorithm for 5 clusters to Figure 2(c) because of the gray levels containing the image.

TABLE 33: Results of the IT2MPE-DMFP validation index to data clustering of image shown in Figure 2(c) using IT2FPCM and IT2FCM algorithm with $m = [1.5, 2.5]$ and $\eta = [1.5, 2.5]$ as parameters.

Image	Clusters	Index of validation IT2MPEDFP					
		IT2FPCM			IT2FCM		
		Defuzz	Lower	Upper	Defuzz	Lower	Upper
	2	**0.65158**	**0.66791**	**0.63526**	**0.65159**	**0.66791**	**0.63528**
	3	1.19860	1.14601	1.25119	1.19862	1.14601	1.25123
	4	1.37171	1.21107	1.53235	1.37174	1.21109	1.53239
	5	2.11610	1.87005	2.36214	2.11613	1.87006	2.36219
Figure 2(c)	6	2.55874	2.21054	2.90695	2.55876	2.21054	2.90699
	7	2.87029	2.40246	3.33812	2.87033	2.40247	3.33820
	8	3.26360	2.68131	3.84588	3.26782	2.68456	3.85108
	9	3.67714	2.97374	4.38054	3.62469	2.92623	4.32314
	10	4.14291	3.29282	4.99301	4.13626	3.28607	4.98644

In Table 40, we can observe that z-values to the hypothesis test of the IT2PE and IT2MPE-DMPF are $-4.95875E+13$ and -3341752881.89, respectively, which are less than the z-critical value that is equal to -1.645 with a significant level (α) of 0.05, whose z-value confirms the acceptance of the alternative hypothesis posed in (15), demonstrating that IT2FPCM algorithm is better than IT2FCM algorithm to z-test of the defuzzification according to IT2PE and IT2MPE-DMFP indices of validation index for the cluster found by the algorithms in image shown in Figure 2(d). Also we can observe that z-value to the IT2XB validation index is 16637443.08, which is greater than the z-critical value that is equal to -1.645; with this information the null hypothesis is accepted demonstrating that the IT2FPCM algorithm is not better than the IT2FCM according to the IT2XB validation index. Figure 6 shows the resulting image clustering performed by the IT2FPCM algorithm for 7 clusters to Figure 2(d) because of the gray levels containing the image.

In Tables 41, 42, and 43 we can observe the averages of the IT2MPE-DFPM, IT2PE, and IT2XB indices of validation, respectively, for 2 to 10 clusters, computed with the results (defuzzification, lower and upper bounds of the interval) obtained by the IT2FPCM and IT2FCM algorithms to Figure 2(e). These tables show the number of clusters for the defuzzification, the upper and lower values that each validation index found like better.

In Table 44, we can observe that z-values to the hypothesis test of the IT2PE and IT2MPE-DMPF are -23529815362

TABLE 34: Results of the IT2PE validation index to the clustering of image shown in Figure 2(c) using IT2FPCM and IT2FCM algorithm with $m = [1.5, 2.5]$ and $\eta = [1.5, 2.5]$ as parameters.

Image	Clusters	Index of validation IT2PE					
		IT2FPCM			IT2FCM		
		Defuzz	Lower	Upper	Defuzz	Lower	Upper
Figure 2(c)	2	**0.19464**	**0.16113**	**0.22815**	**0.19465**	**0.16114**	**0.22817**
	3	0.33148	0.25917	0.40380	0.33150	0.25917	0.40383
	4	0.43944	0.32977	0.54911	0.43947	0.32978	0.54915
	5	0.47479	0.32064	0.62893	0.47481	0.32066	0.62897
	6	0.52781	0.34398	0.71165	0.52784	0.34399	0.71169
	7	0.55923	0.35528	0.76319	0.55927	0.35530	0.76323
	8	0.58906	0.36326	0.81485	0.58911	0.36333	0.81489
	9	0.62291	0.37466	0.87116	0.62580	0.37766	0.87394
	10	0.64876	0.38289	0.91463	0.64964	0.38373	0.91555

TABLE 35: Results of the IT2XB validation index to the clustering of image shown in Figure 2(c) using IT2FPCM and IT2FCM algorithm with $m = [1.5, 2.5]$ and $\eta = [1.5, 2.5]$ as parameters.

Image	Clusters	Index of validation IT2XB					
		IT2FPCM			IT2FCM		
		Defuzz	Lower	Upper	Defuzz	Lower	Upper
Figure 2(c)	2	**0.07255**	**0.04993**	**0.09517**	**0.07255**	**0.04993**	**0.09517**
	3	0.12235	0.07282	0.17187	0.12235	0.07282	0.17187
	4	0.20659	0.11269	0.30048	0.20658	0.11269	0.30048
	5	0.14144	0.07037	0.21252	0.14144	0.07037	0.21252
	6	0.14983	0.07128	0.22839	0.14983	0.07128	0.22838
	7	0.18642	0.08317	0.28966	0.18641	0.08317	0.28965
	8	0.18905	0.08034	0.29777	0.18869	0.08021	0.29718
	9	0.18358	0.07500	0.29216	0.18937	0.07724	0.30150
	10	0.18118	0.07167	0.29068	0.18252	0.07222	0.29283

TABLE 36: Statistical test for the IT2PE, IT2XB, and IT2MPE-DMFP indices of validation for Figure 2(c) clustering.

Dataset	Validation index	Algorithm	N	μ	σ^2	z-value	z-critical value	P value
Figure 2(c)	IT2PE	IT2FCM	30	0.19465271	$6.95E-29$	−7686717373	1.645	0
		IT2FPCM		0.19464102	$1.23E-32$			
	IT2XB	IT2FCM	30	0.07255222	$2.29E-28$	63042445.95	1.645	1
		IT2FPCM		0.0725524	$1.52E-32$			
	IT2MPE-DMFP	IT2FCM	30	0.65159414	$3.67E-27$	−1117084835.71	1.645	0
		IT2FPCM		0.65158179	$2.23E-31$			

TABLE 37: Results of the IT2MPE-DMFP validation index to data clustering of image shown in Figure 2(d) using IT2FPCM and IT2FCM algorithm with $m = [1.5, 2.5]$ and $\eta = [1.5, 2.5]$ as parameters.

Image	Clusters	Index of validation IT2MPEDFP					
		IT2FPCM			IT2FCM		
		Defuzz	Lower	Upper	Defuzz	Lower	Upper
Figure 2(d)	2	**0.15969**	**0.17663**	**0.14274**	**0.15970**	**0.17663**	**0.14276**
	3	0.52574	0.50323	0.54825	0.52576	0.50324	0.54828
	4	0.67185	0.63511	0.70860	0.67188	0.63511	0.70865
	5	0.92726	0.82782	1.02669	0.92729	0.82782	1.02675
	6	0.98403	0.84965	1.11841	0.98406	0.84964	1.11849
	7	1.30797	1.11585	1.50009	1.30804	1.11588	1.50020
	8	1.39216	1.12362	1.66071	1.39200	1.12350	1.66050
	9	1.75595	1.40758	2.10433	1.76502	1.41523	2.11480
	10	1.98371	1.56297	2.40445	1.98233	1.56168	2.40299

TABLE 38: Results of the IT2PE validation index to the clustering of image shown in Figure 2(d) using IT2FPCM and IT2FCM algorithm with $m = [1.5, 2.5]$ and $\eta = [1.5, 2.5]$ as parameters.

Image	Clusters	Index of validation IT2PE					
		IT2FPCM			IT2FCM		
		Defuzz	Lower	Upper	Defuzz	Lower	Upper
	2	**0.25819**	**0.22646**	**0.28993**	**0.25821**	**0.22646**	**0.28995**
	3	0.36130	0.27685	0.44576	0.36132	0.27686	0.44578
	4	0.45157	0.32364	0.57949	0.45159	0.32365	0.57953
	5	0.49465	0.33272	0.65658	0.49468	0.33274	0.65662
Figure 2(d)	6	0.54882	0.35591	0.74172	0.54885	0.35593	0.74177
	7	0.58437	0.36581	0.80292	0.58441	0.36584	0.80298
	8	0.61986	0.37772	0.86200	0.61989	0.37775	0.86203
	9	0.64495	0.38357	0.90634	0.64494	0.38355	0.90633
	10	0.66579	0.38710	0.94449	0.66588	0.38714	0.94462

TABLE 39: Results of the IT2XB validation index to the clustering of image shown in Figure 2(d) using IT2FPCM and IT2FCM algorithm with $m = [1.5, 2.5]$ and $\eta = [1.5, 2.5]$ as parameters.

Image	Clusters	Index validation IT2XB					
		IT2FPCM			IT2FCM		
		Defuzz	Lower	Upper	Defuzz	Lower	Upper
	2	**0.09227**	0.06428	**0.12026**	**0.09227**	0.06428	**0.12026**
	3	0.07539	0.04451	0.10626	0.07539	0.04451	0.10626
	4	0.09492	0.05034	0.13951	0.09492	0.05034	0.13950
	5	0.08523	0.04182	0.12864	0.08523	0.04182	0.12864
Figure 2(d)	6	0.10438	0.04713	0.16164	0.10438	0.04713	0.16164
	7	0.09127	**0.04010**	0.14244	0.09126	**0.04010**	0.14243
	8	0.10893	0.04570	0.17217	0.10897	0.04571	0.17223
	9	0.10584	0.04343	0.16825	0.10532	0.04326	0.16738
	10	0.10343	0.04109	0.16577	0.10347	0.04111	0.16584

TABLE 40: Statistical test for the IT2PE, IT2XB, and IT2MPE-DMFP indices of validation for Figure 2(d) clustering.

Dataset	Validation index	Algorithm	N	μ	σ^2	z-value	z-critical value	P value
	IT2PE	IT2FCM	30	0.25820712	$8.76E-30$	$-4.95875E+13$	1.645	0
		IT2FPCM		0.2581946	$3.35E-30$			
Figure 2(d)	IT2XB	IT2FCM	30	0.07538511	$6.34E-27$	16637443.08	1.645	1
		IT2FPCM		0.07538535	$1.30E-28$			
	IT2MPE-DMFP	IT2FCM	30	0.15969978	$3.39E-30$	−3341752881.89	1.645	0
		IT2FPCM		0.15968716	$4.25E-28$			

TABLE 41: Results of the IT2MPE-DMFP validation index to data clustering of image shown in Figure 2(e) using IT2FPCM and IT2FCM algorithm with $m = [1.5, 2.5]$ and $\eta = [1.5, 2.5]$ as parameters.

Image	Clusters	Index of validation IT2MPEDFP					
		IT2FPCM			IT2FCM		
		Defuzz	Lower	Upper	Defuzz	Lower	Upper
	2	**0.39275**	**0.41011**	**0.37538**	**0.39276**	**0.41011**	**0.37541**
	3	0.82005	0.79494	0.84516	0.82007	0.79494	0.84519
	4	0.93366	0.85748	1.00984	0.93368	0.85749	1.00988
	5	1.46587	1.30791	1.62383	1.46590	1.30792	1.62388
Figure 2(e)	6	1.48807	1.24435	1.73180	1.48811	1.24438	1.73184
	7	1.61212	1.33261	1.89163	1.61217	1.33262	1.89171
	8	2.20927	1.81935	2.59920	2.20931	1.81917	2.59946
	9	2.18831	1.74485	2.63178	2.18867	1.74513	2.63220
	10	2.28302	1.78155	2.78449	2.28147	1.78004	2.78291

FIGURE 6: The resulting image clustering performed by the IT2FPCM algorithm for 7 clusters to Figure 2(d) because of the gray levels containing the image.

TABLE 42: Results of the IT2PE validation index to the clustering of image shown in Figure 2(e) using IT2FPCM and IT2FCM algorithm with $m = [1.5, 2.5]$ and $\eta = [1.5, 2.5]$ as parameters.

Image	Clusters	Index of validation IT2PE					
		IT2FPCM			IT2FCM		
		Defuzz	Lower	Upper	Defuzz	Lower	Upper
Figure 2(e)	2	**0.25722**	**0.22472**	**0.28971**	**0.25723**	**0.22473**	**0.28973**
	3	0.34121	0.25920	0.42321	0.34122	0.25921	0.42324
	4	0.42472	0.30360	0.54583	0.42474	0.30362	0.54586
	5	0.48178	0.32465	0.63891	0.48181	0.32467	0.63894
	6	0.53382	0.34760	0.72005	0.53386	0.34762	0.72010
	7	0.55569	0.34695	0.76443	0.55573	0.34698	0.76448
	8	0.58706	0.35582	0.81830	0.58706	0.35580	0.81831
	9	0.62347	0.37038	0.87656	0.62353	0.37041	0.87665
	10	0.65010	0.37895	0.92124	0.65020	0.37899	0.92142

and −2455999372.25, respectively, which are less than the z-critical value that is equal to −1.645 with a significant level (α) of 0.05, whose z-value confirms the acceptance of the alternative hypothesis posed in (15), demonstrating that IT2FPCM algorithm is better than IT2FCM algorithm to z-test of the defuzzification according to IT2PE and IT2MPE-DMFP indices of validation index for the cluster found by the algorithms in image shown in Figure 2(e). Also we can observe that z-value to the IT2XB validation index is 305500241.9, which is greater than the z-critical value that is equal to −1.645; with this information the null hypothesis is accepted demonstrating that the IT2FPCM algorithm is not better than the IT2FCM according to the IT2XB validation index. Figure 7 shows the resulting image clustering performed by the IT2FPCM algorithm for 7 clusters to Figure 2(e) because of the gray levels containing the image.

5. Conclusions

IT2FPCM is an extension of the FPCM algorithm based on Type-2 Fuzzy Logic concepts, in order to enhance its ability of handling uncertainty and making it less susceptible to noise. This algorithm was tested using the Wine, WDBC, Iris Flower, Ionosphere, Abalone, and Cover type benchmark datasets and a set of images shown in Figure 2. In order to observe if the proposal is better than the IT2FCM algorithm we performed 30 experiments with each dataset and images used for a number of clusters from 2 to 10, in order to make a hypothesis testing with the assumption made in

TABLE 43: Results of the IT2XB validation index to the clustering of image shown in Figure 2(e) using IT2FPCM and IT2FCM algorithm with $m = [1.5, 2.5]$ and $\eta = [1.5, 2.5]$ as parameters.

Image	Clusters	Index validation IT2XB					
		IT2FPCM			IT2FCM		
		Defuzz	Lower	Upper	Defuzz	Lower	Upper
Figure 2(e)	2	**0.09682**	0.06740	**0.12625**	**0.09682**	0.06740	**0.12625**
	3	0.08314	0.05171	0.11457	0.08314	0.05171	0.11457
	4	0.10315	0.05592	0.15038	0.10315	0.05592	0.15038
	5	0.08615	0.04418	0.12812	0.08615	0.04418	0.12811
	6	0.11084	0.05313	0.16855	0.11084	0.05313	0.16855
	7	0.11129	0.05049	0.17209	0.11129	0.05049	0.17209
	8	0.09285	**0.04078**	0.14492	0.09285	**0.04078**	0.14491
	9	0.10428	0.04261	0.16594	0.10423	0.04261	0.16586
	10	0.12035	0.04734	0.19336	0.12045	0.04736	0.19354

TABLE 44: Statistical test for the IT2PE, IT2XB, and IT2MPE-DMFP indices of validation for Figure 2(e) clustering.

Dataset	Validation index	Algorithm	N	μ	σ^2	z-value	z-critical value	P value
Figure 2(e)	IT2PE	IT2FCM	30	0.25722842	$8.46E-31$	−23529815362	1.645	0
		IT2FPCM		0.2572161	$7.37E-30$			
	IT2XB	IT2FCM	30	0.04078364	$1.06E-06$	305500241.9	1.645	1
		IT2FPCM		0.04078258	$1.05E-06$			
	IT2MPE-DMFP	IT2FCM	30	0.39275846	$9.74E-30$	−2455999372.25	1.645	0
		IT2FPCM		0.39274591	$7.73E-28$			

FIGURE 7: Resulting image clustering performed by the IT2FPCM algorithm for 5 clusters to Figure 2(e) because of the gray levels containing the image.

(15), to prove that the proposed method is better with a significant difference with respect to the other existing methods. Statistical tests were performed with the number of clusters that each validation index indicates as the best; in these statistical tests for the datasets and images we can observe that 69.45% of the hypothesis tests performed with the different indices of validation are affirming the alternative hypothesis based on (15), and 30.55% of the hypothesis tests reject the alternative hypothesis.

It is noteworthy that the parameters used in this work are not the optimal ones for both algorithms, and to find the optimal parameters for both algorithms used in this work we can use optimization algorithms like in [33]. We can use the PSO, GSA, and GA algorithms among others, in order to improve the performance and automate the interval type-2 clustering algorithms that were used.

Competing Interests

The authors declare that there is no conflict of interests regarding the publication of this paper.

References

[1] W. E. Phillips, R. P. Velthuizen, S. Phuphanich, L. O. Hall, L. P. Clarke, and M. L. Silbiger, "Application of fuzzy c-means segmentation technique for tissue differentiation in MR images of a hemorrhagic glioblastoma multiforme," *Magnetic Resonance Imaging*, vol. 13, no. 2, pp. 277–290, 1995.

[2] M.-S. Yang, Y.-J. Hu, K. C.-R. Lin, and C. C.-L. Lin, "Segmentation techniques for tissue differentiation in MRI of ophthalmology using fuzzy clustering algorithms," *Magnetic Resonance Imaging*, vol. 20, no. 2, pp. 173–179, 2002.

[3] K. Hirota and W. Pedrycz, "Fuzzy computing for data mining," *Proceedings of the IEEE*, vol. 87, no. 9, pp. 1575–1600, 1999.

[4] J. C. Bezdek, *Pattern recognition with fuzzy objective function algorithms*, Plenum Press, NY, USA, 1981.

[5] N. S. Iyer, A. Kandel, and M. Schneider, "Feature-based fuzzy classification for interpretation of mammograms," *Fuzzy Sets and Systems*, vol. 114, no. 2, pp. 271–280, 2000.

[6] X. Chang, W. Li, and J. Farrell, "C-means clustering based fuzzy modeling method," in *Proceedings of the 9th IEEE International Conference on Fuzzy Systems (FUZZ-IEEE '00)*, vol. 2, pp. 937–940, May 2000.

[7] R. Krishnapuram and J. M. Keller, "A possibilistic approach to clustering," *IEEE Transactions on Fuzzy Systems*, vol. 1, no. 2, pp. 98–110, 1993.

[8] N. R. Pal, K. Pal, and J. C. Bezdek, "A mixed c-means clustering model," in *Proceedings of the 6th IEEE International Conference on Fuzzy Systems*, vol. 1, p. 11, IEEE, Barcelona, Spain, July 1997.

[9] N. N. Karnik and J. M. Mendel, "Operations on type-2 fuzzy sets," *Fuzzy Sets and Systems*, vol. 122, no. 2, pp. 327–348, 2001.

[10] J. Mendel, *Uncertain Rule-Based Fuzzy Logic Systems: Introduction and New Directions*, Prentice-Hall, 2001.

[11] F. C.-H. Rhee and C. Hwang, "A type-2 fuzzy C-means clustering algorithm," in *Proceedings of the Joint 9th IFSA World Congress and 20th NAFIPS International Conference*, pp. 1926–1929, British Columbia, Canada, July 2001.

[12] M. H. Fazel Zarandi, M. Zarinbal, and I. B. Turksen, "Type-II fuzzy possibilistic C-Mean clustering," in *Proceedings of the Joint International Fuzzy Systems Association World Congress (IFSA '09) and European Society of Fuzzy Logic and Technology Conference (EUSFLAT '09)*, pp. 30–35, July 2009.

[13] M. H. F. Zarandi, S. M. Golsefid, and S. Bastani, "Dual centers fuzzy type-2 clustering," in *Proceedings of the 9th Joint World Congress on Fuzzy Systems and NAFIPS Annual Meeting (IFSA/NAFIPS '13)*, pp. 1215–1220, Alberta, Canada, June 2013.

[14] E. Rubio and O. Castillo, "Designing type-2 fuzzy systems using the interval type-2 fuzzy C-means algorithm," in *Proceedings of the International Seminar Computing Intelligence (ISCI '13)*, Tijuana, Mexico, November 2013.

[15] Y.-H. Byeon and K.-C. Kwak, "Knowledge discovery and modeling based on conditional fuzzy clustering with interval Type-2 fuzzy," in *Proceedings of the 7th International Joint Conference on Knowledge Discovery, Knowledge Engineering and Knowledge Management (IC3K '15)*, pp. 440–444, Lisbon, Portugal, November 2015.

[16] L. Yu, J. Xiao, and G. Zheng, "Robust interval type-2 possibilistic c-means clustering and its application for fuzzy modeling," in *Proceedings of the 6th International Conference on Fuzzy Systems and Knowledge Discovery (FSKD '09)*, pp. 360–365, IEEE, Tianjin, China, August 2009.

[17] O. Obajemu, M. Mahfouf, and L. A. Torres-Salomao, "A new interval type-2 fuzzy clustering algorithm for interval type-2 fuzzy modelling with application to heat treatment of steel," *IFAC Proceedings*, vol. 47, no. 3, pp. 10658–10663, 2014.

[18] B.-I. Choi and F. C.-H. Rhee, "Interval type-2 fuzzy membership function generation methods for pattern recognition," *Information Sciences*, vol. 179, no. 13, pp. 2102–2122, 2009.

[19] E. Rubio and O. Castillo, "Interval type-2 fuzzy clustering for membership function generation," in *Proceedings of the 2013 IEEE Workshop on Hybrid Intelligent Models and Applications (HIMA '13)*, pp. 13–18, IEEE, Singapore, April 2013.

[20] L. Tlig, M. Sayadi, and F. Fnaeich, "A new descriptor for textured image segmentation based on fuzzy type-2 clustering approach," in *Proceedings of the 2nd International Conference on Image Processing Theory, Tools and Applications (IPTA '10)*, pp. 258–263, Paris, France, July 2010.

[21] M. H. F. Zarandi and M. Zarinbal, "A new image enhancement method type-2 possibilistic C-mean approach," in *Proceedings of the 2013 Joint IFSA World Congress and NAFIPS Annual Meeting (IFSA/NAFIPS '13)*, pp. 1131–1135, June 2013.

[22] R. Ceylan, Y. Özbay, and B. Karlik, "A novel approach for classification of ECG arrhythmias: type-2 fuzzy clustering neural network," *Expert Systems with Applications*, vol. 36, no. 3, part 2, pp. 6721–6726, 2009.

[23] M. A. Raza and F. C.-H. Rhee, "Interval type-2 approach to kernel possibilistic C-means clustering," in *Proceedings of the IEEE International Conference on Fuzzy Systems (FUZZ-IEEE '12)*, pp. 1–7, Brisbane, Australia, June 2012.

[24] D. D. Nguyen and L. T. Ngo, "Multiple kernel interval type-2 fuzzy c-means clustering," in *Proceedings of the IEEE International Conference on Fuzzy Systems (FUZZ-IEEE '13)*, pp. 1–8, Hyderabad, India, July 2013.

[25] M. U. Nguyen, L. T. Ngo, and T. T. Dao, "Improved interval type-2 fuzzy subtractive clustering for obstacle detection of robot vision from stream of depth camera," in *Proceedings of the 12th International Conference on Intelligent Systems Design and Applications (ISDA '12)*, pp. 903–908, November 2012.

[26] S. M. Golsefid, M. H. F. Zarandi, and S. Bastani, "Fuzzy type-2 c-ellipses clustering," in *Proceedings of the IFSA World Congress*

and NAFIPS Annual Meeting (IFSA/NAFIPS '13), pp. 1221–1226, June 2013.

[27] S. M. M. Golsefid and M. H. Fazel Zarandi, "Dual-centers type-2 fuzzy clustering framework and its verification and validation indices," *Applied Soft Computing Journal*, vol. 47, pp. 600–613, 2016.

[28] E. Rubio, O. Castillo, and P. Melin, "A new validation index for fuzzy clustering and its comparisons with other methods," in *Proceedings of the IEEE International Conference on Systems, Man, and Cybernetics (SMC '11)*, pp. 301–306, October 2011.

[29] J. Yen and R. Langari, "Fuzzy logic in pattern recognition," in *Fuzzy Logic: Intelligence, Control, and Information*, pp. 351–377, Prentice Hall, Upper Saddle River, NJ, USA, 1999.

[30] M. K. Pakhira, S. Bandyopadhyay, and U. Maulik, "A study of some fuzzy cluster validity indices, genetic clustering and application to pixel classification," *Fuzzy Sets and Systems*, vol. 155, no. 2, pp. 191–214, 2005.

[31] B. Rezaee, "A cluster validity index for fuzzy clustering," *Fuzzy Sets and Systems*, vol. 161, no. 23, pp. 3014–3025, 2010.

[32] K.-L. Wu and M.-S. Yang, "A cluster validity index for fuzzy clustering," *Pattern Recognition Letters*, vol. 26, no. 9, pp. 1275–1291, 2005.

[33] E. Rubio and O. Castillo, "Optimization of the interval type-2 fuzzy C-means using particle swarm optimization," in *Proceedings of the World Congress on Nature and Biologically Inspired Computing (NaBIC '13)*, pp. 10–15, IEEE, August 2013.

Permissions

All chapters in this book were first published in AFS, by Hindawi Publishing Corporation; hereby published with permission under the Creative Commons Attribution License or equivalent. Every chapter published in this book has been scrutinized by our experts. Their significance has been extensively debated. The topics covered herein carry significant findings which will fuel the growth of the discipline. They may even be implemented as practical applications or may be referred to as a beginning point for another development.

The contributors of this book come from diverse backgrounds, making this book a truly international effort. This book will bring forth new frontiers with its revolutionizing research information and detailed analysis of the nascent developments around the world.

We would like to thank all the contributing authors for lending their expertise to make the book truly unique. They have played a crucial role in the development of this book. Without their invaluable contributions this book wouldn't have been possible. They have made vital efforts to compile up to date information on the varied aspects of this subject to make this book a valuable addition to the collection of many professionals and students.

This book was conceptualized with the vision of imparting up-to-date information and advanced data in this field. To ensure the same, a matchless editorial board was set up. Every individual on the board went through rigorous rounds of assessment to prove their worth. After which they invested a large part of their time researching and compiling the most relevant data for our readers.

The editorial board has been involved in producing this book since its inception. They have spent rigorous hours researching and exploring the diverse topics which have resulted in the successful publishing of this book. They have passed on their knowledge of decades through this book. To expedite this challenging task, the publisher supported the team at every step. A small team of assistant editors was also appointed to further simplify the editing procedure and attain best results for the readers.

Apart from the editorial board, the designing team has also invested a significant amount of their time in understanding the subject and creating the most relevant covers. They scrutinized every image to scout for the most suitable representation of the subject and create an appropriate cover for the book.

The publishing team has been an ardent support to the editorial, designing and production team. Their endless efforts to recruit the best for this project, has resulted in the accomplishment of this book. They are a veteran in the field of academics and their pool of knowledge is as vast as their experience in printing. Their expertise and guidance has proved useful at every step. Their uncompromising quality standards have made this book an exceptional effort. Their encouragement from time to time has been an inspiration for everyone.

The publisher and the editorial board hope that this book will prove to be a valuable piece of knowledge for researchers, students, practitioners and scholars across the globe.

List of Contributors

Aleksandar Janjic
University of Nis, Faculty of Electronic Engineering, 18 000 Nis, Serbia

Artur Nor and Eugene Korotkov
National Research Nuclear University "MEPhI", Kashirskoe Highway, 31, 115409,Moscow, Russia

Eugene Korotkov
Institute of Bioengineering, Research Center of Biotechnology of the Russian Academy of Sciences, Leninsky Ave. 33, bld. 2, 119071,Moscow, Russia

Aveek Basu
Department of Management, BIT Mesra Kolkata Campus 700107, India

Sanchita Ghosh
Department of Computer Science, BIT Mesra Kolkata Campus 700107, India

Solomon T. Girma
Department of Electrical Engineering, Pan African University Institute of Sciences, Technology and Innovations, Addis Ababa, Ethiopia

Abinet G. Abebe
Ethio Telecom, Radio Access Network Rollout Department, Addis Ababa, Ethiopia

A. A. Abd El-Latif
Department ofMathematics, Faculty of Science and Arts at Balgarn, University of Bisha, Sabt Al-Alaya 61985, Saudi Arabia

A. A. Abd El-Latif
High Institute of Computer King Marriott, Alexandria, Egypt

H. Aygün and V. Çetkin
Department of Mathematics, Kocaeli University, 41380 Kocaeli, Turkey

Krzysztof Piasecki
Department of Investment and Real Estate, Poznań University of Economics and Business, 61-875 Poznań, Poland

Ewa Roszkowska
Faculty of Economics and Management, University of Bialystok, 15-062 Bialystok, Poland

R. Mastani Shabestari and T. Allahviranloo
Department of Mathematics, Science and Research Branch, Islamic Azad University, Tehran, Iran

R. Ezzati
Department of Mathematics, Karaj Branch, Islamic Azad University, Karaj, Iran

Meraj A. Khan and Abdulrahman Aljohani
Department of Mathematics, University of Tabuk, Tabuk, Saudi Arabia

Izhar Ahmad
Department of Mathematics and Statistics, King Fahd University of Petroleum and Minerals, Dhahran 31261, Saudi Arabia

Shuker Mahmood Khalil, Mayadah Ulrazaq, Samaher Abdul-Ghani and Abu Firas Al-Musawi
Department of Mathematics, College of Science, University of Basrah, Basrah 61004, Iraq

Israa Abdzaid Atiyah and Adel Mohammadpour
Faculty of Mathematics and Computer Science, Amirkabir University of Technology, Tehran, Iran

S. Mahmoud Taheri
School of Engineering Science, College of Engineering, University of Tehran, Tehran, Iran

Soumya Ranjan Das and Prakash K. Ray
Department of Electrical Engineering, IIIT Bhubaneswar, Odisha 751003, India

Asit Mohanty
Department of Electrical Engineering, CET Bhubaneswar, Odisha 751003, India

Shahzad Faizi, Tabasam Rashid and Sohail Zafar
Department of Mathematics, University of Management and Technology, Lahore 54770, Pakistan

Sunil Kumar and Ashwani Kumar Dhingra
Mechanical Engineering Department, University Institute of Engineering & Technology, Maharshi Dayanand University, Rohtak, Haryana 124 001, India

Bhim Singh
Mechanical Engineering Department, School of Engineering & Technology, Sharda University, Greater Noida, Uttar Pradesh 201 306, India

Tomoe Entani
Graduate School of Applied Informatics, University of Hyogo, Kobe, Hyogo 6500047, Japan

Aby K. George and Harpreet Singh
Department of Electrical and Computer Engineering,Wayne State University, Detroit, MI 48202, USA

Macam S. Dattathreya and Thomas J. Meitzler
Tank Automotive Research, Development and Engineering Center,Warren, MI 48397, USA

Shazia Kanwal
Department of Mathematics, GC University, Faisalabad-38000, Pakistan

Akbar Azam
Department of Mathematics, COMSATS University,

Abraham Assefa Tsehayae
School of Civil and Environmental Engineering, Addis Ababa Institute of Technology, Addis Ababa University, Room 206, AAiT Main Building, Addis Ababa, Ethiopia Islamabad-44000, Pakistan

Aminah Robinson Fayek
Department of Civil and Environmental Engineering, Hole School of Construction Engineering, University of Alberta, 7-287 Donadeo Innovation Centre for Engineering, Edmonton, AB, Canada T6G 1H9

Katsuhiro Honda, Yurina Suzuki and Seiki Ubukata
Graduate School of Engineering, Osaka Prefecture University, Sakai, Osaka 599-8531, Japan

Akira Notsu
Graduate School of Humanities and Sustainable System Sciences, Osaka Prefecture University, Sakai, Osaka 599-8531, Japan

Cansel Yormaz, Simge Simsek and Serife Naz Elmas
Department of Mathematics, Pamukkale University, Denizli 20070, Turkey

M. Clement Joe Anand and Janani Bharatraj
Department of Mathematics, Hindustan Institute of Technology and Science (Deemed to be University), Chennai 603103, India

Huang Hui, Jia Rong and Wang Songkai
Institute of Water Resources and Hydro-Electric Engineering, Xi'an University of Technology, Xi'an 710048, China

Huang Hui
School of Electric Power, North China University ofWater Resources and Electric Power, Zhengzhou 450011, China

Elid Rubio, Oscar Castillo, Fevrier Valdez, Patricia Melin, Claudia I. Gonzalez and Gabriela Martinez
Tijuana Institute of Technology, Tijuana, BC, Mexico

Index